UCLA Symposia on Molecular and Cellular Biology, New Series

Series Editor, C. Fred Fox

RECENT TITLES

Volume 33
Yeast Cell Biology, James Hicks, *Editor*

Volume 34
Molecular Genetics of Filamentous Fungi, William Timberlake, *Editor*

Volume 35
Plant Genetics, Michael Freeling, *Editor*

Volume 36
Options for the Control of Influenza, Alan P. Kendal and Peter A. Patriarca, *Editors*

Volume 37
Perspectives in Inflammation, Neoplasia, and Vascular Cell Biology, Thomas Edgington, Russell Ross, and Samuel Silverstein, *Editors*

Volume 38
Membrane Skeletons and Cytoskeletal—Membrane Associations, Vann Bennett, Carl M. Cohen, Samuel E. Lux, and Jiri Palek, *Editors*

Volume 39
Protein Structure, Folding, and Design, Dale L. Oxender, *Editor*

Volume 40
Biochemical and Molecular Epidemiology of Cancer, Curtis C. Harris, *Editor*

Volume 41
Immune Regulation by Characterized Polypeptides, Gideon Goldstein, Jean-François Bach, and Hans Wigzell, *Editors*

Volume 42
Molecular Strategies of Parasitic Invasion, Nina Agabian, Howard Goodman, and Nadia Nogueira, *Editors*

Volume 43
Viruses and Human Cancer, Robert C. Gallo, William Haseltine, George Klein, and Harald zur Hausen, *Editors*

Volume 44
Molecular Biology of Plant Growth Control, J. Eugene Fox and Mark Jacobs, *Editors*

Volume 45
Membrane-Mediated Cytotoxicity, Benjamin Bonavida and R. John Collier, *Editors*

Volume 46
Development and Diseases of Cartilage and Bone Matrix, Arup Sen and Thomas Thornhill, *Editors*

Volume 47
DNA Replication and Recombination, Roger McMacken and Thomas J. Kelly, *Editors*

Volume 48
Molecular Strategies for Crop Protection, Charles J. Arntzen and Clarence Ryan, *Editors*

Volume 49
Molecular Entomology, John H. Law, *Editor*

Volume 50
Interferons as Cell Growth Inhibitors and Antitumor Factors, Robert M. Friedman, Thomas Merigan, and T. Sreevalsan, *Editors*

Volume 51
Molecular Approaches to Developmental Biology, Richard A. Firtel and Eric H. Davidson, *Editors*

Volume 52
Transcriptional Control Mechanisms, Daryl Granner, Michael G. Rosenfeld, and Shing Chang, *Editors*

Volume 53
Recent Advances in Bone Marrow Transplantation, Robert Peter Gale and Richard Champlin, *Editors*

Volume 54
Positive Strand RNA Viruses, Margo A. Brinton and Roland R. Rueckert, *Editors*

UCLA Symposia Board

Positive Strand RNA Viruses

Positive Strand RNA Viruses

Proceedings of a UCLA Symposium
held in Keystone, Colorado,
April 20–26, 1986

Editors

Margo A. Brinton
The Wistar Institute
Philadelphia, Pennsylvania

Roland R. Rueckert
Biophysics Laboratory
University of Wisconsin
Madison, Wisconsin

Alan R. Liss, Inc. • New York

Address all Inquiries to the Publisher
Alan R. Liss, Inc., 41 East 11th Street, New York, NY 10003

Printed in the United States of America

Library of Congress Cataloging-in-Publication Data

UCLA Symposium on Positive Strand RNA Viruses
(1986 : Keystone, Colo.)
Positive strand RNA viruses.

(UCLA symposia on molecular and cellular
biology ; new ser., v. 54)
Includes bibliographies and index.
1. Viruses, RNA--Congresses. I. Brinton,
Margo A. II. Rueckert, Roland R. III. Title.
IV. Series. [DNLM: 1. RNA Viruses--congresses.
W3 U17N new ser. v.54 / QW 168 U17p 1986]
QR395.U25 1986 616'.0194 86-33709
ISBN 0-8451-2653-9

Contents

Contributors . . . . **xi**

Preface
Margo A. Brinton and Roland R. Rueckert . . . . **xxi**

I. GENOME STRUCTURE AND VIRAL EVOLUTION

Evolution and Genome Structure
Paul Kaesberg **3**

Replication and Selection Kinetics of Short-Chained RNA Species
Christof K. Biebricher **9**

Comparative Organization and Genome Structure in Picornaviruses
Ann C. Palmenberg **25**

Requirement for Secondary Structure Formation During Coliphage RNA Replication
Donald R. Mills, Christine Priano, and Fred Russell Kramer **35**

Construction of Genomic-Length cDNA Clones of Tobacco Mosaic Virus
D.A. Knorr, D.L. Beck, and W.O. Dawson **47**

II. ANTIGENIC AND FUNCTIONAL ANALYSES OF VIRION PROTEINS AND CELLULAR RECEPTORS

Structure and Function of Human Rhinovirus 14 and Mengo Virus: Neutralizing Antigenic Sites, Putative Receptor Binding Site, Neutralization by Drug Binding
Michael G. Rossmann, Edward Arnold, Greg Kamer, Marcia J. Kremer, Ming Luo, Thomas J. Smith, Gerrit Vriend, Roland R. Rueckert, Anne G. Mosser, Barbara Sherry, Ulrike Boege, Douglas G. Scraba, Mark A. McKinlay, and Guy D. Diana **59**

The Three Dimensional Structure of Poliovirus: Implications for Assembly and Immune Recognition
James M. Hogle, Marie Chow, and David J. Filman **79**

Isolation and Characterization of a Monoclonal Antibody Which Blocks Attachment of Human Rhinoviruses
Richard J. Colonno, Joanne E. Tomassini, and Pia L. Callahan **93**

Photolabeling of the Influenza Hemagglutinin With a Hydrophobic Probe
François Boulay, Robert W. Doms, and Ari Helenius **103**

Epitope Mapping of a Flavivirus Glycoprotein
Franz X. Heinz, Günther Winkler, Christian Mandl, Farshad Guirakhoo, Wolfgang Tuma, and Christian Kunz **113**

III. GENE EXPRESSION

Genomic Organization of Murine Coronavirus MHV
Stuart G. Siddell **127**

Multiple Subgenomic mRNA's Are Involved in the Gene Expression of Equine Arteritis Virus, a Non-Arthropod Borne Togavirus
Bernard A.M. Van der Zeijst, M.F. van Berlo, W.C. Vooys, E.J. Snijder, P. Bredenbeek, M.C. Horzinek, P.J.M. Rottier, and W.J.M. Spaan **137**

Gene Expression in Turnip Yellow Mosaic Virus
Anne-Lise Haenni, Marie-Dominique Morch, Gabrièle Drugeon, Rosaura Valle, Rajiv L. Joshi, and Térèse-Marie Denial **149**

Cotranslational Disassembly of Filamentous Plant Virus Nucleocapsids *In Vitro* and *In Vivo*
T. Michael A. Wilson and John G. Shaw **159**

Sindbis Virus Structural Genes are Expressed in *Saccharomyces cerevisiae*
Milton J. Schlesinger, Duanzhi Wen, and Mingxaio Ding **183**

IV. PROTEOLYTIC PROCESSING OF VIRAL PROTEINS

Polyprotein Processing in the Expression of the Genome of Cowpea Mosaic Virus
Ab van Kammen, Martine Jaegle, Pieter Vos, Joan Wellink, and Rob Goldbach **195**

Replication of Alphaviruses and Flaviviruses: Proteolytic Processing of Polyproteins
James H. Strauss, Ellen G. Strauss, Chang S. Hahn, Young S. Hahn, Ricardo Galler, W. Reef Hardy, and Charles M. Rice **209**

Assembly of Moloney Murine Leukemia Virus: Requirement for Myristylation Site in PR65^{gag}
Alan Rein, Melody R. McClure, Nancy R. Rice, Ronald B. Luftig, and Alan M. Schultz **227**

V. GENOME REPLICATION

Replication and Packaging Sequences in Defective Interfering RNAs of Sindbis Virus
Sondra Schlesinger, Robin Levis, Barbara G. Weiss, Manuel Tsiang, and Henry Huang **241**

Alphavirus Plus Strand and Minus Strand RNA Synthesis
Dorothea Sawicki and Stanley Sawicki **251**

Replication of Flaviviruses
Margo A. Brinton and Janet B. Grun **261**

Mechanism of RNA Replication by the Poliovirus RNA Polymerase, HeLa Cell Host Factor, and VPg
J. Bert Flanegan, Dorothy C. Young, Gregory J. Tobin, Mary Merchant Stokes, Carol D. Murphy, and Steven M. Oberste 273

Leader RNA-Primed Transcription and RNA Recombination of Murine Coronaviruses
Michael M.C. Lai, Shinji Makino, Ralph S. Baric, Lisa Soe, Chien-Kou Shieh, James G. Keck, and Stephen A. Stohlman 285

The Spatial Folding of the 3′Noncoding Region of Aminoacylatable Plant Viral RNAs
Cornelis W.A. Pleij, Jan Pieter Abrahams, Alex van Belkum, Krijn Rietveld, and Leendert Bosch 299

Mutational Analysis of the Functions of the tRNA-Like Region of Brome Mosaic Virus RNA
Theo W. Dreher and Timothy C. Hall 317

Mutational Analysis of the Internal Promoter for Transcription of the Subgenomic RNA4 of BMV
Loren E. Marsh, Theo W. Dreher, and Timothy C. Hall 327

VI. TRANSLATION, PROTEIN MODIFICATION, AND ASSEMBLY

Processing of Coronavirus Proteins and Assembly of Virions
K.V. Holmes, J.F. Boyle, R.K. Williams, C.B. Stephensen, S.G. Robbins, E.C. Bauer, C.S. Duchala, M.F. Frana, D.G. Weismiller, S. Compton, J.J. McGowan, and L.S. Sturman 339

Biosynthesis of the Structural Proteins of SFV—A Recombinant DNA Approach
Henrik Garoff, Paul Melançon, Daniel Cutler, Laurie Roman, Jim Hare, Marino Zerial, and Danny Huylebroeck 351

Structure-Function Relationships in the Glycoproteins of Alphaviruses
Ellen G. Strauss, Alan L. Schmaljohn, Diane E. Griffin, and James H. Strauss 365

Mechanism of RNA Virus Assembly and Disassembly
S.C. Harrison, P.K. Sorger, P.G. Stockley, J. Hogle, R. Altman, and R.K. Strong 379

TMV Encapsidation Initiation Sites on Non-Virion RNA Species
Albert Siegel, Chintamani Atreya, Fumihiro Terami, and D'Ann Rochon 397

VII. PATHOGENESIS AND VIRULENCE

Molecular Determinants of CNS Virulence of the Coronavirus Mouse Hepatitis Virus-4
Michael J. Buchmeier, Robert G. Dalziel, Marck J.M. Koolen, and Peter W. Lampert 409

Responses of Plant Cells to Virus Infection With Special Reference to the Sites of RNA Replication
R.I.B. Francki 423

Study on Virulence of Poliovirus Type 1 Using *In Vitro* Modified Viruses
A. Nomoto, M. Kohara, S. Kuge, N. Kawamura, M. Arita, T. Komatsu, S. Abe, B.L. Semler, E. Wimmer, and H. Itoh **437**

Cellular Receptors in Coxsackievirus Infections
Richard L. Crowell, Kuo-Hom Lee Hsu, Maggie Schultz, and Burton J. Landau **453**

Nucleic Acid Sequence Analysis of Sindbis Pathogenesis and Penetration Mutants
Robert E. Johnston, Nancy L. Davis, David F. Pence, Susan Gidwitz, and Frederick J. Fuller **467**

Natural Distribution of Wild Type I Poliovirus Genotypes
Rebeca Rico-Hesse, Mark A. Pallansch, Baldev K. Nottay, and Olen Kew **477**

Rhinovirus Detection by cDNA: RNA Hybridization
W. Al-Nakib, G. Stanway, M. Forsyth, P.J. Hughes, J.W. Almond, and D.A.J. Tyrrell **487**

Single-Cycle Growth Kinetics of Hepatitis A Virus in BSC-1 Cells
David A. Anderson, Stephen A. Locarnini, Bruce C. Ross, Anthony G. Coulepis, Bruce N. Anderson, and Ian D. Gust **497**

Host Range Determinants of Avian Retrovirus Envelope Genes
Carol A. Bova and Ronald Swanstrom **509**

Rubella Virus Associated With Cytoskeleton (Rubella VACS) Particles—Relevant to Scrapie?
Diane Van Alstyne, Marc DeCamillis, Paul S. Sunga, and Richard F. Marsh **519**

VIII. STRATEGIES FOR CONTROL OF POSITIVE STRAND VIRUS DISEASES

Molecular Basis of Antigenicity of Poliovirus
P.D. Minor, D.M.A. Evans, M. Ferguson, G.C. Schild, J.W. Almond, and G. Stanway **539**

Control of Foot-and-Mouth Disease: The Present Position and Future Prospects
Fred Brown **555**

Status of Hepatitis A Vaccines
Robert H. Purcell **565**

Two Contrasting Types of Host Resistance of Potential Use for Controlling Plant Viruses
B.D. Harrison, H. Barker, D.C. Baulcombe, M.W. Bevan, and M.A. Mayo **575**

Index **587**

Contributors

S. Abe, Japan Poliomyelitis Research Institute, Tokyo 189, Japan **[437]**

Jan Pieter Abrahams, Department of Biochemistry, University of Leiden, Leiden, The Netherlands **[299]**

J.W. Almond, Department of Microbiology, University of Reading, Reading, United Kingdom; present address: National Institute for Biological Standards and Control, Holly Hill, London NW3 6RB, United Kingdom **[487,539]**

W. Al-Nakib, MRC Common Cold Unit, Harvard Hospital, Salisbury, United Kingdom **[487]**

R. Altman, Department of Biochemistry and Molecular Biology, Harvard University, Cambridge, MA 02138; present address: School of Medicine, Stanford University, Stanford, CA 94305 **[379]**

Bruce N. Anderson, Virus Laboratory, Fairfield Hospital, Fairfield 3078, Victoria, Australia **[497]**

David A. Anderson, Virus Laboratory, Fairfield Hospital, Fairfield 3078, Victoria, Australia **[497]**

M. Arita, Department of Enteroviruses, National Institute of Health, Tokyo 190-12, Japan **[437]**

Edward Arnold, Department of Biological Sciences, Purdue University, West Lafayette, IN 47907 **[59]**

Chintamani Atreya, Department of Biological Sciences, Wayne State University, Detroit, MI 48202; present address: Faculty of Medicine, Memorial University, St. John's, Newfoundland A1B 3V6, Canada **[397]**

Ralph S. Baric, Departments of Microbiology and Neurology, University of Southern California School of Medicine, Los Angeles, CA 90033; present address: Department of Parasitology, University of North Carolina, Chapel Hill, NC 27514 **[285]**

H. Barker, Scottish Crop Research Institute, Invergowrie, Dundee DD2 5DA, United Kingdom **[575]**

E.C. Bauer, Department of Pathology, Uniformed Services University of the Health Sciences, Bethesda, MD 20814 **[339]**

D.C. Baulcombe, Plant Breeding Institute, Trumpington, Cambridge CB2 2LQ, United Kingdom **[575]**

D.L. Beck, Department of Plant Pathology, University of California, Riverside, CA 92521 **[47]**

The numbers in brackets are the opening page numbers of the contributors' articles.

M.W. Bevan, Plant Breeding Institute, Trumpington, Cambridge CB2 2LQ, United Kingdom **[575]**

Christof K. Biebricher, Max Planck Institut für Biophysikalische Chemie, D-3400 Göttingen, Federal Republic of Germany **[9]**

Ulrike Boege, Department of Biochemistry, University of Alberta, Edmonton, Alberta T6G 2H7, Canada **[59]**

Leendert Bosch, Department of Biochemistry, University of Leiden, Leiden, The Netherlands **[299]**

François Boulay, Department of Cell Biology, Yale University School of Medicine, New Haven, CT 06510-8002 **[103]**

Carol A. Bova, Department of Biochemistry, Lineberger Cancer Research Center, University of North Carolina, Chapel Hill, NC 27514 **[509]**

J.F. Boyle, Department of Pathology, Uniformed Services University of the Health Sciences, Bethesda, MD 20814 **[339]**

P. Bredenbeek, Institute of Virology, Veterinary Faculty, State University Utrecht, 3508 TD Utrecht, The Netherlands **[137]**

Margo A. Brinton, The Wistar Institute, Philadelphia, PA 19104 **[261]**

Fred Brown, Wellcome Biotechnology Ltd, Beckenham, Kent, United Kingdom **[555]**

Michael J. Buchmeier, Scripps Clinic and Research Foundation, La Jolla, CA 92037 **[409]**

Pia L. Callahan, Merck Sharp and Dohme Research Laboratories, West Point, PA 19486 **[93]**

Marie Chow, Department of Applied Biological Sciences, Massachusetts Institute of Technology, Cambridge, MA 02139 **[79]**

Richard J. Colonno, Merck Sharp and Dohme Research Laboratories, West Point, PA 19486 **[93]**

S. Compton, Department of Microbiology, Uniformed Services University of the Health Sciences, Bethesda, MD 20814 **[339]**

Anthony G. Coulepis, Virus Laboratory, Fairfield Hospital, Fairfield 3078, Victoria, Australia **[497]**

Richard L. Crowell, Department of Microbiology and Immunology, Hahnemann University School of Medicine, Philadelphia, PA 19102 **[453]**

Daniel Cutler, European Molecular Biology Laboratory, D-6900 Heidelberg, Federal Republic of Germany **[351]**

Robert G. Dalziel, Scripps Clinic and Research Foundation, La Jolla, CA 92037 **[409]**

Nancy L. Davis, Department of Microbiology, North Carolina State University, Raleigh, NC 27695-7615 **[467]**

W.O. Dawson, Department of Plant Pathology, University of California, Riverside, CA 92521 **[47]**

Marc DeCamillis, Quadra Logic Technologies Inc, Vancouver, British Columbia V6H 3Z6, Canada **[519]**

Térèse-Marie Denial, Institut Jacques Monod, CNRS, Université Paris VII, 75251 Paris, France **[149]**

Guy D. Diana, Microbiology and Medicinal Chemistry, Sterling-Winthrop Research Institute, Columbia Turnpike, Rensselaer, NY 12144 **[59]**

Mingxaio Ding, Department of Microbiology and Immunology, Washington University School of Medicine, St. Louis, MO 63110 **[183]**

Robert W. Doms, Department of Cell Biology, Yale University School of Medicine, New Haven, CT 06510-8002 **[103]**

Theo W. Dreher, Department of Biology, Texas A&M University, College Station, TX 77843-3258 **[317,327]**

Gabrièle Drugeon, Institut Jacques Monod, CNRS, Université Paris VII, 75251 Paris, France **[149]**

C.S. Duchala, Department of Pathology, Uniformed Services University of the Health Sciences, Bethesda, MD 20814 **[339]**

D.M.A. Evans, National Institute for Biological Standards and Control, Holly Hill, London NW3 6RB, United Kingdom **[539]**

M. Ferguson, National Institute for Biological Standards and Control, Holly Hill, London NW3 6RB, United Kingdom **[539]**

David J. Filman, Department of Molecular Biology, Research Institute of Scripps Clinic, La Jolla, CA 92037 **[79]**

J. Bert Flanegan, Department of Immunology and Medical Microbiology, University of Florida College of Medicine, Gainesville, FL 32610 **[273]**

M. Forsyth, MRC Common Cold Unit, Harvard Hospital, Salisbury, United Kingdom **[487]**

M.F. Frana, Department of Pathology, Uniformed Services University of the Health Sciences, Bethesda, MD 20814 **[339]**

R.I.B. Francki, Department of Plant Pathology, Waite Agricultural Research Institute, University of Adelaide, Glen Osmond, 5064 South Australia **[423]**

Frederick J. Fuller, Department of Microbiology, North Carolina State University, Raleigh, NC 27695-7615 **[467]**

Ricardo Galler, Division of Biology, California Institute of Technology, Pasadena, CA 91125 **[209]**

Henrik Garoff, European Molecular Biology Laboratory, D-6900 Heidelberg, Federal Republic of Germany **[351]**

Susan Gidwitz, Department of Microbiology, North Carolina State University, Raleigh, NC 27695-7615; present address: Department of Medicine, University of North Carolina, Chapel Hill, NC 27514 **[467]**

Rob Goldbach, Department of Molecular Biology, Agricultural University, 6703 BC Wageningen, The Netherlands **[195]**

Diane E. Griffin, Department of Medicine and Neurology, Johns Hopkins University, Baltimore, MD 21205 **[365]**

Janet B. Grun, The Wistar Institute, Philadelphia, PA 19104 **[261]**

Farshad Guirakhoo, Institute of Virology, University of Vienna, Vienna A-1095, Austria **[113]**

Ian D. Gust, Virus Laboratory, Fairfield Hospital, Fairfield 3078, Victoria, Australia **[497]**

Anne-Lise Haenni, Institut Jacques Monod, CNRS, Université Paris VII, 75251 Paris, France **[149]**

Chang S. Hahn, Division of Biology, California Institute of Technology, Pasadena, CA 91125 **[209]**

Young S. Hahn, Division of Biology, California Institute of Technology, Pasadena, CA 91125 **[209]**

Timothy C. Hall, Department of Biology, Texas A&M University, College Station, TX 77843-3258 **[317,327]**

W. Reef Hardy, Division of Biology, California Institute of Technology, Pasadena, CA 91125 **[209]**

Jim Hare, European Molecular Biology Laboratory, D-6900 Heidelberg, Federal Republic of Germany **[351]**

B.D. Harrison, Scottish Crop Research Institute, Invergowrie, Dundee DD2 5DA, United Kingdom **[575]**

S.C. Harrison, Department of Biochemistry and Molecular Biology, Harvard University, Cambridge, MA 02138 **[379]**

Franz X. Heinz, Institute of Virology, University of Vienna, Vienna A-1095, Austria **[113]**

Ari Helenius, Department of Cell Biology, Yale University School of Medicine, New Haven, CT 06510-8002 **[103]**

James M. Hogle, Department of Biochemistry and Molecular Biology, Harvard University, Cambridge, MA 02138; present address: Scripps Clinic and Research Foundation, La Jolla, CA 92037 **[79,379]**

K.V. Holmes, Department of Pathology, Uniformed Services University of the Health Sciences, Bethesda, MD 20814 **[339]**

M.C. Horzinek, Institute of Virology, Veterinary Faculty, State University Utrecht, 3508 TD Utrecht, The Netherlands **[137]**

Kuo-Hom Lee Hsu, Department of Microbiology and Immunology, Hahnemann University School of Medicine, Philadelphia, PA 19102 **[453]**

Henry Huang, Department of Microbiology and Immunology, Washington University School of Medicine, St. Louis, MO 63110 **[241]**

P.J. Hughes, Department of Biology, University of Essex, Essex, United Kingdom **[487]**

Danny Huylebroeck, European Molecular Biology Laboratory, D-6900 Heidelberg, Federal Republic of Germany **[351]**

H. Itoh*, Japan Poliomyelitis Research Institute, Tokyo 189, Japan, **[437]***deceased

Martine Jaegle, Department of Molecular Biology, Agricultural University, 6703 BC Wageningen, The Netherlands **[195]**

Robert E. Johnston, Department of Microbiology, North Carolina State University, Raleigh, NC 27695-7615 **[467]**

Rajiv L. Joshi, Institut Jacques Monod, CNRS, Université Paris VII, 75251 Paris, France **[149]**

Paul Kaesberg, Biophysics Laboratory and Biochemistry Department, University of Wisconsin, Madison, WI 53706 **[3]**

Greg Kamer, Department of Biological Sciences, Purdue University, West Lafayette, IN 47907 **[59]**

N. Kawamura, Department of Microbiology, Faculty of Medicine, University of Tokyo, Tokyo 113, Japan **[437]**

James G. Keck, Departments of Microbiology and Neurology, University of Southern California School of Medicine, Los Angeles, CA 90033 **[285]**

Olen Kew, Division of Viral Diseases, Center for Infectious Diseases, Centers for Disease Control, Atlanta, GA 30333 **[477]**

D.A. Knorr, Department of Plant Pathology, University of California, Riverside, CA 92521 **[47]**

M. Kohara, Department of Microbiology, Faculty of Medicine, University of Tokyo, Tokyo 113, Japan; and Japan Poliomyelitis Research Institute, Tokyo 189, Japan **[437]**

T. Komatsu, Department of Enteroviruses, National Institute of Health, Tokyo 190-12, Japan **[437]**

Marck J.M. Koolen, Scripps Clinic and Research Foundation, La Jolla, CA 92037 **[409]**

Fred Russell Kramer, Institute of Cancer Research and Department of Genetics and Development, College of Physicians and Surgeons, Columbia University, New York, NY 10032 **[35]**

Marcia J. Kremer, Department of Biological Sciences, Purdue University, West Lafayette, IN 47907 **[59]**

S. Kuge, Department of Microbiology, Faculty of Medicine, University of Tokyo, Tokyo 113, Japan **[437]**

Christian Kunz, Institute of Virology, University of Vienna, Vienna A-1095, Austria **[113]**

Michael M.C. Lai, Departments of Microbiology and Neurology, University of Southern California School of Medicine, Los Angeles, CA 90033 **[285]**

Peter W. Lampert, University of California, San Diego, CA 92093 **[409]**

Burton J. Landau, Department of Microbiology and Immunology, Hahnemann University School of Medicine, Philadelphia, PA 19102 **[453]**

Robin Levis, Department of Microbiology and Immunology, Washington University School of Medicine, St. Louis, MO 63110 **[241]**

Stephen A. Locarnini, Virus Laboratory, Fairfield Hospital, Fairfield 3078, Victoria, Australia **[497]**

Ronald B. Luftig, Department of Microbiology, Louisiana State University Medical Center, New Orleans, LA 70112 **[227]**

Ming Luo, Department of Biological Sciences, Purdue University, West Lafayette, IN 47907 **[59]**

Shinji Makino, Departments of Microbiology and Neurology, University of Southern California School of Medicine, Los Angeles, CA 90033 **[285]**

Christian Mandl, Institute of Virology, University of Vienna, Vienna A-1095, Austria **[113]**

Loren E. Marsh, Department of Biology, Texas A&M University, College Station, TX 77843-3258 **[327]**

Richard F. Marsh, Department of Veterinary Science, University of Wisconsin-Madison, Madison, WI 53706 **[519]**

M.A. Mayo, Scottish Crop Research Institute, Invergowrie, Dundee DD2 5DA, United Kingdom **[575]**

Melody R. McClure, LBI-Basic Research Program, NCI-Frederick Cancer Facility, Frederick, MD 21701 **[227]**

J.J. McGowan, Department of Microbiology, Uniformed Services University of the Health Sciences, Bethesda, MD 20814 **[339]**

Mark A. McKinlay, Microbiology and Medicinal Chemistry, Sterling-Winthrop Research Institute, Columbia Turnpike, Rensselaer, NY 12144 **[59]**

Paul Melancon, European Molecular Biology Laboratory, D-6900 Heidelberg, Federal Republic of Germany **[351]**

Donald R. Mills, Institute of Cancer Research and Department of Genetics and Development, College of Physicians and Surgeons, Columbia University, New York, NY 10032 **[35]**

P.D. Minor, National Institute for Biological Standards and Control, Holly Hill, London NW3 6RB, United Kingdom **[539]**

Marie-Dominique Morch, Institut Jacques Monod, CNRS, Université Paris VII, 75251 Paris, France **[149]**

Anne G. Mosser, Biophysics Laboratory, University of Wisconsin, Madison, WI 53706 **[59]**

Carol D. Murphy, Department of Immunology and Medical Microbiology, University of Florida College of Medicine, Gainesville, FL 32610 **[273]**

A. Nomoto, Department of Microbiology, Faculty of Medicine, University of Tokyo, Tokyo 113, Japan **[437]**

Baldev K. Nottay, Division of Viral Diseases, Center for Infectious Diseases, Centers for Disease Control, Atlanta, GA 30333 **[477]**

Steven M. Oberste, Department of Immunology and Medical Microbiology, University of Florida College of Medicine, Gainesville, FL 32610 **[273]**

Mark A. Pallansch, Division of Viral Diseases, Center for Infectious Diseases, Centers for Disease Control, Atlanta, GA 30333 **[477]**

Ann C. Palmenberg, Biophysics Laboratory, University of Wisconsin, Madison, WI 53706 **[25]**

David F. Pence, Department of Microbiology, North Carolina State University, Raleigh, NC 27695-7615 **[467]**

Cornelis W.A. Pleij, Department of Biochemistry, University of Leiden, Leiden, The Netherlands **[299]**

Christine Priano, Department of Genetics and Development, College of Physicians and Surgeons, Columbia University, New York, NY 10032 **[35]**

Robert H. Purcell, Hepatitis Viruses Section, Laboratory of Infectious Diseases, National Institute of Allergy and Infectious Diseases, NIH, Bethesda, MD 20892 **[565]**

Alan Rein, LBI-Basic Research Program, NCI-Frederick Cancer Facility, Frederick, MD 21701 **[227]**

Charles M. Rice, Division of Biology, California Institute of Technology, Pasadena, CA 91125; present address: Department of Microbiology and Immunology, Washington University School of Medicine, St. Louis, MO 63110 **[209]**

Nancy R. Rice, LBI-Basic Research Program, NCI-Frederick Cancer Facility, Frederick, MD 21701 **[227]**

Rebeca Rico-Hesse, Division of Viral Diseases, Center for Infectious Diseases, Centers for Disease Control, Atlanta, GA 30333 **[477]**

Krijn Rietveld, Department of Biochemistry, University of Leiden, Leiden, The Netherlands; present address: Gist Brocades NV, P.O. Box 1, 2600 Delft, The Netherlands **[299]**

S.G. Robbins, Department of Microbiology, Uniformed Services University of the Health Sciences, Bethesda, MD 20814 **[339]**

D'Ann Rochon, Department of Biological Sciences, Wayne State University, Detroit, MI 48202; present address: Agriculture Canada, Vancouver, British Columbia V6T 1X2, Canada **[397]**

Laurie Roman, European Molecular Biology Laboratory, D-6900 Heidelberg, Federal Republic of Germany **[351]**

Bruce C. Ross, Virus Laboratory, Fairfield Hospital, Fairfield 3078, Victoria, Australia **[497]**

Michael G. Rossmann, Department of Biological Sciences, Purdue University, West Lafayette, IN 47907 **[59]**

P.J.M. Rottier, Institute of Virology, Veterinary Faculty, State University Utrecht, 3508 TD Utrecht, The Netherlands **[137]**

Roland R. Rueckert, Biophysics Laboratory, University of Wisconsin, Madison, WI 53706 **[59]**

Dorothea Sawicki, Department of Microbiology, Medical College of Ohio, Toledo, OH 43614 **[251]**

Stanley Sawicki, Department of Microbiology, Medical College of Ohio, Toledo, OH 43614 **[251]**

G.C. Schild, National Institute for Biological Standards and Control, Holly Hill, London NW3 6RB, United Kingdom **[539]**

Milton J. Schlesinger, Department of Microbiology and Immunology, Washington University School of Medicine, St. Louis, MO 63110 **[183]**

Sondra Schlesinger, Department of Microbiology and Immunology, Washington University School of Medicine, St. Louis, MO 63110 **[241]**

Alan L. Schmaljohn, Department of Microbiology, University of Maryland School of Medicine, Baltimore, MD 21201 **[365]**

Alan M. Schultz, LBI-Basic Research Program, NCI-Frederick Cancer Facility, Frederick, MD 21701 **[227]**

Maggie Schultz, Department of Microbiology and Immunology, Hahnemann University School of Medicine, Philadelphia, PA 19102 **[453]**

Douglas G. Scraba, Department of Biochemistry, University of Alberta, Edmonton, Alberta T6G 2H7, Canada **[59]**

B.L. Semler, Department of Microbiology and Molecular Genetics, College of Medicine, University of California, Irvine, CA 92717 **[437]**

John G. Shaw, Department of Plant Pathology, University of Kentucky, Lexington, KY 40546-0091 **[159]**

Barbara Sherry, Biophysics Laboratory, University of Wisconsin, Madison, WI 53706 **[59]**

Chien-Kou Shieh, Departments of Microbiology and Neurology, University of Southern California School of Medicine, Los Angeles, CA 90033 **[285]**

Stuart G. Siddell, Institute of Virology, Versbacher Str.7, D-8700 Wurzburg, Federal Republic of Germany **[127]**

Albert Siegel, Department of Biological Sciences, Wayne State University, Detroit, MI 48202 **[397]**

Thomas J. Smith, Department of Biological Sciences, Purdue University, West Lafayette, IN 47907 **[59]**

E.J. Snijder, Institute of Virology, Veterinary Faculty, State University Utrecht, 3508 TD Utrecht, The Netherlands **[137]**

Lisa Soe, Departments of Microbiology and Neurology, University of Southern California School of Medicine, Los Angeles, CA 90033 **[285]**

P.K. Sorger, Department of Biochemistry and Molecular Biology, Harvard University, Cambridge, MA 02138; present address: MRC Laboratory for Molecular Biology, Hills Road, Cambridge CB2 2QH, United Kingdom **[379]**

W.J.M. Spaan, Institute of Virology, Veterinary Faculty, State Universty Utrecht, 3508 TD Utrecht, The Netherlands **[137]**

G. Stanway, Department of Biology, University of Essex, Essex, United Kingdom; present address: National Institute for Biological Standards and Control, Holly Hill, London NW3 6RB, United Kingdom **[487,539]**

C.B. Stephensen, Department of Pathology, Uniformed Services University of the Health Sciences, Bethesda, MD 20814 **[339]**

P.G. Stockley, Department of Biochemistry and Molecular Biology, Harvard University, Cambridge, MA 02138; present address: Biotechnology Unit, Department of Genetics, University of Leeds, Leeds, LS2 9JT, United Kingdom **[379]**

Stephen A. Stohlman, Departments of Microbiology and Neurology, University of Southern California School of Medicine, Los Angeles, CA 90033 **[285]**

Mary Merchant Stokes, Department of Immunology and Medical Microbiology, University of Florida College of Medicine, Gainesville, FL 32610 **[273]**

Ellen G. Strauss, Division of Biology, California Institute of Technology, Pasadena, CA 91125 **[209,365]**

James H. Strauss, Division of Biology, California Institute of Technology, Pasadena, CA 91125 **[209,365]**

R.K. Strong, Department of Biochemistry and Molecular Biology, Harvard University, Cambridge, MA 02138 **[379]**

L.S. Sturman, New York State Department of Health, Albany, NY 12201 **[339]**

Paul S. Sunga, Quadra Logic Technologies Inc, Vancouver, British Columbia V6H 3Z6, Canada **[519]**

Ronald Swanstrom, Department of Biochemistry, Lineberger Cancer Research Center, University of North Carolina, Chapel Hill, NC 27514 **[509]**

Fumihiro Terami, Department of Biological Sciences, Wayne State University, Detroit, MI 48202 **[397]**

Gregory J. Tobin, Department of Immunology and Medical Microbiology, University of Florida College of Medicine, Gainesville, FL 32610 **[273]**

Joanne E. Tomassini, Merck Sharp and Dohme Research Laboratories, West Point, PA 19486 **[93]**

Manuel Tsiang, Department of Microbiology and Immunology, Washington University School of Medicine, St. Louis, MO 63110 **[241]**

Wolfgang Tuma, Institute of Virology, University of Vienna, Vienna A-1095, Austria **[113]**

D.A.J. Tyrrell, MRC Common Cold Unit, Harvard Hospital, Salisbury, United Kingdom **[487]**

Rosaura Valle, Institut Jacques Monod, CNRS, Université Paris VII, 75251 Paris, France **[149]**

Diane Van Alstyne, Quadra Logic Technologies Inc, Vancouver, British Columbia V6H 3Z6, Canada **[519]**

Alex van Belkum, Department of Biochemistry, University of Leiden, Leiden, The Netherlands **[299]**

M.F. van Berlo, Institute of Virology, Veterinary Faculty, State University Utrecht, 3508 TD Utrecht, The Netherlands **[137]**

Bernard A.M. Van der Zeijst, Institute of Virology, Veterinary Faculty, State University Utrecht, 3508 TD Utrecht, The Netherlands; present address: University of Utrecht, Department of Infectious Diseases, Section Bacteriology, 3508 TD Utrecht, The Netherlands **[137]**

Ab van Kammen, Department of Molecular Biology, Agricultural University, 6703 BC Wageningen, The Netherlands **[195]**

W.C. Vooys, Institute of Virology, Veterinary Faculty, State University Utrecht, 3508 TD Utrecht, The Netherlands **[137]**

Pieter Vos, Department of Molecular Biology, Agricultural University, 6703 BC Wageningen, The Netherlands **[195]**

Gerrit Vriend, Department of Biological Sciences, Purdue University, West Lafayette, IN 47907 **[59]**

D.G. Weismiller, Department of Pathology, Uniformed Services University of the Health Sciences, Bethesda, MD 20814 **[339]**

Barbara G. Weiss, Department of Microbiology and Immunology, Washington University School of Medicine, St. Louis, MO 63110 **[241]**

Joan Wellink, Department of Molecular Biology, Agricultural University, 6703 BC Wageningen, The Netherlands **[195]**

Duanzhi Wen, Department of Microbiology and Immunology, Washington University School of Medicine, St. Louis, MO 63110 **[183]**

R.K. Williams, Department of Pathology, Uniformed Services University of the Health Sciences, Bethesda, MD 20814 **[339]**

T. Michael A. Wilson, Department of Virus Research, John Innes Research Institute, Colney Lane, Norwich NR4 7UH, United Kingdom **[159]**

E. Wimmer, Department of Microbiology, School of Medicine, State University of New York, Stony Brook, NY 11794 **[437]**

Günther Winkler, Institute of Virology, University of Vienna, Vienna A-1095, Austria **[113]**

Dorothy C. Young, Department of Immunology and Medical Microbiology, University of Florida College of Medicine, Gainesville, FL 32610 **[273]**

Marino Zerial, European Molecular Biology Laboratory, D-6900 Heidelberg, Federal Republic of Germany **[351]**

Preface

The classification of viruses by their genome structure and transcriptional strategy was initiated in Baltimore in 1971. According to this system, the positive-strand RNA viruses are those whose genomes function as an mRNA. Seven families of animal viruses, almost all of the plant viruses, and 1 family of bacteriophages are of this type. Diseases caused by these viruses are of medical, agricultural and economic significance.

Positive-strand RNA viruses have been major players in the development of virology. In the late 1890's, it was demonstrated with both tobacco mosaic virus (TMV), and foot and mouth disease virus that disease transmissibility was associated with a bacterial-free filtrate. The proof of serial transmission of TMV by Beijerinck in 1899 represented the beginning of virological research. TMV was also the first virus to be crystallized and the one for which infectivity of the genome RNA was first demonstrated. The discovery by Rous that virus infection could induce tumor formation was made in chickens with the positive-strand RNA leukosis virus. Yellow fever virus was the first virus to be identified as the cause of a human disease. Enders, Weller and Robbins first demonstrated that viruses could multiply to high titers in explanted cell cultures, using poliovirus. The first animal virus plaque test was developed using Western equine encephalitis virus.

Even with this impressive list of "firsts", detailed information about the positive-strand RNA viruses has been slow in coming. One reason for this is that many of these viruses have been difficult to cultivate to high titers. The recent application of molecular biological and x-ray crystalographic techniques to the analysis of positive-strand RNA viruses has led to an explosion of new information. The results of these analyses has indicated previously unsuspected nucleic acid, amino acid and barrel structure homologies among positive-stranded RNA viruses and imply that genetic exchanges have occurred between plant and animal viruses. This exciting realization, together with the recent recognition of positive-strand RNA virus genetic recombination in both plants and animals, have made it clear that plant and animal virologists, who traditionally have conversed infrequently, now have much in common to discuss. It, therefore, seemed timely to organize a meeting of positive-strand RNA virologists to encourage new scientific interactions and technical approaches to the study of these viruses. The Keystone meeting succeeded in accomplishing these goals and it is hoped, also defined a group of scientists who will continue to meet and exchange information.

We gratefully acknowledge substantial support from UCLA Symposia Director's Sponsors—Burroughs-Wellcome Company, Monsanto Corporation Research Laboratories, The Upjohn Company—and Grant Number DAMD 17-86-G-6004 from the United States Army Medical Research and Development Command. We also thank Merieux Institute for contributing to this meeting. Additional support was provided by gifts from contributors to the Director's Fund Biosciences Laboratory, Corporate Technology, and Allied-Signal Corporation; AMOCO Corporation; Merieux Institute; Allelix, Inc.; AMGen, Inc.; Boehringer Ingelheim Pharmaceuticals, Inc.; Hoechst-Roussel Pharmaceuticals, Inc.; New England Biolabs Foundation; Ayerst Laboratories Research, Inc.; Bristol-Myers Company; Celanese Research Company; Wyeth Laboratories, and Codon Corporation.

Margo A. Brinton
Roland R. Rueckert

I. GENOME STRUCTURE AND VIRAL EVOLUTION

Positive Strand RNA Viruses, pages 3–7

EVOLUTION AND GENOME STRUCTURE[1]

Paul Kaesberg

Biophysics Lab. and Biochemistry Dept., U. Wisconsin
Madison, Wisconsin 53706

ABSTRACT Most of our insight into the origin and evolution of viruses comes from comparative studies of viral genomes and their encoded proteins. Positive strand RNA viruses, as we know them, probably originated rather late in evolutionary time and evolved by mechanisms that included RNA recombination.

INTRODUCTION

In looking over the program and the list of participants it's not difficult to come to the conclusion that this is a momentous meeting--a meeting all of us will recall again and again in our writing and in our memory. We are very much indebted to the organizers, Margo Brinton and Roland Rueckert.

You will note that the opening session includes in its title both Evolution and Genome Structure--a rather speculative subject and a very concrete one. The subject of the evolution of viruses invariably generates great interest. Nevertheless, the number of scientists who admit to working in this area is small and the number of noteworthy papers is correspondingly small. There are several reasons for this.

Until recently there has not been a well-defined body of experimental techniques that relate to viral evolution. Now there exist sequencing and the associated computer

[1]This work was supported by NIH Public Health Service Grants AI-1466 and AI-15342 and Career Award AI-21942.

procedures for comparing viruses, genetic engineering techniques for producing defined modifications of viruses, and innovative methods for letting mutation take its course at the molecular level. All these topics will be dicussed in the papers that follow. All their methodologies can provide evidence of evolution but can also provide evidence of other important matters relating to genomes and their expression. Thus "Genome Structure" in the title of this session was meant to imply that all aspects of RNA genomes were fair game.

VIRUS SIMILARITIES

Even though primary, secondary and tertiary structure of genomes and their encoded proteins are intimately related, it is often only one of these aspects that provides striking evidence of virus similarities, similarities that may be related to geometry as well as to function or evolution. Thus the existence and universality of the icosahedral structure of viruses (1) ranging in size from the smallest of plant viruses to tipula iridescent virus, some thousand times more massive, may demonstate what virion structures are possible and likely. The existence of the β-barrel crystallographic structure (See the papers of Hogle and of Rossmann, below) in all viruses examined thus far may signify a common function. A successful search for non-viral β-barrels could have implications for viral evolution. Amino acid sequence comparisons show no evidence of these remarkable similarities in tertiary structure. On the other hand all known primary sequences of the genomes of positive strand viruses translate into a protein region having noteworthy amino acid sequence similarity in the so-called GDD region (2) and signify a commonality in the mechanism of action of all viral RNA replicases. A lesser, though significant, similarity exists with retrovirus reverse transcriptases (2) and is indicative of functional or evolutionary relatedness. In two instances a more extensive similarity has been demonstrated in the amino acid sequences of genome encoded proteins. All of the animal picornaviruses show extensive similarity suggestive of close homology of both structure and function (See the paper of Palmenberg, below). More pointedly, the plant virus cowpea mosaic virus shows extensive similarity to the picornaviruses in amino acid sequence, in encoded functions and in gene order (See the paper of van Kammen, below). The plant bromovirus, alfalfa mosaic virus, tobamovirus and cucumovirus groups and the animal alpha viruses share three domains of amino acid sequence similarity even though these domains are expressed differently at the level of translation (See the paper of Strauss, below). This suggests that divergence of mechanism of genome expression does not necessarily indicate a dissimilarity of encoded functions.

ON THE ORIGIN OF POSITIVE STRAND VIRUSES

A second important reason for the paucity of scientists admitting to studying evolution of viruses is that virus evolution is closely bound to the origin of viruses and there is no consensus on virus origin. Indeed the known diversity of viruses suggests their multiple origin. There is no "big bang" theory of virology.

Several years ago David Baltimore discussed the origin and evolution of RNA-based viruses in a meeting whose Proceedings were published in the Annals of the New York Academy of Sciences (3). In that definitive paper he reached several conclusions that merit repeating (although I won't here give the arguments). 1) The character of the cells from which viruses originally emerged is unknown. A plausible case can be made for either DNA-based cells or primitive RNA-based cells. Possibly RNA viruses are the only remaining vestige of a former world of RNA-based cells. 2) There are cogent reasons for believing in a common origin of the replicating entity of positive strand RNA viruses of eukaryotes. The evidence for a common origin of the replicating entity of all RNA viruses, including negative strand viruses, double-stranded RNA viruses, retroviruses and RNA phages is less secure. 3) Accepting the premise that the four kinds of RNA viruses of eukaryotes arose separately, the existence of RNA recombination nevertheless provides a ready mechanism for exchange of genes among them.

RNA RECOMBINATION

The virus similarities alluded to above suggest that RNA recombination is likely to be intimately involved in the origin of positive strand viruses. Such recombination provides an obvious mechanism for promoting the association of genes from different sources, an association that might be regarded as the defining characteristic of a virus. Such a scenario of an association of functional domains suggests that positive strand RNA viruses (and possibly all viruses) are relative latecomers in evolution and arose from associations of proteins or domains that already possessed well defined functions (4). Modifications of these functions, together with the acquisition and modification of other functions appropriate for proliferation in their particular hosts, would have resulted in what we recognize to be viruses today.

RNA RECOMBINATION IN PLANTS

Although RNA recombination is well known in the case of animal viruses, to date its existence in the large world of plant viruses has not been reported and this has been widely regarded as arguing against the central involvement of recombination in RNA virus origin and evolution.

Josef Bujarski and I have now found quite unequivocal evidence of RNA recombination in brome mosaic virus (BMV) (5). The genome of BMV is made up of three RNAs designated RNA1, RNA2, and RNA3. The 3' terminal region of all three RNAs is similar both in sequence and in tertiary structure. This structure has been implicated in several important viral functions. We have deleted a 20 base long stem-and-loop region in the 3' terminal region of BMV RNA3 to produce a modified virus that is still infectious but is perhaps less viable than wild-type. Upon prolonged infection, we find that RNA3 no longer has the modified sequence. Sequence analysis of several recombinants shows that the region from the original modification to the 3' terminus has been replaced by RNA1 or RNA2 sequence of similar but not always identical length. No evidence remains of the originally modified RNA3. We judge that active RNA recombination has produced a more capable RNA3.

It thus seems that all positive strand RNA viruses have had available to them mechanisms both for changing individual bases and for changing and exchanging substantial chunks of RNA--chunks of RNA that may be genes or groups of genes. And it is one of our tasks to determine how these changes have taken place and how it all started.

REFERENCES

1. Kaesberg P (1956). Structure of Small "Spherical" Viruses. Science 124:626.
2. Kamer G, Argos P (1984). Primary Structural Comparison of RNA-dependent Polymerases from Plant, Animal and Bacterial Viruses. Nucl Acids Res 12:7269.
3. Baltimore D (1980). Evolution of RNA Viruses. Annals N Y Acad Sci 354:492.
4. Rossmann M, Abad-Zapatero C, Murthy M, Lilias L, Jones A, Strandberg B (1983). Structural Comparisons of Some Small Spherical Plant Viruses. J Mol Biol 165:711.
5. Bujarski J, Kaesberg P (1986). Genetic Recombination Between RNA Components of a Multipartite Plant Virus. Nature In Press.

Positive Strand RNA Viruses, pages 9–23

REPLICATION AND SELECTION KINETICS OF SHORT-CHAINED RNA SPECIES

Christof K. Biebricher

Max-Planck-Institut für Biophysikalische Chemie
D-3400 Göttingen, Fed. Rep. of Germany

ABSTRACT In vitro replication of short-chained self-replicating RNA species is a suitable model system for the study of evolution and selection at the molecular level. Phenotypic expression of a genotype is reduced to its ability to be effectively replicated by Qβ replicase added as highly purified preparation and constituting an environmental factor. The high template specificity of Qβ replicase is due to kinetic control of formation of initiation complexes. A minimal replication mechanism can be written by compiling the elementary steps of chain initiation, elongation and termination. The effects of double strand formation between plus and minus strands, of asymmetric synthesis in the plus and the minus cycles and the complicated competition behavior between several RNA species have been determined by mathematical analysis, leading to compact equations, and by computer simulations. The computed concentration profiles agreed with the experimental profiles. A virtually unlimited number of different self-replicating RNA species can be obtained by template-free RNA synthesis with Qβ replicase: after long lag times non-reproducible self-replicating RNA species are produced de novo. When amplification of RNA is suppressed by omission of one triphosphate, a slow condensation of triphosphates to oligonucleotides is observed to be catalysed by Qβ replicase.

INTRODUCTION

Plus-strand viruses have remarkably simple replication systems: one enzyme, the replicase, is (nearly) sufficient to catalyse all necessary steps in amplifying the RNA some 10^5 fold starting from one strand per cell (1). The replicase of coliphage Qβ is a stable enzyme that can be purified readily from infected cells to homogeneity (2,3). For in vitro studies of replication, replicase serves as an extraneous environmental factor. Protein synthesis does not occur and there is no selection for the correct message. The phenotypic expression of the genotype is reduced to its ability to reproduce effectively with Qβ replicase under the prevailing experimental conditions.

RESULTS AND DISCUSSION

Template Specificity of Qβ Replicase.

In vivo, Qβ replicase amplifies viral RNA specifically while ignoring cellular RNA (4). Part of the information for replication must thus reside on the RNA, which thus shares with the replicase a complementary catalytic role. In vitro, other RNA templates are also accepted by Qβ replicase: poly(C) and C-containing nucleotide copolymers direct the transcription into complementary RNA (6-8). The reaction requires high template and nucleoside triphosphate concentrations; no autocatalytic amplification occurs. Synthesis of cRNA from natural RNA by Qβ replicase has been reported (8,9), but yields were low. Directing its own transcription, however, is not sufficient for an RNA to be reproduced; only when the transcript also directs synthesis and when template, replica and replicase are released at the end of each replication round, does an autocatalytic growth occurs in which each strand cross-catalytically directs the synthesis of its complement. In the absence of translation, the assignment of plus and minus strands is arbitrary.

A class of self-replicating RNA species with chain lengths in the range of 70 to 250 nucleotides is amplified autocatalytically by Qβ replicase to large levels. Such 'variants' are found in vivo in infected cells as a heterogeneous RNA sedimenting around 6 S (10), and in vitro as products of template-free incorporations (11-14), probably synthesized by amplifying 6 S RNA impurities present in the replicase preparations (15). Sequence comparisons of several RNA species (13,14,16) revealed no recognition signals except for a C cluster at the 3'-end, which is necessary but not sufficient for replication. Indeed, it can be shown that the information of an RNA for being accepted for replication does not depend on its primary structure alone: One RNA species, SV-11, can be folded into alternative tertiary structures, only one of which is replicated by Qβ replicase (17).

Recognition is not a static but rather a dynamic process controlled kinetically by the life-times of the initiation complexes. Replicase binds rapidly to most single- and double-stranded RNAs and DNAs (18-20). The dissociation rates of most complexes are high, whereas dissociation rates of enzyme complexes with optimized templates are extremely slow, with half-lifes of the complexes on the order of 1 day (21). The 3'-end of the RNA must be recognized and at least two molecules of GTP must be bound to it before priming by phosphodiester formation can take place. Complex formation is favored by high concentrations of template and/or GTP and template specificity of the replicase may be partially overcome by using such conditions (22).

Replication kinetics.

Successive complexation of nucleoside triphosphates followed by phosphodiester formation and pyrophosphate release probably occurs by Watson-Crick base-pairing to a short double-helical region at the replication fork (23). As chain elongation progresses, separation of replica and template must occur by some not yet understood mechanism. After completion of the replica it is adenylated and released first (24,25). The resulting inactive template-replicase complex must be recycled by dissociation. A complete mechanism of replication can be constructed by combining all steps involved in replication. It consists of two cross-catalytic cycles, one for each of to the two complementary strands (26).

Incorporation profiles of template-instructed replication show two clearly distinguishable growth phases: at low RNA concentrations and high enzyme excess, exponential growth is observed until the enzyme is saturated with template. Incorporation continues then linearly until it eventually levels off due to product inhibition. The linear growth rate is defined as the number of strands synthesized per enzyme molecule and minute; it can be calculated from the slope if the enzyme concentration and chain length and nucleotide composition of the RNA are known. For the determination of the growth rate in the exponential phase the displacement of the curve on the time axis must be used (27) because incorporation of radioactive nucleotides is usually not sensitive enough to measure the very low RNA concentrations in the early exponential phase (Fig. 1). The growth rates in the exponential phase are typically three times the growth rates in the linear phase (27). The simultaneous action of several replicase molecules on the same strand (28) can be excluded for the short-chained RNA species (22). Instead the slow reactivation step of the replication is responsible for the higher speed in the exponential phase: the new replica can initiate a new replication round before template and enzyme are recycled. In the linear phase, however, this step is rate-controlling. A new chain is initiated typically in a few seconds, chain elongation including release of the replica is completed in 10 - 30 s, while dissociating the inactive template-replicase complex requires 100 - 300 s. The last step is also dependent on the triphosphate concentration (26,29); however, the role and fate of this triphosphate are unknown.

The replication rate in the linear phase was independent of the enzyme concentration from 50 nM up to the measured maximum of 500 nM; below 50 nM the rate drops. The replication rates in the exponential phase were independent of enzyme concentration above 100 nM; the slow decline in the concentration range below is due to the increasing contribution of the bimolecular step to the overall replication rate (26,27,30).

Interaction of Complementary Strands

Analysis of the incorporation products revealed that in the late linear growth phase the majority of the RNA species is present predominantly as double strand formed by reaction of the free complementary strands with each other. Template is irreversibly lost by that reaction, because double strands were found to be unable to repli-

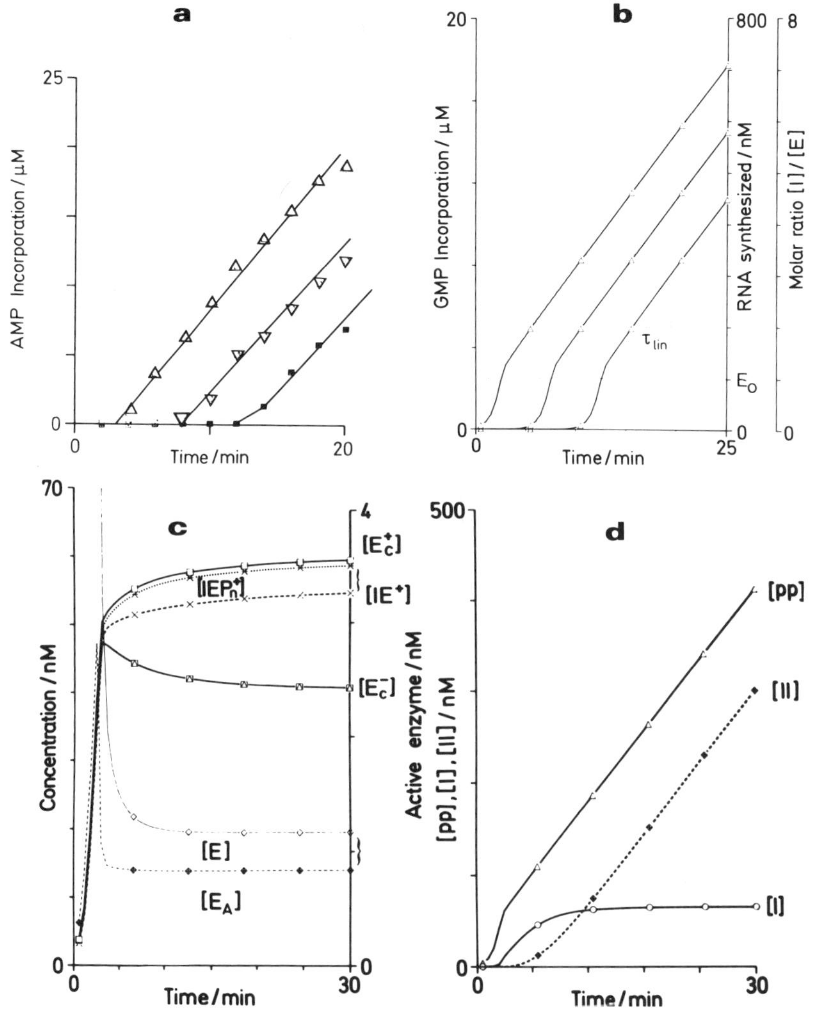

FIGURE 1. Incorporation profiles. a: Experimental determination of the replication rates of MNV-11; b: Simulated incorporation profiles; c: In the linear growth phase, most enzyme is present in inactive complexes $[IE^+]$, $[IEP_n^+]$ (left scale), only a small part is active ($[E_A]$, $[E]$, right scale); d: nucleotides are predominantly incorporated in the exponential growth phase into enzyme bound single strands $[E_c^+]$, in the early linear growth phase into free single strands $[I]$, in the late linear growth phase into double strands $[II]$.

cate (17,25,28). Eventually a steady state is formed where formation of template strands by replication is balanced by loss via double strand formation. In the late linear growth phase, nucleotide incorporation results only in increasing the concentration of double strands (25). The rate of double strand formation depends upon the sequence complexity and the degree of intramolecular base-pairing. All short-chained self-replicating RNA species have evolved unusually strong secondary structures to slow down double strand formation and to prevent replica and template from re-forming a double strand during chain elongation, but large differences among different species remain in the measured steady concentration of templates. For all species with chain lengths around 100, single strands were found to be predominantly complexed to enzyme (where they are more protected from double strand formation) and the steady state concentration of free strands is low. Exclusive synthesis of double strands begins then shortly after saturation of the enzyme with template. Species with chain lengths around 200, most distinctly MDV-1, reach concentrations of free template in the late linear phase exceeding the enzyme concentration. With accumulation of double stranded RNA the incorporation rate drops due to their competition with template for free enzyme; product inhibition is thus observed (29).

The production of double strands masks plus/ minus asymmetry in the synthesis cycles of plus and minus strands; asymmetry ratios between corresponding elementary rates in the cross-catalytic cycles were usually found in the range of 1 to 2 (25).

Competition Between Species

The competition and selection behavior among different species is generally too complicated to make quantitative predictions. Viral RNA species competing in one cell supply each other with their expression products and a defect often becomes noticeable only after recloning. For RNA species replicating in vitro mutations can only work cis, i.e. they affect only the expression of the mutated RNA itself. However, different species in one test tube share the same environment including enzyme, substrate and other growth influencing components. A reasonable quantitative description for selective value is its relative population change $(1/{}^{i}I_0)d^{i}I_0/dt$; species with the highest selective values are enriched in the population (30). At high dilutions of the RNA strands with large excesses of substrate and enzyme, RNA species grow independently as they would in the absence of the other species. The selection value equals the overall replication rate in the exponential growth phase (31). Experiments done by different research groups confirm these results (30,32).

In the linear growth phase, however, when enzyme is saturated with template, different species must compete for the enzyme, thus interfering with each other's growth. The selective value is no longer determined by the overall growth rate. Instead, the rate of template-enzyme binding becomes crucial for selection since, once a template has succeeded in binding a free enzyme molecule, it will replicate, complex formation being nearly irreversible (20,30). If one species reaches steady state by double strand formation, its concentration change and thus its selective value vanish. Another species present in much smaller concentration, however, has a negligible rate of double strand formation and its selective value is positive. As long as its concentration remains low compared to the enzyme concentration it grows exponentially with the rate it would if only the steady state concentration of free enzyme would be present and is thus rapidly enriched in the population. When its concentration rises to levels where double strand formation becomes noticeable its relative growth and its selective value decrease because of the loss of template by double strand formation. Both species compete then for template. Eventually a new steady state is reached where both species coexist in steady state concentrations determined only by their rate constant values of template binding and double strand formation. The agreement between analytical equations, computer simulations and experimental results is excellent (30).

Generally, in the exponential growth phase species with shorter chains are favored due to their higher rates of enzyme reactivation. In the linear growth phase, the overall replication rate of species with shorter chain lengths is also higher, but selection favors instead species with higher chain lengths because their rate constant values are higher for template binding and lower for double strand formation (Fig. 2).

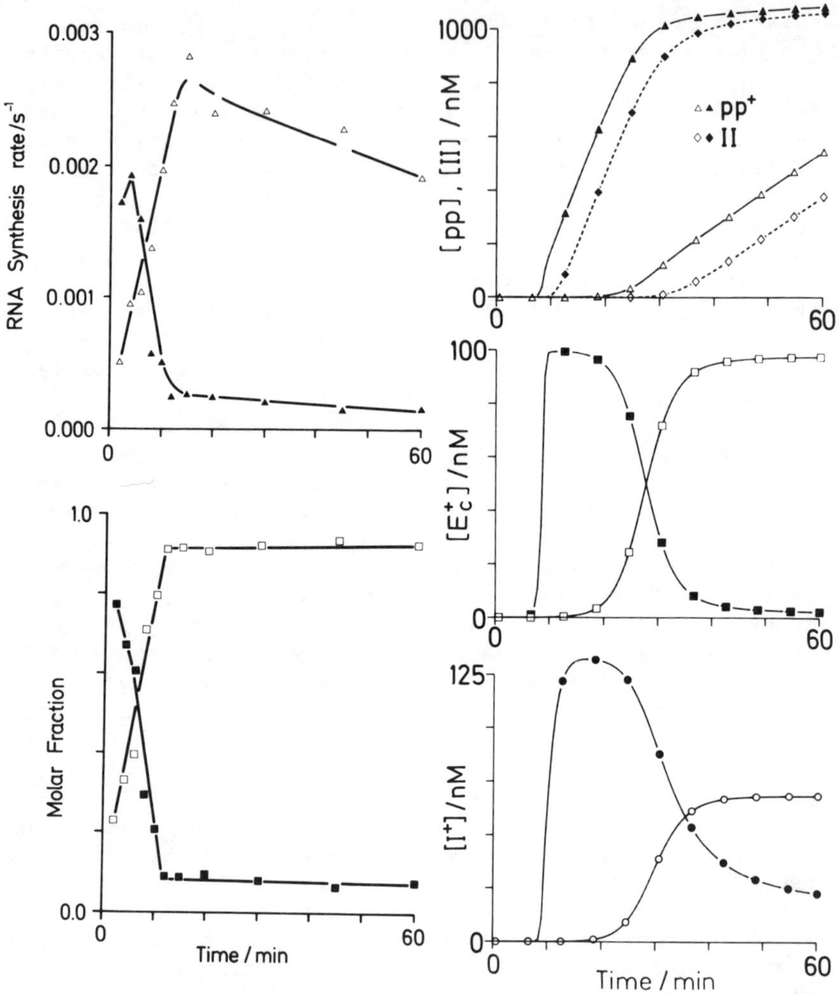

FIGURE 2. Competition between two species. Left: experimental. The more slowly replicating MDV-1 (open symbols) rapidly takes over in the linear growth phase of MNV-11 (filled symbols) by capturing the majority of enzyme; eventially both species coexist.
Right: computer simulation. Species 2 (open symbols) is rapidly outgrown in the exponential phase by species 1 (filled symbols) which has a faster enzyme reactivation rate and saturates the enzyme after 10 min. Species 2 grows up exponentially and captures most of the enzyme because of its higher rate of enzyme binding.

The Quasispecies

Populations of nucleic acids, particularly of RNA, synthesized in vivo or in vitro contain many mutants. Mutation frequencies differ for different sequence positions ('hot spots'), often attributed to different mutation rates. Even though mutation rates do differ, the mutation frequency is mainly controlled by the error propagation during the autocatalytic reproduction (31). Mutations severely affecting reproduction are most frequent, however, the populations of the resulting mutants remain small. In contrast, mutants with nearly neutral mutations accumulate. After an indefinite number of replication rounds a steady state is reached where each mutant has a constant relative abundance determined only by its synthesis rates by mutation and replication and by its decomposition rates (31). Experimental evidence has established the microheterogeneity of viral RNAs (33,34) and of self-replicating RNAs synthesized in vitro (13), but extensive sampling of the genotypes in a population remains to be done.

Point mutation error rates not based on error frequency measurements were determined for some DNA polymerases (35) using viral DNAs; however, insertions and deletions remained undetected due to their high impact of these mutations on gene expression. Measurements with Qβ viral RNAs are less accurate because of the undefined replication conditions in vivo (36). We found high frequencies of slippage errors for short-chained self-replicating RNA; attempts to compare sequences of cloned RNA indicated high mutation rates and high mutation fixation rates, since quasispecies distributions were restored during the 30 generations required to amplify single RNA strands to macroscopic appearance. The selective values calculated as described can be used to describe the competition success of the variants; however, because of the sequence homologies of variants (in contrast to species) the rates of formation of hybrid double strands have also to be taken into account. In most cases members of a quasispecies distribution form hybrid strands with almost the same rate constant values as they have for forming homologous double strands. Even for mutants present in very low concentration formation of double strands can not be neglected when the total concentration of strands is high. Analytical equations and computer simulations show that stable co-existence - as was observed with species - only occurs among variants when formation of homologous strands is markedly favored (30). Experimental confirmation of this result (in progress) is hampered by the difficulty of distinguishing variants. Sensitivity of the double strands against RNases indicates the presence of hybrid double strands. As expected, the heterogeneity is concentrated at a few positions, which could be determined by 5'-end sequencing of the specific fragments (Fig. 3).

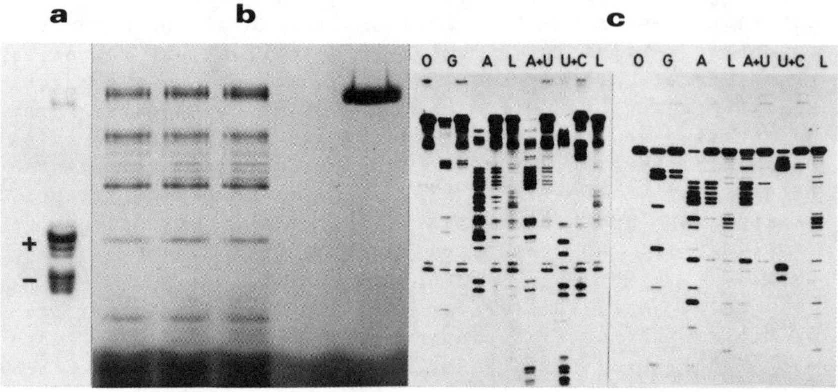

FIGURE 3. Evidence for sequence heterogeneity.
a: Electropherogram of single stranded RNA from cloned MNV-11 (d = double strand).
b: double stranded MNV-11 RNA treated with RNase A (right lane) reveals single strand breaks after melting (left lanes).
c: Sequencing of plus strand MNV-11 is ambiguous; the ambiguity is removed in the sequence of a band from electropherogram b.

Synthesis without template instruction

When high concentrations of Qβ replicase are added to incorporation mixtures in the absence of a template, after long lag times a sudden outgrowth of self-replicating RNA species is observed (3,27,37). The kinetics of this reaction differs clearly from that of template-instructed processes, which excludes amplification of self-replicating RNA present as impurities in the solution (27). The resulting RNA products differ from experiment to experiment in sequence, kinetic parameters and chain lengths (37). In analogy to similar reactions with DNA polymerase I (38,39) and RNA polymerase (40) this reaction was called de novo synthesis (3). Surprisingly, in contrast to those enzymes, Qβ replicase produces RNA species with sequence complexities of around 70. The preparation of Qβ replicase incapable of de novo synthesis has been reported (41). However, these authors apparently overlooked that the reaction requires low ionic strength, high triphosphate and high enzyme concentrations (27,31) to proceed. Enzyme prepared carefully in my laboratory according to the reported method (41) did synthesize RNA de novo under the proper conditions (42).

De novo synthesis proceeds in two phases:
In the first phase, a self-replicating RNA strand is produced. This is a very rare and non-reproducible process and can not be investigated directly. Possibly nucleoside triphosphates are linked more or less at random.
In the second phase, once a self-replicating RNA is produced, it is rapidly amplified to macroscopic levels. The second step is highly selective, since only templates able to grow autocatalytically are amplified. The difference in sequence complexity of the de novo products of DNA polymerase I and Qβ replicase is thus obvious.

The slow condensation of nucleotides to oligomers can be observed readily by omitting one of the pyrimidine triphosphates from the incorporation mixture. Qβ replicase then incorporates nucleoside triphosphates at a rate 4 to 5 orders of magnitude slower than template-directed synthesis into oligonucleotides of chain length 5 - 20 with the sequence $pppGpR(pN)_x$ (42). When initiation of a new chain is suppressed by replacing GTP with ITP, Qβ replicase modifies RNA on the 3'-end by slowly attaching a tail of a heterogeneous nucleotide sequence. The latter reaction may be responsible for a poorly understood optimization process where suddenly new RNA species better adapted to the environment arise, usually accompanied with an increase in chain length (37). These RNA species are not produced de novo, because the sequence is strongly related to the ancestor RNA, while de novo synthesis results in synthesis of unrelated sequences.

Do the RNA synthesis reactions without template instruction play an important biological role? Probably not. The very slow phosphodiester formation in the absence of template is probably only a normal side reaction of the enzyme. It is the selective amplification of some extremely rare products that makes them manifest in in vitro experiments.

ACKNOWLEDGMENTS

The excellent technical assistance by Mr R. Luce and Ms M. Druminski are gratefully acknowledged. I am indebted to Drs M. Eigen and W.C. Gardiner for many contributions to the work and for stimulating discussions.

REFERENCES

1. Haruna I (1965). Autocatalytic synthesis of a viral RNA in vitro. Science 150:884.
2. Kamen R (1972). A new method for the purification of Qβ RNA-dependent RNA polymerase. Biochim Biophys Acta 262:88.
3. Sumper M, Luce R (1975). Evidence for de novo production of self-replicating and environmentally adapted RNA structures by Qβ replicase. Proc Nat Acad Sci USA 72:162.
4. Haruna I, Spiegelman S (1965). Specific template requirements of RNA replicases. Proc Nat Acad Sci USA 54:579.
5. Hori KL, Eoyang L, Banerjee AK, August JT (1967). Template activity of synthetic ribopoymers in the Qβ RNA polymerase reaction. Proc Nat Acad Sci USA 57:1790.
6. Eikhom TS and Spiegelman S (1967). The dissociation of Qβ replicase and the relation of one of the components to a poly(C)-dependent poly(G)-polymerase. Proc Nat Acad Sci USA 57:1833.
7. Mitsunari Y, Hori K (1973). Qβ replicase-associated, poly(C)-dependent poly(G)-polymerase. J Biochem 74:263.
8. Palmenberg A, Kaesberg P (1974). Synthesis of complementary strands of heterologous RNAs with Qβ replicase. Proc Nat Acad Sci USA 71:1371.

9. Vournakis JN, Carmichael GG, Efstradiadis A (1976). Synthesis of RNA complementary to rabbit globin mRNA by Qβ replicase. Biochem Biophys Res Comm 70:774.
10. Banerjee AK, Rensing U, August JT (1969). Replication of RNA viruses. Replication of a natural 6 S RNA by the Qβ RNA polymerase. J Mol Biol 45:181.
11. Kacian DL, Mills DR, Spiegelman S (1971). The mechanism of Qβ replication: Sequence at the 5′ terminus of a 6 S RNA template. Biochem Biophys Acta 238:212.
12. Kacian DL, Mills DR, Kramer FR, Spiegelman S (1972). A replicating RNA molecule suitable for a detailed analysis of extracellular evolution and replication. Proc Nat Acad Sci USA 69:3038.
13. Schaffner W, Ruegg KJ, Weissmann C (1977). Nanovariant RNAs: Nucleotide sequence and interaction with bacteriophage Qβ replicase. J Mol Biol 117:877.
14. Mills DR, Kramer FR, Dobkin C, Nishihara T, Spiegelman S (1975). Nucleotide sequence of microvariant RNA: Another small replicating molecule. Proc Nat Acad Sci USA 72:4252.
15. Weissman C, Billeter MA, Goodman HM, Hindley J, Weber H (1973). Structure and function of phage RNA. Annu Rev Biochem 42:303.
16. Mills DR, Kramer FR, Spiegelman S (1973). Complete nucleotide sequence of a replicating RNA molecule. Science 180:916.
17. Biebricher CK, Diekmann S, Luce R (1982). Structural analysis of self-replicating RNA synthesized by Qβ replicase. J Mol Biol 154:629.
18. August JT, Banerjee AK, Eoyang L, Franze de Fernandez MT, Hori K, Kuo CH, Rensing U, Shapiro L (1968). Synthesis of bacteriophage Qβ RNA. Cold Spring Harbor Symp Quant Biol 33:73.
19. Silverman PM (1973). Analysis by liquid polymer phase partition of the binding of Qβ RNA polymerase to nucleic acid. Arch Biochem Biophys 157:234.
20. Biebricher CK (1983). Darwinian selection of self-replicating RNA molecules. Evol Biol 16:1.
21. Nishihara R, Mills DR, Kramer FR (1983). Localization of the Qβ replicase recognition site in MDV-1 RNA. J biochem (Tokyo) 93:669.
22. Blumenthal T (1980) Qβ replicase specificity: different templates require different GTP concentrations for initiation. Proc Nat Acad Sci USA 77:2601.

23. Weissmann C (1974). The making of a phage. FEBS Lett (Suppl.) 40:S10–S18.
24. Dobkin C. Mills DR, Kramer FR, Spiegelman S (1979). RNA replication: Required intermediates and the dissociation of template, product, and Qβ replicase. Biochemistry 18:2038.
25. Biebricher CK, Eigen M, Gardiner WC (1984). Kinetics of RNA replication: plus-minus asymmetry and double strand formation. Biochemistry 23:3186.
26. Biebricher CK, Eigen M, Gardiner WC (1983). Kinetics of RNA replication. Biochemistry 22:2544.
27. Biebricher, CK, Eigen M, Luce R (1981). Kinetic analysis of template-instructed and de novo RNA synthesis by Qβ replicase. J Mol Biol 148:391.
28. Weissmann C, Feix G, Slor H (1968). In vitro synthesis of phage RNA: The nature of the intermediates. Cold Spring Harbor Symp Quant Biol 33:83.
29. Biebricher CK (1986). Darwinian selection of self-replicating RNA molecules. Chem Script 26:
30. Biebricher CK, Eigen M, Gardiner WC (1985). Kinetics of RNA replication: competition and selection among self-replicating RNA species. Biochemistry 24:6550.
31. Eigen M, Schuster P (1977). The hypercycle - a principle of natural self-organization. Part A: Emergence of the hypercycle. Naturwissenschaften 64:541.
32. Kramer FR, Mills DR, Cole PE, Nishihara T, Spiegelman S (1974). Evolution in vitro: Sequence and phenotype of a mutant RNA resistant to ethidium bromide. J Mol Biol 89:719.
33. Domingo E, Sabo D, Taniguchi T, Weissmann C (1978). Nucleotide sequence heterogeneity of an RNA phage population. Cell 13:735.
34. Domingo E, Martinez-Salas E, Sobrino F, de la Torre JC, Portela A, Ortin J, Lobez-Galindez C, Perez-Brena P, Villanueva N, Najera R, VandePol S, Steinhauer D, DePolo N, Holland J (1986). The quasispecies (extremely heterogeneous) nature of viral RNA genome populations: biological relevance - a review. Gene 40:1.
35. Loeb LA, Kunkel TA (1982). Fidelity of DNA synthesis. Annu Rev Biochem 52:429.
36. Batschelet E, Domingo E, Weissmann C (1976). The proportion of revertant and mutant phage in a growing population, as a function of mutation and growth

rate. Gene 1:27.

37. Biebricher CK, Eigen M, Luce R (1981). Product analysis of RNA generated de novo by Qβ replicase. J Mol Biol 148:369.

38. Schachman HK, Adler J, Radding CM, Kornberg A (1960). Enzymatic synthesis of deoxyribonucleic acid, VII. Synthesis of a polymer of deoxyadenylate and deoxythymidylate. J Biol Chem 235:3242.

39. Kornberg A, Bertsch LL, Jackson JF, Khorana HG (1964). Enzymatic synthesis of deoxyribonucleic acid. Oligonucleotides as templates and the mechanism of their replication. Proc Natl Acad Sci USA 51:315.

40. Krakow JS, Karstadt M (1967) Azotobacter vinelandii RNA polymerase. Unprimed synthesis of rIC copolymer. Proc Natl Acad Sci USA 58:2094

41. Hill D, Blumenthal T (1983) Does Qβ replicase synthesize RNA in the absence of template? Nature 301:350.

42. Biebricher CK, Eigen M, Luce R (1986). Template-free RNA synthesis by Qβ replicase. Nature

Positive Strand RNA Viruses, pages 25–34

COMPARATIVE ORGANIZATION AND GENOME STRUCTURE IN PICORNAVIRUSES[1]

Ann C. Palmenberg

Biophysics Laboratory, University of Wisconsin
Madison, Wisconsin 53706

ABSTRACT Picornaviruses are among the smallest eucaryotic, positive-strand RNA viruses. The genome encodes a single, large protein, which is processed by progressive post-translational cleavage into 11-15 different, mature viral peptides. All picornaviruses share a remarkable degree of homology in their genome organization and virion structure, but nucleotide and amino acid sequence comparsions can be used to subdivide the viruses into four groups or genera, which may represent their phylogenic relationships.

INTRODUCTION

The family of picornaviruses contains a diverse variety of highly virulent human and animal pathogens including: polio, rhino (common cold), hepatitis A, coxsackie, encephalomyocarditis (EMC) and foot-and-mouth disease virus. Classification as a picornavirus is based upon virion structure, genome organization and overall biological character. These properties are remarkably similar among the viruses in spite of wide disparities in host range and tissue pathology.

All virions contain a single-stranded RNA genome enclosed in a protein capsid shell. The capsids are composed of sixty subunits, each of which contains four non-identical polypeptide chains (1,2). Translation of

---

[1]This work was supported by NIH Public Health Service grant AI-17331 to ACP

virion RNA proceeds primarily from a single, strong initiation site, to produce a giant precursor polypeptide which represents most of the theoretical coding capacity of the genome. The polyprotein is processed in a series of proteolytic cleavage steps to yield mature virion capsid proteins, as well as other viral proteins of a non-capsid nature. Figure 1 illustrates a typical genome and the characteristic processing cascade. (For reviews of structure and classification see refs. 3-6.)

Electron microscopy cannot distinguish particle surface differences among the viruses. However, physical properties such as pH lability, buoyant density and thermostability are traditionally used to subdivide the viruses into 4 groups or genera (Table 1).

The enteroviruses are typified by human poliovirus (3 serotypes), but also include murine poliovirus (Theiler's virus), coxsackie A and B, swine vesicular virus, human echoviruses, hepatitis A virus and other human and animal viruses. Entero-particles are stable over a wide range of pH values, but can be denatured by heat (50°, 1hr) in the absence of $MgCl_2$. The average buoyant density is 1.33-1.35 gm/ml.

Rhinoviruses are a major causative agent of the common cold in humans (89 identified serotypes) and cattle (2 serotypes). These viruses are acid labile, and unlike the enteroviruses, can be denatured by heat in the presence of $MgCl_2$. Because rhino particles are permeable

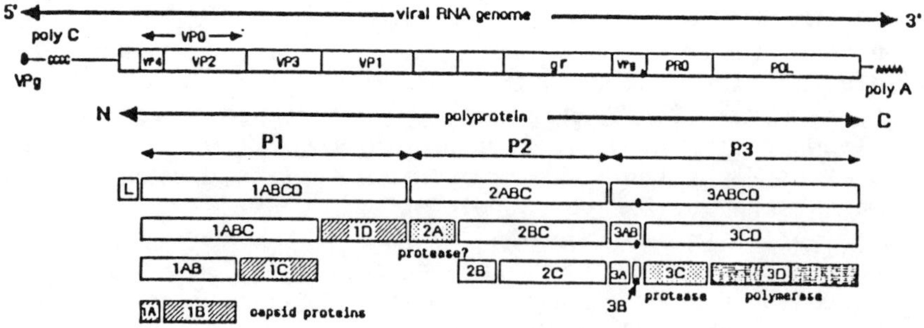

FIGURE 1. PICORNAVIRAL PROCESSING MAP

to CsCl, their buoyant density can range from 1.33-1.41 gm/ml.

The aphthoviruses, or foot-and-mouth disease viruses (FMDV, 7 serotypes), infect many types of cloven-hoofed animals (pigs, goats, sheep, cattle, etc.). The have the largest genomes, the smallest capsid peptides, and correspondingly, the highest buoyant density (1.42 gm/ml) of any picornavirus. The viruses are labile below pH 6.

Cardioviruses such as encephalomyocarditis (EMC), mengo, Maus Elberfeld (ME) and Colombia SK are primarily murine in host range, though humans and swine are sometimes also suceptible. Most isolates are serologically indistinguishable from one another, but can be characterized by unique RNA fingerprint patterns and viral pathology. The buoyant density of cardioviruses is 1.34 gm/ml. The viruses are thermolabile at pH 5-7 in the presence of 0.1M Cl^- or Br^- ions.

Complete nucleotide sequence information is now available for many picornaviruses, including representative members from each of these subgroups (see refs. 7-14 for sequences). Comparative analyses and computer-generated alignments using these data have affirmed that a high degree homology is shared by all the viruses. The comparisons have also brought to light a few discrepencies in the present method for subgroup assignment. A limited summary of these results is presented below, along with a proposal for a somewhat altered subgroup classification system, which may more accurately reflect the phylogenic relationships among the viruses.

THE VIRAL RNA

Picornaviral genomes vary in length from 7102 bases (rhino-2) to 8332 bases (FMDV-O1K), not including the 3' poly(A) tail (50-150 bases) (Table 1). Non-coding bases are distributed (unequally) between the 5' (610-1250 bases) and the 3' (42-126 bases) regions. In contrast to most eucaryotic messages, the 5' ends of picornaviral RNAs are not capped in the usual manner with 5'-5' triphosphate linkages. Instead, these viruses have a small viral-coded protein (VPg, MW 2-3000) attached by a tyrosine-O^4-phosphodiester bond to the 5' nucleotide (pUp) of both plus and minus strands (15). Some evidence suggests that VPg may play a role in the initiation of positive and negative strand RNA synthesis (16).

TABLE 1
PROPERTIES OF THE RNAS[d]

	Genome length bases	5'Non-coding bases	poly C	3'Non-coding bases	Poly protein codons
Polio-1 Sabin	7441	742	-	72	2209
Polio-2 Sabin	7440	747	-	72	2207
Polio-3 Sabin	7432	742	-	72	2206
Coxsackie B3	7400[a]	738	-	101	2200[a]
Hepatitis A	7478	734	-	64	2227
Rhino-14	7208	624	-	47	2179
Rhino-2	7102	610	-	42	2150
EMC[b]	7835	833	+	126	2292
FMDV-A10[c]	8342[a]	1250[a]	+	96	2332
FMDV-O1K[c]	8332	1244	+	92	2332

TABLE 2
RNA BASE COMPOSITION[d] (%)

	Coding Region				Non-Coding Region			
	A	C	G	U	A	C	G	U
Polio-1 Sabin	30	23	23	24	23	24	25	28
Polio-2 Sabin	30	23	23	24	24	24	25	27
Polio-3 Sabin	30	23	23	24	22	26	26	26
Coxsackie B3	29[a]	23[a]	25[a]	23[a]	26	25	24	25
Hepatitis A	30	15	22	33	20	23	24	33
Rhino-14	33	20	20	27	24	24	23	29
Rhino-2	33	19	20	28	26	24	22	28
EMC[b]	26	25	24	25	21	35	23	21
FMDV-A10[c]	26	28	25	21	17[a]	41[a]	21[a]	21[a]
FMDV-O1K[c]	25	28	26	21	20	38	22	20

[a]Based on partial sequence data.
[b]Poly(C) tract determined, 115 bases.
[c]Poly(C) tract estimated, 150 bases.
[d]Sequence refs: 7-14, see also 3.

Individual sequences and distinctive RNA structures (computer-predicted) within the 5' non-coding regions are distinguishing features of particular groups of viruses. For example, all polio, rhino and coxsackie viruses share extensive nucleotide homology (about 60%) and predicted structures within their 5' segments. These sequences align very well for about 600 bases, then polio and coxsackie exibit an "extra" ~140 bases, which are not present in rhino. Hepatitis A shares very little 5' homology with the other viruses. Cardio- and aphthovirus 5' regions are typified by the presence of a homopolymeric poly(C) tract, whose length (50-150 bases) and exact location relative to the 5' end (150-400 bases) vary with different isolates of virus. The biological role of the poly(C) tract remains to be determined. Aphthoviruses have the longest non-coding regions of any picornaviruses (even without the poly(C)) and also have large, stable secondary structures (predicted) near their 5' ends which are peculiar to this group (17).

The nucleotide compositions, especially in the coding regions of the genome, are also characteristic of each variety of picornavirus (Table 2). The rhinoviruses and hepatitis A have a distinct bias towards high A+U content (60%). Polio and coxsackie (52% A+U), EMC (50% A+U) and FMDV (<47%) have progressively lower A+U content. All picornavirus polyproteins have similar amino acid compositions, but codon selection, especially in the third position of the triplet, reflects (or perhaps causes?) the characteristic nucleotide composition bias. Most rhinovirus codons end in A or U (xxA + xxU = 67%) rather than G or C (xxG + xxC = 33%). The preference is reversed for aphthoviruses (xxA + xxU = 36%, xxG + xxC = 64%). Cardio and enteroviruses (except hepatitis A, which is like rhino) are not as discriminatory in the third codon position. The factors which influence codon and nucleotide usage are not very clear. It is possible these features may play a role in host or tissue tropism, though it is interesting to note that hepatitis and polio, both human pathogens, select somewhat different coding patterns.

THE VIRAL PROTEINS

Mature picornaviral peptides are derived by progressive post-translational cleavage of the polyprotein. At least three proteolytic activities are involved in these functions, and all are probably viral-encoded. Peptide 2A (see Fig. 1) is responsible for the first event in poliovirus processing, catalyzing cleavage between the P1/P2 regions while the polyprotein is still nascent on a ribosome (18). The first cleavage within coxsackie, rhino and hepatitis viruses is probably analogous to that of polio, and also catalyzed by 2A. But, in the cardio and aphthoviruses, the nascent cleavage occurs at a different location (2A/2B), and may possibly be catalyzed by a different (yet unidentified) agent (12,19).

Most of the remaining cleavages in all polyproteins are carried out by the viral 3C protease. This enzyme is capable of mono- and bimolecular reactions within its precursor substrates (3ABCD → 3AB+3C+3D), and is also responsible for several processing steps within capsid precursor peptides (P1 → VP0+VP3+VP1). Protease 3C belongs to the thiol class of enzymes (20-23).

The final cleavage within picornaviral polyproteins is maturation processing of VP0 → VP4+VP2 (VP0 is also called 1AB). These reactions are observed in vivo only during the final stages of virion morphogenesis, concomitent with RNA association with provirions. An agent responsible for VP0 proteolysis has never been isolated. However, recent crystallographic resolution of rhino and polio capsid protein structures has suggested that VP0 scission may occur through an unusual serine

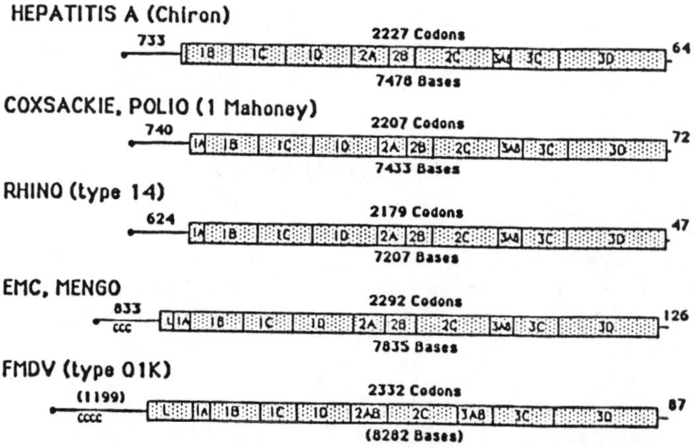

FIGURE 2. REPRESENTATIVE GENOMIC MAPS

protease-like, self-cleavage mechanism, which is catalyzed in part, by virion RNA (12,19).

By convention, picornaviral peptides and peptide precursors are named according to the region of the polyprotein from which they originate (L, P1, P2 or P3) (6). Figure 2 illustrates five viral genomes and their polyproteins, drawn (roughly) to scale. All known, sequenced viruses fit closely into one of these patterns.

The leader or "L" peptides are found only in cardio and aphthoviruses. The EMC leader is about 8kd in size. FMD viruses have two nested L peptides (16 and 23kd), which share common carboxy ends, but have different, in-phase translational start sites (24). The activity of the leader peptide(s) has not been determined.

The four P1 region peptides 1A, 1B, 1C and 1D (in polio, 8, 30, 26 and 33kd) are the capsid structural proteins, and are also called VP4, VP2, VP3 and VP1, respectively. Cell-free processing experiments suggest that capsid peptides which originate from a common precursor molecule stay together as a subunit throughout particle morphogenesis (25). With the exception of hepatitis A, which has a very small (or nonexistant) VP4 peptide, the overall size and shape (structure) of the capsid peptides is similar among the viruses. Individual differences in the primary sequences give rise to variation in particle antigenicity, receptor binding capacity and virion stability.

The middle portion of the polyprotein yields peptides 2A, 2B and 2C (in polio, 12, 14 and 36kd respectively). FMDV genomes have very small or deleted 2A sequences when compared to other viruses. As mentioned above, polioviral 2A is a protease, but the other P2 region peptides have not been assigned definitive biological roles. Protein 2C is the probable genetic locus of the guanidine resistance marker, a compound which effects the initiation of RNA synthesis (26). However, 2C is not a polymerase, and its contribution to the replication cycle remains unclear.

The P3 peptides, especially 3D (51kd in polio), are more closely associated with genome replication. Peptide 3D can catalyze elongation of nascent RNA chains in primer-dependent reactions, an activity which identifies this enzyme as the central element of viral polymerase complexes (16). Protein 3B is VPg (2.5kd), the peptide attached to the 5' end of the genome during initiation of RNA synthesis. Aphthoviruses have three tandemly-linked, different sequences at this position of the polyprotein, making the FMDV P3 segment somewhat longer than in other viruses. Peptide 3C (20kd) is the viral-specific protease responsible for most of the processing events within the polyprotein.

SEQUENCE COMPARISONS (AND PHYLOGENY?)

The compliation of sequence information is not, in itself, a particularly rewarding scientific endeavor. It is only when the data is analyzed from comparative, structural and evolutionary perspectives, that the information becomes meaningful. This is especially true of picornaviral sequences, where more than 300,000 nucleotides from 32 viral strains are now available for comparison. The data presented above represent only a small fraction of this information. Viral similarities and relationships have also been examined by sequence alignments (nucleotide and amino acid) and by dot matrix comparisons. Details of these comparisons will be published elsewhere.

Although analysis is by no means complete, all alignment information strongly suggests that traditional picornaviral subgroup classifications based on physical properties of the virions (rhino-, entero, cardio-, aphtho-) inaccurately represent the viral phylogenic relationships. Figure 3 illustrates a proposal for alternative viral groupings, which are more consistent with the data.

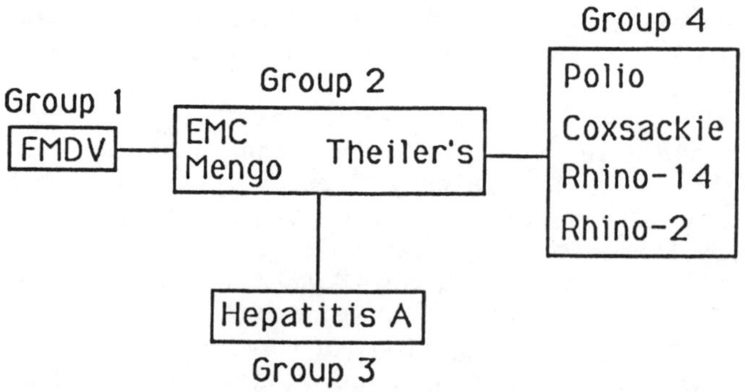

FIGURE 3. PROPOSAL FOR NEW SUBGROUP DIVISIONS

To summarize, genome organization, capsid structure and sequence homology place the rhino, polio and coxsackieviruses as related members of the same group (group 4). The rhinovirus themselves, can be further subdivided into "rhino-14-like" or rhino-2-like" on the basis of overall sequence similarities. Hepatitis A strains (group 3) are dissimilar from all other picornaviruses and appear to belong to a separate group. The FMD viruses, likewise, merit their own subdivision (group 1). Theiler's virus (murine polio), previously characterized as an enterovirus, is much more cardio-like in sequence and organization, and clearly belongs with other members of this category (group 2) such as EMC and mengovirus (27). As a rule, sequence homology is always much stronger between members of the same group, than among viruses from different groups. For example, EMC and mengo have >93% peptide sequence identity, while EMC and polio share <40% common residues (unpublished observations).

A consistent pattern of intergroup relationships is also beginning to emerge from the sequence data. While these relative group placements must still be considered tentative, the group 3 viruses usually seem to occupy a middle position in any comparative scheme (as illustrated in Fig. 3), exibiting transitional sequences and intermediate structural elements which are not held in common by representatives of the other three subgroups (3). With eventual development of a complete phylogenic "tree", we hope to more accurately assess whether the cardio-like viruses (group 3) indeed, represent a central "branch" among picornaviruses.

REFERENCES

1. Hogle JM, Chow M, Filman DJ (1985). Science 229:1358.
2. Rossmann MJ, Arnold E, Erickson JW, Frankenberger EA, Griffith JP, Hecht H-J, Johnson JE, Kamer G, Luo M, Mosser AG, Rueckert RR, Sherry B, Vriend G (1985). Nature 317:145.
3. Palmenberg AC (1986). In Rowlands DJ, Mahy BWJ, Mayo M (eds): "The Molecular Biology of Positive Strand RNA Viruses" London: Academic Press, in press.
4. Rueckert RR (1976). In Fraenkel-Conrat H, Wagner RR (eds): "Comprehensive Virology, Vol 6" New York: Plenum, pp 131-212.

5. Rueckert RR (1985). In Fields B, Knipe DM, Chanock RM, Melnick JL, Roizman B, Shope RE (eds): "Virology" New York: Raven Press, pp 705-738.
6. Rueckert RR, Wimmer E (1984). J Virol 50:957.
7. Toyoda H, Kohara M, Kataoka Y, Suganuma T, Omata T, Imura N, Nomoto A (1984). J Mol Biol 174:561.
8. Tracy S, Liu H-L, Chapman NM (1985). Virus Res 3:263.
9. Najarian R, Caput D, Gee W, Potter S, Renard A, Merryweather J, Van Nest G, Dina D (1985). Proc Natl Acad Sci 82:2627.
10. Stanway G, Hughes P, Mountford R, Minor P, Almond J (1984). Nuc Acid Res 12:7859.
11. Skern T, Sommergruber W, Blass D, Gruendler P, Fraundorfer F, Pieler C, Fogy I, Kuechler E (1985). Nuc Acid Res 13:2111.
12. Palmenberg AC, Kirby EM, Janda MJ, Drake NL, Duke GM, Potratz KF, Collett MS (1984). Nuc Acid Res 12:2969.
13. Boothroyd JC, Highfield PE, Cross GAM, Rowlands DJ, Lowe PA, Brown F, Harris TJR (1981). Nature 290:800.
14. Forss S, Strebel K, Beck E, Schaller H (1984). Nuc Acid Res 12:6587.
15. Rothberg PG, Harris TJR, Nomoto A, Wimmer E (1980). Proc Natl Acad Sci 75:4868.
16. Flanegan JB, Baltimore D (1977). Proc Nat Acad Sci 74:2677.
17. Newton SE, Carroll AR, Campbell RO, Clarke BE, Rowlands DJ (1986). Gene 40:331.
18. Toyoda H, Nicklin MJH, Murray MG, Anderson CW, Dunn JJ, Studier FW, Wimmer E (1986). Cell submitted.
19. Palmenberg AC (1986). J Cell Biochem in press.
20. Palmenberg AC, Pallansch MA, Rueckert RR (1979). J Virol 32:770.
21. Gorbalenya AE, Svitkin YV, Kazachkov YA, Agol VI (1979). FEBS Lett 108:1.
22. Palmenberg AC, Rueckert RR (1982). J Virol 41:244.
23. Svitkin YV, Gorbalenya AE, Kazachkov YA, Agol VI (1979). FEBS Lett 108:6.
24. Beck E, Fross S, Strebel K, Cattaneo R, Feil G (1983). Nuc Acid Res 11:7873.
25. Palmenberg AC (1982). J Virol 44:900.
26. Anderson-Stillman K, Bartal S, Tershak DR (1984). J Virol 50:922.
27. Nitayaphan S, Omilianowski D, Toth MM, Parks GD, Rueckert RR, Palmenberg AC, Roos RP (1986). Intervirology submitted.

Positive Strand RNA Viruses, pages 35–45

REQUIREMENT FOR SECONDARY STRUCTURE FORMATION DURING COLIPHAGE RNA REPLICATION

Donald R. Mills, Christine Priano, Fred Russell Kramer

Institute of Cancer Research
and Department of Genetics and Development,
College of Physicians and Surgeons, Columbia University
New York, New York 10032

ABSTRACT The rate of increase of replicating populations of MDV-1 RNA is greater than that of microvariant RNA, despite the greater length of MDV-1 RNA. Furthermore, Qβ replicase pauses more frequently during MDV-1 RNA chain elongation, and it takes a longer time to complete MDV-1 RNA chains. The explanation for this apparent contradiction is that MDV-1 RNA possesses more extensive secondary structures than microvariant RNA and the formation of secondary structures during chain elongation prevents the lethal collapse of the product strand upon the template strand in the replication complex. These results suggest that the extensive secondary structures that are present in the genomic RNAs of coliphages are required for the maintenance of their single-strandedness during replication.

INTRODUCTION

The genome of bacteriophage Qβ is a single-stranded, linear RNA that contains 4,220 nucleotides (1). The phage genes are encoded in three cistrons that are translated in the same reading frame. These cistrons encode (reading from the 5'-end of the RNA) the maturation protein, the coat protein, and one of the four subunits that comprise the RNA-directed RNA polymerase, Qβ replicase (2).

Qβ replicase is highly selective for its own template RNAs. It promotes the autocatalytic synthesis of Qβ RNA (3), the genomic RNA of the closely related group III RNA bacteriophages (4), and several naturally occurring "variant RNAs" (5-9). It cannot replicate other viral RNAs, nor any Escherichia coli RNA (10).

Two variant templates of Qβ replicase, MDV-1 RNA (6) and microvariant RNA (7), have been isolated from Qβ replicase reactions that were incubated in the absence of exogenous template. Their nucleotide sequences have been completely determined (7,11,12). The mechanism of their replication has been studied in detail (6,13-19) and has been shown to be fundamentally similar to the replication of Qβ RNA. The mechanism of RNA synthesis is illustrated in Figure 1. The replicase first binds to a highly structured internal region of the template RNA (17). The synthesis of the product strand is then initiated at the 3' end of the template, and the product strand is synthesized in the 5' to 3' direction. Although the replicase relies on base pairing for the synthesis of the product strand, the mechanism of replication is such that, in general, both the product and the template are single-stranded. After the completion of product strand elongation, both the product strand and the template strand are released from the replication complex as single-stranded RNAs (14). The replicase is then free to bind to any RNA for use as a template in the next round of synthesis.

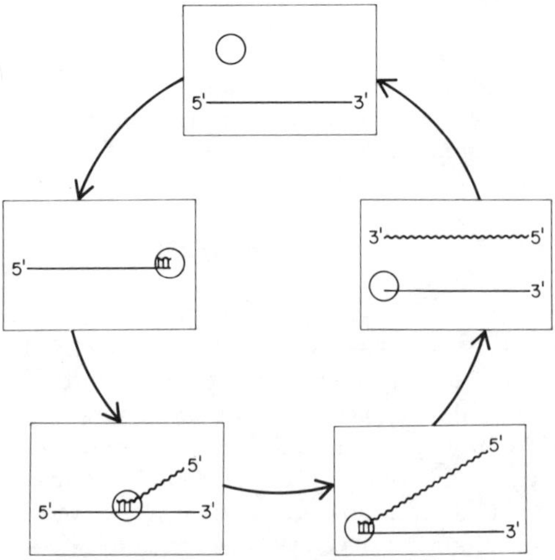

FIGURE 1. The replication cycle of Qβ replicase. The replicase (circle) synthesizes a single-stranded product from a single-stranded template.

In this paper we compare the synthesis of MDV-1 RNA to the synthesis of microvariant RNA. We studied the secondary structures that they can form, the kinetics of their synthesis, the variability in their rates of chain elongation, and the propensity of their complementary strands to reassociate. The results suggest that the formation of secondary structures during chain elongation is required for autocatalytic replication.

RESULTS

Replicable RNAs contain many Secondary Structures.

The sequences of both microvariant RNA and MDV-1 RNA possess many self-complementary regions that are capable of forming hairpin structures. Figure 2 shows the structure of each RNA that is predicted by computer to be most stable (20). These structures have been confirmed experimentally. They have been directly visualized (21), utilizing hollow-cone, dark-field electron microscopy (22). Chemical agents that modify nucleotides in single-stranded regions have been used to identify specific structures (15). Band

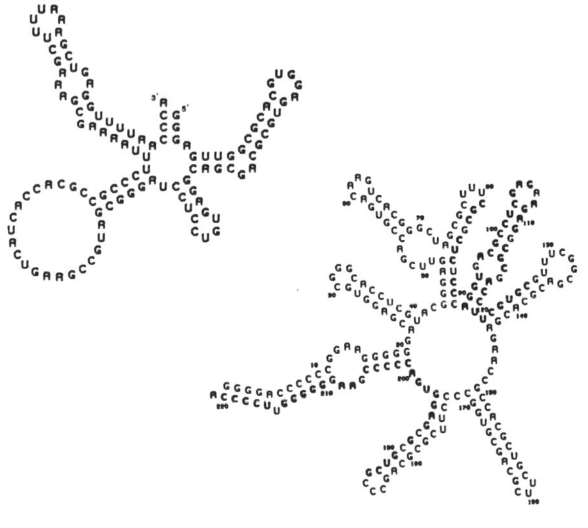

FIGURE 2. Comparison of the nucleotide sequences and secondary structures of microvariant (+) RNA (115 nucleotides) and MDV-1 (+) RNA (221 nucleotides).

compression in RNA sequencing gels (12) has been used to identify regions that form secondary structures (23). And finally, the tertiary structure of each RNA was probed by subjecting it to mild cleavage with ribonuclease T_1, which only attacks single-stranded regions (24). The results of these experiments confirm that both RNAs are highly structured, though MDV-1 RNA has a more extensive set of structures than microvariant RNA.

The Kinetics of RNA Synthesis.

Under conditions of replicase excess, when both the template and the product strands can serve as templates for each round of replication, RNA synthesis proceeds at an exponential rate (25). When enough RNA has been synthesized to saturate all of the replicase with template, synthesis proceeds at a linear rate. The kinetics of synthesis of microvariant RNA were compared to the kinetics of synthesis of MDV-1 RNA (Fig. 3). The results show that during the exponential phase of synthesis it took less time

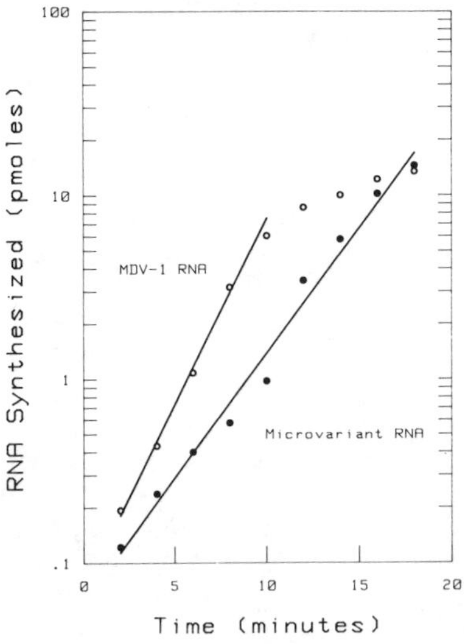

FIGURE 3. Comparison of the kinetics of MDV-1 RNA synthesis and microvariant RNA synthesis. The amount of RNA in each reaction increased exponentially, until the number of RNA strands equaled the number of active replicase molecules.

for the MDV-1 RNA population to double in number than it did for the microvariant RNA population. Table 1 summarizes the results of this comparison.

TABLE 1.

KINETIC ANALYSIS OF RNA REPLICATION

RNA	T_d (sec)	C_s (pmol/ml)
MDV-1	89	145
microvariant	129	375

T_d is the doubling time of the RNA population.
C_s is the RNA concentration at saturation.

These results were surprising, because microvariant RNA is half the length of MDV-1 RNA. All other factors being equal, the microvariant population should have increased at a faster rate than the MDV-1 RNA population. When we compared the ability of each RNA to bind Qβ replicase and the ability of each RNA to initiate the synthesis of the product strand, no differences were observed (19). We therefore expected that the slower rate of increase of the microvariant RNA population would be accounted for by a slower rate of product chain elongation.

Pausing of the Replicase during Chain Elongation.

Electrophoretic analysis of the distribution of partially synthesized product chains indicates that the rate of chain elongation is highly variable. Apparently, the progress of the replicase is temporarily interrupted at a small number of specific sites, and then resumes spontaneously with a finite probability (13). Since the time spent between these pause sites is negligible compared with the time spent at pause sites, the mean time of chain elongation is well approximated by the sum of the mean times spent at each pause site. Nucleotide sequence analysis indicates that pauses occur just after the synthesis of sequences that can form hairpin structures (13). Thus, the formation of secondary structures plays a major role in determining the rate of chain elongation.

We compared the chain elongation of microvariant RNA and MDV-1 RNA, to see whether more pauses occurred during microvariant chain elongation, accounting for its slower population growth rate. Each RNA was synthesized in a two-stage reaction (13). In the first stage, replication was permitted to proceed until the first pyrimidine triphosphate was required. In the second stage, the missing pyrimidine triphosphate was added to the reactions and samples were taken at various times. Electrophoretic analysis of the partially synthesized RNAs (Fig. 4) indicated that more pauses occur during the synthesis of MDV-1 RNA than during the synthesis of microvariant RNA. Contrary to our expectations, it took longer for Qβ replicase to elongate the MDV-1 RNA than microvariant RNA (19). Thus, MDV-1 RNA populations reproduce at a faster rate than microvariant RNA populations, despite their longer chain length and slower chain elongation rate.

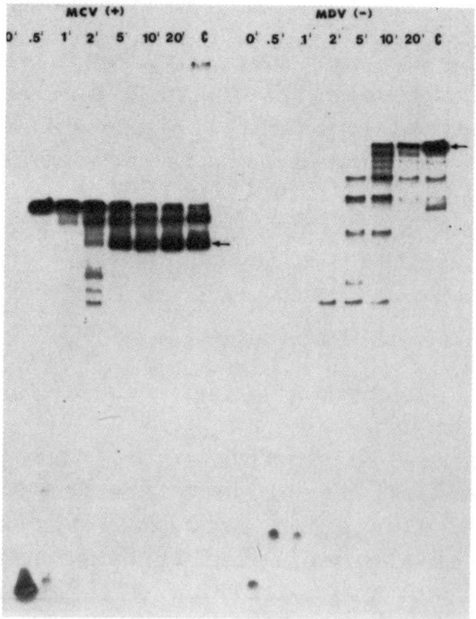

FIGURE 4. Comparison of chain elongation during the synthesis of microvariant RNA (MCV RNA) and MDV-1 RNA. Samples taken at different times were analyzed by gel electrophoresis. Arrows indicate the location of full-length RNA chains. Fewer pauses occur during the synthesis of microvariant RNA (left panel), and the microvariant RNA chains are completed sooner.

Reassociation of Complementary Strands in the Replication Complex.

We found the explanation for these apparent contradictions in a closer examination of the kinetics of synthesis of each RNA species. We discovered that there were more moles of microvariant RNA present at the replicase saturation point than there were moles of MDV-1 RNA when its kinetics became linear (see Table 1). This result indicated that inactive microvariant RNAs were accumulating during the exponential phase of synthesis.

We prepared double-stranded microvariant RNA and MDV-1 RNA, melted each apart, and then determined the reassociation constant (C_ot) of each, under conditions similar to those that occur during RNA synthesis (19). Different concentrations of each RNA were studied in separate reactions, and the fraction of RNA that had reassociated in each, as a function of time, was determined by polyacrylamide gel electrophoresis. The results show that the complementary strands of microvariant RNA are much more likely to reassociate than are the complementary strands of MDV-1 RNA (Fig. 5). These data suggest that

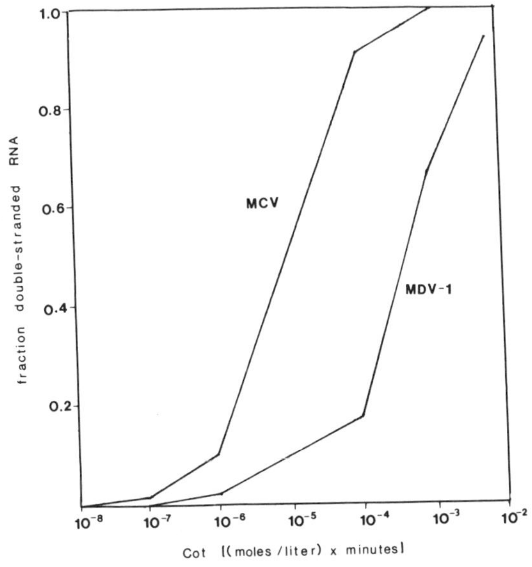

FIGURE 5. Comparison of the reassociation of the complementary strands of microvariant RNA (MCV RNA) and MDV-1 RNA.

there is a higher probability that the template and product strands of microvariant RNA will hybridize to each other while associated with the replicase. Moreover, a direct examination of the RNA present during the exponential phase of synthesis indicated that double-stranded RNAs occur much more frequently in reactions that synthesize microvariant RNA (19). Since double-stranded RNAs cannot serve as templates for Qβ replicase (18), these results provide an explanation for the faster rate of MDV-1 RNA population growth.

DISCUSSION

Because of its more stable structure, MDV-1 RNA is less likely to hybridize to its complement in the replication complex than is microvariant RNA. An MDV-1 RNA population doubles in less time than a microvariant RNA population, despite the shorter length and faster elongation time of microvariant RNA. Thus, the formation of secondary structures during chain elongation prevents the formation of lethal double strands.

Since the formation of secondary structures dramatically slows down chain elongation (13), few structures would be present in a replicating RNA, unless there was a selective advantage to their presence. The maintenance of the single-stranded state, made possible by the formation of secondary structures, provides this advantage. We have observed the results of this selective pressure during in vitro evolution experiments (9,24,26). Most of the nucleotide substitutions that were seen occurred in single-stranded regions of the RNA. The few mutations that were seen in double-stranded regions occurred as base-pair substitutions, preserving the integrity of their hairpin stem.

The RNA coliphages possess highly structured, single-stranded genomes (27-29). Our data suggest that the extensive structures present in these genomes evolved, not only as a means of possessing functional surfaces, but also as a means of maintaining single-strandedness during replication. In spite of the high cost of a reduced replication rate, only those viral RNAs that retain their structures through evolution are able to give rise to viable progeny.

ACKNOWLEDGEMENTS

The work reviewed in this grant was supported by National Science Foundation grant PCM-82-15902, American Cancer Society grant MV-191A, and National Institutes of Health grants GM-32044 and GM-33345.

REFERENCES

1. Mekler P (1981) Ph.D. Thesis, University of Zurich.
2. Blumenthal T, Carmichael GG (1979). RNA replication: function and structure of Qβ replicase. Ann Rev Biochem 48:525.
3. Haruna I, Spiegelman S (1965). Specific template requirements of RNA replicases. Proc Natl Acad Sci USA 54:579.
4. Miyake T, Haruna, I, Shiba T, Itoh TH, Yaname K, Watanabe I (1971). Grouping of RNA phages based on the template specificity of their RNA replicases.
5. Banerjee AK, Eoyang L, Hori K, August JT (1967). Replication of RNA viruses, X. Replication of a natural 6s RNA by the Qβ RNA polymerase. J Mol Biol 45:181.
6. Kacian DL, Mills DR, Kramer FR, Spiegelman S (1972). A replicating RNA molecule suitable for a detailed analysis of extracellular evolution and replication. Proc Natl Acad Sci USA 69:3038.
7. Mills DR, Kramer, FR, Dobkin C, Nishihara T, Spiegelman S (1975). Nucleotide sequence of microvariant RNA: another small replicating molecule. Proc Natl Acad Sci USA 72:4252.
8. Schaffner W, Ruegg KJ, Weissmann C (1977). Nanovariant RNAs: nucleotide sequence and interaction with bacteriophage Qβ replicase. J Mol Biol 117:877.
9. Priano C, Mills DR, Kramer FR (1986). Sequence, structure, and evolution of a 77 nucleotide template for Qβ replicase. In preparation.
10. Haruna I, Spiegelman S (1965). Recognition of size and sequence by an RNA replicase. Proc Natl Acad Sci USA 54:1189.
11. Mills DR, Kramer FR, Spiegelman S (1973). Complete nucleotide sequence of a replicating RNA molecule. Science 180:916.

12. Kramer FR, Mills DR (1978). RNA sequencing with radioactive chain-terminating ribonucleotides. Proc Natl Acad Sci USA 75:5334.
13. Mills DR, Dobkin C, Kramer FR (1978). Template-determined, variable rate of RNA chain elongation. Cell 15:541.
14. Dobkin C, Mills DR, Kramer FR, Spiegelman S (1979). RNA replication: required intermediates and the dissociation of template, product, and Qβ replicase. Biochemistry 18:2038.
15. Mills DR, Kramer FR, Dobkin C, Nishihara T, Cole P (1980). Modification of cytidines in a Qβ replicase template: analysis of conformation and localization of lethal nucleotide substitutions. Biochemistry 19:228.
16. Kramer FR, Mills DR (1978). Secondary structure formation during RNA synthesis. Nucleic Acids Res 9:5109.
17. Nishihara T, Mills DR, Kramer FR (1983). Localization of the Qβ replicase recognition site in MDV-1 RNA. J Biochem 93:669.
18. Bausch JN, Kramer FR, Miele EA, Dobkin C, Mills DR (1983). Terminal adenylation in the synthesis of RNA by Qβ replicase. J Biol Chem 258:1978.
19. Priano C, Mills DR, Kramer FR (1986). The role of secondary structures in the prevention of strand collapse during RNA replication. In preparation.
20. Zuker M, Stiegler P (1981). Optimal computer folding of large RNA sequences using thermodynamics and auxiliary information. Nucleic Acids Res 9:133.
21. Klotz G, Kramer FR, Kleinschmidt AK (1980). Conformational details of partially base-paired small RNAs in the nanometer range. Electron Microscopy 2:530.
22. Klotz G (1982) Dark field imaging of nucleic acid molecules. 10th Int Cong Electron Microscopy, p 57.
23. Mills Dr, Kramer FR (1979). Structure-independent nucleotide sequence analysis. Proc Natl Acad Sci USA 76:2232.
24. Kramer FR, Mills DR, Rudner R (1986). Selection of a mutant RNA resistant to ribonuclease T_1. In preparation.
25. Haruna I, Spiegelman S (1965). Autocatalytic synthesis of a viral RNA _in vitro_. Science 180:884.

26. Kramer FR, Mills DR, Cole PE, Nishihara T, Spiegelman S (1974). Evolution in vitro: sequence and phenotype of a mutant RNA resistant to ethidium bromide. J Mol Biol 89:719.
27. Min Jou W, Haegeman G, Ysebaert M, Fiers W (1972). Nucleotide sequence of the gene coding for the bacteriophage MS2 coat protein. Nature 237:82.
28. Fiers W, Contreras R, Duerinck F, Haegeman G, Merregaert J, Min Jou W, Raeymakers A, Volckaert G, Ysebaert M, Van de Kerckhove J, Nolf F, Van Montagu M (1975). A-protein gene of bacteriophage MS2. Nature 256:273.
29. Fiers W, Contreras R, Duerinck F, Haegeman G, Iserentant D, Merregaert J, Min Jou W, Molemans F, Raeymaekers A, Van den Berghe A, Volckaert G, Ysebaert M (1976). Complete nucleotide sequence of bacteriophage MS2 RNA: primary and secondary structure of the replicase gene. Nature 260:500.

Positive Strand RNA Viruses, pages 47–56

CONSTRUCTION OF GENOMIC-LENGTH cDNA CLONES OF TOBACCO MOSAIC VIRUS[1]

D.A. Knorr, D.L. Beck, W.O. Dawson

Department of Plant Pathology, University of California
Riverside, California 92521

ABSTRACT Genomic-length cDNA of tobacco mosaic virus (TMV) was assembled from a cloned library of overlapping fragments. cDNAs at the 5' and 3' termini of the viral genome were synthesized separately. A synthetic oligonucleotide primer was used to generate a *Pst* I site (unique in the TMV genome) at the 5' terminus, while a different primer was used to generate an *Nde* I site (not unique in TMV) at the 3' terminus. Addition of these restriction sites permitted removal of non-TMV sequences from cloned cDNAs by restriction endonuclease cleavage followed by treatment with exonuclease VII. To aid manipulation of the cloned genome, *Pst* I linkers were added to 3' terminal cDNAs. Thus, genomic-length cDNAs ligated into plasmid cloning vectors could be cleanly excised as single fragments by *Pst* I. Genomic-length TMV cDNA constructs containing the phage lambda P promoter from pPM1 were transcribed *in vitro*. *E. coli* RNA polymerase transcripts from three of four genomic-length cDNA constructs were infectious, even though these contained 6 non-TMV nucleotides at the 3' end. Transcripts of constructs with no additional nucleotides and containing 6 additional nucleotides at the 5' end also were infectious.

INTRODUCTION

Tobacco mosaic virus (TMV) infects ~200 plant species

[1]This research was supported in part by Grants PCM 72921664 from the National Science Foundation and 85-CRCR-1-1795 from the U.S. Department of Agriculture.

worldwide, causing serious economic losses in tobacco, tomato, and other crop plants. The virions are rod-shaped, 18 X 300 nm, encapsidating a single strand of messenger-sense RNA 6.4 kb in length. The virus encodes at least four proteins: two that are translated from intact plus strands, and two others that are translated from subgenomic mRNAs (1).

TMV has provided a useful model system for studying genetics, protein translation, virus structure, and pathogen-host interactions. The first studies on viral protein sequence changes, as well as biological variation, and in vitro mutability, used TMV (2-4). Many strains and biological variants of TMV have been described, and three of these have been completely sequenced (5-7). Numerous TMV-specific functions have been described, including host range, symptomology, elicitation of host defense mechanisms, cell-to-cell spread, encapsidation, and cross-protection, as well as several functions involved in RNA replication. However, in spite of a general knowledge of the organization and replication of this virus, functions of only two viral sequences are known. This situation is due to an inability to accurately map mutations in the TMV genome using traditional genetic techniques.

We have attempted to overcome these limitations to studying the genetics of TMV by extending the virus life cycle through an artificial DNA phase. This approach previously has been successfully utilized for two other RNA plant pathogens, potato spindle tuber viroid (8) and brome mosaic virus (9). Unlike RNA, DNA sequences can be precisely altered, spliced, or otherwise recombined using currently available technology.

METHODS AND RESULTS

The primary goal was to create an exact DNA copy of TMV that could manipulated as a single fragment containing no additional nucleotides. The DNA would then be utilized to initiate TMV infection in plants.

Because the amount of non-infectious or variant RNAs in the parental population could not be estimated, we felt the best approach to cloning TMV was to generate multiple independent clones of each region of the genome. This strategy allows genomic-length cDNAs to be assembled in different combinations, thereby increasing the probability of generating an infectious copy of the original wild-type parental genome.

cDNA synthesis A series of overlapping cDNA clones representing the entire TMV genome was generated. TMV virion RNA, polyadenylylated at the 3' terminus, was used as template for reverse-transcription primed with oligo-d$(T)_{12-18}$ (10). Reaction conditions for synthesis of second-strand have been described elsewhere (11). Double-stranded DNA of TMV was inserted into pBR322 after addition of *Pst* I or *Bam* HI linkers. By this method, an array of cDNA clones representing all except approximately 200 nucleotides at the 5' terminus of TMV was produced. To accurately synthesize and incorporate useful restriction sites into 3' and 5' termini, synthetic oligonucleotide primers were used. The oligomer 5'-d(CATATGGGCCCC-OH) was used to prime first-strand synthesis and create an *Nde* I restriction site at the 3' terminus of TMV cDNAs (Fig 2). However, because this site is not unique in the TMV genome, *Pst* I linkers were added following second-strand synthesis to facilitate manipulation of genomic-length cDNAs. To select for cDNA specific to the 5' terminus of TMV, the synthetic oligomer, 5'-d(CTGCAGTATTTTTA-OH), was used to prime second-strand synthesis on cDNAs containing 5' termini. This primer creates a *Pst* I restriction site at the 5' terminus of TMV cDNA. Following *Pst* I/*Hind* III-digestion, cDNA specific to each terminus was inserted into pUC19.

The strategy for generating genomic-length constructions of TMV cDNA by progressively joining together restriction fragments from smaller clones is illustrated in Figure 1. For example, to construct pTMV100, the 1446 bp *Pst* I/*Hind* III fragment from TK-49 (containing the 5' end of TMV) was ligated to the 1886 bp *Hind* III/*Bam* HI fragment from clone B-61 and inserted into pUC19, producing clone p5F3. In the same manner, p3F1 was made from clone 3-15 (containing the 3' end of TMV) and B-68. Inserts from p5F3 (3332 bp) and p3F1 (3073 bp) were removed by *Pst* I/*Bam* HI digestion and ligated together with *Pst* I-digested pBR322 to give pTMV100. Four additional clones with genomic-length inserts, pTMV101, pTMV102, pTMV103, and pTMV104, were constructed in a similar fashion (11).

Exact DNA copies of TMV with no additional nucleotides were prepared *in vitro* by treating separate batches of *Pst* I- or *Nde* I-digested pTMV103 with exonuclease VII to remove protruding nucleotides, digesting each with *Bam* HI, then ligating 5' and 3' fragments isolated by gel electrophoresis.

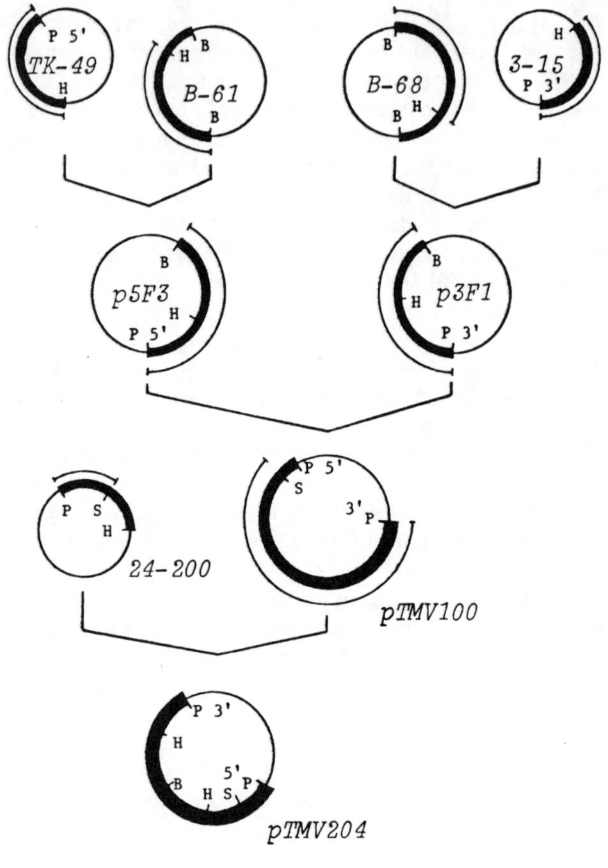

FIGURE 1. Construction of pTMV204, one of three clones used to transcribe infectious TMV RNA. Plasmid vector sequences are shown as single lines, TMV cDNA inserts as heavy lines. Fragments used for ligations are indicated by barred lines. 5' and 3' correspond to termini of the TMV genome. B = Bam HI, H = Hind III, P = Pst I, S = Sma I restriction sites.

To prepare templates for in vitro transcription the phage lambda P_R promoter sequence from pPM1 (a gift from Dr. Paul Ahlquist) was added to the 5' terminus of several genomic-length TMV cDNAs. Clone 24-200 was constructed with the 5' 270 nucleotides of TMV inserted between the <u>Sma</u> I and <u>Eco</u> RI sites of pPM1 so that transcription initiates with the first nucleotide of TMV (12).

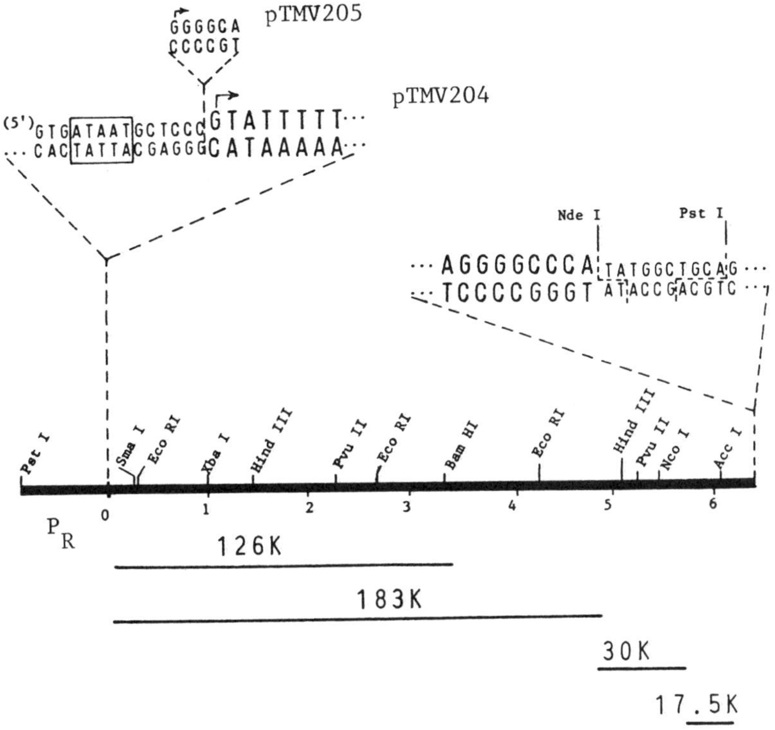

FIGURE 2. Restriction map of genomic-length TMV cDNA/pPM1 promoter fragment of pTMV204. Left sequence shows the promoter-TMV junction sequence, with additional nucleotides of pTMV205 inserted above, as shown. A box surrounds the promoter consensus sequence and arrows show the first nucleotides synthesized during transcription. Sequence on the right shows arrangement of 3' <u>Nde</u> I and <u>Pst</u> I restriction sites. TMV sequences are represented as larger letters with upper strand as plus-sense. Numbers under the map indicate each kb of the TMV genome. Locations and molecular weights of known TMV gene products shown below.

Xcy I-Pst I fragments from genomic-length TMV cDNAs (containing all but the 5' 256 nucleotides of TMV), plus Pst I-Xcy I fragments from 24-200 (containing the 5' 256 nucleotides of TMV joined to the promoter) were ligated into Pst I-digested pBR322 as shown in Figure 1 to generate pTMV201, pTMV202, pTMV203, and pTMV204.

To prepare a transcription template with all 3' non-TMV nucleotides removed, 3' fragments from pTMV204, treated with Nde I and exonuclease VII, followed by Bam HI digestion, were gel-purified then ligated to 5' Pst I/ Bam HI fragments of pTMV204 (Figure 2).

An additional construction, clone pTMV205, contains an insertion at the promoter-TMV junction sequence of pTMV204 such that transcription initiates six nucleotides upstream from the 5' terminus of TMV (Figure 2).

Infectivity cDNA of TMV was not infectious when inoculated directly onto test plants, although cDNAs were prepared in a variety of ways. No plants became infected when 5 µg of either undigested or Pst I-digested plasmid DNA from pTMV101, pTMV103, or pTMV204 was mechanically inoculated onto pinto bean or Xanthi tobacco leaves, whereas under the same conditions, 1 ng of TMV virion RNA was highly infectious. Total RNA isolated from *E. coli* carrying pTMV204 was not infectious, nor was an exact cDNA copy of the TMV genome, prepared *in vitro* from pTMV103.

The infectivity of *in vitro* transcripts of TMV cDNA was also tested. Reaction mixtures for *in vitro* transcriptions contained 40 mM Tris·HCl (pH 7.9), 10 mM $MgCl_2$, 150 mM KCl, 0.1 mM EDTA, 2 mM DTT, 0.5% bovine serum albumin, 150 µM (each) ATP, CTP, UTP, and m^7GpppG, 100 units of RNasin, 4 µg of Pst I-digested plasmid DNA and 1 unit of *E. coli* RNA polymerase in a total volume of 92.5 µl. After 3 min. at 37°C, GTP was added to 30 µM and after 15 min., GTP was increased to 150 µM in a final reaction volume of 100 µl. After 45 min. 100 µl of ice cold RNA assay buffer (1% sodium pyrophosphate, pH 9.0/1% bentonite/1% celite) was added, and the mixture was rubbed immediately onto leaves of pinto bean (*Phaseolus vulgaris* L.) and tobacco (*Nicotiana tabacum* L. var. Xanthi). Transcripts of Pst I-digested plasmids pTMV201-pTMV204 produced under the conditions described should contain the exact, 5'-capped, sequence of TMV RNA with the addition of 6 nucleotides at the 3' end. Three of the four different constructions tested were infectious (Table 1). In addition, pTMV204 prepared with no additional 3' nucleotides and pTMV205 both were infectious.

TABLE 1
INFECTIVITY OF TMV cDNA TRANSCRIPTS[a]

Sample	Infected Xanthi/total	Lesions/8 pinto bean leaves
	Experiment 1	
pTMV201	5/6	4
pTMV202	0/6	0
pTMV203	1/6	7
pTMV204	6/6	320
+ GpppG	6/6	227
Acc I only	0/6	0
Eco RV only	0/6	0
pTMV101	0/6	0
pTMV104	0/6	0
TMV RNA, 100 pg/µl	6/6	73
	Experiment 2	
pTMV204		
complete		88
without cap		0
without incubation		0
RNase A, 0.12 µg/ml added		
Before transcription		23
After transcription		0
DNase, 0.6 µg/ml		
Before transcription		0
After transcription		31
	Experiment 3	
in vitro-prepared exact 3' sequence (~500 ng)		4
pTMV204 (~500 ng)		6
	Experiment 4	
pTMV204		612
pTMV205 (5' insertion mutant)		2

[a] Unless otherwise stated, transcription reaction mixtures contained 4 µg of _Pst_ I-digested plasmid and m^7GpppG.

Several controls demonstrate that infectious RNA resulted from the transcription reaction and not from contaminating RNA. Infectivity of TMV cDNA transcripts resulted only when a synthetic cap structure was present in transcription reactions. Use of the unmethylated cap analog GpppG resulted in almost as many lesions as use of m^7GpppG (Table 1). RNase treatment destroyed infectivity after, but not before transcription. DNase treatment destroyed infectivity before, but not after transcription. By linearizing pTMV204 with Acc I or Eco RV, transcripts shortened by 346, or lengthened by 962 nucleotides respectively, were prepared. Neither of these two treatments yielded infectious transcripts, although TMV transcripts extended at the 3' terminus by 6 additional nucleotides were infectious. The specific infectivity of plasmid transcripts could not be compared to that of parental RNA because the size of *in vitro* transcripts was not homogeneous, making quantification difficult. However, transcripts of pTMV201, pTMV203, and pTMV204 reproducibly infected plants (Table 1), with transcripts of pTMV204 exhibiting the highest infectivity. In 12 separate experiments, pTMV204 transcripts infected every plant inoculated and gave over 200 lesions per pinto bean leaf in several experiments.

Characteristics of the Progeny Virus. TMV RNA produced *in vitro* was biologically identical to the parental virus in host range and symptoms caused on 12 different plant species. Progeny virus isolated from plants infected by cDNA transcripts appeared identical to the original parental virus by electron microscopy and serology in double diffusion tests. The 3' terminal nucleotide sequence of RNA isolated from progeny virus of pTMV203 and pTMV204 was identical to that of the parental virus RNA (data not shown). No differences were observed in specific infectivity of virions or virion RNA isolated from plants infected either with *in vitro* transcripts or with parental virus.

DISCUSSION

We have constructed several independent cDNA clones of the complete genome of TMV from a library of subgenomic cDNA clones. Our strategy was designed to produce genomic-length constructions that could be excised cleanly as single fragments, allowing movement to other vectors, or the addition of promoters to allow transcription of the entire genome. Production of infectious transcripts from three of four TMV cDNA constructs demonstrates the success of this method.

Although infectious RNA transcripts of the exact TMV sequence without additional nucleotides could be produced, this proved to be unnecessary and inconvenient. Transcripts with additional bases at the 3' end also were infectious, and virus isolated from infected plants appeared to be identical to the parental U1 TMV, both physically and biologically. However, addition of extra nucleotides to the 5' terminus of the transcripts reduced infectivity. Yet after replication, the progeny virus regained levels of infectivity equal to that of the parental virus, and the extra nucleotides were not retained. This corroborates earlier observations with poliovirus (13) and brome mosaic virus (9) demonstrating the ability of enzymes involved in replication of RNA viral genomes to recognize and/or replicate only viral sequences.

The principle objective of this work was to extend the life cycle of TMV through a DNA phase, thereby generating an experimental system to effectively examine the molecular genetics of TMV. Accomplishment of this objective allows the viral genome to be manipulated directly and precisely. Viral functions may be accurately mapped and sequences from diverse sources combined to create unique genomes for studying regulation and biological functions of RNA viral genes. TMV genes with confirmed biological activity may be studied individually or expressed in other backgrounds. In addition, preliminary data suggest that biological variants may occur less frequently in virus populations derived from cloned genomes. Such systems may be useful for studying fidelity of replication and evolution in RNA viruses.

ACKNOWLEDGEMENTS

We thank Carol Boyd and George Grantham for technical assistance.

REFERENCES

1. Zaitlin M, and Israel HW (1975). Tobacco mosaic virus. CMI/AAB Descriptions of Plant Viruses No. 151.

2. Kunkel LO (1947). Variation in pathogenic viruses. Ann Rev Microbiol 1:85-100.
3. Mundry KW, Gierer A (1958). Die erzeugung von mutationen des tabakmosaicvirus durch chemischie behandlung seiner nucleinsaure in vitro. Z Vererbungslehre 89:614-630.
4. Wittman HG (1960). Comparison of the tryptic peptides of chemically induced and spontaneous mutants of tobacco mosaic virus. Virology 12:609-612.
5. Goelet P, Lomonossoff P, Butler PJG, Akam ME, Gait MJ, Karn J (1982). Nucleotide sequence of tobacco mosaic virus RNA. Proc Natl Acad Sci USA 79:5818-5822.
6. Ohno T, Aoyagi M, Yamanashi Y, Saito H, Ikawa S, Meshi T, Okada Y (1984). Nucleotide sequence of the tobacco mosaic virus (tomato strain) genome and comparison with the common strain genome. J Biochem 96:1915-1923.
7. Nishiguchi M, Kituchi S, Kiho Y, Ohno T, Meshi T, Okada Y (1985). Molecular basis of plant viral virulence: the complete nucleotide sequence of an attenuated strain of tobacco mosaic virus. Nucleic acids research 13:5585-5590.
8. Cress DE, Kiefer MC, Owens RA (1983). Construction of infectious potato spindle tuber viroid cDNA clones. Nucleic Acids Research 11:6821-6835.
9. Ahlquist P, French R, Janda M, Loesch-Fries LS (1984). Multicomponent RNA plant virus infection derived from cloned viral cDNA. Proc Natl Acad Sci USA 81:7066-7070.
10. Murray MG, Hoffman LM, Jarvis NP (1983). Improved yield of full-length phaseolin cDNA clones by controlling premature anticomplementary DNA synthesis. Plant Molecular Biology 2:75-84.
11. Dawson WO, Beck DL, Knorr DA, Grantham GL (1986). cDNA cloning of the complete genome of tobacco mosaic virus and production of infectious transcripts. Proc Natl Acad Sci USA 83:1832-1836.
12. Ahlquist P, Janda M (1984). cDNA cloning and in vitro transcription of the complete brome mosaic virus genome. Molecular & Cellular Biology 4:2876-2882.
13. Racaniello VR, Baltimore D (1981). Cloned poliovirus cDNA is infectious in mammalian cells. Science 214:916-919.

II. ANTIGENIC AND FUNCTIONAL ANALYSES OF VIRION PROTEINS AND CELLULAR RECEPTORS

Positive Strand RNA Viruses, pages 59–77

STRUCTURE AND FUNCTION OF HUMAN RHINOVIRUS 14 AND MENGO VIRUS: NEUTRALIZING ANTIGENIC SITES, PUTATIVE RECEPTOR BINDING SITE, NEUTRALIZATION BY DRUG BINDING

Michael G. Rossmann[2], Edward Arnold[2], Greg Kamer[2], Marcia J. Kremer[2], Ming Luo[2], Thomas J. Smith[2], Gerrit Vriend[2], Roland R. Rueckert[3], Anne G. Mosser[3], Barbara Sherry[3], Ulrike Boege[4], Douglas G. Scraba[4], Mark A. McKinlay[5] and Guy D. Diana[5]

[2]Department of Biological Sciences, Purdue University, West Lafayette, Indiana 47907;
[3]Biophysics Lab, University of Wisconsin, 1525 Linden Drive, Madison, Wisconsin 53706;
[4]Department of Biochemistry, University of Alberta, Edmonton, Alberta T6G 2H7, Canada;
[5]Microbiology and Medicinal Chemistry, Sterling-Winthrop Research Institute, Columbia Turnpike, Rensselaer, New York 12144

INTRODUCTION

The structure of human rhinovirus 14 (HRV14) (1), poliovirus Mahoney type 1 (2) and Mengo virus (3) have now been determined. These and a variety of RNA plant viruses as well as an insect virus (4) are all based on the same quaternary and tertiary structural principles.

These viruses consist of an icosahedral protein shell (Fig. 1) surrounding an RNA core. The lack of visible structure in the central cavity results from the random orientation of the asymmetrical RNA molecule. VP1, VP2 and VP3 each consists of an eight-stranded anti-parallel β-barrel

[1]The work was supported by a grant from the Sterling-Winthrop Research Institute to M.G.R. and by grants from the National Institutes of Health and the National Science Foundation to M.G.R. E.A. was supported by a National Institutes of Health Postdoctoral Fellowship during part of this work.

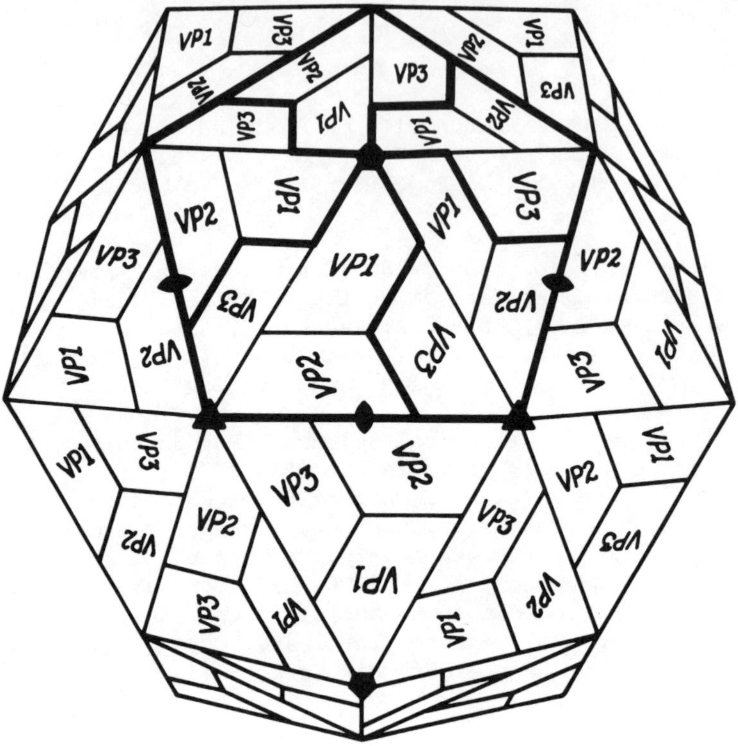

FIGURE 1. Relationship of the pseudo equivalent VP1, VP2 and VP3 subunits in the icosahedral capsid. The thick outline corresponds to the 6S (VP1, VP0, VP3) protomer and the 14S pentamer observed in assembly experiments.

closely similar to that observed in southern bean mosiac virus (SBMV) (5) and other plant viruses (Fig. 2). They have a pseudo threefold relation to each other as if the chains were identical and obeyed the rules of a $\underline{T}$ = 3 virus (6).

HRV14 AND MENGO VIRUS STRUCTURE

In describing the picornavirus structure, this paper will refer primarily to the HRV14 results as the Mengo structure had not yet been fully studied when this manuscript

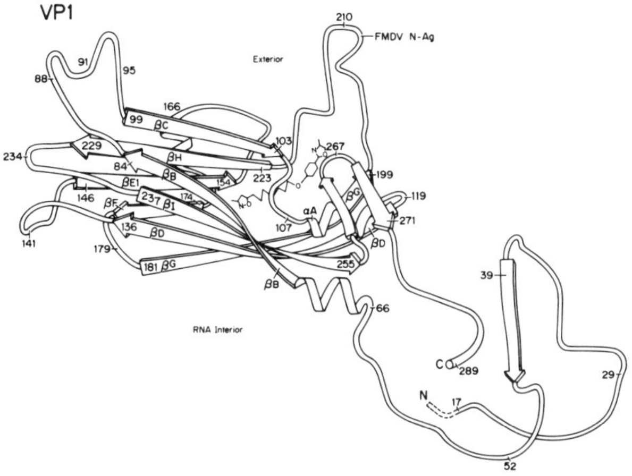

FIGURE 2. Diagrammatic drawing of polypeptide folding of VP1. Similar folds occur in VP2, VP3, plant viruses such as SBMV and other spherical RNA viruses. Shown also is the site of the antiviral WIN compound binding.

had to be written.

The first 64 of the 73 amino-terminal residues of VP1 reside under VP3, whereas the first 42 of the 71 amino-terminal residues of VP3 are under VP1. Thus, the predominant positions of VP1 and VP3 at the RNA-protein interface are exchanged relative to their positions at the exterior surface. The first 25 of the 69 residues of the internal structural protein VP4 are not seen in the electron density map, implying that they lack icosahedral symmetry. VP4 is an extended polypeptide chain, positioned internally below, but in contact with, VP1 and VP2. Its visible amino end surrounds the fivefold axis. The carboxy ends of VP1 and VP3 are external and function in part to associate proteins within a protomer. Large sequence insertions relative to the typical shell domain form protrusions on VP1, VP2 and VP3 and create a deep cleft or "canyon" on the HRV14 viral surface. The canyon separates the major part of five VP1 subunits (in the "north") clustered about a pentamer axis from the surrounding VP2 and VP3 subunits (in the "south"), thus forming a moat around the VP1 protrusions on the fivefold axis. The south canyon walls are lined with the carboxy-terminal ends of VP1 and a large sequence insertion in VP1 corresponding to helices αD and αE in the equivalent SBMV capsid protein. The north canyon wall is partially lined with the carboxy-terminus of VP3. VP2 is hardly associated at all with the canyon, whereas VP1 is the major contributor to the residues lining the canyon. The canyon is 25 Å deep and 12 - 30 Å wide.

Because of the additional elaborations which VP1 has on the surface relative to VP2 and VP3, its overall shape is that of a kidney, with the depression forming a large part of the canyon. The first 16 residues of VP1 are not seen in the electron density map. The shell domain of VP1 in HRV14 starts at residue 74. The small sequence insertion between βB and βC in rhinovirus and poliovirus is not found in foot-and-mouth disease virus (FMDV). This loop forms a major immunogen in HRV14, NIm-IA, and poliovirus. The residues in VP1 of HRV14 that are analogous to αD and αE helices in SBMV protrude to the surface and form part of the south rim of the canyon, but do not form helices in HRV14. The most external portion of this segment contains the major antigenic site of FMDV and has been predicted to form an α-helix (7).

The proximity of Ser 10 in VP2 to the carboxy-terminus of VP4 suggests that the VP0 cleavage which occurs during the virus maturation step is autoproteolytic. There is no histidine next to Ser 10 and the carboxy end of VP4. However, nucleotide bases of the RNA could act as proton acceptors in the autocatalysis. Thus, the insertion of RNA into the growing capsids could trigger the cleavage of VP0 into VP2 and VP4. This "enzyme" is, therefore, composed both of protein and RNA components.

There is a large 43-residue insertion in the VP2 position, corresponding to βE2 of VP1 and VP3, forming an external mushroom-shaped "puff". This is positioned next to the VP1 elaborations, associated with the major antigenic site in FMDV, which line the south canyon wall. The most external residues of this puff correspond to NIm-II of HRV14. In contrast to VP1 and VP3, the carboxy-terminus of VP2 has no extension beyond the shell domain.

All residues of VP3 can be seen in the electron density map. The 26 amino-terminal residues form a fivefold β-cylinder about the pentamer axis (Fig. 3). This fivefold cylinder extends down into the RNA to a radius of 111 Å. The polypeptide emerges from the β-cylinder and circles around the base of the VP1 shell domain, probably making extensive contact with the RNA in the central cavity. It then emerges on the viral surface near residue 61 and enters the shell

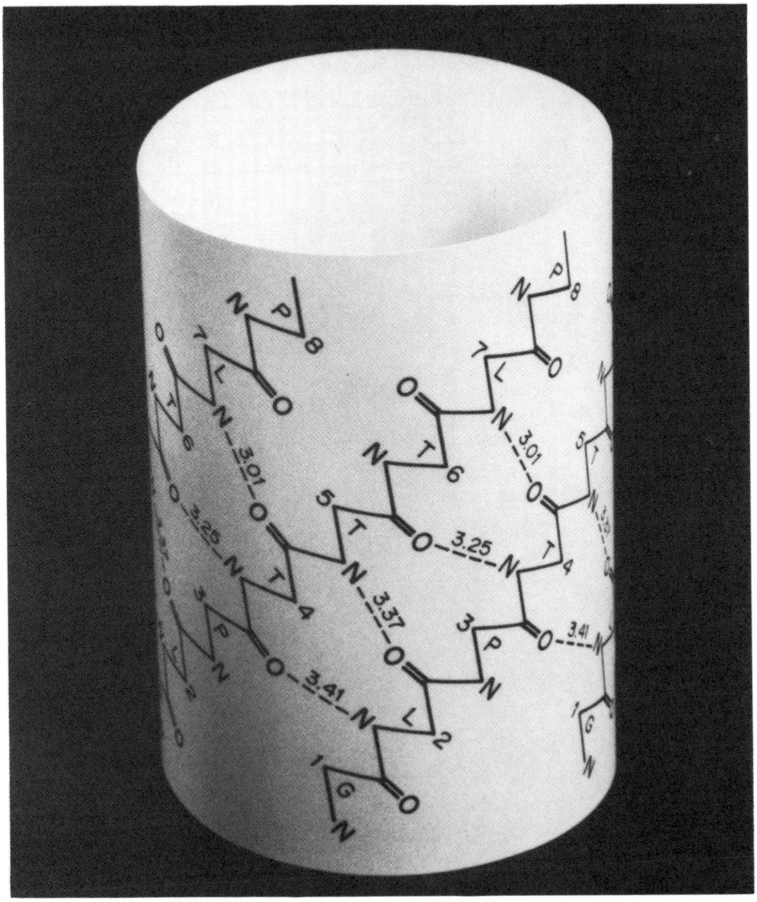

FIGURE 3. The fivefold β-cylinder formed by the amino-terminal ends of VP3 at the protein/RNA interface.

domain at residue 72. The top corner of the VP3 shell domain, between βB and βC, is the NIm-III site of HRV14, structurally equivalent to NIm-IA in VP1.

The Mengo virus structure shows greatest similarity to HRV14 in VP3. In VP2 the puff is deleted as was correctly predicted by Ann Palmenberg's sequence alignments (8). The major differences occur, however, in VP1. Here residues at the top corner corresponding to NIm-IA are deleted. The

Table 1

Alignment of HRV14, VP1, VP2, VP3, SBMV and TBSV based on Structural Superpositions

	HRV14								
	VP1	VP2	VP3	SBMV	TBSV	STNV	ConA	Flu	Exceptional Physical Properties
	66 D	55 K	42 N						13+
	67 V	56 P	43 L						
	68 E	57 D	44 L						12+,22-
αZ	69 C	58 T	45 E						
	70 F	59 S	46 I						
	71 L	60 V	47 I						3+,19+
	72 G	61 C	48 Q						
	73 R		49 V	71 H	108 H	29 F			6+,8-
	74 A		50 D	72 C	109 R	30 A	2 D		6+
	75 A	63 F	51 T	73 E	110 E	31 L	3 T		22-
	76 C	64 Y	52 L	74 L	111 Y	32 I	4 I	164 L	3+,4+,13-,17-,18-,20+
	77 V	65 T	53 L	75 S	112 L	33 N	5 V	165 N	
βB	78 H	66 L		76 T	113 T	34 S	6 A	166 V	
	79 V						7 V	167 T	0,19+
	80 T	69 K	69 I	79 L	116 N	37 T	40 W	168 M	
	81 E	70 T	70 P	80 A	117 N	38 N	41 N	169 P	
	82 I	71 W	71 L	81 V	118 S		42 M	170 N	
	83 Q	72 T	72 N	82 T	119 S		43 Q	171 N	
	84 N	73 T	73 A		120 G		44 D	172 D	
	85 K						45 G		
	86 D						46 K		
	87 A						47 V		
	88 T						48 G		
	89 G								
	90 I								
Top	91 D								
corner	92 N								
	93 H								
	94 R								
	95 E								
	96 A								
	97 K								
	98 L								
	99 F	77 G	80 V				49 T	177 L	
βC	100 N	78 W	81 F	87 T		43 T	50 A	178 Y	23+
	101 D	79 C	82 G	88 S		44 V	51 H	179 I	
	102 W	80 W	83 T	89 E		45 Q	52 I	180 W	14+,23+
	103 K						53 I	181 G	
	104 I			92 M		48 S	54 L	182 I	
	105 N			93 P		49 N	55 F		
αA_0	106 L			94 F		50 G			
	107 S			95 T		51 I			
	108 S			96 V	141 L				
	109 L			97 G	142 F				
	110 V	90 G	95 T	98 T	143 S				
	111 Q	91 V	96 L	99 W	144 W				14+,23+
	112 L	92 F	97 L	100 L	145 L				0,3+,7+,9-,17-,21+
	113 R	93 G	98 G	101 R	146 P	54 G			15+
αA	114 K	94 Q	99 E	102 G	147 A	55 D			13+
	115 K	95 N	100 I	103 V	148 L				
	116 L	96 M	101 V	104 A	149 A	58 N			
	117 E	97 F	102 Q	105 Q	150 S	59 Q			
	118 L	98 F	103 Y	106 N	151 N	60 R			
	119 F	99 H	104 Y	107 W	152 F	61 S			14+,23+
	120 T	100 S	105 T	108 S	153 D	62 G			4-,7-
	121 Y	101 L	106 H	109 K	154 Q	63 D			6+
	122 V	102 G	107 W	110 Y	155 Y				18-,20+,23+
	123 R	103 R	108 S	111 A	156 S				0,6+,10-,15+,16+,18+, 22+
βD	124 F	104 S	109 G						
	125 D	105 G	110 S	114 A	159 S	69 S			7-,14-
	126 S	106 Y	111 L	115 I	160 V	70 H	89 V	198 A	
	127 E	107 T	112 R	116 R	161 V	71 K	90 R	199 S	6+,10-,15+,16+,18+,22+
	128 Y	108 V	113 F	117 Y	162 L	72 L	91 V	200 G	2+,3+,4+,6-,7+,14+, 17+,18-,20+,23+
	129 T	109 H	114 S	118 T	163 D	73 H	92 G	201 R	

Table 1 (continued)

	HRV14 VP1	HRV14 VP2	HRV14 VP3	SBMV	TBSV	STNV	ConA	Flu	Exceptional Physical Properties
βD	130 I	110 V	115 L	119 Y	164 Y	74 V	93 L	202 V	2+,3+,4+,7+,13-,17-,18-,19+,20+
	131 L	111 Q	116 M	120 L	165 V	75 R	94 S	203 T	
	132 A	112 C	117 Y	121 P	166 P	76 G	95 A		9+
	133 T	113 N	118 T	122 S	167 L	77 T	96 S		
	134 A	114 A	119 G	123 C	168 C	78 A	97 T		9+,14-
	135 S	115 T	120 P	124 P	169 G		98 G		
	136 Q	116 K	121 A	125 T	170 T				
	143 S	117 F	122 L	126 T	171 T	80 T			
	144 S	118 H	123 S	127 T	172 E	81 V		206 T	
βE	145 N	119 S	124 S	128 S	173 V	82 S	104 N	207 R	4-
	146 L	120 G	125 A	129 G	174 G	83 Q	105 T	208 R	7-
	147 V	121 C	126 K	130 A	175 R	84 T	106 I	209 S	
	148 V	122 L	127 L	131 I	176 V	85 F	107 L	210 Q	3+,13-,17-,19+,21+
	149 Q	123 L	128 I	132 H	177 A	86 R	108 S	211 Q	
	150 A	124 V	129 L	133 M	178 L	87 F	109 W		
	151 M	125 V	130 A	134 G	179 Y	88 I	110 T		
	152 Y	126 V	131 Y	135 F	180 F	89 W	111 F		2+,4+,13-,14+,17-,18-,20+,23+
	153 V	127 I	132 T	136 Q	181 D	90 F	112 T		
	154 P	128 P	133 P	137 Y	182 K	91 R	113 S		10-,11+,15+,16+
	155 P	129 E	134 P	138 D	183 D				9+,11+,12+
	156 G	130 H	135 G						7-
	157 A	131 Q	136 A						
	158 P	132 L	137 R	142 T	187 D				
	159 N		138 G	143 I	188 E				
	160 P	PUFF	139 P	144 P	189 P				0,1-,2-,4+,9+,11+,20-
	161 K		140 Q	145 V	190 A	109 A			
αB	162 E			146 S	191 D	110 N			
	163 W			147 V	192 R	111 F			
	164 D			148 N	193 V	112 M			
	165 D			149 Q	194 E	113 S			
	166 Y								
	167 T								
	168 W								
	169 Q			150 L	195 L	115 Y			
	170 S			151 S	196 F	116 N			
	171 A			152 N	197 N	117 P			
	172 S			153 L	198 F	118 I	127 H		17-
	173 N			154 K	199 G		128 F		
βF	174 P	185 P	149 T	155 G	200 V	129 K	129 M		
	175 S	186 H	150 H	156 Y	201 L	130 D	130 F		23+
	176 V	187 Q	151 V	157 V	202 K	131 V	131 N		19+
	177 F	188 F	152 V	158 T	203 E	132 T	154 L		
	178 F	189 I	153 W	159 G	204 T	133 L	155 E		23+
	179 K	190 N	154 D	160 P	205 A	134 N	156 L		2-,20-
	180 V	191 L	155 I	161 L	206 P	135 C	157 T		9+
Fourth corner down	181 G								
βG1	182 D		159 S			143 K	171 G		7-
	183 T	196 T	160 T			144 D	172 R	212 T	6+,12+,16+,18+
	184 S	197 A	161 I	183 I	210 A	145 R	173 A	213 I	
	185 R	198 T	162 V	184 T	211 M	146 I	174 L	214 I	
	186 F	199 I	163 M	185 I	212 L	147 I	175 F	215 P	0,2+,3+,4+,7+,10+,11-,12-,13-,14+,16+,17-,18-,19+,20+,21+
	187 S	200 V	164 T	186 A	213 R	148 N	176 Y	216 N	
	188 V	201 I	165 I	187 L	214 I	149 L	177 A	217 I	1+,3+,4+,5+,7+,8+,10+,11-,13-,16-,17-,19+,20+,21+
	189 P	202 P	166 P	188 D	215 P	150 P	178 P	218 G	0,1-,2-,3-,4+,8-,9+,11+,14-,19-,20-
	190 Y	203 Y	167 W	189 T	216 T	151 G	179 V	219 S	23+
	191 V	204 I	168 T			152 Q	180 V	220 R	
	192 G	205 N	169 S	192 V			181 I	221 P	
	193 L	206 S	170 G	193 S				222 W	
	194 A	207 V	171 V	194 E	218 K			223 V	

Table 1 (continued)

	HRV14								
	VP1	VP2	VP3	SBMV	TBSV	STNV	ConA	Flu	Exceptional Physical Properties
	195 S	208 P	172 Q	195 K	219 V				
	196 A	209 I	173 F	196 R	220 K				15+,22+
↑	197 Y	210 D	174 R	197 Y	221 R	154 V			6+,10+,23+
βG2	198 N	211 S	175 Y	198 P	222 Y	155 N			23+
↓	199 C	212 M	176 T	199 F	223 C	156 Y			8−
	200 F			200 K	224 N	157 N			
	201 Y			201 T	225 D				
	FMDV loop								
	217 V		180 T	217 N	232 K			224 T	
	218 L		181 Y	218 I	233 L			225 G	17−
	219 N		182 T	219 L	234 I			226 L	
	220 H		183 S	220 V	235 D		189 S	227 S	
	221 M		184 A	221 P	236 L		190 A	228 S	
↑	222 G		185 G	222 A	237 G	171 I	191 F	229 R	5+
	223 S	219 S	186 F	223 R	238 Q	172 F	192 E	230 I	
	224 M	220 L	187 L	224 L	239 L	173 M	193 A	231 I	0,1+,3+,5+,11−,13−,16−,17−,21+
βH	225 A	221 M	188 S	225 V	240 G	174 L	194 T	232 I	
	226 F	222 V	189 C	226 T	241 I	175 Q	195 F	233 Y	
	227 R	223 I	190 W	227 A	242 A	176 I	196 A	234 W	
	228 I			228 M	243 T	177 G		235 T	
↓	229 V			229 E	244 Y	178 D			12+,15−,22−
	230 N			230 G	245 G	179 S			7−,11+,14−
	231 E			231 G	246 G				
Second	232 H			232 S	247 A		199 I		
corner	233 D						200 K		
down	234 E						201 S		
	235 H						202 P		
	236 K						204 S	240 G	
↑	237 T	236 P	203 G	236 A	249 R		205 H	241 D	
	238 L	237 S	204 Q	237 V	250 L		206 P	242 V	
	239 V	238 L	205 V	238 N	251 A	182 G	207 A	243 L	
	240 K	239 P	206 Y	239 T	252 V	183 L	208 D	244 V	
	241 I	240 I	207 L	240 G	253 G	184 W	210 I	245 I	3+,4+,8+,17−,19+
	242 R	241 T	208 L	241 R	254 E	185 D	211 A	246 N	6+,16+
	243 V	242 V	209 S	242 L	255 L	186 S	212 F	247 S	
	244 Y	243 T	210 F	243 Y	256 F	187 S	213 F	248 N	23+
	245 H	244 I	211 I	244 A	257 L	188 Y	214 I	249 G	19+
βI	246 R	245 A	212 S	245 S	258 A	189 E	215 S		
	247 A	246 P	213 C	246 Y	259 R	190 A	216 N		
	248 K	247 M		247 T	260 S	191 V			
	249 H	248 C				192 Y			23+
	250 V	249 T				193 T			
	251 E	250 E	218 K	249 R	262 T	194 D			5−,6+,10−,12+,13+,16+,18+,21−
	252 A	251 F	219 L	250 L	263 L				17−
	253 W	252 S	220 R	251 I	264 Y				23+
↓	254 I	253 G	221 L	252 E	265 F				
	255 P	254 I	222 M	253 P	266 P				4+,9+
	256 R	255 R	223 K	254 I	267 Q				6+,10−,15+,16+,18+,22+
	257 A	256 S	224 D	255 A	268 P				
	258 P	257 K	225 T	256 A	269 T				
	259 R	258 S	226 Q	257 A	270 N				
	260 A	259 I	227 T	258 L	271 T				
	261 L								
	262 P								
	263 Y								

Notes:

(1) Comments indicate physical properties dominant at aligned positions which may be the determining factors that are required for producing the virus β-barrel fold.

(2) The column headed "Exceptional physical properties" gives the particular property which is unusually large (+) or small (−) for the aligned set of amino acids.

0 Small minimum base change per codon
1 Helix forming
2 Sheet forming
3 Hydrophobicity (1) (Ponmanvolan)

Table 1 (continued)

Notes (continued):

4	Hydrophobicity (2) (Tanford)
5	Hydration potential
6	Polarity (1)
7	Bulkiness
8	pK (1)
9	pK (3)
10	Transfer of free energy
11	Turn forming
12	Hydrophilicity (Kuntz)
13	Polarity (2)
14	Volume
15	pI
16	Transfer free buried energy
17	Surface tension transfer energy
18	Hydrophilicity
19	Parallel sheet forming
20	Anti-parallel sheet forming
21	Hydropathy (Kyte & Doolittle)
22	Charge
23	Aromaticity

(3) All alignments were performed by superposition of the three-dimensional structures.
(4) Coordinates for the hemagglutinin spike of influenza virus ("Flu") were kindly supplied by Don Wiley.

whole of the top βB-βC corner is folded over into the canyon between VP2 and VP1. The "FMDV-loop" is deleted as is also the carboxy-terminal end lining the south canyon wall. The result is a totally changed virion surface (no puff, no FMDV loop, no NIm-IA protrusion), and a filling in of part of the canyon leaving only a deep "pit" into the deepest part by the highly conserved PPGAYP sequence of VP1.

ASSEMBLY

Assembly of picornaviruses (9) proceeds from 6S protomers of VP1, VP3 and VP0, via 14S pentamers of five 6S protomers, to mature virions. The final step involves inclusion of the RNA into empty capsids or partially assembled shells with simultaneous cleavage of VP0 into VP2 and VP4. Conversely, in vitro disassembly, produced by mild denaturation, proceeds via the expulsion of VP4 followed by the RNA (9).

Both the amino and carboxy ends of VP1 and $VP3_5$ are intertwined with each other. Furthermore, if VP4 and VP2 are considered as VP0, then VP0 is also intertwined with VP1 and $VP3_5$, which strongly suggests that the 6S protomer is as shown in Figure 1. These protomers are themselves intertwined because of the fivefold β-cylinder formed by the amino ends of the VP3's and the proximity of the observed amino ends of VP4's to the fivefold axis. Thus, the 14S pentamers closely correlate with the observed structure, shown diagrammatically in Figure 1.

The protein VP2, once cleaved from VP4, is globular and does not contact the other proteins extensively, although there are extensive solvent accessible regions around VP2 continuing into VP1. This, as well as the extraordinarily internal heavy atom sites on VP2, is consistent with loose binding of VP2 to the capsid and the channels that lead to the interior through a pore in the canyon floor. Disruption of pentamer-pentamer contacts, mediated by a slight reorientation of VP2 or its complete removal, could provide a port by which the VP4 and RNA can exit. Binding of a cell receptor in the canyon next to VP3 could facilitate this process, possibly accompanied by an isoelectric change.

SEQUENCE ALIGNMENTS

The three-dimensional coordinates of the three larger viral proteins have been superimposed on each other and on the structures of SBMV and tomato bushy stunt virus (TBSV) using techniques developed by Rossmann (10,11). The resultant structural alignments are shown in Table 1. The aligned residues have been systematically examined for common physical properties and some of these are also shown in Table 1. These characteristics may be essential for providing the common virus folding pattern in the absence of any obvious sequence homology. This "fingerprint" of the virus β-barrel fold can be used (12) to detect similar folds in other viral sequences.

NEUTRALIZING IMMUNOGENIC SITES

The immunological response to a virus is a major defense against disease in animals. Antibodies can bind to viruses but they do not necessarily neutralize infectivity. Despite the 60-fold equivalence of each potential binding site on the virus, as few as four neutralizing antibodies per virion can be sufficient to inhibit infectivity of poliovirus (13). Neutralizing antibodies usually change the isoelectric point of the picornavirions (14,15), indicating that a conformational change frequently accompanies neutralization. Antibodies may neutralize by interfering with cell attachment (antibodies to NIm-II and IA), membrane penetration or virus uncoating (antibodies to NIm-III) (cf. also 16). Antibodies that bind to poliovirus may require bivalent attachment for neutralization of the virus (13,17).

Amino acid residues within the major neutralization immunogens of HRV14 have been identified by Sherry and Rueckert (18) and by Sherry et al. (19). They found four major immunogenic neutralization sites NIm-IA, NIm-IB, NIm-II and NIm-III, each one composed of overlapping epitopes where a given mutant was resistant to many or all of the antibodies directed against that site. The residues, corresponding to escape mutants, invariably faced outward towards the viral exterior.

Many of the methods that have been used to determine antibody binding sites depend on the use of synthetic peptides as antigens. Peptides associated with neutralizing antigenic regions do, in general, elicit antibodies that can neutralize the intact virus. In a significant number of cases, however, the sequence in question lies far below the viral surface or is even buried in the RNA. This suggests that some peptides can elicit antibodies which subsequently bind to totally unrelated portions of the native virus.

THE CANYON AS RECEPTOR BINDING SITE

Despite the sequence and surface similarities of picornaviruses, they have different host and tissue specificity. This is particularly obvious in a comparison of HRV14 and Mengo virus. The presence of the 25 Å deep canyon, circulating around each of the 12 pentamer vertices, suggests that this is the site for cell receptor binding. An antibody molecule, whose Fab fragment would have a diameter of the order of 35 Å, would have difficulty in reaching the canyon floor, its entrance being blocked by the canyon rim. Thus, the residues in the deeper recesses of the canyon would not be under immune selection and could remain constant, permitting the virus to retain its ability to seek out the same cell receptor.

Although retention of the canyon structure for all picornaviruses is to be expected, variation in the residues lining the canyon should be anticipated between viruses that attach themselves to different host cell receptors. Those parts of the carboxy-terminal ends of VP1 and VP3 which line the canyon walls are some of the least conserved amino acids among picornaviruses. As the topology of the canyon should be retained, the highly conserved, structurally equivalent sequences (MYVPPGAPNP starting at 151 of VP1 and AYTPPGARGP

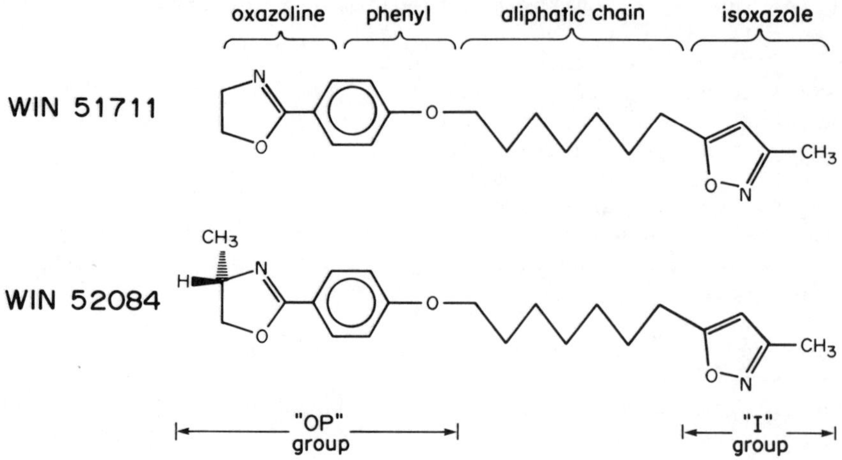

FIGURE 4. Structural formula of the antiviral compounds WIN 51711 and WIN 52084.

starting at 130 of VP3 for HRV14) in rhino, polio and FMD viruses situated near the floor of the canyon may be significant. The conserved sequence in VP1 in part provides the drug binding site mentioned below.

ANTIVIRAL DRUG BINDING

Arildone (20-23) and WIN 51711 (24-26) are examples of compounds (27,28) which inhibit picornavirus replication by inhibiting viral uncoating without affecting cellular attachment. These compounds stabilize the virion against alkaline and heat denaturation, loss of VP4 and induce a significant pI change. The effect of these substances on the virion properties in some ways resembles that of neutralizing antibodies or of sulfhydryl reagents (29,30).

WIN 51711 represents a class of compounds which inhibits picornavirus replication in tissue culture and in animal infected models. A number of compounds related to WIN 51711 have been synthesized at Sterling-Winthrop Research Institute and tested against several different picornaviruses. The end point used in these studies is the minimal inhibition concentration (MIC), defined as the concentration which

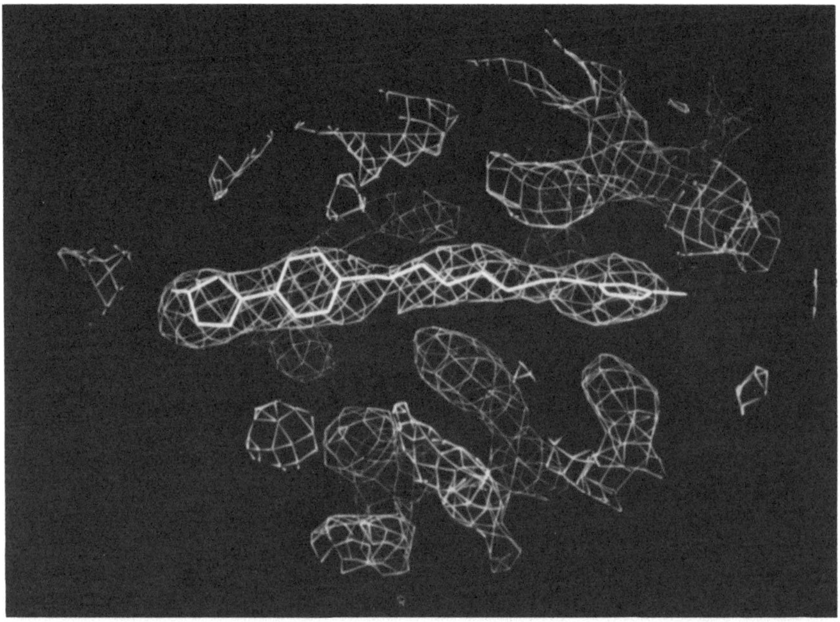

FIGURE 5. Model of WIN 52084 in its electron density.

produces a 50% reduction of plaque count in cell culture. Although many of these compounds have a fairly wide spectrum of anti-picornavirus activity, the MIC values can vary considerably between picornavirus pathogens. The MIC values are dependent upon a number of factors: the binding affinity of the compound to the virus, the ability of the bound compound to interfere with the uncoating process and the incorporation of the compound into the host cell. The compounds that have been selected for testing on the basis of their efficacy against human rhinoviruses, polioviruses and coxsackie A and B viruses were inactive against encephalomyocarditis and hepatitis A virus.

The site of binding on HRV14 has been determined for two such compounds, namely WIN 51711 and WIN 52084 (Fig. 4). These compounds consist of a hydrophobic phenoxazole head, a

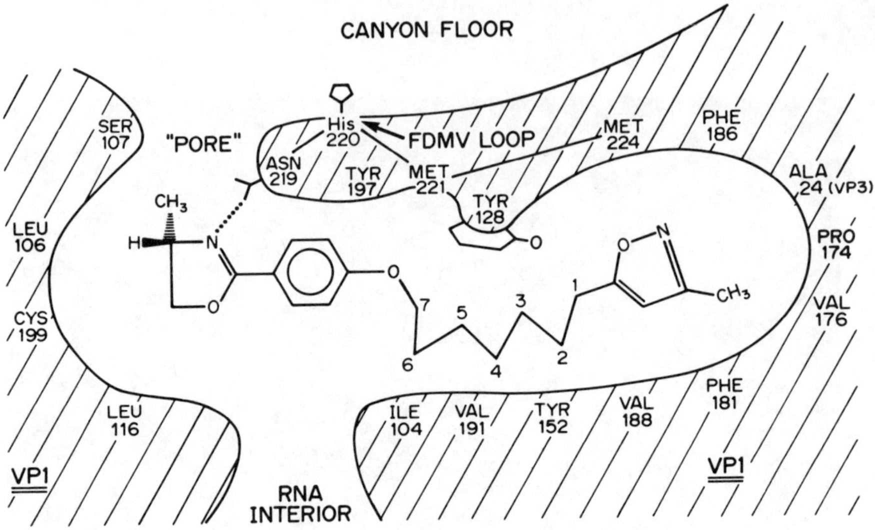

FIGURE 6. Diagrammatic view of WIN 52084 binding to VP1 of HRV14.

seven-membered aliphatic chain and an isoxazole functional tail. They are two of close to a thousand such compounds that have been tested against a screen of various picornaviruses by the Sterling-Winthrop Company.

There were close to 60 drug molecules per virion as estimated both by the height of the electron density and radioactively labelled WIN 51711 (31). The electron density for both WIN 51711 and WIN 52084 when bound to HRV14 was unequivocal (Fig. 5). The isoxazole (I) end binds into a hydrophobic pocket formed by the interior of the VP1 β-barrel (Fig. 2), in part created by the displacements of Phe 152 and Met 221. The oxazoline-phenolic (OP) end of the compounds binds into a hydrophilic pore on the base of the canyon. The pore opens into large channels leading to the RNA interior (Fig. 6). The residues involved in drug binding are given in Table 2. A hydrogen bond between the nitrogen atom of the oxazoline ring and Asn 219 orients this end of the compound. Steric displacement of some of the residues by the drug affect considerable portions of βH leading from the FMDV loop across the canyon floor in VP1. The main chain is

Table 2

Residues within 3.6 Å of the Bound WIN Compounds

VP	Residue Number	HRV14 Amino Acid Type	Interaction	HRV2, HRV39, & HRV49	PV1 Mahoney	PV1 Sabin	PV2 Lansing	PV3 Leon	Mengo	FMDV (A10)
1	104	I	phenolic oxygen	I	I	I	I	I	L	L
1	106	L	phenyl	L	Y	Y	Y	Y	-	-
1	107	S	methyl on oxazoline	Q	K	K	K	K	-	-
1	116	L	oxazoline	F	L	L	L	L	S	A
1	128	Y	phenolic oxygen, C_3 in aliphatic chain	I	L	F	F	F	V	I
1	152	Y	C_4, C_2 & C_1 of aliphatic chain	Y	Y	Y	Y	Y	C	V
1	174	P	isoxazole	A	P	P	P	P	P	P
1	176	V	methyl of isoxazole	V	I	I	V	I	V	A
1	186	F	isoxazole	F	I	I	I	I	F	L
1	188	V	C_2 and C_3 of aliphatic chain	L	V	V	V	V	V	L
1	191	V	C_6 of aliphatic chain	L	V	V	V	V	N	T
1	197	Y	phenol C_7 & C_5 of aliphatic chain	Y	Y	Y	Y	Y	L	L
1	199	C	oxazoline	M	H	H	H	H	A	T
1	219	N	phenyl; N of oxazoline	N	N	N	N	D	S	G
1	221	M	phenyl	M	F	F	F	F	F	M
1	224	M	C_1 of aliphatic chain	L	L	L	L	L	L	A
3	24	A	isoxazole	A	A	A	A	A	I	V

Compound	MIC (μM)	HRV2	HRV39	HRV49	PV1 Mahoney	PV1 Sabin	PV2 Lansing	PV3 Leon	Mengo	FMDV (A10)
WIN 51711	0.4	3.8	9.3	4.7	8.7	1.1	0.8	1.1	inactive	inactive
WIN 52084	0.06	0.1	2.2	0.8	>9.3	6.9	5.7	1.2	no data	no data

Notes:

(i) Abbreviations for viruses: HRV, human rhinovirus; PV, poliovirus; FMDV (A10), foot-and-mouth disease virus strain A10.

(ii) Sequence alignments are based on work by A. Palmenberg (unpublished).

displaced as much as 3 Å in places and the side chains by even larger distances.

Table 2 lists not only the residues at the drug binding site, but also minimum inhibition concentrations (MIC) required to reduce the plaque counts by one-half. The apparent insensitivity of HRV2 to WIN 51711 is explained in terms of the steric hindrance caused by the leucine, corresponding to valine 188 in HRV14.

Numerous substitutions of these compounds have been tested and the resultant MIC values can be explained on a qualitative basis in terms of the drug's potential binding ability. The most important observation is that the drug becomes inactive on replacing the nitrogen and oxygen atoms in the oxazoline ring by carbon. Similarly, the compounds are inactive in those viruses (namely encephalomyocarditis, Mengo, hepatitis A and FMDV) which do not conserve Asn 219 of HRV14 as Asn or Asp (although the entrance to the pocket is far more constricted in Mengo virus). The hydrogen bond between the drug nitrogen or oxygen atoms and Asn 219 is, therefore, probably essential to the drug function.

The effect of the compounds on the virion can be compared to that of the binding of an NAD co-factor to a dehydrogenase (32,33). The isoxazole group (compared to the adenine end of NAD) binds into a hydrophobic pocket and stabilizes the virions against denaturation by heat or alkaline treatment. The binding of the aliphatic chain, corresponding to the pyrophosphate in NAD, causes essential conformational changes to permit the binding of the functional oxazoline-phenyl group (or nicotinamide group of NAD). The OP group covers a pore, thereby changing the pI, in the canyon floor which admits anions (e.g. $Au(CN)_2$ as used in the structure determination of HRV14) to the RNA, causing swelling and disassembly.

The structure of poliovirus Mahoney type I (2) shows electron density at the precise site of the WIN compounds. Thus a "co-factor" is binding to the same site as is occupied by the WIN compounds. This co-factor (perhaps a peptide or lipid component) would permit the virion to penetrate the membrane as a complete virion. The WIN compound might compete with the co-factor but binds more tenaciously, thus inhibiting disassembly.

ACKNOWLEDGMENTS

We are grateful to Rich Colonno (Merck, Sharp and Dohme)

for the initial analysis of $Au(CN)_2$ treated HRV14 and for stimulating discussions, Stan Lemon (UNC) for MIC measurements with respect to hepatitis A virus and R. K. Kulnig (Sterling-Winthrop Research Institute) for providing the crystal structure of WIN 51711.

REFERENCES

1. Rossmann MG, Arnold E, Erickson JW, Frankenberger EA, Griffith JP, Hecht HJ, Johnson JE, Kamer G, Luo M, Mosser AG, Rueckert RR, Sherry B, Vriend G (1985). Structure of a human common cold virus and functional relationship to other picornaviruses. Nature (London) 317:145.
2. Hogle JM, Chow M, Filman DJ (1985). Three-dimensional structure of poliovirus at 2.9 Å resolution. Science 229:1358.
3. Luo M, Vriend G, Kamer G, Minor I, Arnold E, Rossmann MG, Boege U, Scraba DG, Duke G, Palmenberg A (1986). The structure of Mengo virus at atomic resolution. Manuscript in preparation.
4. Johnson JE, Hosur MV (1986). Unpublished results.
5. Abad-Zapatero C, Abdel-Meguid SS, Johnson JE, Leslie AGW, Rayment I, Rossmann MG, Suck D, Tsukihara T (1980). Structure of southern bean mosaic virus at 2.8 Å resolution. Nature (London) 286:33.
6. Caspar DLD, Klug A (1962). Physical principles in the construction of regular viruses. Cold Spring Harbor Symp Quant Biol 27:1.
7. Pfaff E, Mussgay M, Böhm HO, Schulz GE, Schaller H (1982). Antibodies against a preselected peptide recognize and neutralize foot and mouth disease virus. EMBO J 1:869.
8. Palmenberg AC (1986). Unpublished results.
9. Rueckert RR (1976). On the structure and morphogenesis of picornaviruses. In Fraenkel-Conrat H, Wagner RR (eds): "Comprehensive Virology," New York: Plenum Press, Vol 6, p 131.
10. Rao ST, Rossmann MG (1973). Comparison of super-secondary structures in proteins. J Mol Biol 76:241.
11. Rossmann MG, Argos P (1976). Exploring structural homology of proteins. J Mol Biol 105:75.
12. Rossmann MG, Palmenberg A (1986). Unpublished results.
13. Icenogle J, Shiwen H, Duke G, Gilbert S, Rueckert R, Anderegg J (1983). Neutralization of poliovirus by a

monoclonal antibody: kinetics and stoichiometry. Virology 127:412.

14. Mandel B (1976). Neutralization of poliovirus: a hypothesis to explain the mechanism and the one-hit character of the neutralization reaction. Virology 69:500.
15. Emini EA, Jameson BA, Wimmer E (1983). Priming for and induction of anti-poliovirus neutralizing antibodies by synthetic peptides. Nature (London) 304:699.
16. Diamond DC, Jameson BA, Bonin J, Kohara M, Abe S, Itoh H, Komatsu T, Arita M, Kuge S, Nomoto A, Osterhaus ADME, Crainic R, Wimmer E (1985). Antigenic variation and resistance to neutralization in poliovirus type 1. Science 229:1090.
17. Emini EA, Ostapchuk P, Wimmer E (1983). Bivalent attachment of antibody onto poliovirus leads to conformational alteration and neutralization. J Virol 48:547.
18. Sherry B, Rueckert R (1985). Evidence for at least two dominant neutralization antigens on human rhinovirus 14. J Virol 53:137.
19. Sherry B, Mosser AG, Colonno RJ, Rueckert RR (1986). Use of monoclonal antibodies to identify four neutralization immunogens on a common cold picornavirus, human rhinovirus 14. J Virol 57:246.
20. Caliguiri LA, McSharry JJ, Lawrence GW (1980). Effect of arildone on modifications of poliovirus in vitro. Virology 105:86.
21. McSharry JJ, Caliguiri LA, Eggers HJ (1979). Inhibition of uncoating of poliovirus by arildone, a new antiviral drug. Virology 97:307.
22. Diana GD, Salvador UJ, Zalay ES, Johnson RE, Collins JC, Johnson D, Hinshaw WB, Lorenz RR, Thielking WH, Pancic F (1977). Antiviral activity of some β-diketones. 1. Aryl alkyl diketones. In vitro activity against both RNA and DNA viruses. J Med Chem 20:750.
23. Diana GD, Salvador UJ, Zalay ES, Carabateas PM, Williams GL, Collins JC, Pancic F (1977). Antiviral activity of some β-diketones. 2. Aryloxyl alkyl diketones. In vitro activity against both RNA and DNA viruses. J Med Chem 20:757.
24. McKinlay MA, Steinberg BA (1986). Oral efficacy of WIN 51711 in mice infected with human poliovirus. Antimicrob Agents Chemother 29:30.
25. Otto MJ, Fox MP, Fancher MJ, Kuhrt MF, Diana GD, McKinlay MA (1985). In vitro activity of WIN 51711, a

new broad-spectrum antipicornavirus drug. Antimicrob Agents Chemother 27:883.
26. Fox MP, Otto MJ, McKinlay MA (1986). The prevention of rhinovirus and poliovirus uncoating by WIN 51711: a new antiviral drug. Antimicrob Agents Chemother, manuscript submitted for publication.
27. Ninomiya Y, Ohsawa C, Aoyama M, Umeda I, Suhara Y, Ishitsuka H (1984). Antivirus agent, Ro 09-0410, binds to rhinovirus specifically and stabilizes the virus conformation. Virology 134:269.
28. Lonberg-Holm K, Gosser LB, Kauer JC (1975). Early alteration of poliovirus in infected cells and its specific inhibition. J Gen Virol 27:329.
29. Koch F, Koch G (1985). "The Molecular Biology of Poliovirus." New York: Springer-Verlag.
30. Fenwick ML, Cooper PD (1962). Early interactions between poliovirus and ERK cells: some observations on the nature and significance of the rejected particles. Virology 18:212.
31. Smith TJ, Kremer MJ, Luo M, Vriend G, Arnold E, Kamer G, Rossmann MG, McKinlay MA, Diana GD, Otto MJ (1986). The site of attachment in human rhinovirus 14 for antiviral agents that inhibit uncoating. Manuscript in preparation.
32. Adams MJ, Buehner M, Chandrasekhar K, Ford GC, Hackert ML, Liljas A, Rossmann MG, Smiley IE, Allison WS, Everse J, Kaplan NO, Taylor SS (1973). Structure-function relationships in lactate dehydrogenase. Proc Natl Acad Sci US 70:1968.
33. Holbrook JJ, Liljas A, Steindel SJ, Rossmann MG (1975). Lactate dehydrogenase. In Boyer PD (ed): "The Enzymes," New York: Academic Press, 3rd edn, Vol XI, p 191.

Positive Strand RNA Viruses, pages 79–92

THE THREE DIMENSIONAL STRUCTURE OF POLIOVIRUS: IMPLICATIONS FOR ASSEMBLY AND IMMUNE RECOGNITION

James M. Hogle*, Marie Chow**, and David J. Filman*,

*Department of Molecular Biology
Research Institute of Scripps Clinic
La Jolla, CA 92037

**Department of Applied Biological Sciences
Massachusetts Institute of Technology
Cambridge, MA 02139

First described in 1908 as the causative agent of poliomyelitis (1), poliovirus has been the subject of intensive study since the demonstration 35 years ago that the virus could be grown in cultured cells (2). Consequently, poliovirus has become perhaps the best understood viral pathogen, making it a good model system for studying properties of other more complex viruses. We have now solved the structure of the Mahoney strain of type 1 poliovirus at high (2.9 A) resolution by x-ray crystallographic methods (3). The structure of a related virus, rhinovirus 14, has also been solved recently (4). These structures make it possible for the first time to investigate the structural basis of biological functions in an intact animal virus. The structures have already been shown to have important implications for viral assembly and for immune recognition (both of which are discussed below). In the future, these structures are also expected to provide additional insight into the mechanisms of viral neutralization by antibodies, and into the structural basis for the host and tissue specificity of these viruses. In particular, the structure of poliovirus provides a unique opportunity to study the structural basis for neurotropism and its attenuation.

POLIOVIRUS

Poliovirus is a member of the Picornavirus family, which also includes the rhinoviruses, the coxsackieviruses, foot and mouth disease virus, and hepatitis A virus. The virus is approximately 300 A in diameter, with a molecular mass of 8.4 million daltons. The capsid is comprised entirely of 60 copies each of the four coat protein subunits VP1, VP2, VP3, and VP4 (33,000, 30,000, 26,000, and 7,400 daltons respectively) arranged on a T=1 icosahedral surface (5). The capsid encloses a single-stranded molecule of RNA (2.5 million daltons, 7400 nucleotides) that is of positive polarity, is polyadenylated at its 3' end and is covalently linked to a 22 amino acid protein (VPg) at its 5' end. The RNA is translated as a single large open reading frame to yield a large polyprotein (molecular mass 220,000 daltons). The capsid sequences are located at the amino terminus of the polyprotein in the order VP4-VP2-VP3-VP1. The capsid sequences are excised in an early cleavage as the precursor P1-1a (100,000 daltons). Subsequent processing (which occurs concurrently with the early stages of capsid assembly) yields VP0, VP3, VP1. A final cleavage (which occurs late in assembly after encapsidation of the viral RNA) yields the mature virion proteins VP4, VP2, VP3, VP1 (6).

Poliovirus Structure.

Capsid Protein Subunits. The structures of the four capsid proteins are shown in figure 1. VP1, VP2, and VP3 display an obvious structural homology. Each of the three is composed of a conserved core with variable elaborations. The cores consist of an eight-stranded antiparallel beta barrel with two flanking helices. Four of the beta strands (B, I, D, and G in figure 1d) make up a large twisted beta sheet that forms the front and bottom surfaces of the barrel. The remaining four strands (C, H, E, and F in figure 1d) make up a shorter flatter beta sheet that forms the back surface of the barrel. The strands of the front and back surfaces are connected at one end by short loops, giving the barrel the shape of a triangular wedge. Although the cores of the three large capsid proteins are very similar, the connecting loops and amino and carboxyl terminal extensions are all dissimilar. In particular, VP1 has a large insertion (residues 207-237) in the loop connecting the G and H strands of the beta barrel and a re-

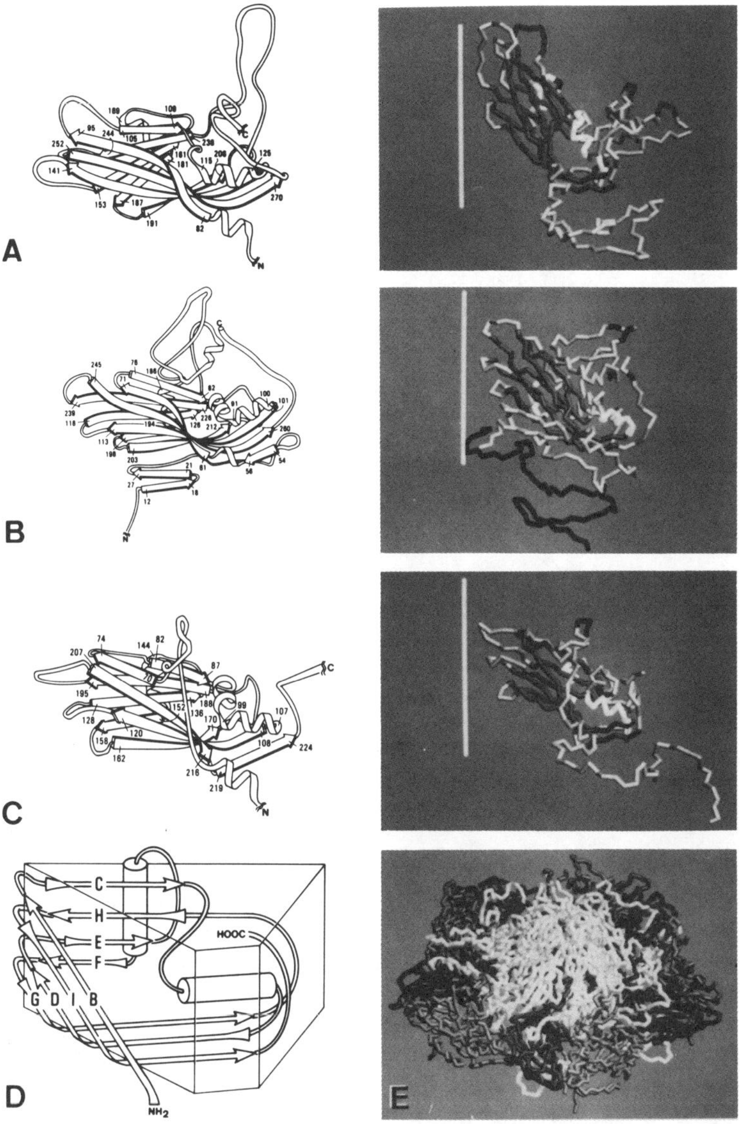

FIGURE 1. Alpha carbon models of the poliovirus capsid proteins. (a) VP1; (b) VP2 and VP4; (c) VP3; (d) diagrammatic representation of the conserved folding pattern of the major capsid proteins; (e) the organization of the major capsid proteins around the fivefold axis. The structurally conserved "cores" of the proteins each consists of a radial

FIGURE 1, cont.: (back helix), a tangential (front) helix, and an eight-stranded antiparallel beta barrel. In the left-most panels, strands of the beta barrel are indicated by arrows, and are labelled either by residue number, or by single strand designations used for the icosahedral plant viruses (23). In a and c, terminal extensions have been removed from the ribbon drawings for clarity. In a, b, and c helices and beta strands of the core are white and dark gray, respectively; variable loops and terminal extensions are medium gray, except for the sites of monoclonal release mutations, and VP4, which are black. Vertical bars extend from 110 to 160 A along the fivefold axis (in a), or the threefold axis (in b and c). In e, five copies of each VP1 (light gray), VP2 (medium gray), and VP3 (dark gray) are shown arranged in a pentamer.

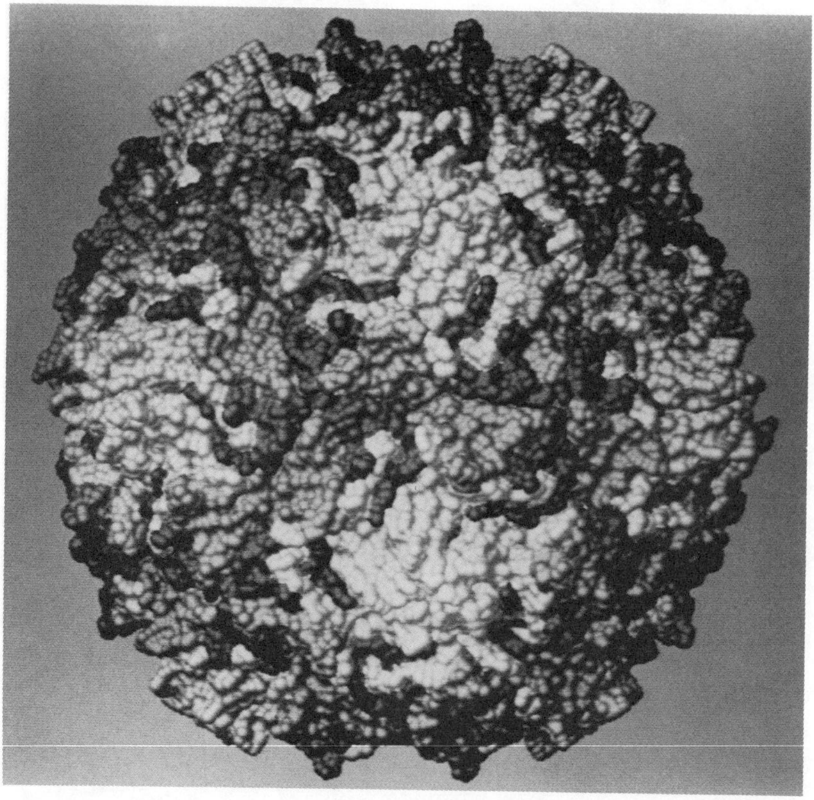

FIGURE 2. The outer surface of poliovirus. VP1 is light gray; VP2 is medium gray; VP3 is dark gray.

latively large loop (residues 96-104) connecting the top two strands (B and C) of the beta barrel. VP2 has a large insertion (residues 127-185) in the loop connecting the E strand of the barrel with the "back" helix. The largest insertion in VP3 (residues 53-69) occurs in the middle of the B strand of the beta barrel.

The structural description of the three large proteins in terms of cores, connecting loops, and terminal extensions is also relevant to function, since each of these units appears to play a different role in the structure of the intact virus. The cores make up the continuous shell of the virus, the connecting loops and carboxyl terminal extensions make up many of the major features on the external surface of the virion, and the amino terminal extensions cover much of the internal surface of the capsid. The small protein VP4 (shown along with VP2 in figure 1b) is similar in many respects to the amino terminal extensions of VP1 and VP3. Like them, it is rather extended in structure (except for a short two stranded antiparallel beta sheet near its amino terminus), and it occupies a totally internal position, forming a significant portion of the internal surface of the capsid. Since VP4 is covalently linked to VP2 until very late in the assembly of the virions, it may be considered to be the detached amino terminal extension of VP2.

Virion structure. The arrangement of the major capsid proteins in the virion is depicted in figures 1e and 2. Five copies of VP1 are clustered around the particle fivefold axis, with the narrow ends of the beta barrels closest to the axis. A pronounced tilt of the beta barrels outward along the fivefold axis results in the exposure of the top three connecting loops at the narrow end of the barrel, and the formation a prominent surface protrusion extending to 160 A radius. VP2 and VP3 alternate around the particle threefold axis, also with the narrow ends of the barrels closest to the axis. The tilts of the barrels of VP2 and VP3 are less pronounced. As a result, only the top two loops at the narrow end of the barrels of VP2 and VP3 are exposed, and the surface protrusion at the threefold extends to a lesser radius (150 A). This threefold protrusion is ringed by two sets of outward projections (or promontories). The larger projection is dominated by the large insertion in VP2 (residues 127-185) with contributions from the carboxyl terminus of VP2, the large insertion in VP1 (residues 207-237) and residues 271-295 near the carboxyl terminus of VP1. This large projection extends to a radius of 165 A. The smaller projection is formed by the insertion in VP3 (residues 53-70) and extends to 155 A radius. The peaks at the fivefold axes are surrounded by broad deep valleys, while the threefold plateaus are separated by saddle surfaces at the twofold axes.

Implications for Assembly.

The amino terminal extensions (including VP4) form an extensive network of interactions on the inner surface of the capsid. Within the protomer (the set of subunits VP4-VP2-VP3-VP1 that were derived from a single P1-1a precursor molecule) the amino terminus of VP1 folds extensively under the inner surface of VP3 and the amino terminus of VP3 folds around the bottom of VP1. The interactions in this network are particularly strong between fivefold related protomers. However, the interactions between adjacent pentamers are less extensive. This observation is consistent with a proposal (that was originally based on the characterization of assembly intermediates within infected cells) that the pentamer is an important intermediate in the assembly of virions (5). One of the interactions found in the network is particularly striking. The amino termini of five VP3 subunits wrap around the fivefold axis, forming a twisted tube that is stabilized by parallel beta interactions. This structure can form only upon pentamer formation, so that it may well serve to direct and/or stabilize the formation of the pentameric intermediate.

The disposition of the termini generated by the processing of the P1-1a protomer also has interesting implications for assembly. The carboxyl terminus of VP4 is very close to the the first ordered residue at the amino terminus of VP2 (residue 5). The four disordered residues could comfortably fill the gap. However, both termini are located in the interior of the virion, inaccessible to exogenous proteases. Since this cleavage is thought to occur late in assembly (subsequent to encapsidation of RNA), either there must be a substantial conformational rearrangement subsequent to the cleavage, or the cleavage must be "autocatalytic". In contrast, the amino terminus of VP2 is spatially distant from the carboxyl terminus of VP3, and the amino terminus of VP3 is distant from the carboxyl terminus of VP1; indeed the carboxyl termini are located on the exterior, while the amino termini are on the interior of the virion. The termini must, therefore, undergo substantial rearrangement subsequent to the cleavages. The apparent role of the amino terminal extensions in directing and stabilizing the formation of pentamers confirms that proteolytic processing and subsequent rearrangement play an important role in the control of virion assembly.

Similarities with Other Viruses.

As it had been predicted from earlier sequence comparison, the structures of the Mahoney strain of type 1 poliovirus and of rhinovirus 14 are strikingly similar. In fact, the "cores" of the capsid proteins are nearly identical, and the major structural differences occur in loops that are exposed on the surface of the virions (especially the top loop (residues 96-104) in VP1, and the large insertions in VP1 and VP2). The degree of sequence and structural similarity clearly indicates that the two viruses are very closely related, and points to the danger of classifying viruses within families on the basis of pathology, or physical characteristics such as acid stability or permeability to ions.

A much more surprising observation is that poliovirus (and hence the rhinovirus) capsid proteins are similar in structure to the capsid proteins of the icosahedral plant viruses whose structures are known. The capsid proteins of the plant viruses have an identical folding pattern (illustrated in figure 1d), and a similar organization into cores (that pack together to form the capsid shell), connecting loops (that decorate the outer surface), and amino terminal strands (that are located on the inner surface of the capsid, and play a role in directing assembly). Overall, the packing of capsid proteins in poliovirus is similar to the packing of proteins in the "S-domains" of the T=3 plant viruses (although it differs significantly in detail, since the constraints imposed by the quasi-equivalent packing of chemically identical subunits have been relaxed). Although these structural similarities strongly indicate an evolutionary relationship, the exact nature of the relationship is unclear. Despite the structural similarity, plant viruses and picornaviruses differ significantly in their gene order, their mechanisms for the control of gene expression, and in the details of their assembly processes. Therefore, alternatives to direct descent from a common ancestor must be considered. One alternative possibility is that these viruses have "borrowed" a common protein and its message from their respective hosts.

Location of Neutralizing Antigenic Sites.

Two methods have been used to identify amino acid residues in the neutralizing antigenic sites of poliovirus: 1) mapping the sequence changes in mutant viruses selected for resistance to neutralization by specific monoclonal antibodies (monoclonal release mutants) (7-10), and 2) identification of synthetic peptides capable of either eliciting neutralizing antibodies or priming for a neutralizing response upon subsequent injection of subimmunizing doses of intact virus (11-16).

Monoclonal release mutants. The sites of sequence changes found in monoclonal release mutants are shown in figure 3a. When the sequence changes are grouped into areas that might be included within the "footprint" of a single Fab binding site (approximately 40 A in diameter), they are seen to cluster into three groups. Cluster 1 (figure 3b) consists of the mutations located around the fivefold axes, including residues in the top two loops of the VP1 barrel (residues 89-103, and 254) and a residue from the loop connecting strand E with the back helix (residue 168). Cluster 2 (figure 3c) contains mutations located near the twofold axes and includes residues from the large insertion in VP1 (residues 222-224), from the large insertion of VP2 (residues 166, 169, and 170), and from the carboxyl terminus of VP2 (residue 270). Cluster 3 (figure 3d) consists of mutations situated near the threefold axes, including residues from the top loop of the VP2 barrel (residue 72), from the top loop of the VP3 barrel (residues 71-73), from the insertion in VP3 (residues 58-60), and residues near the carboxyl terminus of VP1 (residues 284-287).

There are several interesting findings concerning the distribution of monoclonal release mutants. 1) The sites of monoclonal release mutations are all located in exposed loops that are readily accessible to antibody binding. Thus, in all probability, the mutation sites are parts of

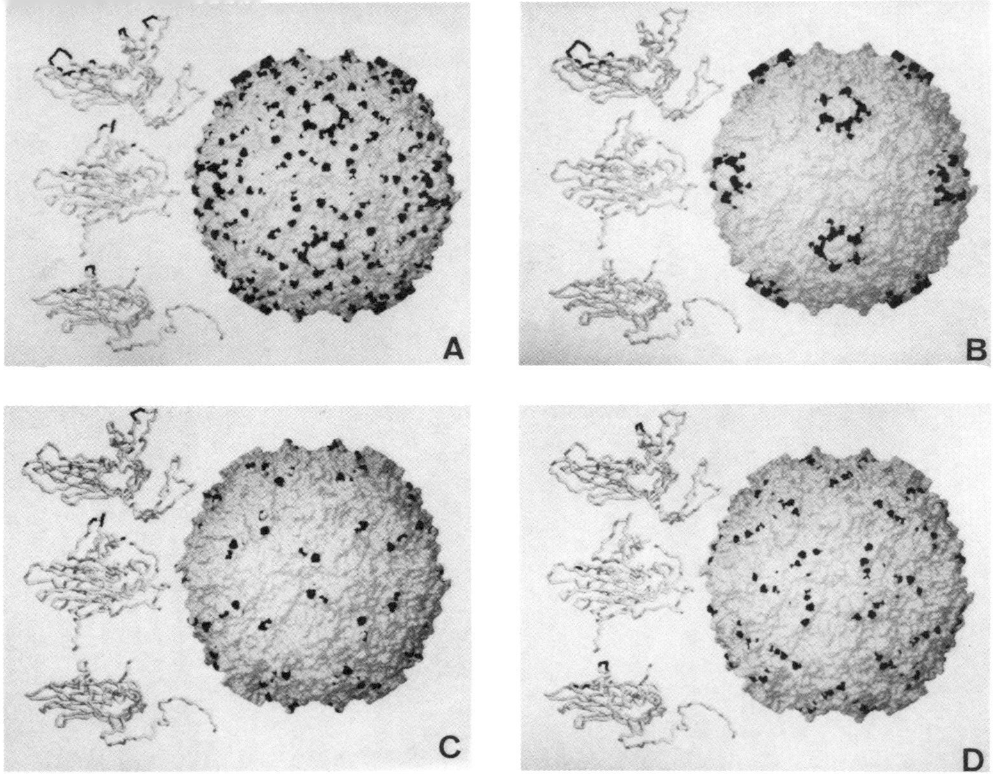

FIGURE 3: Clusters of monoclonal release mutations in poliovirus. (a) all mutation sites. (b) cluster 1. (c) cluster 2. (d) cluster 3. The sites of individual mutations are shown in black on alpha carbon models of VP1, VP2, and VP3, and on the virion surface.

direct antibody binding sites; and none of the mutations identified to date would be required to act by disrupting a spatially distant antibody binding site. The situation of antibody binding sites on exposed loops at the surface of an antigen has also been observed in the haemagglutinin and neuraminidase of influenza virus (17,18). The decoration of a virus surface with exposed (and mutable) loops seems likely to be a common mechanism by which animal viruses Escape immune surveillance. 2) All of the regions contain residues from sequentially distant areas of the coat protein, and two of the regions (2 and 3) include residues from more than one protein. In at least one case there is direct evidence that the antibody binding site itself is composed of residues from more than one protein, since mutations in either the insertion in VP1 or the insertion in VP2 allow the virus to escape neutralization by the same monoclonal antibody. Consequently, these sites fit the definition of a discontinuous or conformational determinant. 3) The three clusters of mutation sites are not equally important in all strains of poliovirus. For reasons that are not yet clear, the overwhelming majority of mouse monoclonal antibodies to type 3 poliovirus map to cluster 1, whereas the majority of mouse monoclonal antibodies to type 1 poliovirus map to clusters 2 and 3. We have recently used a unique trypsin cleavage site within cluster 1 (19) to determine the extent to which cluster 1 dominates the immune response in sera (20). These investigations have shown that the degree of immunodominance of cluster 1 varies with the strain of mouse, and that in some animals (including monkeys and perhaps humans) cluster 1 is of little importance.

Synthetic peptides. Most of the synthetic peptides that have been shown to elicit or prime for a neutralizing response map in (or very near to) the clusters of monoclonal release mutations. Specifically, synthetic peptides have been used to identify sites in the top three loops of the VP1 beta barrel (in cluster 1), in the loop connecting beta strand E to the back helix of VP1 (cluster 1), in the major insertions of VP1 and VP2 (cluster 2), in the top loop of VP3 (cluster 3), and in the region near the carboxyl terminus of VP1 (cluster 3). However, some of the peptides that elicit or prime for a neutralizing response do not map into any of the previously identified clusters. It is interesting that these exceptions (including a peptide corresponding to the bottom loop of VP1, and several peptides from the amino terminal extension of VP1) are all located in the interior of the mature virion. One possible explanation for the action of these buried peptides is that a conformational change in the virus could result in the exposure of these residues, along with a loss of infectivity. Such a conformational change could be induced or trapped by the anti-peptide antibodies, and perhaps by the binding of other antibodies as well. Experiments to provide direct experimental evidence for such a conformational change are in progress.

The antigenic sites identified by antibodies to synthetic peptides are generally considered to be continuous or sequential determinants. Poliovirus provides an interesting counterexample, in that most of the sequential sites are included in larger conformational sites defined by monoclonal release mutations. At least in the case of poliovirus, there may be no clear distinction between sequential and conformational sites.

FUTURE PROSPECTS

Knowledge of the three dimensional structure of poliovirus has already provided insight into the distribution of neutralizing antigenic sites in an intact virion. In the future, we expect the structure to facilitate the design and interpretation of additional studies, leading to an understanding of the structural basis for antigen-antibody interactions, and of the mechanisms by which antibodies neutralize virus infectivity. We also expect the structure to serve as a valuable tool for the design and interpretation of experiments using infectious DNA copies of the genome and recombinant methods to characterize the regions of the virion responsible for receptor recognition and the structural basis for neurotropism and its attenuation. On a practical side, the structure may also assist in the design of improved vaccines for poliovirus and for related viruses such as hepatitis A and foot-and-mouth disease. For example, knowledge of the three dimensional distribution of antigenic sites may permit the design of synthetic vaccines that use several synthetic peptides to mimic a conformational site. Alternatively, the structure may be used to identify suitable sites for the application of recombinant DNA methods to produce stable (and consequently, safer) attenuated strains of the viruses.

ACKNOWLEDGEMENTS

This work was supported by NIH grant AI-20566 to J. M. Hogle, and in part by a grant from the Rockefeller Foundation and by NIH grant AI-22346 to D. Baltimore (Whitehead Institute for Biomedical Research). D.J. Filman was supported in part by training grants NS-07078 and NS-12428. Preliminary crystallographic studies that led to this work were supported by NIH grant CA-13202 to S.C. Harrison. We would like to thank Richard Lerner, David Baltimore, and Stephen Harrison for their interest and support throughout various phases of this work; and Joseph Icenogle, Carl Fricks, Michael Oldstone and Ian Wilson for helpful comments and discussions. Ribbon drawings in figure 1 were produced with the assistance of E. Getzoff. The molecular surface representation in figure 2 was calculated by A. J. Olson, using the programs AMS and RAMS (21,22).

REFERENCES

1) Landsteiner, K and Popper, E (1908). Mikroskopische Praparate von einem menschlichen und zwei Affenruckenmarken. Wien. Klin. Wochenshr. 21:1830.

2) Enders, JF, Weller, TH, and Robbins, FC (1949) Cultivation of the Lansing strain of poliomyelitis virus in cultures of various human embryonic tissues. Science 109:85-87.

3) Hogle, JM, Chow, M, and Filman, DJ (1985). Three-dimensional structure of poliovirus at 2.9 A resolution. Science 229:1358-1365.

4) Rossmann, MG, Arnold, E, Erickson, JW, Fankenberger, EA, Griffith, JP, Hecht, H-J, Johnson, JE, Kamer, G, Luo, M, Mosser, AG, Rueckert, RR, Sherry, B, and Vriend, G (1985). Structure of a human common cold virus and functional relationship to other picornaviruses. Nature 317:145-153.

5) Rueckert, RR (1976) On the structure and morphogenesis of picornaviruses. In Comprehensive Virology, vol. 6, Fraenkel-Conrat, H and Wagner, RR eds., Plenum, pp 131-213.

6) Pallansch, MA, Kew, OM, Semler, BL, Omilianowski, DR, Anderson, CW, Wimmer, E, Rueckert, RR (1984). Protein processing map of poliovirus. J. Virol. 49:873-880.

7) Minor, PD, Schild, GC, Bootman, J, Evans, DMA, Ferguson, M, Reeve, P, Spitz, M, Stanway, G, Cann, AJ, Hauptmann, R, Clarke, LD, Mountford, RC, and Almond, JW (1983). Location and primary structure of a major antigenic site for poliovirus neutralization. Nature 301:674-679.

8) Minor, PD, Evans, DMA, Ferguson, M, Schild, GC, Westorp, G, and Almond, JW (1985). Principal and subsidiary antigenic site of VP1 involved in the neutralization of poliovirus type 3. J. Gen. Virol. 65:1159-1165.

9) Diamond, DC, Jameson, BA, Bonin, J, Kohara, M, Abe, S, Itoh, H, Komatsu, T, Arita, M, Kuge, S, Nomoto, A, Osterhaus, ADME, Crainic, R, and Wimmer, E (1985). Antigenic variation and resistance to neutralization in poliovirus type 1. Science 229:1090-1093.

10) Minor, PD, Ferguson, M, Evans, DMA, and Icenogle, JP (1986). Antigenic structure of polioviruses of serotypes 1, 2, and 3. J. Gen. Virol, in press.

11) Chow, M, Yabrov, R, Bittle, J, Hogle, J, and Baltimore, D (1985). Synthetic peptides from four separate regions of the poliovirus type 1 capsid protein VP1 induce neutralizing antibodies. Proc. Natl. Acad. Sci. USA 82:910-914.

12) Emini, EA, Jameson, BA, and Wimmer, E (1983). Priming for and induction of anti-poliovirus neutralizing antibodies by synthetic peptides. Nature 304:699-703.

13) Emini, EA, Jameson, BA, and Wimmer, E (1984). Identification of a new neutralization antigenic site on poliovirus coat protein VP2. J. Virol. 52:719-721.

14) Emini, EA, Jameson, BA, and Wimmer, E (1984). Identification of multiple neutralization antigenic sites on poliovirus type 1 and the priming of the immune response with synthetic peptides. In Modern Approaches to Vaccines, Chanock, R. M. and Lerner, R. A. eds.

Cold Spring Harbor Laboratory, pp 65-75.

15) Ferguson, M, Evans, DMA, Magrath, DI, Minor, PD, Almond, JW, and Schild, GC (1985). Induction by synthetic peptides of broadly reactive, type-specific neutralizing antibody to poliovirus type 3. Virology 143:505-515.

16) Jameson, BA, Bonin, J, Murray, MG, Wimmer, E, and Kew, O (1985). Peptide-induced neutralizing antibodies to poliovirus. In Vaccines 85, Lerner, RA, Chanock, RM, and Brown, F eds, Cold Spring Harbor Laboratory, pp. 191-198.

17) Wiley, DC, Wilson, IA, and Skehel, JJ (1981) Structural identification of the antibody-binding sites of Hong Kong influenza haemagglutinin and their involvement in antigenic variation, Nature, 289:373-378.

18) Colman, PM, Varghese, JN, and Laver, WG (1983) Structure of the catalytic and antigenic sites in influenza virus neuraminidase, Nature, 305:41-44.

19) Fricks, CE, Icenogle, JP, and Hogle, JM (1985) Trypsin sensitivity of the Sabin strain of type 1 poliovirus: cleavage sites in virions and related particles. J. Virol. 54:856-859.

20) Icenogle, JP, Minor, PD, Ferguson, M, and Hogle, JM (1986) Modulation of the humoral response to a 12 amino acid site on the polivirus virion. J. Virol., In press.

21) Connolly, ML (1983). Analytical molecular surface calculation. J. Appl. Crystallogr. 16:548-558.

22) Connolly, ML (1985). Depth-buffer algorithms for molecular modelling. J. Mol. Graphics 3:19-24.

23) Rossmann, MG, Abad-Zapatero, C, Murthy, MRN, Liljas, L, Jones, TA and Strandberg, B (1983). Structural comparisons of some small spherical plant viruses. J. Mol. Biol. 165:711-736.

Positive Strand RNA Viruses, pages 93–102

ISOLATION AND CHARACTERIZATION OF A MONOCLONAL ANTIBODY WHICH BLOCKS ATTACHMENT OF HUMAN RHINOVIRUSES

Richard J. Colonno, Joanne E. Tomassini, and Pia L. Callahan

Merck Sharp and Dohme Research Laboratories
West Point, Pennsylvania 19486

ABSTRACT Previous studies have suggested that the vast majority of human rhinovirus serotypes initiate infection by attachment to a single cell surface receptor. Using HeLa cells as an immunogen, a mouse monoclonal antibody was isolated which specifically inhibited infection of 78 of 88 human rhinovirus serotypes tested. The receptor monoclonal antibody was subsequently used to isolate a membrane protein believed to be important for rhinovirus attachment. The normal cellular function of this protein remains to be determined. The antibody was found to have a very high binding affinity for the cellular receptor protein and was capable of displacing bound virions. No human rhinovirus variants were found which were capable of bypassing the receptor antibody block. The ability of the antibody to protect primates from human rhinovirus infection was demonstrated in a chimpanzee animal model system.

INTRODUCTION

The human rhinoviruses (HRVs) represent a subgroup of picornaviruses and contain some 115 antigenically distinct serotypes (1,2). They are of particular medical importance because they have been identified as the major causative agent of the common cold in humans (3). Recent reviews of receptors utilized during picornavirus infections indicate that distinct receptor families exist for attachment of specific groups of picornaviruses (4,5). Receptor specificity has been further demonstrated by the isolation of monoclonal antibodies which specifically block attachment of poliovirus and group B coxsackieviruses (6-9).

Traditional approaches to controlling HRV infections do not appear to be feasible, since there are a vast number of serotypes. Therefore, it becomes of greater importance to define specific areas of the virus or steps in its replicative cycle that are common to a multitude of serotypes. Lonberg-Holm et al. suggested that HRVs could be divided into at least two receptor families which did not compete with other picornaviruses except coxsackievirus A21 (10). We recently expanded these competition binding studies to include 24 randomly selected HRV serotypes and found that 20 of the 24 HRVs tested competed for a single receptor on the surface of susceptible cells (11). We refer to this group as the major group of HRVs. The other four serotypes competed with each other for a second and different cellular receptor and are referred to as the minor group of HRVs. Subsequent studies have led to the isolation of a monoclonal antibody which blocks attachment of 90% of HRV serotypes. The receptor antibody was used to isolate a receptor protein involved in HRV attachment. Characterization of the antibody and the cellular receptor protein are summarized in this paper and define a novel target that may be used in the control of HRV infection.

RESULTS AND DISCUSSION

Since our initial studies using 24 serotypes of HRVs in a competition binding assay indicated that HRVs could be divided into only 2 groups based on receptor specificity, we reasoned that it should be possible to isolate a monoclonal antibody to the major HRV group receptor and that such an antibody would be capable of protecting cells from infection by a large number of serotypes. Mice were immunized with HeLa cells and HeLa cell membranes and were used to generate some 8720 hybridomas (12). Supernatant fluid from each of these hybridoma cultures was used to treat HeLa cell monolayers prior to a challenge infection with HRV-14 (13). Only 2 hybridoma cultures were identified which produced antibody capable of protecting the HeLa cell monolayers from virus induced cytopathic effect. These cells were recloned twice to ensure the production of monoclonal specific antibodies. Analysis and characterization of the two monoclonal antibodies indicated that both were of the IgG-1 isotype and each was capable of blocking the binding of the other in cell binding assays (data not shown). These two monoclonal antibodies, referred to as receptor antibody, apparently recognize the same epitope and were used interchangeably in the experiments described below.

As detailed elsewhere (13), three criteria were used to show that the receptor antibody binds to the major HRV receptor on cells. First, cell protection assays involving the 24 HRV serotypes used in our initial competition binding assays resulted in protection of cells from infection by the 20 HRV serotypes belonging to the major group while the receptor antibody showed no protective effect against the 4 minor group serotypes. This is precisely what one would predict if the antibody were directed against the major HRV cellular receptor. Testing of an additional 64 HRV serotypes in the cell protection assay confirmed that 90% of HRV serotypes share a single cellular receptor, since the receptor antibody could protect cells from infection by 78 of the 88 serotypes tested (Table 1). The second criterion involved cell protection assays using 19 other viruses, both RNA and DNA, to determine if the receptor antibody would block infection by other groups of viruses. Previous studies predicted that the antibody should inhibit only coxsackievirus A serotypes, since coxsackievirus A21 was shown to share a common receptor with HRV-14 (10). Again results (Table 1) were as predicted and demonstrated that the receptor antibody could block infection by the 3 coxsackievirus A serotypes tested in addition to the major group of HRVs. No inhibitory effect was observed when cells were infected with poliovirus, echovirus or coxsackievirus B serotypes. In addition to those listed in Table 1, the antibody could not block infection by vesicular stomatitis virus, Newcastle disease virus, parainfluenza virus, influenza A virus, respiratory syncytial virus, human coronavirus, vaccinia virus, and adenovirus.

The last criterion used to prove that the receptor antibody was directed to the major HRV receptor took advantage of the narrow host range exhibited by this

TABLE 1
CELL PROTECTION ASSAYS USING RECEPTOR ANTIBODY

Protection Against Infection By:				
HRV-3	HRV-19	HRV-38	HRV-58	HRV-75
HRV-4	HRV-20	HRV-39	HRV-59	HRV-76
HRV-5	HRV-21	HRV-40	HRV-60	HRV-77
HRV-6	HRV-22	HRV-41	HRV-61	HRV-78
HRV-7	HRV-23	HRV-42	HRV-63	HRV-79
HRV-8	HRV-24	HRV-43	HRV-64	HRV-80
HRV-9	HRV-25	HRV-45	HRV-65	HRV-81
HRV-10	HRV-26	HRV-46	HRV-66	HRV-83
HRV-11	HRV-27	HRV-48	HRV-67	HRV-84
HRV-12	HRV-28	HRV-50	HRV-68	HRV-85
HRV-13	HRV-32	HRV-51	HRV-69	HRV-86
HRV-14	HRV-33	HRV-52	HRV-70	HRV-88
HRV-15	HRV-34	HRV-54	HRV-71	HRV-89
HRV-16	HRV-35	HRV-55	HRV-72	HRV-Hanks
HRV-17	HRV-36	HRV-56	HRV-73	
HRV-18	HRV-37	HRV-57	HRV-74	
CV-A13	CV-A18	CV-A21		
No Protection Against Infection By:				
HRV-1A	HRV-2	HRV-30	HRV-44	HRV-49
HRV-1B	HRV-29	HRV-31	HRV-47	HRV-62
PV-1	PV-3	CV-B3	EV-6	
PV-2	CV-B2	EV-1		

HeLa R-19 cell monolayers (3 x 10^5 cells) were treated with 30 μg receptor antibody for 30 minutes at 37°C prior to low multiplicity infection with virus. Cells were incubated for 16-24 hours and examined for cytopathic effect (13). Absence of any cytopathic effect was indicative of protection.
HRV = human rhinovirus; CV = coxsackievirus; PV = poliovirus; and EV = echovirus.

group of viruses. The major group of HRVs can only attach to cells of human or higher primate origin (13). This is the major reason why an animal model system does not exist to study the clinical aspects of a HRV infection. Parallel binding studies using radiolabeled receptor antibody and HRV-15 on 29 different cell lines demonstrated that the antibody could only bind to cells of human or chimpanzee origin. The receptor antibody could not bind to any cell lines which were incapable of binding virus. The results of this study clearly indicated that the receptor antibody and HRV-15 display identical cell tropisms. The experiments described above clearly demonstrate that the receptor antibody is directed against the cellular receptor structure utilized by the major group of HRVs for attachment.

Competition binding studies were then performed to ascertain whether all 10 of the HRVs which were able to infect cells in the presence of the receptor antibody shared a single receptor. Results (13) showed that the binding of all of the minor HRV serotypes could be effectively competed with HRV-2, indicating that these 10 serotypes share a common receptor which is different from the major HRV group receptor. In addition, all ten of these minor HRV serotypes were able to attach to mouse L cells which clearly distinguishes these viruses from the major group of HRVs (13). We conclude from these results that there are only two cellular receptors utilized by the vast number of HRVs for attachment to susceptible cells.

Characterization studies showed that the receptor antibody had a high affinity for the HRV receptor as evidenced by its rapid binding to receptors and ability to displace previously-bound virions (13). Infectious virus particles do not appear to be capable of bypassing the receptor antibody block. Plaque assays involving infection of cells with over 10^9 plaque forming units of three different major HRV serotypes failed to result in the formation of any viral plaques or other evidence of infection of the cell monolayer (13). This is an amazing result in light of the fact that the multiplicity of infection was over 3000 infectious virions per cell. If just one virion had been able to enter each cell, the entire monolayer would have been destroyed.

The receptor protein was isolated using an immunoaffinity column in which the receptor antibody was covalently linked to an Affi-gel resin (Bio-Rad). Cell membranes were treated with 0.3% sodium deoxycholate to solubilize surface proteins. Chromatography of this material over the immunoaffinity column resulted in at least a 4000-fold purification of a protein with an apparent molecular weight of 90 Kd (14). An identical protein could be isolated from several different cells which was able to bind the major group of HRVs but could not be isolated from cell lines unable to bind this group of HRVs. To prove that this 90 Kd protein was indeed involved in HRV attachment, rabbits were immunized with the purified membrane protein and the resulting antiserum assayed to determine if it were capable of blocking virus binding. Results (14) demonstrated that the polyclonal antiserum had the same virus receptor specificity as the receptor antibody and could also block binding of the receptor antibody. Initial characterization of this protein has indicated that it is an acidic glycoprotein and contains sialic acid as its terminal oligosaccharide. Removal of the sialic acid with neuraminidase did not affect receptor antibody recognition.

The normal cellular function of this human receptor remains to be determined. To ascertain whether blockage of this receptor results in any inhibition of cell growth and division, sparsely plated monolayers of HeLa cells were treated with a large excess of either receptor antibody or polyclonal IgG antibody prepared against purified receptor protein. Cells were incubated for 6 days and cell counts determined on days 3 and 6. Results (Fig. 1) show that extended exposure to antibodies directed at the major HRV receptor protein had no effect on cell growth and division. The 6-day time period represented 7 cell doublings and a 30-fold increase in cell number. Analysis of DNA, RNA, and protein synthesis also indicated that antibody binding to this surface protein had no inhibitory or toxic effect on cells (unpublished data). Since the HRV major receptor protein was found on virtually all human cells tested (13), we assume it must have some functional role in humans which is probably not essential in tissue culture.

A crude animal model was established using adult chimpanzees to test the concept of receptor blockage as a method of inhibiting HRV infection. While chimpanzees possess the major HRV receptor and are capable of replicating the virus, they do not develop the clinical symptoms associated with the common cold in man (15). However, they will seroconvert following HRV infection. An experiment was set up in which six adult chimps were divided into 3 groups of 2. The first group received an intranasal administration of placebo followed by a

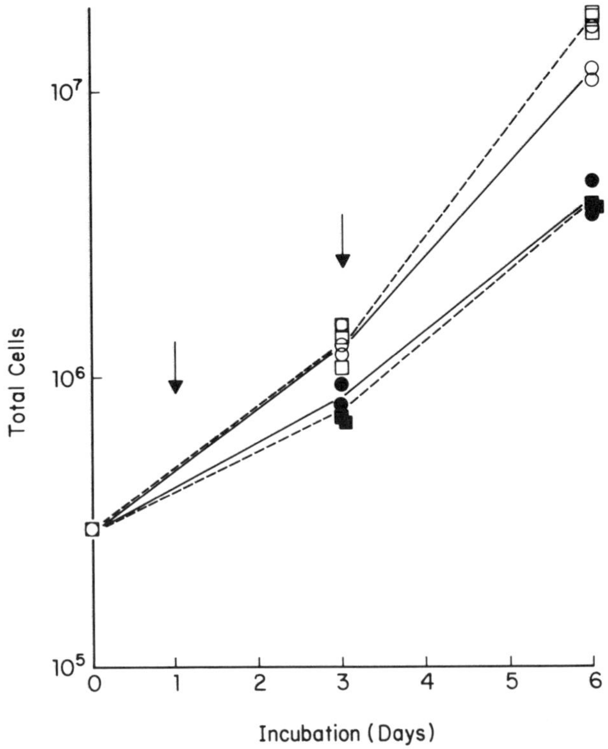

FIGURE 1. Effect of receptor antibody on HeLa cell growth. Sparse HeLa cell monolayers in 25 cm^2 flasks were treated on days 1 and 3 with 165 μg of receptor antibody (□), 3 mg of IgG receptor polyclonal antibody (■), or no antibody (○, ●) and incubated at 37°C. On days 3 and 6, cells were removed from flasks with 50 mM EDTA and cell numbers determined (14).

challenge inoculum of HRV-15. The second group also received the placebo treatment, but were challenged with UV-irradiated HRV-15. The third group was treated intranasally with receptor antibody and then challenged with HRV-15. The animals were bled as indicated and each serum sample analyzed for neutralizing antibody against HRV-15 in a cell protection assay. Table 2 shows that neutralizing titers developed within two weeks in the group one animals. No neutralizing titers were detected in the second group which indicated that viral replication is required to generate an immune response. The animals in the antibody treated group failed to develop an immune response comparable to the control animals in group one. Instead, their titers were nearly identical to the animals in group two and suggest that virus replication was blocked by the receptor antibody.

TABLE 2
HRV CHIMPANZEE ANIMAL MODEL

	Titer in Cell Protection Assay					
Treatment	Placebo		Placebo		Antibody	
Virus	HRV-15		HRV-15(UV)		HRV-15	
Chimp #	1	2	3	4	5	6
Serum Sample						
Day 0	-	-	-	-	-	-
Day 14	25	25	-	-	-	-
Day 21	5	125	-	-	-	-
Day 28	9	27	-	-	-	9

Adult chimpanzees were sedated and intranasally administered 0.2 ml placebo (phosphate buffered saline containing 0.2% w/v albumisol) or receptor antibody (1.25 mg/ml in identical buffer) per nostril. Serum samples were taken on days 0, 14, 21, and 28 and titrated for neutralizing antibody in a cell protection assay (13). - indicates a titer of <2.

The results discussed above clearly indicate that HRVs attach to cells through only two possible receptors. Attachment to this major HRV receptor protein appears to be an absolute requirement for successful infection of cells, since high multiplicity infections of receptor antibody treated cells failed to yield any virus. While we have been able to isolate a receptor antibody which recognizes the major HRV receptor on cell membranes, we have not been able to isolate any monoclonal antibodies capable of binding to the receptor attachment site on virions. Presence of the cellular attachment site on the surface of the virus capsid would result in production of cross-reactive antibodies.

Recently, neutralizing monoclonal antibodies have been mapped to 4 distinct epitopes on HRV-14 (16). Analysis of 18 neutralizing monoclonal antibodies, representing all 4 neutralizing groups, failed to identify an antibody capable of cross reacting with other HRV serotypes (unpublished data). Recent determination of the atomic structure of HRV-14 (17) has revealed the presence of a deep canyon on the surface of the viral capsid. It has been proposed that this canyon harbors the viral attachment site and is inaccessible to antibody molecules (17). Clinically, this is a reasonable assumption, since infection with a given HRV serotype does not result in immunity against other HRV serotypes. This also is supported by results obtained with the neutralizing antibodies and the failure to generate idiotypic antibodies, using our receptor antibody, which are capable of binding to HRV-14 (data not shown). Future studies involving site directed mutagenesis of the HRV-14 genome using an infectious cDNA system may prove this concept. Clearly, HRV receptors represent a novel target for the development of antivirals to prevent HRV infection.

REFERENCES

1. Cooney MK, Fox JP, Kenny GE (1982). Antigenic groupings of 90 rhinovirus serotypes. Infect Immun 37:642-647.
2. Melnick JL (1980). Toxonomy of viruses. Prog Med Virol 26:214-232.
3. Gwaltney JM (1975). Medical reviews, rhinoviruses. Yale J Biol Med 48:17-45.
4. Crowell RL, Landau BJ (1983). Receptors in the initiation of picornavirus infections. Compr Virol 18:1-42.
5. Lonberg-Holm K, Philipson L (1980). Molecular aspects of virus receptors and cell surfaces. In Blough HA, Tiffany TM (eds): "Cell Membranes and Viral Envelopes," New York: Academic Press, Inc.
6. Minor PD, Pipkin PA, Hockley D, Schild GC, Almond JW (1984). Monoclonal antibodies which block cellular receptors of poliovirus. Virus Res 1:203-212.

7. Campbell BA, Cords CE (1983). Monoclonal antibodies that inhibit attachment of group B coxsackieviruses. J Virol 48:561-564.
8. Crowell RL, Field AK, Schleif WA, Long WJ, Colonno RJ, Mapoles JE, Emini EA (1986). Monoclonal antibody that inhibits infection of HeLa and rhabdomyosarcoma cells by selected enteroviruses through receptor blockade. J Virol 57:438-445.
9. Nobis P, Zibirre R, Meyer G, Kuhne J, Warnecke G, Koch G (1985). Production of a monoclonal antibody against an epitope on HeLa cells that is the functional poliovirus binding site. J Gen Virol 66:2563-2569.
10. Lonberg-Holm K, Crowell RL, Philipson L (1976). Unrelated animal viruses share receptors. Nature (London) 259:679-681.
11. Abraham G, Colonno RJ (1984). Many rhinovirus serotypes share the same cellular receptor. J Virol 51:340-345.
12. Long WJ, McGuire W, Palombo A, Emini EA (1986). Enhancing the establishment efficiency of hybridoma cells: Use of irradiated human diploid fibroblast feeder layers. J of Immun Methods 86:89-93.
13. Colonno RJ, Callahan PL, Long WJ (1986). Isolation of a monoclonal antibody that blocks attachment of the major group of human rhinoviruses. J Virol 57:7-12.
14. Tomassini JE, Colonno RJ (1986). Isolation of a receptor protein involved in attachment of human rhinoviruses. J Virol 58:290-295.
15. Dick EC (1968). Experimental infections of chimpanzees with human rhinovirus types 14 and 43. Proc Soc Exp Biol Med 127:1079-1081.
16. Sherry B, Mosser AG, Colonno RJ, Rueckert RR (1986). Use of monoclonal antibodies to identify four neutralization immunogens on a common cold picornavirus, human rhinovirus 14. J Virol 57:246-257.
17. Rossmann MG, Arnold E, Erickson JW, Frankenberger EA, Griffith JP, Hecht H-J, Johnson JE, Kamer G, Luo M, Mosser AG, Rueckert RR, Sherry B, Vriend G (1985). Structure of a human common cold virus and functional relationship to other picornaviruses. Nature 317:145-153.

Positive Strand RNA Viruses, pages 103–112

PHOTOLABELING OF THE INFLUENZA HEMAGGLUTININ WITH A HYDROPHOBIC PROBE

François Boulay, Robert W. Doms, and Ari Helenius[1]

Department of Cell Biology, Yale University School of Medicine, 333 Cedar St., New Haven, CT 06510-8002 USA

ABSTRACT The lipid-soluble photoactivatable carbene generator [^{125}I] TID was used to covalently label both influenza hemagglutinin (HA) and its water soluble ectodomain (BHA). Regardless of whether the virus had been treated with acid to induce the conformational change in HA that triggers membrane fusion activity, the C-terminal membrane anchor of HA2 was labeled. Although BHA, the soluble ectodomain fragment of HA, lacks the C-terminal portion of HA2, it was also labeled. In neutral BHA, the label was in the BHA2 chain whereas in acid treated BHA it was in both HA1 and BHA2. The differences between the ectodomain of intact HA and its bromelain fragment as revealed by TID labeling, is discussed, as well as the implications for HA's interaction with the target membrane during fusion.

INTRODUCTION

The viral factor responsible for the fusion activity of the influenza virus with membranes of the host cell is the hemagglutinin (HA) spike glycoprotein (1,2). After the virus is internalized by the cell, the mildly acidic pH of the endosome triggers an irreversible conformational change in the HA which catalyzes fusion between the viral and endosomal membranes (3-5). This reaction is of importance because it may bear analogy to a variety of biological fusion reactions whose mechanisms are, at present, obscure.

1. Supported by National Institutes of Health Grant - AI 18582.

The HA is well suited for detailed functional analysis because it is so well characterized. Each HA spike is composed of three identical subunits consisting of two disulfide-linked glycoproteins, HA1 and HA2 (6). The HA is anchored in the membrane by the C-terminal domain of the HA2 polypeptide. Exhaustive digestion with bromelain cleaves HA2 from its hydrophobic membrane anchoring segment and releases the trimeric water soluble ectodomain, which comprises 95% of the molecule's mass (7). The structure of bromelain solubilized hemagglutinin (BHA) has been determined by X-ray crystallography (8). As a result, it has proved to be a useful model with which to examine the acid-induced conformational change in HA which leads to membrane fusion.

A significant finding has been that BHA acquires amphiphilic properties under conditions which elicit virus:membrane fusion, presumably due to the exposure of the highly conserved, hydrophobic amino terminus of HA2 (9). As a consequence, BHA binds to liposomes and non-ionic detergents (10). Here we use the lipid-soluble photoactivatable carbene generator [^{125}I]TID (11) to label the hydrophobic domains of BHA and HA both before and after acid treatment. Our results reveal important differences between HA and BHA in terms of exposure of hydrophobic domains and suggest that BHA's interaction with target membranes is more superficial than originally suspected.

METHODS

Preparation of virus The X:31 (A/Hong Kong/1968) recombinant strain of influenza was grown in the allantoic cavity of 11-day embryonated eggs infected with 0.1 ml of phosphate saline buffer, pH 7.5, containing 0.1 hemagglutinating units. Allantoic fluid was collected 48 hr post infection and virus was then purified as previously described (10).

Preparation of BHA The BHA was obtained by exhaustively digesting purified X:31 virus with bromelain (7). After digestion, the viral cores were pelleted at 100,000 x g for 1 hr and the BHA containing supernatant concentrated under pressure in a collodion bag (UH 100/25 from Schleicher and Schull). The concentrated supernatant was loaded on the top of a 5-25% continuous sucrose density gradient (w/v in NaCl 100 mM, TRIS-base 50 mM, MES 20 mM, pH7.5)in a Beckman SW40 Rotor and centrifuged for 16h at 200,000xg. Fractions were collected from the bottom and protein in the first peak was pooled, concentrated to 0.5 - 1.0 mg/mL, and stored at -20°C in small aliquots.

Photolabeling Procedures Photolabeling experiments were carried out on BHA in the presence or absence of liposomes before and after incubation at pH 5.0. Liposomes were freshly prepared before each experiment as described by White and Helenius (12) except that 20mM MES/130 mM NaCl buffer (MES-Saline) containing 20μM tryptophan adjusted to pH 7.0 was used as a buffer. BHA (25μg) was incubated at pH 5.0 in the presence of a large excess of liposomes (>3000 lipid molecules/BHA trimer) (10) for 15 min. at 37°C and then re-neutralized. Routinely, 5μL of TID in ethanol (corresponding to 8-10 x 10^7 DPM) were added to the samples. After 15 min equilibration with the labeling reagent, each sample was transferred into the cap of an Eppendorf tube and exposed to a beam of light from a germicidal mercury lamp. The light beam was cooled by passage through a reservoir of circulating cold water and was directed through 1.5cm of a saturated solution of $CuSO_4$. After 10 min of photolysis, the [^{125}I]TID labeled BHA-liposome mixture was transferred to an Eppendorf tube containing 30μg bovine serum albumin to trap the non-bound reagent. The liposomes were solubilized by 0.5% Triton X-100 and the protein precipitated overnight in 80% acetone at -20°C. In the case of intact virus, the photolabeled virus (30μg) was concentrated by precipitation in TCA (10% w/v) and washed in cold acetone to remove the unbound [^{125}I]TID. The pellets were dissolved in a buffer containing 2% (W/V) SDS, 100mM TRIS at pH 6.8 and 10% (V/V) glycerol, and then analyzed by SDS-polyacrylamide gel electrophoresis according to Laemmli (13). When reduction was required, 10mM DTT was added to the samples. For autoradiography, dried slab gels were exposed to Kodak XAR5 film at -70°C.

RESULTS

Photolabeling of X31 Virus

Virus was photolabeled with [^{125}I]TID in both the neutral and acid forms in order to label bilayer embedded protein domains as well as hydrophobic pockets unmasked after acid treatment. SDS-polyacrylamide gel electrophoresis under non-reducing conditions showed that regardless of pH, the photoprobe was covalently linked to the hemagglutinin

(HA) and, to a lesser extent, on a protein that is probably the matrix protein (M). (Figure 1, lanes 1 and 2). After reduction, the vast majority of the radiolabel was found to be in the HA2 polypeptide (Figure 1, lanes 3 and 4).

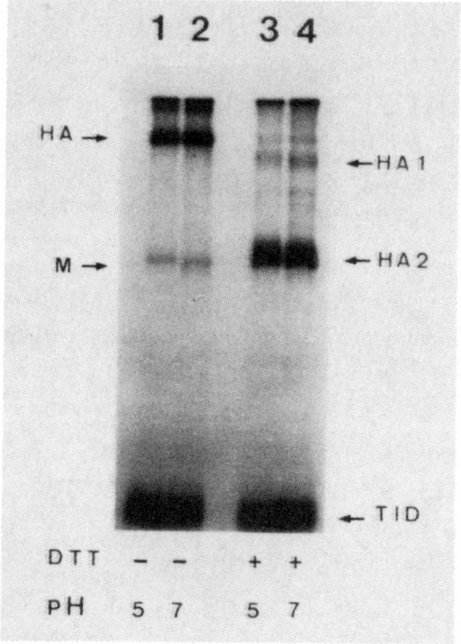

Figure 1 - Autoradiography following SDS-PAGE of X:31 virus photolabeled by [^{125}I] TID in the acid (Lanes 1 and 3) or neutral forms (Lanes 2 and 4).

To determine whether the label in HA2 was confined to the transmembrane domain, TID-labeled virus was digested with bromelain. Bromelain cleaves HA2 close to the point where it enters the viral membrane, thus separting the molecule into transmembrane/cytoplasmic and ectodomain fragments. Following digestion, we found the radiolabel in a small peptide that in all probability was the anchor peptide of HA2 (Figure 2, lane 2) This small peptide was not detected under non-reducing conditions even though it must be present. Most likely, the presence of dithiol bridges shifted the migration of this peptide into the free TID band (Figure 2, lane 1). No strongly radiolabeled BHA and BHA2 fragments appeared in response to the bromelain digestion, demonstrating that the label was confined to the transmembrane domain of HA2. In the case of acid-treated virus, bromelain digestion led to the complete proteolysis of HA, and all radiolabeled peptides migrated with the free [^{125}I]TID or the photolabeled phospholipids (not shown). Thus, we were unable to determine if the label was confined solely to the membrane anchor following acid treatment or if other regions of HA2 were labeled as well.

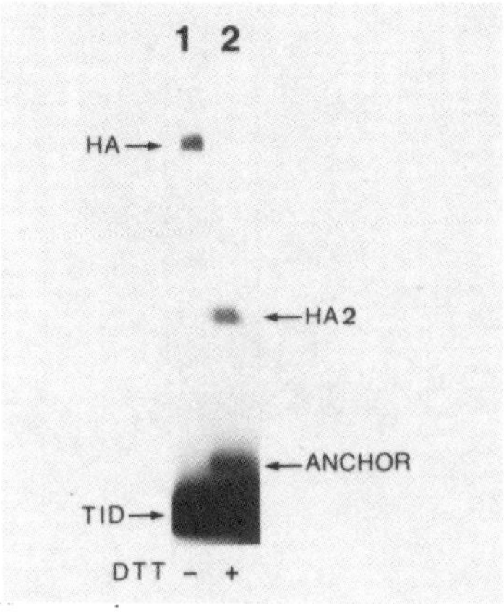

Figure 2 - Autoradiograph following SDS-PAGE of X:31 virus photolabeled by [^{125}I]TID at neutral pH and bromelain digested for 16 hours (7).

Photolabeling of BHA

One of the major consequences of the acid induced conformational change in BHA is its acquistion of amphiphilic properties. The acid conformation binds to both nonionic detergents and to liposomes. Binding is mediated by the BHA2 polypeptide chain, and is hydrophobic in nature (10). Thus, by hydrophobic labeling we could expect to label the domains imbedded in lipid bilayers. We found that after attachment to liposomes at pH 5.0, BHA was labeled on both the BHA1 and BHA2 polypeptide chains. Surprisingly, the neutral form, which is water soluble and does not bind to liposomes, was also labeled, though only on BHA2 (Figure 3, lanes 1 and 2). To obtain further insights on the nature of the [^{125}I]TID labeling, we photoirradiated the neutral and acid treated BHA in the absence of liposomes. A similar distribution of the probe was found in response to pH treatment. (Figure 3, lanes 3 and 4).

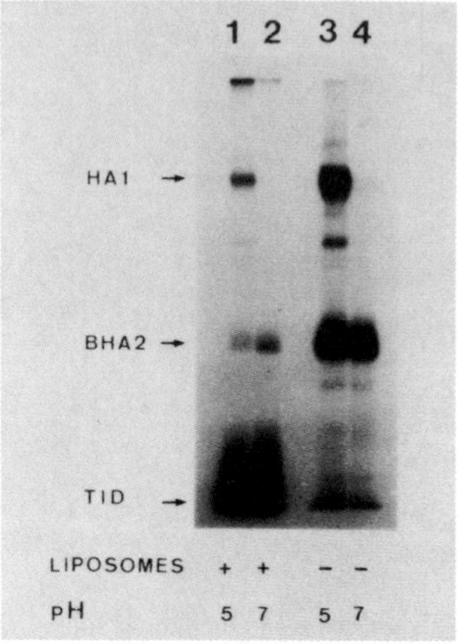

Figure 3 - Autoradiograph following SDS-PAGE under reducing conditions of neutral and acid treated BHA. Photolabeling with [^{125}I]TID was in the presence (Lanes 1 and 2) or absence (Lanes 3 and 4) of liposomes.

It should be noted that in the absence of liposomes, the relative concentration of the probe having access to the hydrophobic domains of BHA is higher than in the presence of liposomes, into which the probe partitions with high efficiency. As a result, we cannot compare the extent of labeling obtained in the presence of liposomes with that obtained in their absence. However, it seems reasonable to assume that the distribution of [^{125}I]TID between BHA1 and BHA2 reflects the degree of exposure of the hydrophobic domains in either the presence or absence of liposomes. This distribution was quantified by counting the radioactive bands from the dry gel, using the autoradiograph as a reference.

Table I.

Quantification of [^{125}I]TID incorporated in BHA.

Liposomes	+	+	-	-	
pH	5	7	5	7	
BHA1	2450	120	15590	680	counts/min
BHA2	1330	2164	20945	15300	counts/min

As can be seen in Table I, the ratio of labeling between BHA1 and BHA2 at neutral pH was quite similar regardless of the presence or absence of liposomes. For acid BHA, however, the extent of BHA2 labeling was decreased relative to BHA1 when liposomes were present.

DISCUSSION

Membrane fusion is an important cellular process, being involved in events such as gamete fusion, muscle cell development, and endo- and exocytosis. Despite its widespread occurrence, little is known about the mechanisms by which fusion occurs. We have used enveloped animal viruses (such as influenza) as paradigms for membrane fusion (4). Being structurally simple and well defined, they represent the best characterized biological fusion systems.

Influenza enters cells by endocytosis (3). The virus is delivered to endosomes where acidic pH triggers the fusion activity of the hemagglutinin spike glycoprotein, allowing the viral nucleocapsid to penetrate into the cytoplasm of the host cell (5). Current evidence suggests that acid pH disrupts intersubunit contacts in HA, leading to dissociation of the trimeric ectodomain with subsequent exposure of the highly conserved, uncharged amino terminus of the HA2 subunit. It has been postulated that this newly exposed domain interacts with target membranes in a hydrophobic manner and plays a critical role in the fusion reaction (10,15). To characterize the interactions of HA and its water soluble ectodomain fragment, BHA, with membranes in more detail, we employed the lipid soluble photoactivatable carbene generator [^{125}I] TID. TID has been successfully used to probe for polypeptide segments buried in the lipid bilayer (11) as well as for hydrophobic pockets in water soluble proteins (14). In this study we used TID to label HA and BHA in order to identify hydrophobic domains unmasked by the acid induced conformational change which leads to membrane fusion.

Our results revealed important differences between HA and BHA at both neutral and acid pH. When allowed to react with virus in the neutral conformation, TID labeled the transmembrane domain of the HA2 subunit. HA1 was not labeled, confirming that TID specifically labels membrane associated protein segments. Water soluble BHA, which lacks the membrane anchor, was labeled nevertheless, indicating the presence of a hydrophobic pocket in the BHA2 subunit. This pocket was not present or accessible to TID in intact HA since digestion with bromelain did not generate [^{125}I] TID labeled BHA2, as would be the case if regions other than the membrane anchor had been labeled. Most probably, bromelain digestion induces a conformational change in HA2 which creates in the molecule one or more sites susceptible to TID labeling. This implies that the structure of BHA, which is known to 3A resolution (8), differs from that of HA in the HA2 region. Such a conformational change need not be large, given the small size of TID which, theoretically, enables it to label small hydrophobic pockets. In fact, no significant differences have been observed between the structure of BHA and HA as inferred indirectly from a variety of genetic, biochemical, and immunological experiments (10,15,16). We can exclude the possibility of potential hydrophobic pocket(s) in the ectodomain of HA which are not labeled due to the partitioning of the probe into the viral membrane, since BHA2 was labeled by TID even in the presence of liposomes which served as a "sink" for the hydrophobic label.

Differences between BHA and HA were also observed at acid pH. Incubation of virus at pH 5 did not alter the distribution of label, which remained exclusively HA2 associated. Interestingly the inclusion of liposomes also did not affect the distribution of labeling (not shown). Owing to the protease sensitivity of acid HA, we were unable to determine if regions other than the membrane anchor were labeled. However, the extent of labeling of HA2 was not increased by acid treatment, making it unlikely that TID bound to segments outside the transmembrane domain. In the case of acid-BHA, both BHA1 and BHA2 were labeled, in the presence or the absence of liposomes, even though previous studies have shown that BHA1 remains water soluble at acid pH whereas BHA2 becomes amphiphilic (9,10). Thus, BHA1 was labeled following acid treatment whereas viral HA1 was not. The absence of labeling in viral HA1 is most probably related to a difference in the destabilization of the trimeric structures of HA's ectodomain and BHA under conditions that trigger membrane fusion. We have observed that BHA can dissociate to dimers and monomers following acid treatment (17). It is plausible that due to the compact packing of HA in the viral membrane, the interface regions between the HA1 subunits are less accessible to TID than in the case of BHA where the interactions between BHA1 subunits are broken following acid treatment.

Taken together, our results show fundamental discrepancies between BHA and the native viral HA at neutral and acid pH. They suggest that the interaction of BHA with a target bilayer, as well as its structure at acid pH, might not faithfully reflect what happens to HA during membrane fusion. In addition, our inability to exclusively label BHA2 at acid pH, as well as the decrease in labeling observed when liposomes were present, indicates that the interaction of the BHA2 subunit with the target membrane may be more superficial than originally suspected. Experiments employing a variety of photoactivatable probes with defined positions in the lipid bilayer will be needed to gain further insight into this important process.

REFERENCES

1. Maeda, T. and Ohnishi, S. (1980) FEBS Lett. 122:283-287.
2. White, J., Helenius, A. and Gething, M.-J. (1982) Nature 300:658-659.
3. Matlin, K.S., Reggio, H., Helenius, A. and Simons, K. (1981) J. Cell Biol. 91:601-613.
4. White, J., Kielian, M. and Helenius, A. (1983) Q. Rev. Biophys. 16:151-195.
5. Yoshimura, A., Kuroda, K., Kawasaki, K., Yamashina, S. Maeda, T. and Ohnishi, S. (1982) J. Virol. 43:284-293.
6. Wiley, D.C., Skehel, J.J. and Waterfield, M. (1977) Virology 79:446-448.

7. Brand, C.M. and Skehel, J.J. (1972) Nat. New Biol. 238:145-147.
8. Wilson, I.A., Skehel, J.J. and Wiley, D.C. (1981) Nature 289:366-373.
9. Skehel, J.J., Bayley, P.M., Brown, E.B., Martin, S.R., Waterfield, M.D., White, J.M., Wilson, I.A. and Wiley, D.C. (1982) Proc. Natl. Acad. Sci. USA 79:968-972.
10. Doms, R.W., Helenius, A. and White, J. (1985) J. Biol. Chem. 260:2973-2981.
11. Brunner, J. and Semenza, G. (1981) Biochemistry 20:7174-7182.
12. White, J. and Helenius, A. (1980) Proc. Natl. Acad. Sci. USA 77:3273-3277.
13. Laemmli, U.K. (1970) Nature (London) 227:680-685.
14. Krebs, J., Buerkler, J., Guerini, D., Brunner, J. and Carafoli, E. (1984) Biochemistry 23:400-403.
15. Doms, R.W., Gething, M.-J., Henneberry, J., White, J. and Helenius, A. (1986) J. Virol. 57:603-613.
16. Daniels, R.S., Downie, J.C., Hay, A.J., Knossow, M., Skehel, J.J., Wang, M.L. and Wiley, D.C. (1985) Cell 40:431-439.
17. Doms, R.W., Agnew, W. and Helenius, A. (submitted).

Positive Strand RNA Viruses, pages 113–123

EPITOPE MAPPING OF A FLAVIVIRUS GLYCOPROTEIN

Franz X. Heinz, Günther Winkler, Christian Mandl, Farshad Guirakhoo, Wolfgang Tuma, and Christian Kunz

Institute of Virology, University of Vienna
A-1095 Vienna, Austria

ABSTRACT The antigenic structure of the glycoprotein (E) of Tick-borne encephalitis virus was analyzed by the use of monoclonal antibodies. At least 14 epitopes were identified which differ in either serological specificity, functional activity or topological location. As revealed by mutual blocking experiments most of these epitopes cluster to form three nonoverlapping antigenic domains, A, B and C, which also have different structural properties. Domain A is very sensitive to denaturation and low pH induced conformational changes. Also domain B is highly conformation-dependent but it is stabilized by disulfide bridges which have a strong tendency for renaturation. Epitopes of domain C are unaffected by reduction and carboxymethylation and therefore seem to represent sequential antigenic determinants. The TBE virus E protein contains a single asparagine-linked high mannose type oligosaccharide side chain. Complete removal of carbohydrate by endoglycosidase F digestion has no effect on infectivity, HA activity, and the antigenic reactivity with monoclonal antibodies or polyclonal immune sera. A characteristic pattern of one way and mutual cooperative interactions occur during the simultaneous binding of monoclonal antibodies to the glycoprotein, which results in an up to sixfold enhancement of antibody avidity. Cooperative effects between functional and nonfunctional antibodies do also occur in neutralization and passive protection assays. TBE virus seems to be quite stable under natural ecological conditions, since neither polyclonal sera nor any of the monoclonal antibodies were capable of differentiating between strains isolated from different hosts in different European countries over a period of 28 years.

INTRODUCTION

Flaviviruses represent a unique family of positive strand enveloped RNA viruses which comprises about 60 distinct virus types. Several flaviviruses are known as important human pathogens which pose a significant public health problem in epidemic and endemic areas (1). Until recently, flaviviruses had been classified as a genus within the family togaviridae, however differences with respect to structural and nonstructural proteins and recent data on the organization of flavivirus genomes (2, 3, 4) led to the establishment of a new virus family (5).

Flaviviruses contain only 3 structural proteins termed E, C and M and have estimated molecular weights of about 8, 15, and 50-60 kilodaltons (kD), respectively. C represents the only protein constituent of the isometric nucleocapsid, whereas both E and M are associated with the viral envelope. The nonglycosylated M protein is derived from a glycosylated precursor protein (NV-2) by a proteolytic cleavage event which seems to occur late during virus maturation (6). In mature virions of most flaviviruses E represents therefore the only glycosylated structural protein. There is however evidence that in certain flaviviruses E may not be glycosylated at all (4, 7).

Experiments using defined viral subunits revealed that the E protein represents the viral hemagglutinin which induces neutralizing antibodies and protects from disease. Recent evidence suggests that E may not be the only virusspecified protein involved in immune protection, since it was shown that monoclonal antibodies to a nonstructural protein (NV-3) as well as the purified protein itself could mediate protection against Yellow Fever virus in animals (8, 9). The E proteins of all flaviviruses share common antigenic determinants as revealed by crossreactivities in hemagglutination inhibition assays. Other parts of the molecule however seem to be more variable resulting in the formation of several flavivirus serocomplexes containing different types, subtypes and strains. This article deals with aspects of the antigenic structure of the E protein of Tick-borne encephalitis (TBE) virus, which is endemic in several European countries, Russia and China.

RESULTS AND DISCUSSION

Epitope Mapping by Monoclonal Antibodies.

Monoclonal antibodies against the E protein of TBE virus were obtained by immunization of Balb/c mice with glycoprotein 'rosettes' . To select those antibodies which define different epitopes, the serological specificities (reactivity with closely related and more distantly related flaviviruses), functional activities (hemagglutination inhibition (HI), neutralization (NT), passive protection), and spatial relationships as determined by competitive binding studies were used as selection criteria:

By these criteria 14 different epitopes were identified (10, Guirakhoo et al, in preparation). These cluster in at least 3 nonoverlapping domains, termed A, B, and C. Two epitopes do not overlap with any other epitope but seem to be structurally related to domain A.

The following conclusions can be deduced from this epitope map: 1. The antigenic domains obtained are inhomogeneous with respect to functional activity and serological specificity of individual epitopes. Domain A for instance contains epitopes which are either broadly flavivirus crossreactive, crossreactive at a reduced level, complex specific, type- or subtype specific. These epitopes also reveal different functional activities in HI, neutralization and protection and some epitopes have no measurable function at all. 2. There is a dissociation of functional activities at the epitope level. HI activity is not necessarily associated with neutralizing activity and protection. All possible combinations of involvement in HI and NT have been observed, i.e. HI+/NT+, HI+/NT-, and HI-/NT+. Similar findings have also been obtained in studies with other flaviviruses, including protective activity for nonneutralizing antibodies (9, 11), a phenomenon which was first described for Sindbis virus (12). These results demonstrate that the exact location of each epitope determines, whether and which viral function is affected by antibody binding. These critical functional sites can be different for HI, NT, and passive protection. 3. The specific functional activity may not only be determined by the epitope itself but also by the type of virus on which the epitope is located. Compared to TBE virus the HA-activity of West Nile (WN) and Murray Valley encephalitis (MVE) virus is apparently more sensitive to the binding of flavivirus cross-reactive antibodies within domain

A. This is not an isolated finding but the same phenomenon has been observed with many other flaviviruses, both with respect to HA and NT (reviewed in 13). The same monoclonal antibody which inhibits HA without neutralization of the homologous virus, may neutralize a heterologous virus without inhibiting its HA activity. Apparently, the functional activity of some monoclonal antibodies is not only determined by the epitopes itself but also by the rest of the molecule and represents an intrinsic property of the protein involved.

Cooperative Interactions between Monoclonal Antibodies.

In the course of competitive binding studies of antibody pairs for the topological mapping of epitopes it became apparent that antibodies may not only block but may even enhance the binding of other antibodies (14). This was a quite frequent phenomenon revealing a complex network of either one way or mutual interactions between pairs of monoclonal antibodies. There seem to be sites of predilection, which bind enhanceable antibodies, and others which bind enhancing antibodies. These sites of predilection correspond to the major antigenic domains thus resulting in a characteristic pattern of cooperative effects. i. Antibodies to epitopes within domains B and C enhance the binding of antibodies to domain A without becoming enhanced by themselves. ii. There is mutual enhancement between antibodies directed against nonoverlapping epitopes within domain A.

For a quantitative evaluation of these binding data Scatchard analyses with selected antibody pairs were performed (14). These revealed that enhanced binding was due to an upto sixfold increase in avidity, depending on the antibody pair analyzed. Since similar cooperative interactions were also demonstrated with purified Fab fragments we assume that antibody binding to certain epitopes can induce conformational changes which result in increased avidity of antibodies to distant sites on the glycoprotein. Enhancement of avidity is most characteristic for antibodies against that antigenic domain (A) which is also most sensitive to denaturation and low pH induced conformational changes.

Cooperative interactions have also been found with monoclonal antibodies against the E protein of all four dengue viruses, Japanese encephalitis, and Yellow Fever virus (reviewed in 13). In all cases bidirectional enhancement involving a flavivirus group-reactive antibody like that within domain A of TBE virus has been demonstrated, suggesting thc presence of structurally homologous domains which participate in these interactions.

Of course one of the crucial questions points to the relevance of these phenomena in functional terms. Functional cooperativity between antibodies to different epitopes can indeed be demonstrated in neutralization and passive protection assays (Heinz et al, in preparation). Under the conditions of a kinetic neutralization assay, no significant neutralization was observed with two antibodies from domain A, which however exerted a strong neutralizing effect when used in combination. This effect also holds true for passive protection as revealed by immunization of mice. The effect of a protective antibody from domain A was at least doubled, when it was applied in combination with a nonprotective antibody. We have therefore to keep in mind that the evaluation of functionally active antibodies and mutants selected in the presence of these antibodies may not give us the whole picture of antigenic sites that are involved in immune protection. For a complete understanding of antiviral immunity we have also to consider that nonfunctional antibodies can increase the functional activity of others, that nonneutralizing antibodies can be protective, and that even nonstructural proteins can participate in immune protection. We also have to take into account that the functional activity of the same epitope may be different on different viruses and even virus strains.

Structural Characterization of Epitopes.

We have first studied the conformation dependence of individual epitopes, the involvement of disulfide bridges in stabilizing antigenic determinants, and low pH induced conformational changes, in view of identifying those sites which might be involved in membrane fusion. The native glycoprotein was treated with SDS only, reduced and carboxymethylated in the presence of SDS, or incubated at pH 5.0 and then analyzed for its reactivity with each monoclonal antibody (15; Guirakhoo et al; in preparation). Characteristically, all epitopes of domain A, except one, are sensitive to pH 5.0, indicating that low pH induced conformational changes take place in this region. Most of these epitopes are also sensitive to denaturation by SDS and therefore are strongly conformation dependent. All epitopes of domain B and C are resistant to SDS, but only domain C is also resistant to reduction and carboxymethylation. This suggests that the native conformation of domain B is stabilized by disulfide-bridges whereas the epitopes of domain C probably represent conformation independent, sequential antigenic determinants. These studies demonstrate that epitopes within the clusters determined by competitive binding assays also have similar structural properties.

For the identification of epitopes on fragments of the E glycoprotein we have analyzed the antigenic reactivity of proteolytic digests obtained by trypsin, α-chymotrypsin and thermolysin treatment of the native protein (15; Guirakhoo et al, in preparation). All epitopes except those of domain B were destroyed upon proteolysis. By Western blotting a 9 kD fragment was identified after trypsin, α-chymotrypsin as well as thermolysin digestion which reacted with polyclonal immune sera and all antibodies defining domain B.

For a more detailed proteinchemical analysis the proteolytic digests were separated by reversed phase HPLC (16) and the immunoreactive peak corresponding to the 9 kD fragment was identified by dot immunoassays using polyclonal and monoclonal antibodies. The identification of several potential protease cleavage sites (16) suggested that this fragment represented a stabilized core structure which is resistant to denaturation and further proteolysis. Although the antigenic reactivity of this domain was lost upon reduction and carboxymethylation it was retained in Western blots, which also involve reduction and SDS-treatment of the protein. This could be due to renaturation by reformation of disulfide bridges during the blotting procedure. To test this hypothesis the following renaturation experiment was carried out (Winkler et al; in preparation). The glycoprotein was isolated by gelpermeation HPLC in the presence of SDS, and either a) carboxymethylated b) reduced and carboxymethylated or c) reduced, dialyzed to allow for renaturation, and then carboxymethylated. The antigenic reactivity of these preparations was assessed by dot immunoassays using monoclonal antibodies to domain B, which revealed that antigenicity was regained after dialysis of the reducing agent. By determination of the cystine/cysteine ratio it was confirmed that the reaquisition of antigenic reactivity within domain B was accompanied by the reformation of disulfide bridges. The strong tendency for renaturation apparently also accounts for the reactivity of these epitopes in Western blots.

Glycosylation and Role of Carbohydrate.

In order to analyze the extent of glycosylation (Winkler et al; in preparation) the E proteins of tick-borne flaviviruses (TBE virus Western and Far Eastern subtype; Louping Ill) as well as mosquito-borne flaviviruses (WN, MVE) and Rocio virus were digested with endoglycosidase F. This enzyme cleaves N-linked glycans of both the high mannose and the complex type (17). The removal of carbohydrate was monitored by a reduction of the migration rate in SDS-PAGE and by the loss of Concanavalin-A binding activity of the E proteins, which was assessed after blotting onto nitrocellulose (18). The results obtained indicate the presence of a single asparagine-linked carbohydrate side chain on the tick-borne and the mosquito-borne viruses analyzed, with the exception of West Nile virus, which is apparently not glycosylated at all. This is in accordance with the sequence data obtained by Wengler et al. (4) which revealed the absence of potential N-glycosidically linked glycosylation sites in the West Nile virus E protein.

To identify the type of carbohydrate side chains digestions were carried out with endoglycosidase H, which is specific for the high mannose type, and endoclycosidase D, which is specific for the complex type (Winkler et al; in preparation). Quite interestingly, the E protein of the tick-borne viruses analyzed contains high mannose type glycans, whereas that of the mosquito-borne viruses is of the complex type. At least for TBE virus (Western subtype) the type of glycan is not dependent on the host cell, since virus harvests derived from chick-embryo cells, PS cells (a porcine kidney cell line), and mouse brain all contain a single high mannose type oligosaccharide side chain.

As shown for several viral glycoproteins (19), glycosylation can strongly influence the reactivity of these proteins with immune sera. Carbohydrate may be directly involved as part of antigenic determinants, it may also stabilize a certain conformation which is lost upon deglycosylation, or it may even shield sites of potential antigenic reactivity (20). We have addressed this question for the TBE virus E glycoprotein by comparing the reactivity of endo F deglycosylated and native virus with monoclonal antibodies and polyclonal immune sera in blocking enzyme immunoassays. Neither monoclonal antibodies to the antigenic domains A, B and C nor polyclonal immune sera from mouse, rabbbit and man revealed any differences between these preparations suggesting that carbohydrate does not play a major role in the antigenic structure of the TBE virus E protein.

Antigenic Variation in Nature.

Studies with many RNA viruses have revealed mutation frequencies 10^4 to 10^6 fold greater than these found with DNA genomes. The potential for such high mutation frequencies is based upon the lack of error correcting mechanisms in RNA replication (21, 22). Stability or variability of a given RNA virus however seems to depend on the conditions under which the virus is allowed to replicate and may be different for each virus. The degree of variation found for different RNA viruses under natural conditions is obviously strongly influenced by environmental and ecological constraints. With respect to vaccination and the use of a certain vaccine strain the question of variation in nature is of paramount importance. We have addressed that question by comparing isolates of TBE virus from different European countries (Austria, Switzerland, Germany, Finland, Czechoslovakia) (23). Neither polyclonal sera nor any of the monoclonal antibodies were capable of distinguishing between these isolates. Remarkably homogenous patterns were also characteristic for peptide maps obtained by limited proteolysis. From these data we have concluded that TBE virus does not seem to be subject to major antigenic variations in nature. In 1985 we have isolated the virus in two of the same natural foci where the virus had also been isolated in 1973 and in a newly established focus in Western Austria (Tyrol) where the first cases were recorded in 1984. Again, none of the monoclonal antibodies revealed antigenic differences between any of these strains compared to the old isolates (Guirakhoo et al; in preparation), corroborating the stability of this virus in nature, at least as the E glycoprotein is concerned.

It is conceivable that the potential for variation of flavivirus genomes is counteracted by the stringent conditions of the ecosystem necessary for the maintenance of these viruses in their natural environment. These stabilizing constraints may be different for individual flaviviruses, due to differences in the number and nature of vectors and hosts involved in their natural cycles (24).

ACKNOWLEDGEMENTS

We thank Heide Dippe, Angela Dohnal and Melby Wilfinger for excellent technical assistance and Susanne Pfauser for expert secretarial help.

REFERENCES

1. Shope RE (1980). Medical significance of togaviruses. An overview of diseases caused by togaviruses in man and in domestic and wild vertebrate animals. In Schlesinger RW (ed):"The togaviruses: Biology, Structure, Replication", New York: Academic Press, p 47.
2. Rice MC, Lenches EM, Eddy SR, Shin SJ, Sheets RL, Strauss JH (1985). Nucleotide sequences of yellow fever virus: Implications for flavivirus gene expression and evolution. Science 229:726.
3. Castle E, Nowak T, Leidner U, Wengler G, Wengler G (1985). Sequence analysis of the viral core protein and the membrane-associated protein V1 and NV2 of the flavivirus West Nile virus and of the genome sequence for these proteins. Virology 145:227.
4. Wengler G, Castle E, Leidner U, Nowak T, Wengler G (1985). Sequence analysis of the membrane protein V3 of the flavivirus West Nile virus and of its gene. Virology 147:264.
5. Westaway EG, Brinton MA, Gaidamovich SY, Horzinek MC, Igarashi A, Kääriäinen L, Lvov DK, Porterfield JS, Russell PK, Trent DW (1985). Flaviviridae. Intervirology 24:183.
6. Shapiro D, Brandt WE, Russell PK (1972). Change involving a viral membrane glycoprotein during morphogenesis of group B arboviruses. Virology 50:906.
7. Wright PJ (1982). Envelope protein of the flavivirus Kunjin is apparently not glycosylated. J Gen Virol 59:29.

8. Schlesinger JJ, Brandriss MW, Walsh EE (1985). Protection against 17 D yellow fever encephalitis in mice by passive transfer of monoclonal antibodies to the nonstructural glycoprotein gp 48 and by active immunization with gp 48. J Immunol 135:2805.
9. Gould EA, Buckley A, Barrett ADT, Cammack N (1986). Neutralizing (54 K) and non-neutralizing (54 K and 48 K) monoclonal antibodies against structural and non-structural yellow fever virus proteins confer immunity in mice. J Gen Virol 67:591.
10. Heinz FX, Berger R, Tuma W, Kunz C (1983). A topological and functional model of epitopes on the structural glycoprotein of tick-borne encephalitis virus defined by monoclonal antibodies. Virology 126:525.
11. Brandriss MW, Schlesinger JJ, Walsh EE, Briselli M (1986). Lethal 17 D yellow fever encephalitis in mice. I. Passive protection by monoclonal antibodies to the envelope proteins of 17 D yellow fever and dengue 2 virus. J Gen Virol 67:229-234.
12. Schmaljohn AJ, Johnson EB, Dalrymple JH, Cole GA (1982). Nonneutralizing monoclonal antibodies can prevent lethal alphavirus encephalitis. Nature 297:70.
13. Heinz FX (1986). Epitope mapping of flavivirus glycoproteins. In Maramorosch K, Murphy FA, Shatkin AJ (eds). "Advances in Virus Research, volume 31" New York:Academic Press, p. 103.
14. Heinz FX, Mandl C, Berger R, Tuma W, Kunz C (1984). Antibody-induced conformational changes result in enhanced avidity of antibodies to different antigenic sites on the tick-borne encephalitis virus glycoprotein. Virology 133:25.
15. Heinz FX, Berger R, Tuma W, Kunz C (1983).Location of immunodominant antigenic determinants on fragments of the tick-borne encephalitis virus glycoprotein: Evidence for two different mechanisms by which antibodies mediate neutralization and hemagglutination inhibition. Virology 130:485.
16. Winkler G, Heinz FX, Kunz C (1984). Exclusive use of high-performance liquid chromatographic techniques for the isolation, 4-dimethylaminoazobenzene-4'-sulphonylchloride amino acid analysis and 4-N,N-dimethylaminoazobenzene-4'-isothiocyanate-phenyl isothiocyanate sequencing of a viral membrane protein. J Chromat 297:63.

17. Elder JH, Alexander S (1982). Endo-ß-N-Acetylglucosaminidase F: Endoglycosidase from Flavobacterium meningosepticum that cleaves both high-mannose and complex glycoproteins. Proc Natl Acad Sci USA 79:4540.
18. Hawkes R (1982). Identification of Concavalin A-binding proteins after sodium dodecyl sulfate-gel electrophoresis and protein blotting. Anal Biochem 123:143.
19. Alexander S, Elder JH (1984). Carbohydrate dramatically influences immune reactivity of antisera to viral glycoprotein antigens. Science 226:1328.
20. Skehel JJ, Stevens DJ, Daniels RS, Douglas AR, Knossow M, Wilson IA, Wiley DC (1984). A carbohydrate side chain on hemagglutinins of Hong Kong influenza viruses inhibits recognition by a monoclonal antibody. Proc Natl Acad Sci USA 81:1779.
21. Holland J, Spindler K, Horodyski F, Grabau E, Nichol S, Vande Pol S (1982). Rapid evolution of RNA genomes. Science 215:1577.
22. Reanny DC (1982). The evolution of RNA viruses. Ann Rev Microbiol 36:47.
23. Heinz FX, Kunz C (1981). Homogeneity of the structural glycoprotein from European isolates of tick-borne encephalitis virus: Comparison with other flaviviruses,. J Gen Virol 57:263.
24. Chamberlain RW (1980) Epidemiology of arthropod-borne togaviruses: The role of Arthropodes as hosts and vectors and of vertebrate hosts in natural transmission cycles. In Schlesinger RW (ed):"The Togaviruses:Biology, Structure, Replication. New York: Academic Press p 175.

III. GENE EXPRESSION

Positive Strand RNA Viruses, pages 127–135

GENOMIC ORGANIZATION OF MURINE CORONAVIRUS MHV[1]

Stuart G. Siddell

Institute of Virology
Versbacher Str. 7
D - 8700 Würzburg

The coronaviruses are enveloped positive-stranded RNA viruses which infect vertebrates and are associated with diseases of economic importance (1, 2). The murine hepatitis viruses induce a wide spectrum of disease patterns and have been used extensively in the laboratory for the study of viral pathogenesis. In particular, the MHV strain, JHM, has attracted increasing interest as a component in a model for virus-induced demyelination (3, 4, 5). The MHV genome is an infectious single-stranded RNA of approximately 20 kilobases. The genome encodes three major virion proteins, the nucleocapsid (N), the membrane (M) and the surface (S) protein, as well as several non-structural proteins (6).

The organization and expression of the MHV genome has been studied in some detail. MHV replicates exclusively in the cytoplasm and in infected cells six subgenomic mRNAs, as well as genome-sized RNA, are produced. These mRNAs are polyadenylated and capped and are synthesized in unequal but constant proportions throughout the infection (7,8,9). The size of the subgenomic mRNAs has been estimated by gel electrophoresis as mRNA 7, 1.9 Kb; mRNA 6, 2.5 Kb; mRNA 5, 3.2 Kb; mRNA 4, 3.6 Kb; mRNA 3 7.8 Kb; mRNA 2, 10.8 Kb (10).

[1]This work was supported by the Deutsche Forschungsgemeinschaft (SFB 165).

The structural relationships of the MHV genomic and subgenomic mRNAs have been determined by RNAse T_1-resistant oligonucleotide fingerprinting (10,11,12,13). These data showed that the intracellular RNAs comprised an overlapping or "nested" set. Furthermore, by relating the mRNA fingerprints to a partial T_1 oligonucleotide map of the genomic RNA it was possible to deduce that this nested set was 3' coterminal. Each mRNA contains the nucleotide sequences of the next smaller RNA plus additional sequences at the 5' end, which are referred to as the unique regions. Additionally, the data indicated that each mRNA possessed common 5' sequences derived from the 5' end of the genome. More detailed oligonucleotide mapping, primer extension, hybridization and sequence data (14,15,16,17,18) confirmed that the mRNAs shared an approximately 70 nucleotide leader sequence. It is thought that during mRNA synthesis this leader RNA is transcribed from the 3' end of the negative-stranded template and is translocated to specific internal reinitiation sites where it acts as a primer for the transcription of individual mRNA bodies (16,19,20,21). At the positions corresponding to the 5' end of the unique regions of mRNAs 7,6,5,4 and 3 (17,22,23,24,25) a region of homology, with the motif UC U/C AAAC has been identified in the genomic RNA. These regions are thought to be involved in regulating the process of discontinuous transcription.

The mRNAs which specify the major virion structural polypeptides of MHV have been identified by in vitro translation of RNA fractions enriched by electrophoresis or velocity sedimentation for each of the subgenomic species (26,27,28,29). mRNA 7 was found to code for the nucleocapsid polypeptide, mRNA 6 for the membrane glycopolypeptide and mRNA 3 for the surface projection glycopolypeptide. Additionally, the synthesis of a 30-35,000 mol.wt. polypeptide was directed by mRNA 2 and a 14-16,000 mol. wt. product was synthesized in fractions containing approximately equal amounts of mRNAs 4 and 5. Genomic RNA from purified MHV virions has also been translated in vitro to produce a series of structurally related polypeptides of more than 200,000 mol.wt. (28). The identities of these products are not known but given the infectious nature of the MHV genome (29) it seems reasonable to assume that they are related to components of the virus-encoded RNA polymerase, which is likely to be translated from the 5' unique region of the genome.

The analysis of the nucleotide sequence of cDNA clones obtained from MHV RNA has further clarified the details of the genomic organization. The nucleotide sequence of the unique regions of the MHV mRNAs 7,6 and 3 have been determined (18,22,23,24). In each case, a single ORF can be identified and the predicted translation products have the characteristics of the N, M and S polypeptides respectively. In each case translation is initiated at the first AUG codon within the mRNA and the expressed ORF does not overlap with any downstream ORF present in the 3' coterminal sequences. In mRNAs 7, 6 and 3 the initiating codon is found in a preferred context (i.e. frequently found amongst functional eucaryotic initator sequences, (31)) and follows closely (8, 4 and 1 nucleotides respectively) the "region of homology" sequences which define the 5' ends of the mRNA bodies.

The unique regions of two MHV mRNAs encoding nonstructural proteins have also been cloned and sequenced. The 5' unique region of MHV mRNA 4 (32) reveals a single ORF and the predicted polypeptide has the size and charge indicative of the 14-16,000 mol.wt. polypeptide detected in infected cells or in vitro translations (27,33). Antiserum from rabbits immunized with a bacterially synthesized polypeptide, encoded by the ORF within the unique region of mRNA 4 specifically immunoprecipitates the 14-16,000 mol.wt. polypeptide from infected cells (Ebner and Siddell, unpublished). Again, the mRNA 4 translation product is initiated at the 5' proximal AUG, in a preferred context, although in this case the initiating codon is approximately 130 nucleotides (as compared to 70-80 in mRNAs 7, 6 and 3) from the 5' terminus.

In contrast to the results obtained with other MHV mRNAs, sequencing of the 5' unique region of mRNA 5 has identified two potential ORFs (34), capable of coding for polypeptides of 12,400 (upstream ORF) and 10,200 (downstream ORF) mol.wt. At present no candidate translation products for these ORFs have been identified in infected cells or by in vitro translations. However, the organization of the ORFs in the unique region of mRNA 5 displays a number of unusual features. Firstly, the upstream ORF overlaps with the expressed ORF in mRNA 4 by 19 nucleotides and with the downstream ORF within the unique region of mRNA 5 by 8 nucleotides. This downstream mRNA 5 ORF itself overlaps by one nucleotide the M protein ORF in mRNA 6. Secondly, the initiator codon context of both mRNA 5 ORFs are amongst those least preferred in eucaryotic mRNAs. Also, the upstream ORF is preceeded by a 5'-proximal AUG which has a context indicative of an upstream non-functional initiator codon. Finally, with the exception of the AUG codon that initiates the downstream ORF, the upstream ORF is devoid for over 300 nucleotides of internal initiation codons, either in or out of frame.

These features themselves do not provide direct evidence for a functionally bicistronic mRNA, but clearly the organization of this region would allow for this possibility. Also, it is striking that many of these features have been found in the mRNAs B and D of the avian infectious bronchitis coronavirus (IBV) (35,36) which are again thought to encode non-structural proteins.

These data taken together have led to a model describing the organization and expression of the MHV genome which is depicted in figure 1. The model is incomplete but it has proven to be an extremely useful framework in which to study the replication of this virus. Essentially MHV genes are expressed from a series of 3' coterminal nested RNAs through a non-overlapping translation strategy. Only the 5' unique regions of the mRNAs are translationally active and only a single polypeptide is translated from each mRNA. The translated polypeptide may, as is known for the S gene product, undergo post-translational cleavage. The viral RNA polymerase activity is encoded in the 5' proximal region of the genome but the number and arrangement of ORFs in this region is unknown.

This strategy appears to have several advantages. Firstly, the genomic location of the polymerase and structural genes, together with the production of a 3' coterminal nested set of mRNAs, facilitates the temporal requirements for the gene products during the infection. Secondly, the production of multiple subgenomic mRNAs allows for the expression of a large viral genome without the synthesis and processing of a corresponding large primary translation product. Thirdly, this strategy has the major advantage that it permits the independent transcriptional regulation of different regions of the viral genome. Although this model appears to be essentially correct with

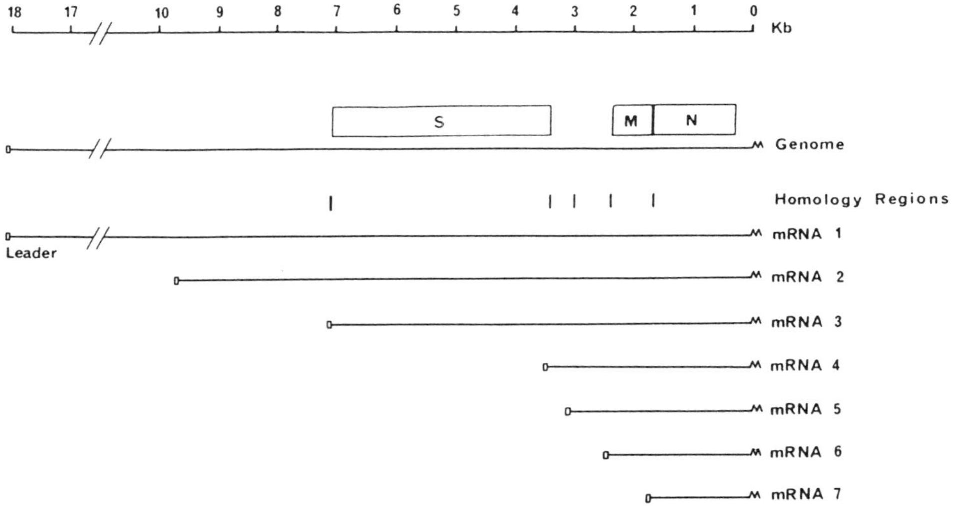

FIGURE 1. The genomic organization of MHV. The relationship between the 3' coterminal nested set and the viral genome is shown, together with the coding regions for the nucleocapsid (N), membrane (M) and surface (S) proteins. The 'regions of homology' are also indicated.

regard to the virion structural proteins, the more recent sequencing data suggests that it may be an oversimplification with regard to the non-structural proteins. However, a considerable amount of work will be necessary to determine which of the non-structural ORFs are translated, their function and the regulation of their expression. This, together with the analysis of the RNA polymerase gene and its products will be one of the most intensive areas of coronavirus research in the near future.

REFERENCES

1. Wege H, Siddell S, ter Meulen V (1982). The biology and pathogenesis of coronaviruses. Curr Top Microbiol Immunol 99: 165.

2. Sturman LS, Holmes KV (1983). The molecular biology of coronaviruses. Adv Virus Res 28: 35.

3. Knobler RL, Lampert PW, Oldstone MBA (1982). Virus persistence and recurring demyelination produced by a temperature sensitive mutant of MHV-4. Nature 298: 279.

4. Watanabe R, Wege H, ter Meulen V (1983). Adoptive transfer of EAE-like lesions from rats with coronavirus-induced demyelinating encephalomyelits. Nature 305: 150.

5. Massa P, Dörries R, ter Meulen V (1986). Virus particles induce Ia antigen expression on astrocytes. Nature 320: 543.

6. Siddell SG, Wege H, ter Meulen V (1983). The biology of coronaviruses. J Gen Virol 64: 761.

7. Spaan WJM, Rottier PJM, Horzinek MC, van der Zeijst BAM (1981). Isolation and identification of virus-specific mutants in cells infected with mouse hepatitis virus (MHV-A59). Virology 108: 424.

8. Wege H, Siddell S, Sturm M, ter Meulen V (1981). Coronavirus JHM: Characterization of intracellular virus RNA. J Gen Virol 54: 213.

9. Lai MMC, Patton CD, Stohlman SA (1982). Further characterization of viruses of mouse hepatits virus: Presence of common 5' and nucleotides. J Virol 41: 557.

10. Leibowitz JL, Wilhelmsen KC, Bond CW (1981). The virus-specific intracellular RNA species of two murine coronaviruses MHV-A59 and MHV-JHM. Virology 114: 39.

11. Lai MMC, Brayton PR, Armen RC, Patton CD, Pugh C, Stohlman SA (1981). Mouse hepatits virus A59: mRNA structure and genetic localization of the sequence divergence from hepatotropic strain MHV-3. J Virol 39: 823.

12. Spaan WJM, Rottier PJM, Horzinek MC, van der Zeijst BAM (1982). Sequence relationship between the genome and the intracellular RNA species 1, 3, 6 and 7 of mouse hepatitis virus strain A59. J Virol 42: 432.

13. Makino S, Taguchi F, Hirano N, Fujiwara K (1984). Analysis of genomic and intracellular viral RNAs of small plaque mutants of mouse hepatitis virus JHM strain. Virology 139: 138.

14. Lai MMC, Patton CD, Baric RS, Stohlman SA (1983). Presence of leader sequences on the mRNA of mouse hepatitis virus. J Virol 46: 1027.

15. Lai MMC, Baric RS, Brayton PR, Stohlman SA (1984). Characterization of leader RNA sequences on the virion and mRNAs of mouse hepatitis virus, a cytoplasmic virus. Proc Natl Acad Sci USA 81: 3626.

16. Spaan W, Delius H, Skinner M, Armstrong J, Rottier P, Smeekens S, van der Zeijst BAM, Siddell SG (1983). Coronavirus mRNA synthesis involves fusion of non-contiguous sequences. EMBO J 2: 1839.

17. Siddell S (1986). The organization and expression of coronavirus genomes. In: Rowlands et. al. (ed.), The Molecular Biology of the Positive-Stranded RNA Viruses Academic Press. Inc. New York.

18. Armstrong J, Niemann H, Smeekens, S, Rottier, P, Warren G (1984). Sequence and topology of a model

intracellular membrane protein, El Glycoprotein from a coronavirus. Nature 308: 751.

19. Jacobs L, Spaan WJM, Horzinek MC, van der Zeijst BAM (1981). Synthesis of subgenomic mRNAs of mouse hepatitis virus is initated independently evidence from UV transcription mapping. J Virol 39: 401.

20. Baric RS, Stohlman SA, Lai MMC (1983). Characterization of replication intermediate RNA of mouse hepatitis virus: Presence of leader RNA sequences on nascent chains. J Virol 48: 633.

21. Lai MMC, Patton CD, Stohlman SA (1982b). Replication of mouse hepatitis virus: Negative-stranded RNA and replicative form RNA are of genome length. J Virol 44: 487.

22. Schmidt I, Skinner M, Siddell SG (1986). Nucleotide sequence of the gene encoding the surface projetion glycoprotein of coronavirus MHV-JHM. submitted for publication.

23. Pfleiderer M, Skinner MA, Siddell SG (1986). Coronavirus MHV-JHM nucleotide sequence of the mRNA that encodes membrane protein. Nucl Acid Res in press.

24. Armstrong J, Smeekens S, Spaan W, Rottier P, van der Zeijst BAM (1984). Cloning and sequencing of the nucleocapsid and El Genes of coronavirus MHV-A59. Adv Exp Med Biol 173: 155.

25. Budzilowicz CJ, Wilczynski SP, Weiss SR (1985). Three intergenic regions of coronavirus mouse hepatitis virus strain A59 genome RNA contain a common nucleotide sequence that is homologous to the 3' end of the viral mRNA leader sequence. J Virol 53: 834.

26. Siddell SG, Wege H, Barthel A, ter Meulen V (1980). Coronavirus JHM. Cell-free synthesis of structural protein p60. J Virol 33: 10.

27. Siddell SG (1983). Coronavirus JHM: Coding assignments of subgenomic mRNAs. J Gen Virol 64: 113.

28. Leibowitz JL, Weiss SR, Paavola E, Bond CW (1982). Cell-free translation of murine coronavirus RNA. J Virol 43: 905.

29. Rottier PJM, Spaan WJM, Horzinek MC, van der Zeijst BAM (1981). Translation of 3 mouse hepatitis virus (MHV A59) subgenomic mRNAs in Xenopus laevis oocytes. J Virol 38: 20.

30. Wege H, Müller R, ter Meulen V (1978). Genomic RNA of the murnie coronavirus JHM . J Gen Virol 41: 217.

31. Skinner MA, Siddell SG (1983). Coronavirus JHM: Nucleotide sequence of the mRNA that encodes nucleocapsid protein. Nucl Acids Res 11: 5045.

32. Kozak M (1983). Comparison of initiation of protein synthesis in procaryotes, eucaryotes and organelles. Microbiol Rev 47: 1.

33. Skinner MA, Siddell SG (1985). Coding sequence of coronavirus MHV-JHM mRNA 4. J Gen Virol 66: 593.

34. Siddell SG, Wege H, Barthel A, ter Meulen V (1981). Coronaviurs JHM: Intracellular protein synthesis. J Gen Virol 53: 145.

35. Skinner MA, Ebner D, Siddell SG (1985). Coronavirus MHV-JHM mRNA 5 has a sequence arrangement which potentially allows translation of a second, downstream open reading frame. J Gen Virol 66: 581.

36. Boursnell MEG, Brown TDK (1984). Sequencing of coronavirus IBV genomic RNA: A 195 base open reading frame encoded by mRNA B. Gene 29: 87.

37. Boursnell MEG, Binns MM, Brown TDK (1985). Sequencing of coronavirus IBV genomic RNA: Three open reading frames encoded by mRNA D. J Gen Virol 66: 2253.

Positive Strand RNA Viruses, pages 137–147

MULTIPLE SUBGENOMIC mRNA's ARE INVOLVED IN THE GENE EXPRESSION OF EQUINE ARTERITIS VIRUS, A NON-ARTHROPOD BORNE TOGAVIRUS.

Bernard A.M. Van der Zeijst[1], M.F.van Berlo, W.C.Vooys, E.J.Snijder, P.Bredenbeek, M.C. Horzinek, P.J.M.Rottier and W.J.M.Spaan

Institute of Virology, Veterinary Faculty, State University Utrecht, Yalelaan 1, 3508 TD Utrecht, The Netherlands.

ABSTRACT Equine arteritis virus (EAV), although structurally similar , differs completely from other togaviruses in its mode of replication. In addition to the genome (RNA1; 4.3×10^6 mol.wt) 5 subgenomic RNA's are present in infected cells with the following molecular weights: 1.3×10^6 (RNA2), 0.9×10^6 (RNA3), 0.7×10^6 (RNA4), 0.3×10^6 (RNA5) and 0.2×10^6(RNA6). The RNA's have a messenger activity but so far only a few viral proteins have been mapped to the RNA's. There are a number of parallels with coronavirus replication. In both cases the genomic and subgenomic RNA's form a nested set sharing 3' sequences. However, in the case of coronaviruses the sequence extensions in the larger RNA's are unique; in EAV this is only true up to RNA2. RNAse-T1 fingerprinting reveals 19 "unique" oligonucleotides in the genome, 6 more than in RNA2. But still considerably less than the expected number of 44, suggesting large repeats in RNA1 of regions also present in RNA2. The mechanism by which the RNA's are generated is also different in both cases. UV-transcription mapping experiments excluded that coronavirus RNA's are processed or spliced from a common precursor of genome length, but the same approach demonstrated that the EAV RNA's do arise from

[1]Present address: University of Utrecht, Dept of Infectious Diseases, Section Bacteriology, P.O. Box 80.171, 3508 TD Utrecht, The Netherlands

such a precursor. Splicing therefore seems to be a possible mechanism. Since EAV is an RNA virus, replicating in the cytoplasm, such a splicing process should be an unorthodox one. Sequence studies are in progress to establish more precisely the homologies between the various RNA's and the points of sequence divergence.

INTRODUCTION

Equine Arteritis Virus (EAV) was discovered in 1953 in Bycurus, Ohio (1). In 1964 there was a great outbreak in Europe among the horses of the Swiss cavalry(2). More recently, in 1977 and 1984 the virus has been causing problems among racehorses in Kentucky (3,4). The clinical signs of EAV are extremely variable. Most infections go unnoticed and are only traced back by the presence of antibodies against the virus (5). With the possible exception of Japan and Australia the virus is endemic all over de world.

EAV is a small, enveloped, positive-stranded RNA virus. The virion has a diameter of 60 $\pm$ 13 nm and consists of an isometric core of about 35 nm which is surrounded by an envelope carrying ringlike subunits (6). On the basis of the morphology, substructure and size of the virion and the properties of the RNA genome, EAV has been originally classified as a non-arthropod borne member of the family Togaviridae, not belonging to one of the existing genera (7-9). EAV is not a widely studied virus. It was introduced in Utrecht by Marian Horzinek in 1972. Over the last years the virus has been characterized. This paper focuses on recent results on its replication.

THE PROTEINS OF EAV

Virions contain three proteins with mol. wts. of 12k (VP1), 14k (VP2) and 21k (VP3). VP1 is a major phosphorylated core protein. VP2, the other major protein is removed by detergent treatment together with the envelope but does not incorporate glucosamine; VP3 is the only glycoprotein, (10). Fig. 1 gives a cartoon of the virion.

The structural proteins of EAV are of lower mol. wt. than those of the other togaviruses. Only LDV which has been considered as another Togaviridae family member on

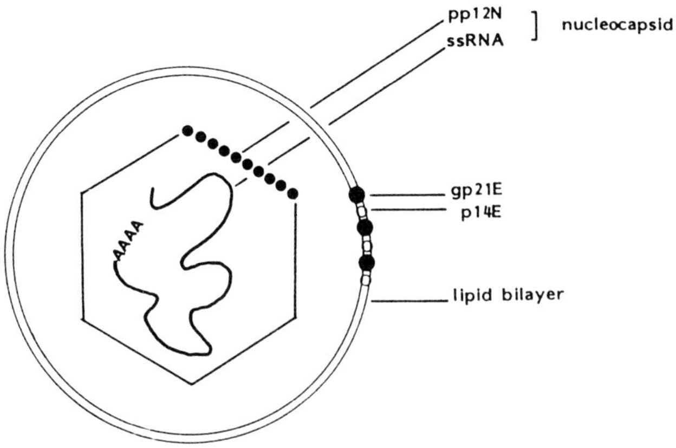

FIGURE 1. A schematic model of EAV. The viral nucleocapsid is an isometric particle (of probably icosahedral symmetry) composed of the positive stranded genomic RNA (mol.wt. 4 x 10^6 and many molecules of the phosphorylated nucleocapsid protein, pp12N(mol.wt. 12k). The viral envelope includes a lipid bilayer derived from host intracellular membranes and two viral proteins, p14E(mol.wt. 14k), the predominant protein, and the glycosylated species gp21 (mol.wt. 21k).

morphological grounds (11) has polypeptides in the same range (12, 13). However, no antigenic relationship at the level of the envelope and nucleocapsid proteins was detected (14), confirming that EAV and LDV belong to separate clusters.

Pulse-chase experiments indicate that the viral proteins do not arise by processing from larger precursors. In addition to the viral structural proteins (VP1 and VP2) three polypeptides with mol. wts. of 60k, 42k and 30k were found in membrane-containing fractions from infected but not from mock-infected cells. The glycosylated envelope protein (VP3) could not be demonstrated in infected cells (15).

THE RNAs OF EAV

The genome of EAV consists of an infectious, colinear molecule of single stranded (ss) RNA, with an S20,w value of 48 S and a mol. wt. of about 4 x 10^6 (16).

Analysis of the RNAs synthesized in EAV-infected BHK-cells revealed the existence of six virus-specific species with mol. wts. of 4.3 x 10^6 (RNA1, comigrating with the viral genome), 1.3 x 10^6 (RNA2), 0.9 x 10^6 (RNA3), 0.7 x 10^6 (RNA4), 0.3 x 10^6 (RNA5) and 0.2 x 10^6 (RNA6) (17; Fig.2).

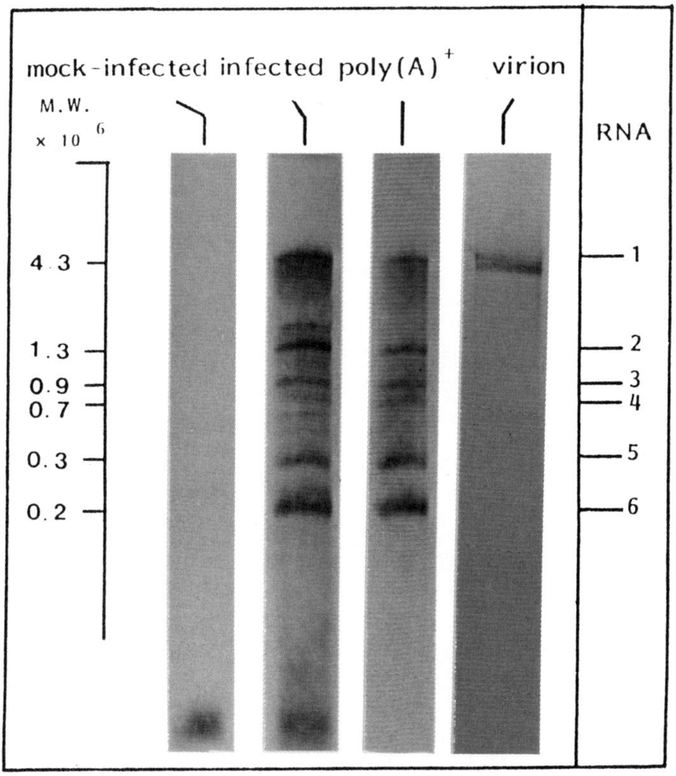

FIGURE 2. Agarose gel electrophoresis of RNAs from equine arteritis virus infected and mock-infected BHK-cells; RNA from purified virions was analyzed for comparison. RNA was labeled with ^{3}H-uridine in the presence of actinomycin D. Polyadenylated RNA from EAV-infected cells was purified by oligo(dT)-cellulose chromatography.

All RNA species are polyadenylated; translation of the unfractionated polyadenylated RNAs in an mRNA dependent reticulocyte cell-free system gave rise to the nucleocapsid protein (VP1) and to a 30kd protein. Of the fractionated molecules, only RNA6 could be translated; its translation product corresponded with VP1. This observation indicates that the nucleocapsid protein is a primary translation product (15).

The sum of the mol. wts. of the subgenomic RNAs amounts to 3.4×10^6 daltons, which is about 20% less than the molecular weight of the genome. The information in the subgenomic RNAs could therefore be adjacent in the genome. On the other hand, the subgenomic RNAs might also contain common sequences, as has been found e.g. in alphaviruses, coronaviruses and several plant viruses (for a review see 18).

To obtain a first impression of the sequence relationships between the EAV-specific RNAs, a northern blot analysis was carried out using a radio-labeled DNA probe complementary to RNA6. It hybridized with all the RNAs indicating that they possess common sequences; the experiment also demonstrates the virus-specificity of the subgenomic RNAs (19).

Using this cDNA probe, homology between the subgenomic RNAs of EAV was established. This technique does not allow to decide, however, whether the positive signals are due to small stretches or extensive regions of homology. Therefore we compared the subgenomic molecules by RNase T1 oligonucleotide fingerprinting. Polyadenylated ^{32}P-labeled RNA was isolated from infected cells, and the molecules were separated by agarose-urea gel electrophoresis. The purified EAV RNAs were digested with RNase T1 and the resistant oligonucleotides were separated by electrophoresis in two dimensions. The fingerprints showed that the intracellular viral RNAs form a nested set. However, some anomalies concerning discrete oligonucleotides were observed; we interpret these as sequence rearrangements, which may occur during the mRNA synthesis (19).The collection of spots found in the fingerprints of the subgenomic RNAs are roughly in agreement with their expected frequency (20; Table 1).

In contrast, RNA1 should have contained about three times the number of oligonucleotides of RNA2. However, instead of about 44 spots only 19 spots were found. This could be due to a high G-content of RNA1, but other explanations (like the occurence of polyploid sequences) are

TABLE 1

RNAse T1-RESISTANT OLIGONUCLEOTIDES IN SUBGENOMIC RNAs OF EAV AS COMPARED WITH THE EXPECTED FREQUENCIES (Ref 20).

	M.W.(10^6)	kbases	T1-spots	
			chain length^{3}15 nucleotides	
			found	expected
RNA1	4.3	13.2	19	44
RNA2	1.3	4.0	13	13
RNA3	0.9	2.8	9	9
RNA4	0.7	2.2		7
RNA5	0.3	0.9	3	3
RNA6	0.2	0.6	4	2

possible.

Thus the RNAs form a nested set. Recent restriction enzyme analyses of cDNA clones and sequence analysis of genomic RNA and RNA6 have shown that the RNA's contain common 3'-terminal ends (data not shown).

REPLICATION STRATEGY, PARALLELS WITH CORONAVIRUSES

The existence of subgenomic messenger RNAs prompted us to compare the translation and transcription strategy of EAV with that of other positive- and single-stranded RNA viruses whose genes are expressed with the aid of subgenomic RNAs. The subgenomic (26S) RNAs of alphaviruses (21) and rubella virus (22) are translated into continuous polypeptide chains, which are further processed to bring forth the viral proteins. The expression of the subgenomic RNAs of coronaviruses (23) and of some plant viruses (24) suggests some similarities with the translation strategy of EAV.

These subgenomic RNAs form a 3' coterminal nested set and only the 5'-terminal parts, which are absent in the next smaller RNAs are translated. In vitro translation of RNA6 of EAV resulted in the synthesis of a 14kd protein. As discussed above, RNA5 must be regarded as an extension of RNA6; in that case the unique region of RNA5 would be only 0.3 kb in length which is insufficient to encode any of the known proteins of EAV. These results suggest that overlapping reading frames exist in both RNAs.

Of central importance to the study of gene expression are the mechanisms by which mRNAs are produced. Upon infection a negative-stranded RNA is synthesized from the positive-stranded virion RNA, which serves as the template for novel positive-strand RNA synthesis. Until now two mechanisms are known for positive-stranded RNA viruses by which subgenomic RNAs can arise: i) internal initiation by the RNA-polymerase on the negative strand of genomic RNA (e.g. alphaviruses, 21, 22), ii) fusion of the leader and body sequences, which are non-contiguous in the genome and are joined in the cytoplasm (e.g. coronaviruses, 25, 26). A number of plant viruses have structural features in common with plus sense animal viruses and they share similarities in their replication strategies, too (27, 28). They also use different ways to produce subgenomic messenger RNAs (24), one of them involving premature termination during negative strand synthesis, followed by independent replication of the subgenomic negative strand (29). Nucleolytic cleavage has been suggested as another possibility for subgenomic RNA production (30).

Further research has focused on the mechanism by which the EAV subgenomic RNAs are synthesized. UV-target sizes for the templates of the EAV specific RNAs were determined and compared to those of mouse hepatitis virus, a coronavirus. The values for RNAs 2-5 were rather uniform and very close to the physical size of RNA1 (4.3×10^6 daltons). Only the target size of the template of RNA6 was smaller (2.8×10^6 daltons), although still much larger than its physical size. The data (17) are consistent with a model in which the individual RNAs are derived from a larger precursor RNA molecule and exclude the possibility that these RNAs are synthesized via independent initiation on one or more template molecules, as described earlier for coronaviral subgenomic RNAs (30, 31). In this system the UV target sizes were almost identical with the physical sizes of the RNAs (Table 2).

TABLE 2 UV TARGET SIZES

Equine arteritis virus			Mouse hepatitis virus	
Target size of template	RNA size	RNA	RNA size	Target size of template
(4.3)	4.3	1	5.6	(5.6)
4.5	1.3	2	4.0	4.0
4.8	0.9	3	3.0	3.0
3.8	0.7	4	1.4	1.4
4.3	0.3	5	1.2	1.2
2.8	0.2	6	0.97	0.9
-	-	7	0.75	0.6

The mechanism of the EAV mRNA synthesis seems to be unique among the positive stranded animal RNA viruses; subgenomic RNAs probably arise by nucleolytic cleavage of a genome-sized RNA. Sequence rearrangements can occur if several cleavage products are fused. Although we have no direct evidence for splicing, the anomalies detected in the T1 fingerprints of the individual mRNAs suggest that rearrangements do exist. However, splicing has only been found to occur in the nucleus and in EAV replication nuclear functions have not yet been identified. Although a universal mechanism does not seem to be involved in RNA splicing, highly conserved sequences located at the intron/exon junctions are required for accurate and efficient splicing (33).

Data on the nucleotide sequence of the genome and of the subgenomic RNAs are needed to unravel the details of the transcription of EAV, which is now a separate genus in the togavirus family (34). These data will be soon available from our laboratory.

REFERENCES

1. Doll ER, Bryans JT, McCollum WH, Crowe M (1957). Isolation of a filterable agent causing arteritis of horses and abortion by mares. Its differentiation from the equine abortion (influenza) virus. Cornell Vet 47:3.
2. Bürki F, Gerber H (1966). Ein virologisch gesicherter Groszausbruch von Equiner Arteritis. Berl & München Tierärztl Wschr 79:391.
3. McCollum WH, Swerczek TW (1978). Studies of an epizootic of equine viral arteritis in racehorses. Equine vet J 2:293.
4. Timoney PJ (1985). Epidemiological features of the 1984 outbreak of equine viral arteritis in the thoroughbred population in Kentucky, USA. Proc Soc Vet Epidem Prev Med: 84.
5 Mumford JA (1985). Preparing for equine arteritis. Equine vet J 17:6.
6. Horzinek MC (1981). Non-arthropod borne Togaviruses. Academic Press, London.
7. Horzinek MC (1973). The structure of Togaviruses. Progr Med Virol 16:109.
8. Horzinek MC (1973). Comparative aspects of Togaviruses. J gen Virol 20:87.
9. Porterfield JS, Casals J, Chumakov MP, Gaidamovich SYa, Hannoun C, Holmes IH, Horzinek MC, Mussgay M, Oker-Blom N, Russel PK and Trent DW. (1978). Togaviridae. Intervirology 9:129.
10. Zeegers JJW, Van der Zeijst BAM and Horzinek MC (1976). The structural proteins of equine arteritis virus. Virology. 73:200.
11. Horzinek MC, Van Wielink PS and Ellens DJ (1975). Purification and electron microscopy of lactic dehydrogenase virus of mice. J. gen. Virol. 26:217.
12. Michaelides MC and Schlesinger S (1973). Structural proteins of lactic dehydrogenase virus. Virology 55:211.
13. Brinton-Darnell MB and Plagemann PGW (1975). Structure and chemical-physical characteristics of lactate dehydrogenase-elevating virus and its RNA. J.Virol. 16:420.
14. Van Berlo MF, Zeegers JJW, Horzinek MC and Van der Zeijst BAM (1983). Antigenic comparison of equine arteritis virus (EAV) and lactic dehydrogenase virus (LDV); binding of Staphylococcal protein A to the

nucleocapsid protein of EAV. Zbl. Vet. Med. B, 30:297.
15. Van Berlo MF, Rottier PJM, Spaan WJM and MC Horzinek (1986a). Equine arteritis virus (EAV) induced polypeptide synthesis. J. gen Virol., in press.
16. Van der Zeijst BAM, Horzinek MC and Moennig V (1975). The genome of equine arteritis virus. Virology. 68:418.
17. Van Berlo MF, Horzinek MC and Van der Zeijst BAM (1982). Equine arteritis virus-infected cells contain six polyadenylated virus-specific RNAs. Virology. 118:345.
18. Strauss EG and Strauss JH. (1983). Replication strategies of the single stranded RNA viruses of eukaryotes. Curr. Top. Microbiol. Immunol. 105:1.
19. Van Berlo MF, Rottier PJM, Horzinek MC and Van der Zeijst BAM. (1986b). Intracellular equine arteritis virus (EAV) specific RNAs contain common sequences. Virology. in press.
20. Beemon KL (1978). Oligonucleotide fingerprinting with RNA tumor virus RNA. Curr. Top. Microbiol. Immunol. 79:73.
21. Kennedy SIT (1980). Synthesis of alphavirus RNA. In: The Togaviruses. p 351-369. ed. R.W. Schlesinger. Academic Press. London.
22. Oker-Blom C, Ulmanen I, Kaariainen L and Petterson RF. (1984). Rubella virus 40S genome RNA specifies a subgenomic 24S mRNA that codes for a precursor to structural proteins. J.Virol.49:403.
23. Siddell S, Wege H and ter Meulen V. (1983). The biology of coronaviruses. J. gen. Virol. 64:761.
24. Joshi S and Haenni A. (1984). Plant RNA viruses: strategies of expression and regulation of viral genes. FEBS letters. 177:163.
25. Spaan W, Delius H, Skinner M, Armstrong J, PJM Rottier, Smeekens S, Van der Zeijst BAM and Siddell SG. (1983). Coronavirus mRNA synthesis involves fusion of non-contiguous sequences. The EMBO J. 2:1839.
26. Lai MM, Baric RS, Brayton PR and Stohlman SA. (1984). Characterization of leader RNA sequences on the virion and mRNAs of mouse hepatitis virus, a cytoplasmic RNA virus. Proc. Natl. Acad. Sci. USA. 81:3626.
27. Haseloff J, Goelet P, Zimmern D, Ahlquist P, Dasgupta R and Kaesberg P. (1984). Striking similarities in amino acid sequence among nonstructural proteins encoded by RNA viruses that have dissimilar

genomic organization. Proc. Natl. Acad. Sci. USA. 81:4358.
28. Ahlquist P, Strauss EG, Rice CM, Strauss JH, Haseloff J and Zimmern D. (1985). Sindbis virus proteins nsP1 and nsP2 contain homology to nonstructural proteins from several RNA plant viruses. J.Virol. 53:536.
29. Goelet P and Karn J. (1982). Tobacco mosaic virus induces the synthesis of a family of 3'coterminal messenger RNAs and their complements. J. Mol. Biol. 154:541.
30. Gonda TJ and Symons RH. (1983). Cucumber mosaic virus replication in cowpea protoplasts: time course of virus, coat protein and RNA synthesis. J. gen. Virol. 45:723.
31. Jacobs L, Spaan WJM, Horzinek MC and Van der Zeijst BAM (1981). Synthesis of subgenomic mRNAs of mouse hepatitis virus is initiated independently: evidence from UV transcription mapping. J.Virol. 39:401.
32. Stern DF and Sefton BM. (1982). Synthesis of coronavirus mRNAs: Kinetics of inactivation of infectious bronchitis virus RNA synthesis by UV light. J.Virol. 42:755.
33. Rogers JH (1985). The origin and evolution of retroposons. Part 1: Mechanisms of RNA splicing. Int. Rev. Cytol. 93:187.
34. Westaway EG, Brinton MA, Gaidamovich S.Ya, Horzinek MC, Igarashi A, Kaariainen L, Lvov DG, Porterfield JS, Russell PK and Trent DW. (1985). Togaviridae. Intervirology. 24:125.

Positive Strand RNA Viruses, pages 149–158

GENE EXPRESSION IN TURNIP YELLOW MOSAIC VIRUS[1]

Anne-Lise Haenni, Marie-Dominique Morch, Gabrièle Drugeon,
Rosaura Valle, Rajiv L. Joshi and Térèse-Marie Denial

Institut Jacques Monod, CNRS - Université Paris VII,
2 Place Jussieu, 75251 Paris Cedex 05

ABSTRACT This paper provides an up-dated account of certain aspects of the strategies employed by turnip yellow mosaic virus for its development, with particular emphasis on the modulation of expression of the virus and on the capacity of the 3' region of its RNA to be recognized by several tRNA-specific enzymes.

INTRODUCTION

Turnip yellow mosaic virus (TYMV) was first described some 40 years ago as a small plant virus composed of ribonucleoprotein particles. Since then, it has been one of the most extensively studied plant viruses and it ranks probably second after tobacco mosaic virus (TMV) in its contribution to our knowledge of virus structure and function.

TYMV was one of the first isometric viruses to be examined by X-ray cristallographic analyses and by electron microscopy using negative staining. It was also the first virus for which it was reported that its RNA genome can be aminoacylated in vitro and in vivo, and in this respect, it has largely led the way in unravelling the analogies and differences that exist between viral RNAs and tRNAs. As template for in vitro translation studies, TYMV RNA has revealed the immense diversity of strategies that a virus can develop to produce its structural and non-structural proteins starting from a small RNA genome.

A recent review on TYMV has focused on the historical background and on the physico-chemical characteristics as

[1]This work was supported partly by a grant from the "ATP: Interaction entre Plantes et Microorganismes".

well as the cytopathological effects of the virus, in addition to describing its structure and molecular biology (1).

After an outline of the characteristics of TYMV, this paper will concentrate on the translation strategies and on the amino acid-accepting activity of the viral RNA.

RESULTS

Structure of TYMV.

TYMV is the type member of the tymovirus group. This group comprises 14 to 17 members, characterized by small icosahedral particles possessing a monopartite RNA genome of Mr 2 X 10^6 of "+" polarity.

TYMV preparations contain at least two types of particles of ∿29 nm in diameter: empty shells devoid of RNA and infectious ribonucleoprotein particles. The viral capsids are constructed of 180 copies of a single protein species (2), the coat protein of Mr 20 000 (20K; K = kilodalton) arranged in 32 morphological subunits.

The great stability of the empty shells indicates that TYMV particles depend on protein-protein hydrophobic interactions for their integrity. TYMV is indeed the prototype of viruses with strong protein-protein interactions and weak protein-RNA interactions (reviewed in ref. 3).

Besides the major infectious nucleoprotein particles that encapsidate the genomic RNA, TYMV preparations contain minor amounts of particles of intermediate densities that harbor various-sized RNAs (4-6), the predominant species being an RNA of Mr 0.24 X 10^6 that codes for the coat protein but is not required for infectivity (7-9). Both genomic and coat protein RNAs are capped (8), and both can be specifically esterified at their 3' end with valine (10-12). The complete nucleotide sequence of the subgenomic RNA has been established (13); it corresponds to the 3' region of the genomic RNA.

Translation of TYMV RNA.

A schematic representation of the genomic organization of TYMV and of the translational strategies employed by this virus is presented in figure 1.

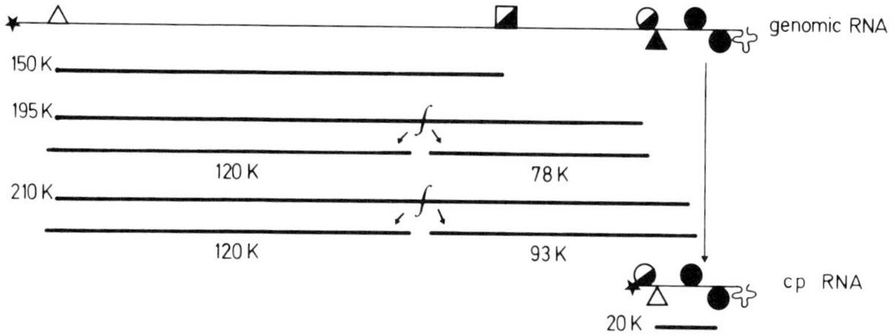

FIGURE 1. Genetic map of TYMV RNA. Viral RNAs and corresponding translation products are schematized by thin and thick lines respectively. △, ▲: accessible and inaccessible initiation codons respectively; ●: termination codon; ◕: leaky termination codon; ◪: signal for alternative arrest of translation; ✶: cap structure; ⚘: tRNA-like structure; ʃ: site of proteolytic cleavage. The molecular weight of the products of proteolytic cleavage is indicated below each fragment (13).

The coat protein gene is silent in the genomic RNA, and is only expressed from the subgenomic RNA where its position becomes 5' proximal (6-8,13). In vitro translation studies have revealed that the affinity of the subgenomic RNA for the translation machinery is greater than that of the genomic RNA (ref. 15). This may account for the efficient production of coat protein at late stages of infection.

In various cell-free translation systems (16,17), TYMV genomic RNA directs the synthesis in about equal amounts of two high molecular weight proteins of 150K and 195K. The 195K protein results from the elongation of the 150K protein beyond a "stop" signal which dictates the arrest of the 150K protein. This "stop" signal does not correspond to a conventional termination site since addition of yeast amber, opal or ochre suppressor tRNAs does not enhance synthesis of the 195K protein to the detriment of the 150K protein (18). Under conditions in which certain tRNA species are limiting in the translation system, the 150K protein is the major high molecular weight protein synthesized. A tRNA species from beef liver has been purified that leads, in such a system, to the appearance of the 195K protein and restores the one to one ratio of 195K versus 150K proteins. This tRNA capable of bypassing the "stop" signal is acceptor of serine. Nucleotide sequence data at the "stop" signal site on the RNA as well as knowledge of the amino acid sequence at the C terminus of the 150K protein should provide valuable clues to our understanding of the characteristics of this unusual "stop" signal.

Addition of yeast amber suppressor tRNA to the translation system, does however modify the pattern of high molecular weight proteins programmed by TYMV RNA: a 210K protein appears, resulting from readthrough of the 195K protein gene (18). The UAG codon terminating the 195K protein gene is present in the untranslated leader sequence of the coat protein RNA (ref. 18). Its position in the corresponding region of the genomic RNA has been confirmed by our recent sequence data: the silent region between the 195K and coat protein cistrons is only 14 nucleotides long. Thus, the C-terminal region of the 210K and the coat protein share the same information on the RNA, but in different reading frames.

In the genomic RNA of TYMV, the UAG codon terminating the 195K protein gene is flanked on either side by CAA, a Gln codon. The resulting nonanucleotide sequence is also present in two other plant viral RNAs: TMV RNA and beet necrotic yellow vein virus (BNYVV) RNA-2 (ref. 19,20). A major $tRNA^{Tyr}$ species isolated from tobacco leaves can overcome the UAG codon in TMV RNA (reviewed in ref. 21) and in BNYVV RNA-2 (ref. 22) in vitro. A similar major $tRNA^{Tyr}$ species isolated from wheat leaves and Drosophila melanogaster can suppress the UAG codon in TMV RNA (ref. 21). We have isolated from calf liver a tRNA species that can overcome the UAG codon in TMV RNA but is not aminoacylated with tyrosine. The sequence of this novel, natural suppressor tRNA in under way. It will be interesting to test whether it also leads to readthrough in TYMV RNA and in BNYVV RNA-2.

An additional strategy employed by TYMV to generate its non-structural proteins, is post-translational cleavage (23). The 195K protein is cleaved in vitro, producing a 120K and a 78K fragment corresponding respectively to the N-terminal and the C-terminal part of the 195K protein. The 210K protein undergoes a similar cleavage. On the other hand, the 150K protein is stable even though it presumably contains the cleavage site; this must reflect a difference in the conformation of the 150K protein as compared to that of the 195K and 210K proteins. Cleavage is not an artefact of the in vitro translation system used since it occurs in three different translation systems (17). A close examination of the kinetics of cleavage after dilution of the in vitro translation products has demonstrated that cleavage is at least in part dilution-insensitive. An intramolecular mechanism of proteolytic maturation has been proposed in which the proteolytic activity would reside in the C-terminal part of the 195K protein (24).

To date, information regarding the production of the viral proteins in vivo is scant. Only the coat protein has been unambiguously detected in TYMV-infected Chinese cabbage leaves (our unpublished observations). However, since the two major non-structural proteins are produced in various in vitro translation systems programmed with TYMV RNA (17), we are confident that they are also produced in vivo.

The function of the non-structural proteins of TYMV are unknown. Nonetheless, two pieces of information suggest a role in replication: 1) the TYMV RNA-dependent RNA polymerase (RNA replicase) isolated from infected tissue appears to contain two major subunits, a virus-coded 115K protein and a host-coded 45K protein (25). The 115K protein could correspond to one of the cleavage products of the 195K protein; 2) in the region of the RNA corresponding to the C-terminal part of the 195K protein, a consensus sequence coding for Gly-Asp-Asp is observed. This sequence is present in the RNA replicase of poliovirus and in the non-structural proteins of several plant and animal viruses (26). Furthermore, in TYMV, alfalfa mosaic virus and brome mosaic virus, this consensus sequence can be extended to a hexapeptide: Ser-Gly-Asp-Asp-Ser-Leu. This homology may highlight a common nucleic acid recognition site in various replicases (26).

tRNA-like Structure.

For the first time in 1970, reports appeared that TYMV RNA can be aminoacylated in vitro at its 3' terminus with valine in conditions identical to those leading to the valylation of tRNA (10,27). The 3' region of the viral RNA capable of being aminoacylated is designated the "tRNA-like" region. The RNAs of four other tymoviruses also accept valine in vitro (reviewed in ref. 28).

Aminoacylation of plant viral RNAs has since been reported for several other virus groups, the RNAs of the members of a given group generally accepting the same amino acid (28). However, comparisons between the tRNA-like region and tRNAs have been the most thoroughly examined in the case of

TYMV RNA. In addition to valyl-tRNA synthetase, TYMV RNA is recognized by several tRNA-specific enzymes including tRNA nucleotidyltransferase and the elongation factors EF-Tu or EF-1α. TYMV RNA is aminoacylated *in vivo* when injected into *Xenopus laevis* oocytes (12), or when it has infected its host (29). Such results are not unexpected since the kinetic parameters of TYMV RNA and tRNA valylation are comparable (11).

More recently, foot-printing experiments have been performed to determine the minimum length of the viral RNA required for valylation: the last 3'-terminal 86 nucleotides are sufficient (reviewed in ref. 28). This information, together with studies on the accessibility of the bases within the 86 nucleotide-long fragment to chemical modifications or nuclease digestion (28) have led to the model schematized in figure 2. An interesting novel feature of this model is that the aminoacyl RNA domain (corresponding in tRNAs to the continuous stacking of the acceptor stem over the T stem) is made up exclusively of nucleotides from the 3' end of the tRNA-like region. Furthermore, the formation of the aminoacyl RNA domain in TYMV as well as in other viral RNAs (28, 30) involves the participation of small bridges linking the two sides of the arm. Similar foot-printing experiments have revealed that the 3'-terminal 55 nucleotides are required for adenylation (28) and the 3'-terminal 47 nucleotides for ternary complex formation with elongation factors (31,32). Thus, the tRNA nucleotidyltransferase and the elongation factor require only the aminoacyl RNA domain for interaction.

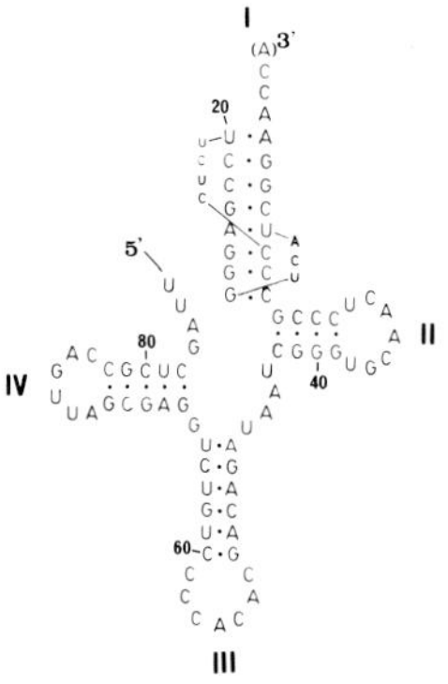

FIGURE 2. Folding of the tRNA-like region of TYMV RNA. I to IV correspond to stems and loops analogous in tRNAs to the acceptor stem, and the T, anticodon and D stems and loops respectively. The aminoacyl RNA domain is composed of regions I and II. Numbering is from the 3' end. The 3'-terminal A residue is bracketed since it is absent from the RNA as extracted from TYMV (27,33; adapted from 32).

It is remarkable that the affinity of wheat germ EF-1α for yeast Val-tRNA and for TYMV Val-RNA is the same. This has revived the previously postulated role of tRNA-like structures in RNA replication (34), by analogy with the RNA replicase of the bacteriophages Qβ and f2 which contain EF-Tu as one of their subunits (reviewed in ref. 35). However, no elongation factor can be detected in the TYMV RNA replicase preparation, and addition of EF-1α does not enhance the activity of the replicase (32).

Even if this leaves us without a clue as to the role of tRNA-like structures in virus development, experiments using the RNA replicase indicate that the integrity of the tRNA-like region is necessary for optimal RNA synthesis (32). Consequently, the intact 3'-terminal structure of the viral RNA is a recognition signal for the replicase. This conclusion is further strengthened by experiments using the in vitro transcribed tRNA-like region. A cDNA clone has been engineered that corresponds to the last 100 nucleotides from the 3' end of TYMV RNA, followed by poly$(A{:}T)_{50}$. After insertion into the Gemini system, in vitro transcription leads to the production of the tRNA-like region of 100 nucleotides followed by the poly(A) tail. When this construct is added to the RNA replicase system programmed with TYMV RNA, it efficiently competes with the viral RNA, a one to one ratio of viral RNA to construct leading to a 50% reduction in replication.

DISCUSSION

Our knowledge of the translational strategies of TYMV and of the tRNA-like properties of its RNA has greatly extended in the past few years. This should help us in further probing aspects of this virus that are still poorly understood. Determination of the complete nucleotide sequence of the genomic RNA which is under way in our group, will bring new dimensions to our understanding of the development of this virus. The availability of antibodies against synthetic peptides corresponding to selected regions of the non-structural proteins should facilitate detection of these proteins in vivo. This will allow us to verify the translational strategies deduced from in vitro studies and possibly to assign a role to these proteins in the life-cycle of the virus. Replication studies in vitro and in vivo are also being pursued to define how the coat protein RNA is produced and what cellular elements are involved in RNA replication. Site-directed mutagenesis on viral cDNA associated to *in vitro* transcription may help pin-point the role of the tRNA-like structure. Finally, production of transgenic plants expressing parts of the viral genome may provide a means of obtaining plants resistant to the virus.

ACKNOWLEDGEMENTS

RV and RLJ are indebted to the "Ministère de la Recherche et de la Technologie"for fellowships. TMD benefited from a grant from the Philippe Foundation and from an anonymous gift.

REFERENCES

1. Matthews REF (1981). Portraits of viruses: turnip yellow mosaic virus. Intervirology 15:121.
2. Klug A, Longley W, Leberman R (1966). Arrangement of protein subunits and the distribution of nucleic acid in turnip yellow mosaic virus. J Mol Biol 15:315.
3. Kaper JM (1975). The Chemical Basis of Virus Structure, Dissociation and Reassembly. In "Frontiers of Biology" 39. Amsterdam: North Holland.
4. Matthews REF (1974). Some properties of TYMV nucleoproteins isolated in cesium chloride density gradients. Virology 60:54.
5. Mellema JR, Benicourt C, Haenni AL, Noort A, Pleij CWA, Bosch L (1979). Translational studies with turnip yellow mosaic virus RNAs isolated from major and minor virus particles. Virology 96:38.
6. Keeling J, Collins ER, Matthews REF (1979). Behavior of turnip yellow mosaic virus nucleoproteins under alkaline conditions. Virology 97:100.
7. Pleij CWA, Neeleman A, Van Vloten-Doting L, Bosch L (1976). Translation of turnip yellow mosaic virus RNA *in vitro*: a closed and an open coat protein cistron. Proc Natl Acad Sci USA 73:4437.
8. Klein C, Fritsch C, Briand JP, Richards KE, Jonard G, Hirth L (1976). Physical and functional heterogeneity in TYMV RNA: evidence for the existence of an independent messenger coding for coat protein. Nucl Acids Res 3:3043.

9. Ricard B, Barreau C, Renaudin H, Mouches C, Bové MJ (1977). Messenger properties of TYMV-RNA. Virology 79: 231.
10. Pinck M, Yot P, Chapeville F, Duranton HM (1970). Enzymatic binding of valine to the 3' end of TYMV RNA. Nature 226:954.
11. Giégé R, Briand JP, Mengual R, Ebel JP, Hirth L (1978). Valylation of the two RNA components of turnip yellow mosaic virus and specificity of the tRNA aminoacylation reaction. Eur J Biochem 84:251.
12. Joshi S, Haenni AL, Hubert E, Huez G, Marbaix G (1978). In vivo aminoacylation and 'processing' of turnip yellow mosaic virus RNA in Xenopus laevis oocytes. Nature 275: 339.
13. Guilley H, Briand JP (1978). Nucleotide sequence of turnip yellow mosaic virus coat protein in RNA. Cell 15:113.
14. Morch MD, Haenni AL (1986). Organization of plant virus genomes that comprise a single RNA molecule. In Mayo M (ed): "The Molecular Biology of the Positive Strand RNA Viruses". London: Academic Press, in press.
15. Benicourt C, Haenni AL (1978). Differential translation of turnip yellow mosaic virus mRNAs in vitro. Biochem Biophys Res Commun 84:831.
16. Benicourt C, Péré JP, Haenni AL (1978). Translation of TYMV RNA into high molecular weight proteins. FEBS Lett 86:268.
17. Zagorski W, Morch MD, Haenni AL (1983). Comparison of three different cell-free systems for turnip yellow mosaic virus RNA translation. Biochimie 65:127.
18. Morch MD, Drugeon G, Benicourt C (1982). Analysis of the in vitro coding properties of the 3' region of turnip yellow mosaic virus genomic RNA. Virology 119:193.
19. Goelet P, Lomonossoff GP, Butler PJG, Akam ME, Gait MJ, Karn J (1982). Nucleotide sequence of tobacco mosaic virus RNA. Proc Natl Acad Sci USA 79:5818.
20. Bouzoubaa S, Ziegler V, Beck D, Guilley H, Richards K, Jonard G (1986). Nucleotide sequence of beet necrotic yellow vein virus RNA-2. J Gen Virol, in press.
21. Hatfield D (1985); Suppression of termination codons in higher eukaryotes. Trends Biochem Sci 10:201.
22. Ziegler V, Richards K, Guilley H, Jonard G, Putz C (1985). Cell-free translation of beet necrotic yellow vein virus: readthrough of the coat protein cistron. J Gen Virol 66:2079.

23. Morch MD, Benicourt C (1980). Post-translational proteolytic cleavage of in vitro-synthesized turnip yellow mosaic virus RNA-coded high-molecular-weight proteins. J Virol 34:85.
24. Morch MD, Zagorski W, Haenni AL (1982). Proteolytic maturation of the turnip-yellow-mosaic-virus polyprotein coded in vitro occurs by internal catalysis. Eur J Biochem 127:259.
25. Mouches C, Candresse T, Bové JP (1984). Turnip yellow mosaic virus RNA-replicase contains host and virus-encoded subunits. Virology 134:78.
26. Kamer G, Argos P (1984). Primary structural comparison of RNA-dependent polymerases from plant, animal and bacterial viruses. Nucl Acids Res 12:7269.
27. Yot P, Pinck M, Haenni AL, Duranton HM, Chapeville F (1970). Valine-specific tRNA-like structure in turnip yellow mosaic virus RNA. Proc Natl Acad Sci USA 67:1345.
28. Joshi S, Joshi RL, Haenni AL, Chapeville F (1983). tRNA-like structures in genomic RNAs of plant viruses. Trends Biochem Sci 8:402.
29. Joshi S, Chapeville F, Haenni AL (1982). Turnip yellow mosaic virus RNA is aminoacylated in vivo in Chinese cabbage leaves. EMBO J 1:935.
30. Rietveld K, Linschooten K, Pleij CWA, Bosch L (1984). The three-dimensional folding of the tRNA-like structure of tobacco mosaic virus RNA. A new building principle applied twice. EMBO J 3:2613.
31. Joshi RL, Faulhammer H, Chapeville F, Sprinzl M, Haenni AL (1984). Aminoacyl RNA domain of turnip yellow mosaic virus Val-RNA interacting with elongation factor Tu. Nucl Acids Res 12:7467.
32. Joshi RL, Ravel JM, Haenni AL (1986). Interaction of turnip yellow mosaic virus Val-RNA with eukaryotic elongation factor EF-1α. Search for a function. EMBO J in press.
33. Litvak S, Carré DS, Chapeville F (1970). TYMV RNA as a substrate of the tRNA nucleotidyltransferase. FEBS Lett 11:316.
34. Litvak S, Tarrago A, Tarrago-Litvak L, Allende JE (1973). Elongation factor-viral genome interaction dependent on the aminoacylation of TYMV and TMV RNAs. Nature NB 241:88.
35. Blumenthal T (1982). Qβ replicase. In Boyer H (ed): "The Enzymes". New York: Academic Press p 267.

Positive Strand RNA Viruses, pages 159–181

COTRANSLATIONAL DISASSEMBLY OF FILAMENTOUS PLANT VIRUS NUCLEOCAPSIDS IN VITRO AND IN VIVO[1]

T. Michael A. Wilson[2], and John G. Shaw[3]

[2]Department of Virus Research, John Innes Research Institute, Colney Lane, Norwich NR4 7UH, U.K.
[3]Department of Plant Pathology, University of Kentucky, Lexington, Kentucky 40546-0091, U.S.A.

ABSTRACT The cotranslational disassembly of tobacco mosaic virus (TMV) was first demonstrated in vitro two years ago. Several details of the phenomenon have been studied more recently and are described in the context of their implications for ribonucleocapsid disassembly in vivo during the early stages of host cell infection. Translationally-active, intermediate complexes, dubbed "striposomes" in vitro, have recently been detected in TMV-infected epidermal cells from tobacco. These complexes appear to contain nascent, virus-specific polypeptides, 80S host ribosomes and partially uncoated TMV particles, in which we presume the 3'-end of the positive-strand TMV RNA remains encapsidated. We are currently engaged in experiments to characterize these intermediate complexes further. In addition to tobacco, preliminary work with epidermal strips from TMV-inoculated pea leaves [*Pisum sativum* L. cv. Argenteum] at best a "subliminal" host for the virus, suggests that the early structural events which lead to virus uptake, uncoating and primary gene expression will occur irrespective of the origin of the cell. Complexes with the physicochemical characteristics of "striposomes" have also been detected in tobacco epidermal cells inoculated with fresh preparations of other rod-shaped plant viruses, including tobacco vein

[1]This work was supported by a grant-in-aid from the Agricultural and Food Research Council to the John Innes Research Institute, and by Grant 58-7B30-3-538 from the United States Department of Agriculture.

mottling virus (a member of the Potyvirus group), potato virus X (type member of the Potexvirus group) or tobacco rattle virus (type member of the Tobraviruses). These results are significant, not only because they broaden the scope of the phenomenon first observed with TMV, but also because paradoxically, no member of any of these virus groups has been found amenable to cotranslational disassembly *in vitro*.

INTRODUCTION

Over 76% of all known plant viruses contain one or more positive-strand RNA molecules as the encapsidated form of their genetic material. Teleologically, this is considered a desirable feature, which facilitates the immediate expression of virus-specific polypeptides in the newly-infected host cell. However, to achieve this, the ribonucleocapsid must disassemble to a greater or lesser extent, during or shortly after the infection process, thereby releasing the viral genome to recruit and subvert available components of the cellular protein synthetic machinery. Herein lies one long-standing paradox of molecular virology. Having evolved, in most cases, an extremely robust, protective, yet structurally fairly simple protein coat (the capsid); to shield the undoubtedly-sensitive, single-stranded RNA (ssRNA) genome from the (assumed) rigors of the extracellular environment during transfer between hosts (often *via* some convenient vector); the virus must also be able to release its genome from this same protein coat under the comparatively mild physicochemical conditions likely to exist within the newly-invaded host cell. In this respect, the architectural problems confronting rod-shaped ribonucleocapsids are especially serious, for release of the RNA from its intimate contacts with every coat protein subunit along the filament requires the complete disassembly of the nucleoprotein structure. Furthermore, the conditions under which this must occur are likely to be similar, if not identical, to those under which the virus was originally assembled. Progress in understanding these so-called "early events" of infection has been slow, the subject is glossed-over in most reviews and textbooks on virology, and receives little attention even in the primary research literature. The questions which we seek to address in this work include: What is(are) the mechanism(s) for virus disassembly and RNA release in vivo? How does the virus ensure that its RNA is uncoated at a "suitable" time and/or place within the plant cell? These questions are not unique to plant virology, but for reasons discussed in recent reviews (1,2), plant virologists seem particularly far from even the most preliminary answers, amenable model systems or testable hypotheses.

Against this background, we were excited by an initial, fortuitous observation (3) made during cell-free translation experiments with preparations of intact tobacco mosaic virus (TMV) particles. TMV RNA-encoded polypeptides were detected in vitro, apparently synthesized during the progressive, stepwise disassembly of the otherwise-stable, rod-shaped nucleoprotein helix. Uncoating of the two large, 5'-proximal, overlapping, cistrons on TMV RNA appeared to occur cotranslationally and was thought to be due to the translocation of active 80S ribosomes (3). Intermediate translation complexes, colloquially called "striposomes" (Figure 1), were easily detected in cycloheximide-arrested translations programmed with TMV particles.

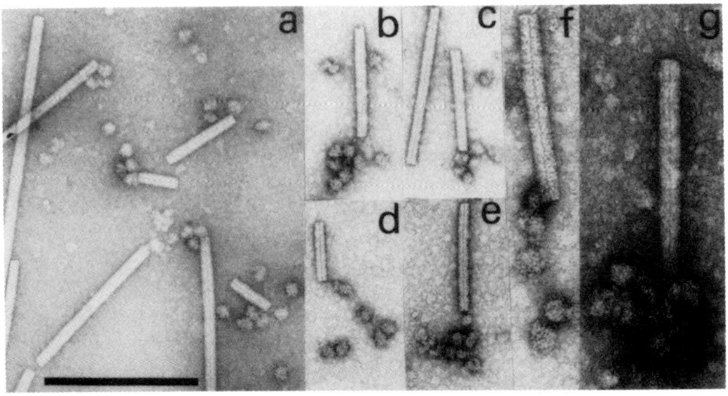

FIGURE 1. Electron micrograph of "striposomes" from a rabbit reticulocyte lysate incubation programmed with pH8-washed TMV particles for 15 min. before cycloheximide treatment (3). Black bar = 300nm at the magnification of panels a-e. Panels f and g were at twice this magnfication.

Several predictions, derived from the hypothesis of cotranslational disassembly were soon confirmed experimentally. First, the rate of polypeptide chain elongation *in vitro* was seen to be reduced from about one amino acid polymerized per second, in incubations conventionally programmed with naked TMV RNA, to approximately one residue per 2-3 seconds with encapsidated RNA templates, as shown in Figure 2 (from ref.3). Second, as the RNA template is exposed sequentially, only during ribosome translocation, one might expect to see higher and more uniform yields of full-length, virus-coded, polypeptides than with corresponding unencapsidated RNA molecules.

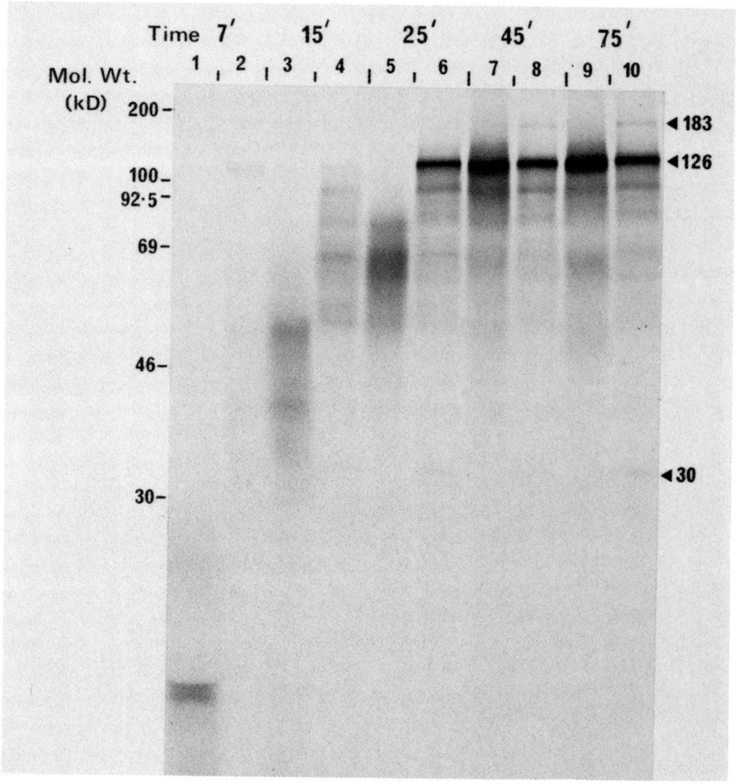

FIGURE 2. Time-course experiment in rabbit reticulocyte lysate incubations programmed with either pH8-washed TMV particles (odd-numbered tracks) or naked TMV RNA (even-numbered tracks), at 10mg/ml or 100µg/ml respectively. Equal amounts of radiolabelled products (10^5 cpm) were loaded on each track. ^{14}C-marker protein positions are shown on the left. TMV RNA encoded products are defined on the right (in kDa).

This would be attributed to the continued protection against ribonucleases of the 3'-terminal, untranslated portion of the viral RNA by viral coat protein. In wheat germ extracts (4) and, more surprisingly, in prokaryotic cell-free translation systems (see below), this prediction was seen to be upheld.

Upon this foundation, we set out to examine the possible involvement of cotranslational disassembly during the natural virus infection process in vivo (5). Three classical reports gave us cause to be optimistic. In 1970, Kiho (6) suggested that partially-uncoated TMV particles were associated with tobacco leaf ribosomes, and in the same year Helms and Zaitlin reported (7) that TMV particles which had lost only 1.7-3.3% of their coat protein from the 5'-end (8) of the RNA were more infectious than untreated virus. They proposed that the pretreatment had overcome the failure of most virus particles to initiate uncoating early during the cell-infection process. The converse phenomenon, that excess free coat protein reduces the specific infectivity of TMV was reported many years ago (9).

In this article we have restricted our results and discussion to the filamentous, positive-strand RNA plant viruses which, for several practical reasons (described below), have proved especially amenable to our experimental approach. Nevertheless, in the broader context of this book and plant virology in general, the reader's attention is drawn to additional data, recently published by Wilson and colleagues (10-12), which suggest that an analogous cotranslational release mechanism may occur with the RNA of preswollen isometric or bacilliform plant virus nucleocapsids, at least in vitro.

RESULTS

Recent Advances in the Cotranslational Disassembly of TMV In Vitro.

To underline the longevity and extreme stability of TMV preparations stored at 4°C, Figure 3A is a recent electron micrograph of an original preparation of TMV made in 1936 by F.C.Bawden and N.W.Pirie (a generous gift from Dr. S.Pierpoint, Rothamsted Experimental Station, Harpenden, Herts, U.K.). In 1986, this virus still produced a normal, severe systemic infection on Nicotiana tabacum cv. White Burley (data not shown) and, when pH8-washed (4) particles were incubated in a rabbit reticulocyte lysate, they were found to be as translationally active as fresh preparations of TMV made this decade (Figure 3B)! This 50-year-old TMV was also used in vivo, in epidermis-inoculation experiments, as described in the appropriate section below.

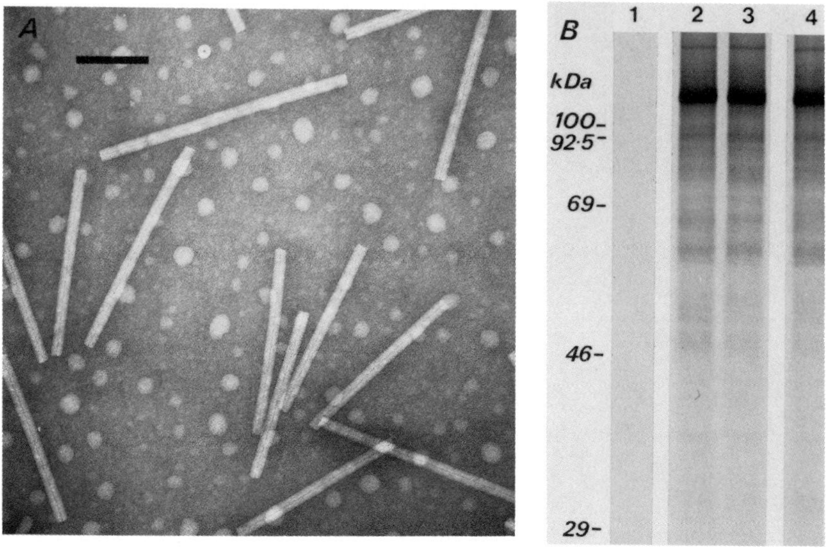

FIGURE 3. A. Electron micrograph (taken January 1986) of a 1936 preparation of TMV (common strain; see text), negatively stained with 1% (w/v) uranyl acetate. Black bar = 100nm. B. In vitro translation products from pH8-washed TMV particles prepared in 1983 (track 2) or 1936 (track 3), or naked TMV RNA from the 1983 virus stock (track 4). Track 1 shows the template-free, control incubation of rabbit reticulocyte lysate. ^{14}C-marker protein positions are shown on the left.

As mentioned above, one prediction from the cotranslational disassembly hypothesis concerned the possible advantage of protecting the 3'-terminal untranslated portion of the RNA, at least until the first ribosome had completed its translocation. This qualitative effect was first observed in wheat germ extracts (4), which are prone to ribonuclease (RNase) "contamination". It was of interest to examine this phenomenon in an even more "RNase-contaminated" cell-free translation system, the 30,000xg supernatant from an Escherichia coli cell lysate. Efficient, but apparently illegitimate, expression of naked TMV RNA in E. coli lysates was first reported over twenty-five years ago (13), and has been a controversial subject ever since (14, 15). Equally controversial and historical have been reports of the association of TMV with the chloroplasts of infected cells (16-18). In vitro translation products from naked or encapsidated TMV RNA after just 20 min. in a 70S-ribosome-based translation system are shown in Figure 4.

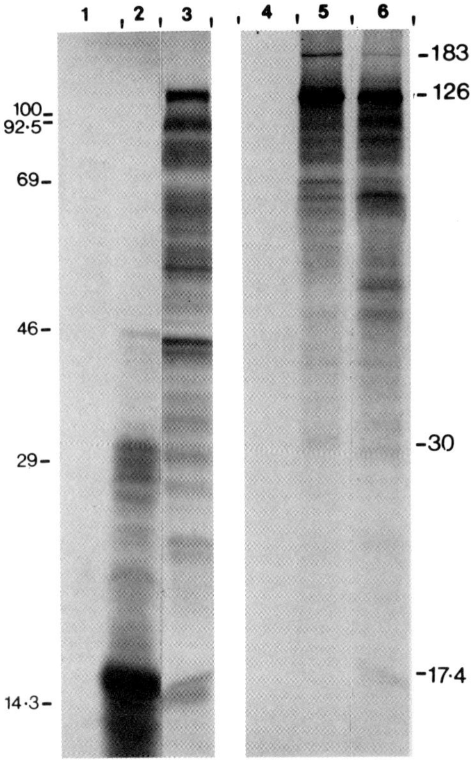

FIGURE 4. L-[^{35}S]-methionine-labelled translation products encoded by naked TMV RNA (100μg/ml; tracks 2,5) or pH8-washed TMV particles (4mg/ml; tracks 3,6) in either an E. coli (tracks 1-3) or rabbit reticulocyte (tracks 4-6) cell-free system (19). Template-free control incubations are shown in tracks 1 and 4. ^{14}C-marker protein positions and significant TMV-encoded products (in kDa) are shown on the left and right respectively.

This completely unexpected result confirmed the protective aspect of the cotranslational disassembly process and also illustrated the very different functional behaviour of the same template when presented either naked or encapsidated. However it also raised a serious question concerning the mechanism of gene expression. The 70 nucleotide-long, 5'-leader sequence for the 126 kilodalton (kDa) gene product of TMV RNA is completely devoid of G residues (20), hence no approximation to the consensus Shine-Dalgarno (21) purine-rich sequence found in prokaryotic mRNAs seems possible.

From the earliest experiments it was noted that some sub-classes of shorter-than-full-length rodlets, containing 3'-coterminal subgenomic RNA species, apparently failed to express efficiently in vitro by cotranslational disassembly. To investigate the cause of this, two strains of TMV were examined (22) for their patterns of encapsidated gene expression. The results (Figure 5) indicated that, at least in vitro, the strong RNA-protein interactions presumed to occur at the origin of assembly sequence inhibited ribosome translocation. The significance of this conclusion for the expression of subgenomic rodlets in vivo and for the mechanism of complete uncoating of the viral RNA beyond residue 4916 [i.e. the 3'-end of the 183kDa readthrough protein gene (20)], during infection, are discussed below.

To date, cotranslational disassembly has not been demonstrated in vitro with filamentous plant viruses for which available evidence suggests a monodirectional self-assembly mechanism [reviewed in (5)] beginning at the 5'-end of the RNA. Potexviruses [potato virus X (PVX), papaya mosaic virus (PMV)], Potyviruses [potato virus Y (PVY), tobacco vein mottling virus (TVMV)] and possibly also Tobraviruses [tobacco rattle virus (TRV), pea early browning virus (PeBV)] fall in this category, and the presumed strength of RNA-protein interactions at the 5'-end of these virus particles may account for our inability [(3); Thornbury and Wilson, unpublished results)] and the failure of others [(23); M.A.Mayo, personal communication], to detect significant levels of gene expression reproducibly in vitro using native or pH8-washed preparations of virions. Clearly these viruses must uncoat in vivo during infection and it is reassuring in part that, despite our failure to mimic these physiological destabilizing conditions in vitro, we have extracted active intermediate translation complexes, with many of the physicochemical properties of striposomes, from PVX-, TVMV-, or TRV-inoculated tobacco leaves (see below).

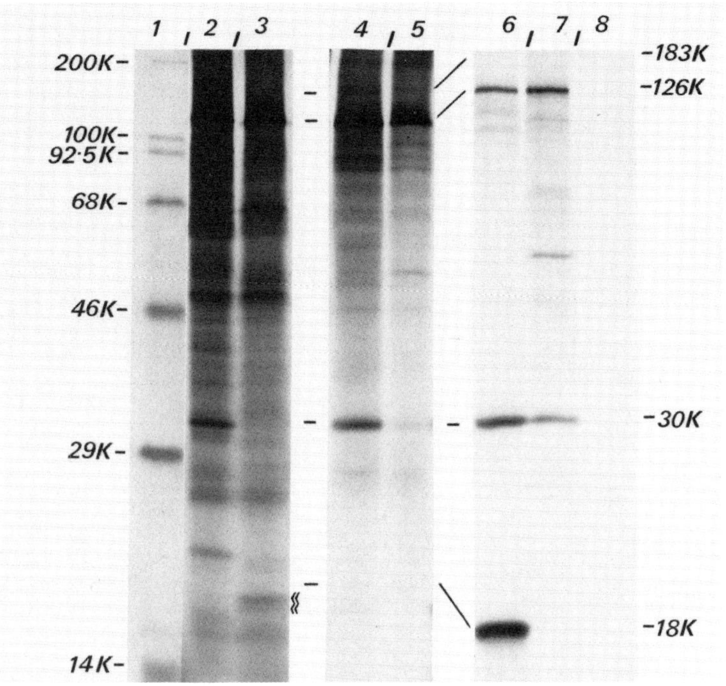

FIGURE 5. L-[^{35}S]-methionine-labelled translation products from rabbit reticulocyte incubations programmed with pH 8.5-washed particles of TMV vulgare (track 3), pH 7.2-washed "long"- (track 5) or "short-rods" (track 7) of the cowpea strain of TMV (Cc) or naked TMV RNAs, phenol-extracted from these same rod(let) preparations (tracks 2, 4 and 6 respectively). Track 8 = template-free control incubation. Track 1 = ^{14}C-marker proteins (sizes given in K= kDa). Strong RNA-protein interactions at the origin-of-assembly sequence apparently block expression of the 30kDa (K) or 18kDa (K) genes in vulgare or Cc TMV (see ref. 22; tracks 3 and 7) respectively.

Following the early work of Bawden and Pirie (9) alluded to above, and in view of our current concept of the effect of pH8-treatment on TMV particles (3), it was of interest to examine the effect of additional free TMV coat protein subunits on the cotranslational disassembly process in vitro. The results were completely unexpected (24) and are summarized in Table 1.

TABLE 1
INHIBITORY EFFECT OF VARIOUS EXOGENOUS PROTEINS ON THE TRANSLATIONAL ACTIVITIES OF SELECTED VIRAL TEMPLATES *IN VITRO*

	Inhibition (%) of total protein synthesis[a]		
Additions (final conc^{s})	TMV RNA (100μg/ml)	pH8-washed TMV particles (2mg/ml)	TRosV RNA (100μg/ml)
2mM Tris-HCl pH 7.5	0	0	0
0.1mg/ml TMVP	17.5	38.0	0
1.0mg/ml TMVP	35.9	63.0	10.9
1.0mg/ml BSA	16.6	4.1	0

[a]Measured as average TCA-insoluble cpm ^{35}S-Met in duplicate 3μl samples from replicate, standard (25μl) rabbit reticulocyte lysate incubations.

The preferential inhibition (approximately 2-fold) of the cotranslational disassembly of pH8-washed virus particles by TMV coat protein, compared with its effect on naked RNA-programmed incubations, adds some support to our views of the "triggering" events surrounding the start of striposome formation (3). More controversially, it lends some credence to one recent hypothesis (25) on the molecular mechanism for cross-protection between viruses in plants. Substantially more work needs to be done in this area, and we are currently examining the effects of various protein fractions, from healthy or TMV-infected leaves, on the cotranslational disassembly process *in vitro*. We believe that our *in vitro* model system provides an amenable and novel approach with which to study this long-standing area of controversy [see refs. in (24)].

Evidence for Cotranslational Disassembly *In Vivo*.

The possibility that a cotranslational disassembly

mechanism might be involved in the early stages of virus infection was investigated.

Preliminary experiments (Shaw and Wilson, unpublished data) with TMV-inoculated tobacco mesophyll protoplasts, or healthy (control) protoplast extracts and TMV, confirmed earlier reports (26) of artifactual virus disassembly in the presence of poly-L-ornithine, membrane phospholipid and Triton X-100. Several additional technical disadvantages led us to abandon this obvious experimental system and adopt the use of peeled strips of TMV-inoculated tobacco leaf (lower) epidermis. This minimized the necessary sample size, the number of uninoculated cells and the amount of chloroplastic material which had proved a serious problem in the protoplast samples. Working with filamentous plant viruses confers two major technical advantages which proved essential for the success of our approach. First, the virus or virus-fragments can be easily identified in the electron microscope. Second, these viruses contain only approx. 5% (w/v) RNA which results in a uniquely low buoyant density, far removed from ribosomes or polysomes (50% (w/v) RNA). Putative striposome complexes, of the type shown in Figure 1, would be expected to have an intermediate density, thus aiding their isolation, purification and characterization.

Double-labelled TMV containing ^{32}P-RNA and ^{3}H-coat protein was prepared, at high specific activity ($3\text{-}4.5 \times 10^5$ dpm ^{32}P and $1\text{-}1.5 \times 10^5$ dpm ^{3}H per μg virus), to follow the fate of the so-called "parental" virus particles (i.e. those derived from the inoculum) within the first hour, or less, following inoculation.

Figure 6 summarizes the results from TMV-inoculated, epidermal cell samples collected 10-40 and 40-70 min. post-infection and subjected to our extraction and fractionation procedure (27). Two new peaks of radiolabelled material, with buoyant densities greater than TMV itself, are seen in both panels A and B. The increased ratio of $^{32}P/^{3}H$ also suggests some degree of uncoating of the parental viral RNA. Control experiments in which double-labelled intact virus (panel C) or partially-uncoated virus (panel D) were added to healthy epidermal cell extracts rule out several trivial explanations for the labelling patterns seen in A and B. Material from zones 1-3 of gradients A and B were collected separately, dialysed and viewed in an electron microscope by negative staining (in 1% (w/v) uranyl acetate) after immune-trapping onto anti-TMV coat protein antibody-coated grids. The results are shown in Figure 7.

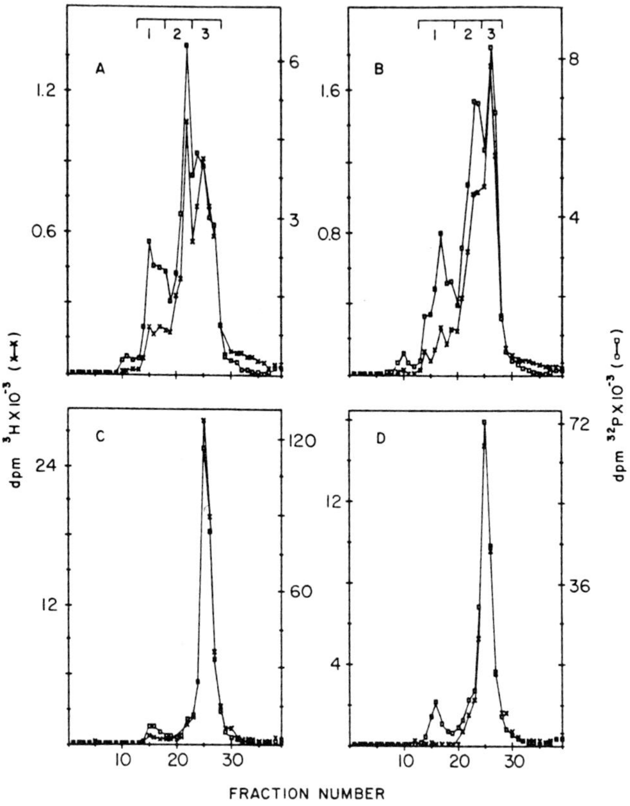

FIGURE 6. Distribution of radioactivity in Cs_2SO_4 gradients layered with a subcellular fraction from tobacco leaf epidermal tissue. Tissue was collected 10-40 (A) or 40-70 (B) min after inoculation with ^{32}P- and ^{3}H-leucine-labelled TMV or after mock-inoculation (C, D). Before extraction 4µg double-labelled TMV (C) or 4µg double-labelled, partially stripped TMV (D) was added to mock-inoculated epidermal samples. After centrifugation, the distribution of radioactivity (X = ^{3}H; □ = ^{32}P) in 50µl aliquots of each 300µl fraction was determined. The right side of each panel represents the top (least dense part) of the gradient. Zones in A and B designated 1, 2 and 3 were collected and examined in the electron microscope (Fig. 7).

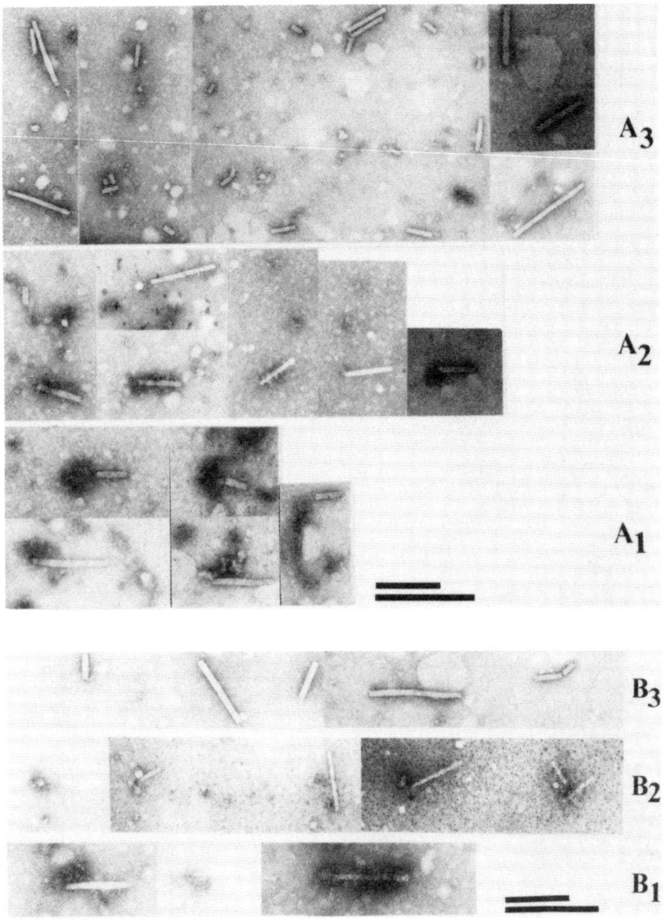

FIGURE 7. Electron micrographs of particles from Cs_2SO_4 gradient zones A1-A3 or B1-B3 presented in Fig. 6. Grids were first coated with anti-TMV antibody and later negatively stained with 1% (w/v) uranyl acetate. Black bars indicate the unit length of TMV (300nm) at the two magnifications used for printing.

As one moves from the least-dense zone 3 (the position of native TMV) to the most-dense zone 1, the electron microscope reveals increasing amounts of cellular components associated predominantly with one end of the clearly distinguishable viral rodlets. We have not yet confirmed, unequivocally, that these components are 80S tobacco ribosomes, but this work is underway. The structural resolution was disappointingly poor and probably results from the harsh salt conditions used during fractionation (see below). We observed a similar reduction in structural detail (c.f. Figure 1) in a control experiment in which ^{35}S-methionine-labelled in vitro striposomes were added to, and processed with, healthy epidermal cell extracts (27). More importantly perhaps, this control experiment revealed two major peaks of ^{35}S-labelled material below the zone of TMV itself in the final Cs_2SO_4 gradient. These peaks were at positions equivalent to zones 1 + 2 in Figure 7 (see ref. 27). Although this result supported our belief that the complexes shown in Figure 7 (panels A1, A2, B1 and B2) were probably "in vivo striposomes", we felt it necessary to demonstrate conclusively that they were also translationally-active. To achieve this, unlabelled virus and relatively large amounts of ^{35}S-methionine (200μCi/leaf) were co-inoculated onto the lower surface of tobacco leaves. Epidermal strips, removed between 15-45 min p.i., were processed as described (27) and the ^{35}S-labelled translation complexes fractionated on Cs_2SO_4 gradients as shown in Figure 8. Peaks designated 1 and 2 co-migrated with authentic in vitro striposomes and were seen to be sensitive to inhibitors of translation such as cycloheximide. These results added further support to the proposed nature of these in vivo complexes containing partially-uncoated TMV particles. This rapid and convenient technique (leaf + ^{35}S-Met + TMV), which avoided the need for double-labelled ($^{32}P/^{3}H$) parental virus, has been used extensively in our most recent experiments, some of which are described below.

Other Filamentous Viruses and New Experimental Plant Systems.

One legitimate criticism of all the work described so far is that, for convenience, it relates only to TMV. Consequently we have recently extended our rapid in vivo approach (Figure 8) to other filamentous plant virus particles. Freshly isolated preparations of TVMV, TRV and PVX were used and in each case (Figure 9) one or more ^{35}S-labelled peaks were observed at a density intermediate between endogenous, labelled tobacco polysomes (the sharper peak in the central panel of Figure 9) and the non-specifically labelled cell

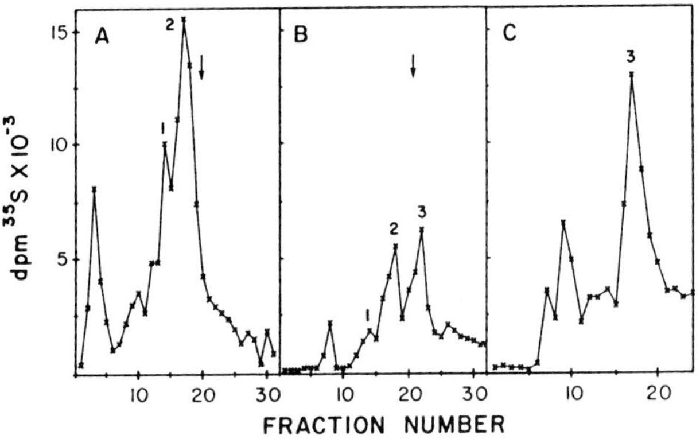

FIGURE 8. Distribution of radioactivity in Cs_2SO_4 gradients layered with subcellular fractions of tobacco epidermal tissue inoculated with unlabelled TMV in the presence of [^{35}S]-methionine (A), or with [^{35}S]-methionine and cycloheximide (B). Leaves were also mock-inoculated with [^{35}S]-methionine (C). The numbers (1, 2 and 3) designate peaks in corresponding positions in different gradients (1 and 2 correspond to putative striposome peaks; see ref. 27). The arrows denote the banding position of intact, free TMV particles. The right side of each diagram is the top of the gradient.

debris (the upper peak in the central panel of Figure 9 and peak 3 in Figures 8B/C). Extrapolating from our results with TMV (Figure 8), we expect that these intermediate-density peaks will reveal analogous striposome-like complexes containing partially-uncoated TVMV, TRV or PVX particles, however this remains to be confirmed. If so, the phenomenon of cotranslational disassembly *in vivo* would seem more widespread than the available *in vitro* data would suggest.

For reasons discussed above, the experimental plant tissue of choice for our work is the infected leaf epidermis. The John Innes Research Institute possesses a mutant line of *Pisum sativum* L., "Argenteum" [JI1397;(28)] so-called because of the silver appearance of its leaves and bracts. This phenotype results from poor attachment of the epidermis to the underlying mesophyll cells, which enables almost the entire epidermis to be removed as a single sheet

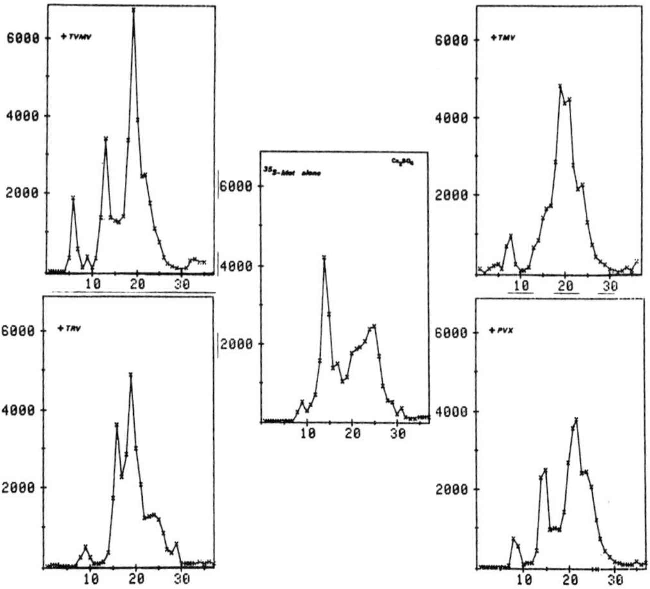

FIGURE 9. Distribution of radioactivity in Cs_2SO_4 gradients layered with subcellular fractions of tobacco epidermal tissue mock-inoculated with [^{35}S]-methionine (centre panel) or inoculated with [^{35}S]-methionine and either unlabelled TMV (top left), TVMV (top right), PVX (bottom left), or TRV (bottom right). Peaks at or below fraction number 14 probably arise from plant polysome material. Peaks at or above fraction number 24 arise by non-specific entrapment of label in low density plant debris. In each case, free virus would band in either fractions 23 or 24. The right side of each diagram is the top of the gradient.

of cells - experimentally, an extremely attractive feature. However *P. sativum* is often considered either completely immune to TMV infection or, at best, is the poorest subliminal host for the virus (29). Nevertheless there are many reports [e.g. (30)(31)] which suggest that host-range classifications of viruses depend upon the late events of virus infection and that the early events, which we are studying, would occur in a much wider variety of cell types. For these reasons, we have performed experiments

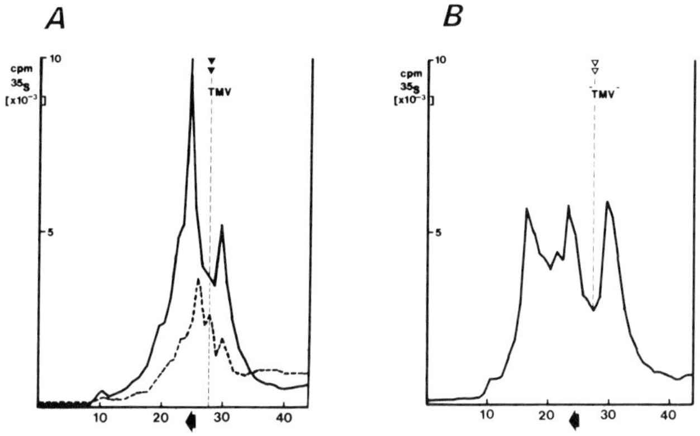

FIGURE 10. Distribution of radioactivity in Cs_2SO_4 gradients layered with subcellular fractions of epidermal tissue from P. sativum L. cv. Argenteum inoculated with unlabelled TMV and ^{35}S-methionine (A) in the presence (broken line) or absence (solid line) of cycloheximide, or mock-inoculated with ^{35}S-methionine alone (B). The real (A) or equivalent (B) banding position of free TMV is marked. The right side of each diagram is the top of the gradient.

analogous to those depicted in Figures 8 and 9, using TMV-inoculated leaves of P. sativum cv. Argenteum (10 leaves/sample). The results are shown in Figure 10 and bear a striking resemblance to those shown in previous Figures, with a TMV-infection-specific peak of ^{35}S-labelled material banding between the unlabelled virus zone and the higher-density plant polysomes. This preliminary result suggests that the early events of infection are host-independent and that, practically, the pea-epidermis may prove a useful model system, particularly for time-course experiments which would require adequate amounts of tissue to be collected within a few minutes, at intervals following inoculation.

In one experiment, using pea-epidermis co-inoculated with ^{35}S-methionine and the 1936-TMV preparation described above, we obtained results directly comparable to those shown in Figure 10A, again confirming the longevity of this virus in storage and its rapid transformation (15-45 min p.i.) into a putative, biologically-active pathogen.

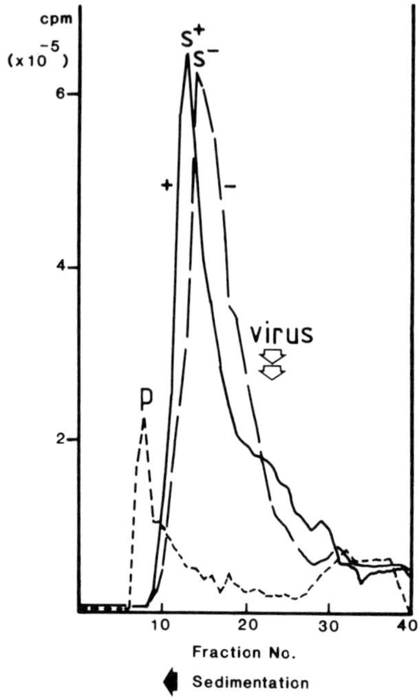

FIGURE 11. Distribution of radioactivity in Nycodenz™ gradients (see text) layered with TMV-specific, ^{35}S-methionine-labelled polyribosomes (P; broken line) or striposomes (S; solid lines) prepared in vitro in rabbit reticulocyte lysate incubations. Striposomes were layered either directly (S^-) or after mixing with an equal volume of S10 (10,000xg supernatant) from homogenized, healthy *N. tabacum* cv. White Burley (S^+). The banding position of free TMV is shown by arrows. The right side of the diagram was the top of each gradient.

We have recently examined the use of isopycnic gradients composed of Nycodenz™ (32) for fractionating our ^{35}S-labelled complexes. In a pilot experiment shown in Figure 11, ^{35}S-methionine-labelled TMV-specific polysomes (P) or striposomes (S) made in rabbit reticulocyte lysates (3) were separated on 5ml Nycodenz gradients. Gradients were prepared from 1.5ml each of 60% (w/v)/50% (w/v)/40% (w/v) Nycodenz in 20mM Tris-HCl, 20mM $MgCl_2$, pH 7.5, and equilibrated at 4°C for 18 hours before centrifugation at

10°C for 16 hours at 44,000 rpm in a Beckman SW50.1 rotor. Addition of a crude 10,000xg supernatant (S10), from Nicotiana tabacum cv. White Burley, to the in vitro striposome sample prior to isopycnic centrifugation had little effect on its banding position. Nycodenz is substantially less chaotropic than Cs_2SO_4 and we therefore anticipate that the structural resolution of striposomes observed in electron micrographs will be improved. If so, Nycodenz will replace Cs_2SO_4 as our fractionation medium for studying in vivo components of the cotranslational disassembly mechanism.

DISCUSSION

Originally and fortuitously observed in vitro, and only with TMV, the phenomenon of cotranslational disassembly appears more widely relevant to other filamentous plant viruses when studied in vivo. A related process seems to occur in vitro with swollen isometric nucleocapsids (10-12) including the animal picornavirus which causes foot-and-mouth disease (Sangar, D.V., AVRI, Pirbright, Surrey, U.K., personal communucation).

We believe that we have detected intermediates in the obligatory uncoating/RNA release mechanism of positive-strand RNA viruses. An intimate association between RNA uncoating and translation seems attractive, simple and desirable from the virus' point-of-view, to assure complete expression of the essential early genes. Our hypthesis is that, following attachment to, or penetration of, the plasmalemma, a short region at the 5'-end of the viral RNA becomes exposed and is sufficient to recruit any available components of the cellular protein synthetic machinery (e.g. cap-binding protein/40S subunits). Ribosome assembly and translocation would follow, resulting in the sequential, stepwise uncoating of the viral RNA. This simple mechanism could account for uncoating of the first continuous coding region in the RNA. In most plant viruses this is a relatively long open-reading frame and is thought to encode either the complete virus-specific RNA-dependent RNA polymerase (replicase) or a subunit thereof. Exactly how any remaining 3'-terminal portions of nucleocapsid are disassembled is open to conjecture. Some colleagues have even proposed a subsequent co-replicational 3'►5' disassembly process!!

Although regions of strong RNA-protein interaction

apparently present a barrier to translocating 80S rabbit reticulocyte ribosomes *in vitro* (22), these regions must disassemble *in vivo*, if only for the complete replication of parental (+) into (-) strand RNA. Filamentous viruses with putative assembly nucleation regions at or near their 5'-termini (e.g. TVMV, PMV, PVX and TRV) failed to initiate cotranslational disassembly *in vitro*, yet they behave like TMV *in vivo* (Figure 9). Examples like this serve to illustrate the gulf between our *in vitro* observations and possible events inside an infected plant cell. Do parental subgenomic rodlets (22) such as I_2 (TMV *vulgare*) or LMC (TMV *Cc*) actually participate in the infection process? If so, they are not essential since purified genome-length rods (or even full-length genomic RNA) alone are sufficient to establish an infection.

We have gone some way towards identifying and speculating upon a novel class of intermediate, virus infection-specific, complex - the striposome. However, one pivotal question remains. Which, if any, of these translationally-active complexes, actually establishes the cellular infection - or do as many virus particles as possible proceed as far as they can along this route?

To characterize these *in vivo* complexes further and to confirm their assumed composition we are currently using immunogold-labelling techniques with antibodies to the nascent polypeptide chains. For TMV itself this requires antiserum to the amino-terminal domains of the 126kDa protein, which unfortunately appears not to be particularly immunogenic (Wilson, unpublished results; Zaitlin, personal communication).

To circumvent this problem; namely that good antisera to the 5'-gene products of most plant viruses are not readily available (since the 5'-gene is usually not the coat protein gene); we have constructed TMV-like, pseudovirus particles. In these, an open-reading-frame for a polypeptide completely alien to plant cells is positioned at the 5'-end of a recombinant RNA molecule which also bears the origin-of-assembly sequence from TMV RNA. Genes for chicken prelysozyme, calf preprochymosin (rennin) and chloramphenicol acetyltransferase have been used since we either have good antisera or a convenient enzymatic assay for their uncoating and expression *in vivo* [Gallie and Wilson, unpublished data; (33)]. We expect that these new tools will further assist in elucidating the cotranslational disassembly phenomenon *in vivo*.

ACKNOWLEDGEMENTS

The authors are grateful to Susan Ballard, Neil Fannin, Paula Podhasky, and Peter Watkins for their excellent technical assistance. We also thank our many colleagues for their advice, encouragement and critical discussions. Special thanks go to Kitty Plaskitt for her patience and skill in dealing with our seemingly endless specimens for the electron microscope.

REFERENCES

1. Shaw JG (1985). Early events in plant virus infections. In Davies JW (ed): "Molecular Plant Virology Vol II", Boca Raton:CRC press, p 1.
2. Wilson TMA (1985). Nucleocapsid disassembly and early gene expression by positive-strand RNA viruses. J gen Virol 66:1201.
3. Wilson TMA (1984). Cotranslational disassembly of tobacco mosaic virus in vitro. Virology 137:255.
4. Wilson TMA (1984). Cotranslational disassembly increases the efficiency of expression of TMV RNA in wheat germ cell-free extracts. Virology 138:353.
5. Wilson TMA, Shaw JG (1985). Does TMV uncoat cotranslationally in vivo? Trends Biochem Sci 10:57.
6. Kiho Y (1970). Polysomes containing infecting viral genome in tobacco leaves infected with tobacco mosaic virus. Japan J Microbiol 14:291.
7. Helms K, Zaitlin M (1970). Enhancement of infectivity of tobacco mosaic virus particles partially uncoated by alkali. Virology 41:549.
8. Perham RN, Wilson TMA (1976). The polarity of stripping of coat protein subunits from the RNA in tobacco mosaic virus under alkaline conditions. FEBS Lett 62:11.
9. Bawden FC, Pirie NW (1957). The activity of fragmented and reassembled tobacco mosaic virus. J gen Microbiol 17:80.
10. Brisco MJ, Hull R, Wilson TMA (1985). Southern bean mosaic virus-specific proteins are synthesized in an in vitro system supplemented with intact, treated virions. Virology 143:392.
11. Brisco MJ, Hull R, Wilson TMA (1986). Swelling of isometric and of bacilliform plant virus nucleocapsids is required for virus-specific protein synthesis in vitro. Virology 148:210.

12. Brisco MJ, Haniff C, Hull R, Wilson TMA, Sattelle DB (1986). The kinetics of swelling of southern bean mosaic virus: a study using photon correlation spectroscopy. Virology 148:218.
13. Nirenberg MW, Matthaei JH (1961). The dependence of cell-free protein synthesis in E. coli upon naturally occurring or synthetic polyribonucleotides. Proc Natl Acad Sci USA 47:1588.
14. Glover JF, Wilson TMA (1982). Efficient translation of the coat protein cistron of tobacco mosaic virus in a cell-free system from Escherichia coli. Eur J Biochem 122:485.
15. Asselin A, Robitaille G, Bellemare G (1985). A note on the translation of cowpea strain tobacco mosaic virus RNA in a prokaryotic cell-free system and possible implications concerning tobamovirus origin. Phytoprotection 66:59.
16. Shalla TA, Peterson LJ, Giunchedi L (1975). Partial characterization of virus-like particles in chloroplasts of plants infected with the U5 strain of TMV. Virology 66:94.
17. Siegel A (1971). Pseudovirions in tobacco mosaic virus. Virology 46:50.
18. Rochon D'A, Siegel A (1984). Chloroplast DNA transcripts are encapsidated by tobacco mosaic virus coat protein. Proc Natl Acad Sci USA 81:1719.
19. Wilson TMA (1986). Expression of the large 5'-proximal cistron of tobacco mosaic virus by 70S ribosomes during cotranslational disassembly in a prokaryotic cell-free system. Virology in press.
20. Goelet P, Lomonossoff GP, Butler PJG, Akam ME, Gait MJ, Karn J (1982). Nucleotide sequence of tobacco mosaic virus RNA. Proc Natl Acad Sci USA 79:5818.
21. Shine J, Dalgarno L (1975). Determinant of cistron specificity in bacterial ribosomes. Nature 254:34.
22. Wilson TMA, Watkins PAC (1985). Cotranslational disassembly of a cowpea strain (Cc) of TMV: evidence that viral RNA-protein interactions at the assembly origin block ribosome translocation in vitro. Virology 145:346.
23. Bendena WG, Abouhaidar M, Mackie GA (1985). Synthesis in vitro of the coat protein of papaya mosaic virus. Virology 140:257.
24. Wilson TMA, Watkins PAC (1985). Influence of exogenous viral coat protein on the cotranslational disassembly of tobacco mosaic virus (TMV) particles in vitro. Virology 149:132.

25. Sherwood JL, Fulton RW (1982). The specific involvement of coat protein in tobacco mosaic virus cross protection. Virology 119:150.
26. Kiho Y, Abe T (1980). Modification of tobacco mosaic virus by polyornithine and lecithin. Microbiol Immunol 24:617.
27. Shaw JG, Plaskitt KA, Wilson TMA (1986). Evidence that tobacco mosaic virus particles disassemble cotranslationally in vivo. Virology 148:326.
28. Marx GA (1982). Argenteum mutant of Pisum: genetic control and breeding behaviour. J Hered 73:413.
29. Cheo PC, Gerard JS (1971). Differences in virus-replicating capacity among plant species inoculated with tobacco mosaic virus. Phytopathology 61:1010.
30. Kiho Y, Machida H, Oshima N (1972). Mechanism determining the host specificity of tobacco mosaic virus. I. Formation of polysomes containing infecting viral genome in various plants. Japan J Microbiol 16:451.
31. Sulzinski MA, Zaitlin M (1982). Tobacco mosaic virus replication in resistant and susceptible plants: in some resistant species virus is confined to a small number of initially infected cells. Virology 121:12.
32. Gugerli P (1984). Isopycnic centrifugation of plant viruses in Nycodenz^R density gradients. J Virol Meth 9:249.
33. Sleat DE, Wilson TMA, Turner PC, Finch JT, Butler PJG (1986). Packaging of recombinant RNA molecules into pseudovirus particles directed by the origin-of-assembly sequence from tobacco mosaic virus RNA. EMBO J, submitted.

Positive Strand RNA Viruses, pages 183–191

Sindbis virus structural genes are expressed in *Saccharomyces Cerevisiae*

Milton J. Schlesinger, Duanzhi Wen and Mingxaio Ding
Department of Microbiology and Immunology
Washington University School of Medicine
St. Louis, MO 63110

ABSTRACT A cDNA containing the genes encoding the three structural proteins of Sindbis virus has been inserted into a yeast *E. coli* shuttle vector such that expression of the virus genes is regulated by the yeast GAL-1 promoter. Cells carrying this vector form virus structural proteins only when cells are grown in galactose-inducing media. The virus polyprotein is proteolytically processed to yield capsid and glycosylated forms of the envelope glycoproteins E1 and p62 - the precursor to virion E2. The glycoproteins form high molecular weight disulfide-linked aggregates in the yeast ER and are not transported to the yeast surface membrane. Neither glycoprotein is acylated with fatty acid and very little p62 is processed to E2. Aberrant forms of the glycoproteins are detected; these appear to be the result of incomplete processing. Cells are arrested in growth during the time of virus protein induction. About 2-3% of yeast protein synthesis appears as virus protein and the glycoproteins can be rapidly obtained in 50% pure form by high speed centrifugation of extracts prepared with SDS in the absence of reducing agents.

Supported by a grant from the Public Health Service, A119454.

The three genes encoding Sindbis virus structural proteins (capsid, E1 and E2) are clustered at the 3' end of the virion RNA but are expressed only from a subgenomic mRNA of 4108 nucleotides (1). This mRNA (noted 26S) has a single site for initiating translation and the three gene products arise by proteolytic cleavages that occur during polypeptide chain synthesis. A cDNA containing the entire Sindbis virus genome (11.7 kb) has been constructed by C.M. Rice and H. Huang. We were interested in expressing the 26S mRNA sequences (the structural genes) in *S. cerevisiae* and especially in ts mutants that were isolated and characterized in R. Scheckman's lab (2). Rice and Huang kindly supplied the virus cDNA which we truncated so that only 26S RNA sequences were present and the first ATG of the clone was the true initiation start site for translation of the polyprotein. This cDNA fragment was inserted immediately 3' to a yeast GAL-1 promoter so that expression would only occur in cells grown in galactose (3). A yeast alcohol dehydrogenase transcription termination sequence (kindly supplied by Dr. Valenzuela, Chiron Corp.) was placed at the 3' end of 26S cDNA. This set of sequences was then placed in the high-copy plasmid pS1-4 which contains 2μ elements, a leaky leu 2 gene and *E. coli* pBR322 sequences (ampicillin resistance and origin of replication). A Leu 2^- strain of yeast was transformed with this plasmid (noted pYMS-2, Fig. 1) and prototrophs were selected. A colony was tested for expression of the virus proteins when cells were grown on galactose. Antivirus antibodies were employed to isolate ^{35}S-labeled virus proteins and these immuneprecipitates were analyzed by SDS/PAGE (Fig. 2). Only cells containing the plasmid and grown in galactose produced the proteins which accounted for 2 to 3% of the total yeast protein synthesized. We used anticapsid, anti-E1 and anti-E2 to show that each of the proteins were made. The capsid appeared in its normal 33 kDa subunit form, indicating that the capsid autoprotease activity functioned in yeast. Very little E2 was detected; instead, its precursor p62 accumulated (Fig. 2). Both p62 and E1 were glycosylated but were not fatty acylated. We showed the former by noting an increase in protein mobility in SDS/PAGE after endoglycosidase F treatment and the latter by labeling with ^{3}H-palmitic acid under conditions that

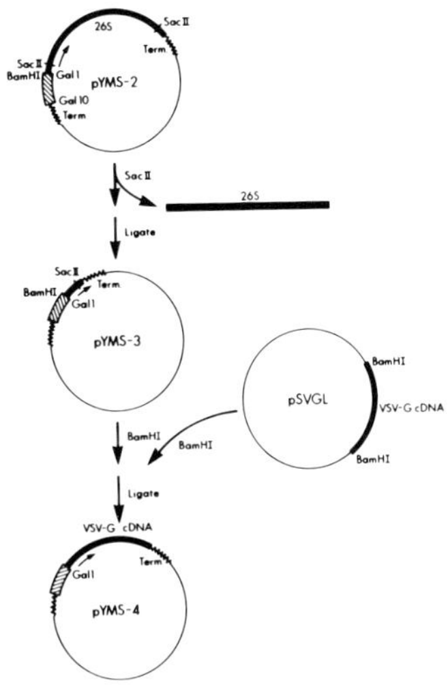

FIGURE 1. The yeast expression vector. The methods used for constructing these plasmids are in ref. 6.

effectively acylate several yeast glycoproteins (4). Two other virus polypeptides were formed in significant amounts in yeast and, based on their reactivity to the antibodies, these were postulated to be non-glycosylated p62-6K-E1 "readthrough" protein and a p62-6K polypeptide. The former would appear if the p62 signal sequence functioned poorly; the latter would form if the E1 signal sequence in 6K functioned better than that of p62. Neither p62 nor E1 were detected in substantial amounts at the yeast surface membrane - based on a surface iodination assay, and they were not found in high mannose forms - a type of oligosaccharide modification occurring in the yeast Golgi (5). This latter result may indicate that the glycoproteins do not move beyond the endoplasmic reticulum. However, another virus glycoprotein - the G of vesicular stomatitis virus - was expressed from a cDNA in these same yeast strains and was transported through the yeast secretory pathway without addition of high mannan structures (6).

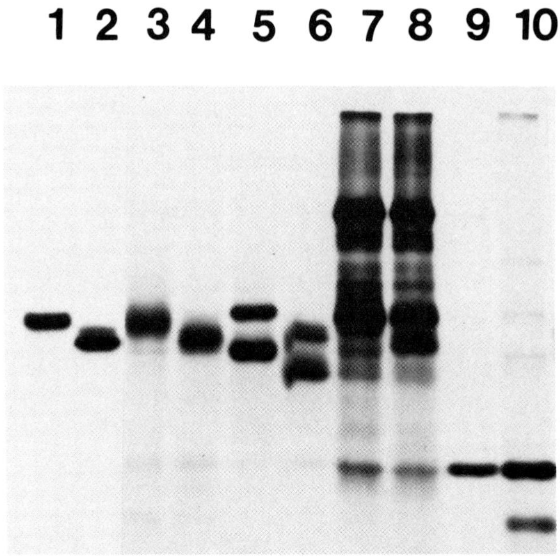

FIGURE 2. Expression of Sindbis virus proteins in yeast. Yeast cells carrying the pYMS-2 were grown to log phase in glucose medium, shifted to glycerol for overnight culture, and then placed in 5% galactose medium. Cultures were labeled with radioactive methionine for 4 hrs. Extracts were prepared by vortexing washed cells with glass beads in 1% SDS. Samples were diluted with "RIPA" buffer and immunoprecipitated with specific anti-virus protein antibodies. Precipitates were treated with Endo F for 18 hrs. 37°C and were analyzed by SDS/PAGE (see ref. 6 for experimental details). Lanes 1, Lysate from infected chicken embryo fibroblasts - anti-E-1 was used for immunoprecipitation; 2. Samples of (1) treated with Endo F; 3. Yeast extract - anti E1 was used for immunoprecipitation; 4. Sample of (3) treated with Endo F; 5. Sample of (1) but anti-E2 was used for immunoprecipitation; 6. Sample of (5) treated with Endo F; 7. Yeast extract - anti-E2 was used for immunoprecipitation; 8. Sample of (7) treated with Endo F; 9. Sample of (1) but anti-capsid was used for immunoprecipitation; 10. Yeast extract - anti-capsid was used for immunoprecipitation.

Production of the foreign virus proteins as a result of galactose induction inhibited cell growth (Fig. 3) and when growth did resume at 36 to 48 hr post induction, the expression of virus cDNA had stopped. We have not examined these cells to determine what type of "mutation" might have allowed a clone to grow out.

We next examined procedures for extracting the virus proteins from yeast and discovered that the glycoproteins formed high molecular weight disulfide linked aggregates (Fig. 4). When synthesized in infected vertebrate and insect cells, p62 and E1 glycoproteins make only intra-molecular disulfide bonds and form a weakly-interacting heterodimer shortly after

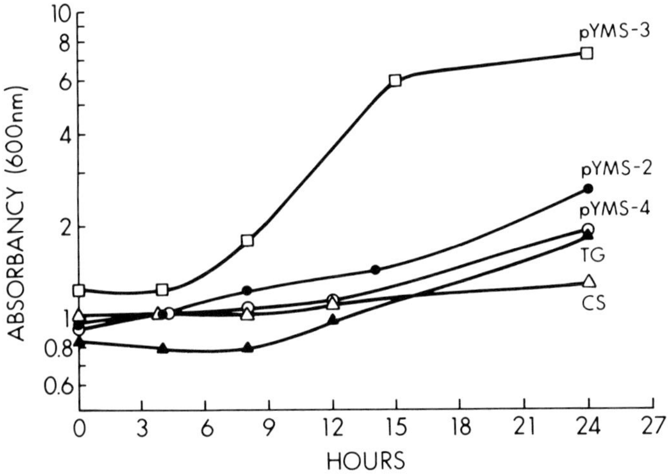

FIGURE 3. Growth of yeast strains expressing virus proteins. Strains carrying the different plasmids were resuspended in galactose medium and their growth followed by absorbancy at 600 nm. pYMS-2 carries the Sindbis genes; pYMS-3 carried no virus genes; pYMS-4 carries the intact VSV G cDNA; TG is a cDNA of VSV G lacking the membrane anchoring sequences; CS is a cDNA of VSV G in which a cysteine in the cytoplasmic domain of the protein has been mutagenized to a serine. Experimental details are in ref. 6.

synthesis. They then move through the Golgi apparatus where p62 is proteolytically cleaved to E2, leading to a much more stable E1.E2 heterodimer (7). The

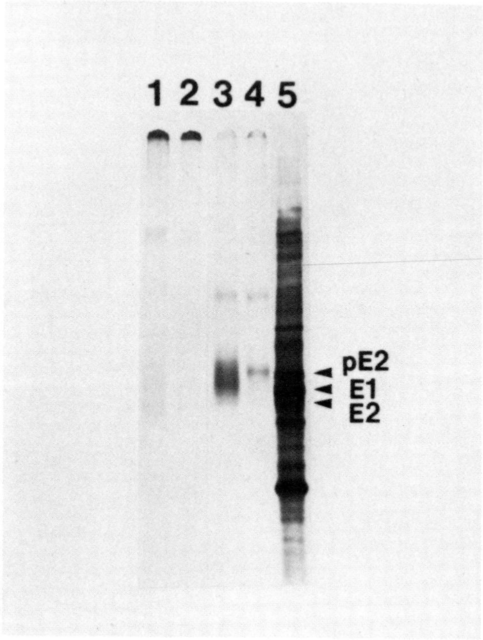

FIGURE 4. Sindbis virus proteins form disulfide-linked oligomers in yeast. Extracts of induced yeast carrying the pYMS-2 plasmid were immunoprecipitated and analyzed by SDS/PAGE (see ref. 8 for experimental details). Lanes 1. Anti-E1 was used and immuneprecipitates were treated with gel-loading buffer in the absence of mercaptoethanol; 2. As in (1) but anti-E2 antibodies were used; 3. As in (1)) but gel-loading buffer contained mercaptoethanol; 4. As in (2) but gel-loading buffer contained mercaptoethanol; 5. Extract of CEF-infected cells, no antibodies were used; but gel-loading buffer did not have mercaptoethanol.

Note: After isolating and reducing the radioactive band from the top of the gel in lanes 1 or 2, all the radioactivity appeared in the bands noted in lanes 3 and 4, respectively.

stoichiometry between E1 and p62 in the yeast system has not been measured, although there appears to be much more glycosylated E1 than p62.

The formation of oligomers in yeast cells allows for a rapid purification step in which the bulk of the yeast proteins are solubilized by hot SDS. The virus aggregates remain insoluble and are separable by brief high speed centrifugation. They are subsequently solubilized by treatment with sulfhydryl-reducing reagents. We are interested in testing these purified preparations as immunogens and for their ability to be renatured and analyzed with regard to their interactions with lipid bilayers. The capsid does not appear in the high molecular-weight disulfide-linked aggregates; this protein has no cysteines. In virus-infected cells capsid has never been detected "free" but only in nucleocapsids bound to virion RNA or, transiently, bound to ribosomes. What form capsid takes in these yeast cells is currently under investigation. This expression system can be further engineered to induce viral RNAs and these could be tested for their binding to capsid.

We do not understand why the virus proteins are not acylated with fatty acids in the yeast cell. Perhaps the abnormal conformation and aggregation of newly made polypeptides block acylation sites. Nothing is known about the yeast glycoproteins that are acylated - but the final acyl-proteins have a stable O-ester linkage. In contrast most of the virus glycoproteins are acylated to yield thioester bonds and the yeast acyl transferases may not recognize a virus transmembranal glycoprotein as substrate.

The E1 and p62 glycoproteins were stable in yeast cells - based on a 15 min pulse-3 hour chase labeling experiment; however, the aberrantly processed forms were unstable.

Thus far we have studied in detail the expression of one other virus glycoprotein in this yeast expression system. cDNA clones encoding the G protein of vesicular stomatitis virus were inserted into the shuttle-vector (pYMS-3) after removal of the Sindbis sequences. Levels of glycosylated G, similar to those of Sindbis virus, were detected in the strains induced with galactose but not in cells grown on glycerol. The G appeared also as disulfide-bonded oligomers with half in a 270,000 mol. wt form (8). In contrast to Sindbis virus glycoproteins, half of G was detected by surface labeling with radiolabeled iodine. No high mannose structures were found on G and no fatty acylation occurred. Production of G arrested cell growth. Forms of G lacking the membrane domain at the carboxyl end of the protein - or with a serine substituted for cysteine in the cytoplasmic domain were induced in yeast. Both produced disulfide-linked oligomers and both arrested growth.

A cDNA coding for the G protein of respiratory syncytial virus is currently being tested in our yeast system. This glycoprotein is distinct from the others in that it appears to have a large amount of O-glycosylation (9). Yeast carry out O-glycosylation so it will be interesting to see if this foreign protein is processed in yeast.

ACKNOWLEDGEMENTS

We gratefully acknowledge cDNA clones provided by M. Johnston, H. Huang and C. Rice (Washington University), J. Rose (Salk Institute) and P. Valenzuela (Chiron Corporation) and antibodies provided by C. Rice (Washington University) and A. Schmaljohn (Frederick, Cancer Center).

REFERENCES

1. Strauss EC, Strauss JH (1986). Structure and replication of the alphavirus genome. In Schlesinger S, Schlesinger MJ (ed): "The Togaviridae and Flaviviridae," New York: Elsevier Biomedical, p 35.
2. Novick P, Field C, Schekman R (1980). Identification of 23 complementation groups required for post-translational events in the yeast secretory pathway. Cell 21: 205.
3. Johnston M, Davis RW (1984). Sequences that regulate the divergent GAL-1-GAL-10 promoter in *Saccharomyces cerevisiae*. Mol Cell Biol 4: 1440.
4. Wen D, Schlesinger MJ (1984). Fatty-acid acylated proteins in secretory mutants of *Saccharomyces cerevisiae*. Mol Cell Biol 4: 688.

5. Schekman R (1982). The secretory pathway in yeast. Trends in Biochem Sci 7: 243.
6. Wen D, Schlesinger MJ (1986). Regulated expression of Sindbis and vesicular stomatitis virus glycopropteins in S. cerevisiae. Proc Natl Acad Sci USA 83: (in press).
7. Rice CM, Strauss JH (1982). Association of Sindbis virion glycoproteins and their precursors. J Mol Biol 154: 325.
8. Wen D, Ding M, Schlesinger MJ (1986) Expression of genes encoding vesicular stomatitis and Sindbis virus glycoproteins in yeast leads to formation of disulfide-linked oligomers. Virology
9. Wertz GW, Collins PL, Huang Y, Gruber C, Levine S, Ball LA (1985). Nucleotide sequence of the G protein gene of human respiratory syncytial virus reveals an unusual type of viral membrane protein. Proc Natl Acad Sci USA 82: 4075.

IV. PROTEOLYTIC PROCESSING OF VIRAL PROTEINS

Positive Strand RNA Viruses, pages 195–207

POLYPROTEIN PROCESSING IN THE EXPRESSION OF THE GENOME OF COWPEA MOSAIC VIRUS

Ab van Kammen, Martine Jaegle, Pieter Vos, Joan Wellink and Rob Goldbach

Department of Molecular Biology, Agricultural University, De Dreijen 11, 6703 BC Wageningen, The Netherlands

ABSTRACT The polyproteins encoded by CPMV B RNA and M RNA contain three types of cleavage sites, Gln-Ser, Gln-Gly and Gln-Met. Cleavage at the Gln-Met sites is achieved by a 32K protease and those at Gln-Ser and Gln-Gly sites by a 24K protease, both encoded by B RNA. VPg is released from the B-RNA encoded polyprotein at a Gln-Ser site at its N- and a Gln-Met site at its C-terminus and is found to be linked by its N-terminal Ser residue to the 5'ends of the viral RNAs. The existence of M-RNA encoded 48K protein, the 60K capsid protein precursor and B-RNA encoded 24K protein, postulated before on basis of _in vitro_ experiments have been demonstrated _in vivo_. These results supplement earlier data and provide a detailed picture of the processing of the CPMV polyproteins.

INTRODUCTION

The genome of cowpea mosaic virus (CPMV) consists of two positive-sense RNA molecules that are encapsidated in separate icosahedral particles each with a diameter of 28 nm (for recent reviews see 1. and 2.). The two nucleoprotein particles referred to as B and M components have similar capsids built up of 60 copies of each of two different coat proteins, but differ in nucleic acid content. B components contain a single RNA molecule (B-RNA) with a mol.weight of

2.04×10^6 and M components a RNA molecule (M-RNA) with a mol.weight of 1.22×10^6. Both RNAs are required for infectivity in plants but B-RNA can support its own replication in protoplasts.

In the main, the structure and organization of the two genomic RNAs of CPMV is similar (1, 2). Each RNA has a small protein, VPg, covalently linked to the 5'end and a poly(A) tail at the 3'end. The sequence of B RNA, 5889 nucleotides excluding the polyA tail, contains a single open reading frame 5598 nucleotides long preceded by a non-translated leader sequence of 206 nucleotides, while the 3'-end non-translated region consists of 85 nucleotides. In M-RNA the open reading frame is 3138 nucleotides, the non-translated leader 160 nucleotides and the 3'-end non-translated region 183 nucleotides long.

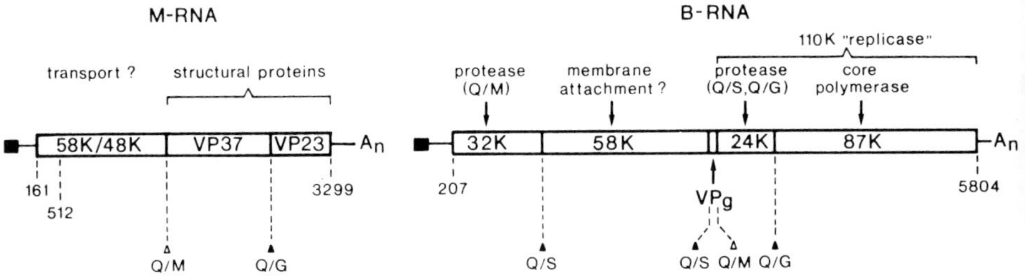

FIGURE 1. Genetic map of CPMV.
The open bars indicate the single open reading frames in M-RNA and B-RNA. The cleavage sites in the polyproteins are marked by open and closed triangles. Translation of M-RNA in vitro starts at the AUG codon at position 161, but more efficiently at the AUG codon at position 512.

STRATEGY OF THE BIPARTITE GENOME OF CPMV

Both CPMV RNAs are translated in vitro and in vivo into large polyproteins that are subsequently cleaved through a number of steps into smaller functional proteins. B-RNA is translated into a single 200K (= kilodalton) protein, which generates upon processing five final proteins in the order NH_2-32K-58K-4K(= VPg)-24K-87K-COOH (Fig. 1). M-RNA is translated into two overlapping proteins of approximately 105K and 95K, which are processed to give the two capsid proteins, VP37 and VP23, together with proteins of 58K and 48K in the order NH_2-58K/48K-VP37-VP23-COOH (Fig. 1).

To most of the viral coded proteins functions have now been ascribed although some of these functions are yet ill-defined. Since B-RNA replicates independently from M-RNA B-RNA must encode information for viral RNA replication. Although B-RNA can dispense with M-RNA for replication, it cannot move to adjacent cells in plants in the absence of M-RNA. Therefore it has been suggested that M-RNA encodes in addition to the two capsid proteins, a protein that effects cell-to-cell transport. The 58K and/or 48K proteins (Fig. 1) are possible candidates for such a function.

Replication of CPMV RNA in vivo is associated with vesicular membrane structures the proliferation of which is induced upon virus infection. From membrane fractions of infected cells containing these cytopathic structures a CPMV-RNA replication complex has been isolated and purified. Highly purified viral RNA replication complex contains three major polypeptides of 110K, 68K and 57K of which the 68K and 57K proteins are probably host-encoded whereas the 110K polypeptide represents a B-RNA encoded protein containing the sequences of both the 24K and 87K B-RNA encoded proteins. The core RNA polymerase activity resides in the 87K polypeptide chain, based on the significant (more than 20%) amino acid sequence homology between this protein and the RNA polymerases from animal picornaviruses. Since RNA replication appears to take place at membranes other B-RNA encoded proteins may be important in anchoring the replication complex to the plant membranes. The B-RNA encoded 58K protein and its direct 60K precursor which also contains the sequence of VPg, are the most distinct membrane proteins and may therefore be candidates for such a function. The release of VPg from the 60K protein may be an important event in the initiation of viral RNA replication where it may play a role in priming RNA replication. A role of VPg or of the generation of VPg in viral RNA replication is substantiated by the finding that complementary minus RNA strands in replicative form RNA are also provided with VPg linked to their 5' ends (3). In addition B-RNA encodes two different proteolytic activities, located in the 32K and 24K proteins which together accomplish all cleavages in the viral polyproteins. In the following sections of this paper recent results on the cleavage specificity of these two proteases and their role in the processing will be discussed.

The genome strategy of CPMV is strikingly similar to that of animal picornaviruses (Figure 2). Apart from their host range and other biological properties the major difference between CPMV and animal picornaviruses is that in CPMV the genome is divided in two RNA molecules whereas that in picornaviruses is a single RNA molecule. Not only have CPMV and picornaviruses a VPg and poly(A) tail and produce

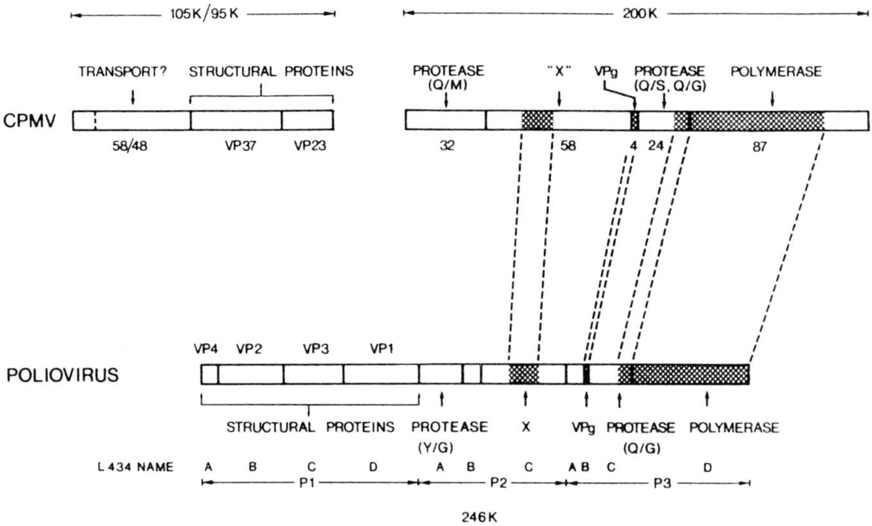

FIGURE 2. Comparison of the functional domains in the polyproteins encoded by CPMV and poliovirus. In addition to a striking analogy in the positions of corresponding functions, there are three regions (hatched boxes) in the 200K B-RNA encoded polyprotein with > 20% sequence homology to regions in the P2 and P3 parts of the poliovirus polyprotein.

functional polypeptides from polyprotein precursors but both viruses encode proteins with similar functions that are found on similar positions in the genetic maps.

Moreover there is significant sequence homology between four non-structural proteins encoded by CPMV B RNA, (58K, VPg, 24K and 87K) and corresponding non-structural proteins of picornaviruses with presumably similar functions in viral RNA replication (4). The relationship between CPMV and picornaviruses is further enhanced by the recent finding that after CPMV, picornaviruses also encode a second protease (5) (protein 2A in the case of poliovirus) the position of which on the genetic map of poliovirus notably corresponds with the position of the 32K protease on the genetic map of CPMV (Fig. 2).

PROTEOLYTIC CLEAVAGE SITES IN THE POLYPROTEINS ENCODED BY B- AND M-RNA

To gain deeper understanding of the proteolytic processing of the B-RNA encoded polyprotein Wellink et al. (6) have recently determined the cleavage sites in the primary 200K translation product of B-RNA. This was achieved by determining partial amino-terminal sequences of the various B-RNA-encoded proteins whereupon the coding regions for these proteins could be located on the B-RNA sequence. Previously Zabel et al. (7) had already mapped the amino-terminal end of VPg on the open reading frame of B-RNA. These sequence analyses revealed the presence of three types of proteolytic cleavage sites in the 200K polyprotein: a glutamine-serine pair (2x), a glutamine-methionine pair (1x) and a glutamine-glycine pair (1x) (Table 1). The latter two sites also occur in the overlapping 105K and 95K polyproteins translated from M-RNA (8).

The polyproteins encoded by B-RNA and M-RNA therefore contain three different types of cleavage sites (Table 1). All cleavage sites share a glutamine residue in the first position while the residue in the second position have varying properties. Methionine has a rather large, non-polar side chain whereas serine and glycine both have small polar

TABLE 1. Amino acids surrounding the proteolytic cleavage sites in the polyproteins encoded by CPMV B- and M-RNA.

Cleavage sites[a)]	Amino acid position							
	-4	-3	-2	-1	1	2	3	4
in B polyprotein:								
32K - 58K	asp	asn	ala	gln	ser	ser	pro	val
58K - VPg	ala	glu	pro	gln	ser	arg	lys	pro
VPg - 24K	ala	asp	ala	gln	met	ser	leu	asp
24K - 87K	ala	gln	ala	gln	gly	ala	glu	glu
in M polyproteins:								
58K - VP37	ala	phe	pro	gln	met	glu	gln	asn
VP37 - VP23	ala	ile	ala	gln	gly	pro	val	cys

Three types of cleavage sites are present in CPMV polyproteins. Gln-Ser is unique for the B-polyprotein. Gln-Met and Gln-Gly sites occur both in B- and M-polyproteins. Five out of six cleavage sites have Ala in position -4. There are 11 Gln-Gly pairs, 7 Gln-Ser pairs and 4 Gln-Met pairs in the polyproteins which are not cleaved. Secondary and tertiary structure are probably important in determining the sites used for proteolytic processing.

side chains. For that reason it seems unlikely that a single protease is responsible for all three types of cleavages. Indeed the available evidence indicates that two different proteases, both encoded by B-RNA are involved in the cleavages. Previously we (9) have proposed that the 32K protein bears the proteolytic activity for cleaving at glutamine-methionine. Direct proof for the 24K protein bearing the second proteolytic activity involved in the cleavages at glutamine-serine and glutamine-glycine, has recently been provided (11, 12) and will be discussed furtheron in this paper.

THE LINKAGE OF VPg TO THE 5'END OF B-RNA AND M-RNA

The determination of the cleavage sites in the 200K polyprotein encoded by B-RNA implies that VPg is released from the precursor as a polypeptide with a length of 28 amino acids. Even though its occurrence at the 5'end is one of the striking features of the structure of the genomic RNAs the function of VPg has not been established and is still subject of speculative discussion. In view of the genetic relationship between CPMV and picornaviruses as outlined above it appears plausible that VPg of CPMV, as for picornaviruses, is somehow involved in the initiation of viral RNA replication, possibly as protein primer. To gain further insight in the role of VPg of CPMV, the linkage of VPg to the virus RNAs has recently been determined by our group (13). This linkage as well as the primary structure of VPg is shown in Figure 3.

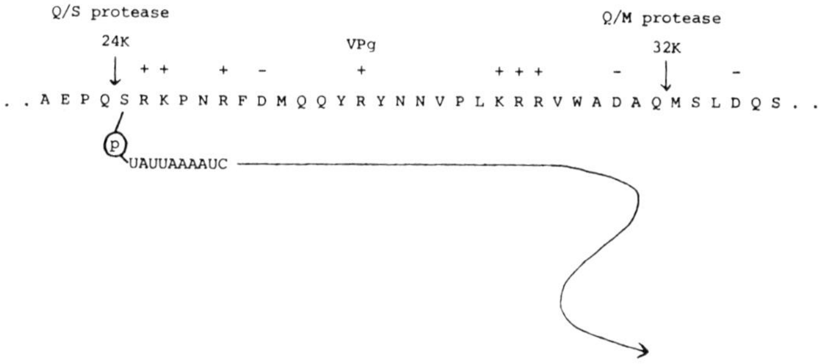

FIGURE 3. VPg consisting of a chain of 28 amino acid residues is released from its precursors by cleavages at a Gln-Ser and a Gln-Met site. VPg is linked to the 5'end of the CPMV RNAs by a phosphodiesterlinkage with the side-chain of the amino terminal serine.

Wherease in picornaviruses VPg is linked to the RNA by a phosphodiester bond with the OH group of a tyrosine residue in the third position from the amino terminus, VPg of CPMV

RNA is linked by the amino terminal serine residue. As in both cases an amino acid residue close to the amino terminal end is involved in the linkage it is suggested that the cleavage of VPg from its precursor and the formation of the phosphodiester linkage with RNA are connected.

DETECTION OF VIRAL PROTEINS IN INFECTED PROTOPLASTS

In section 2 we have described that M-RNA is translated in vitro into two overlapping 105K and 95K polypeptides which are cleaved into products of 58K, 48K and the two capsid proteins. In vivo the capsid proteins are the only M-RNA encoded products readily detectable. We have searched for the in vivo occurrence of 58K, 48K as well as for the 60K precursor to both capsid proteins to establish the model derived from the in vitro observations. Indeed it has now been demonstrated by Western blotting using antiserum against the VP23 capsid protein that the 60K capsid precursor is detectable in the S30 supernatant of CPMV infected cowpea protoplasts incubated in the presence of Zn^{++}.

To investigate whether the M-RNA encoded 58K and 48K non-capsid proteins are produced in vivo an octapeptide with a sequence corresponding to that of the common carboxy termini of both proteins was synthesized. Using antibodies raised against this peptide the 48K protein could be detected in CPMV infected protoplasts but a 58K protein was not found (14). This result demonstrates that in any case the 95K polyprotein is synthesized in vivo which is then rapidly cleaved into 48K and 60K products. The 105K polyprotein is either not produced in vivo or in amounts which are below the level of detection.

The existence of the B-RNA encoded 24K protein postulated by the translation model described in section 2 (see also Fig. 1) has also recently been verified (14). For its detection antibodies against a synthetic peptide of 25 amino acid residues corresponding to the C-terminal part of the hypothetical 24K protein, were used. The antiserum made the detection of virus specific 24K protein possible both in cowpea protoplasts infected with complete virus (M + B) and in protoplasts infected with B components (Fig. 4). The occurrence

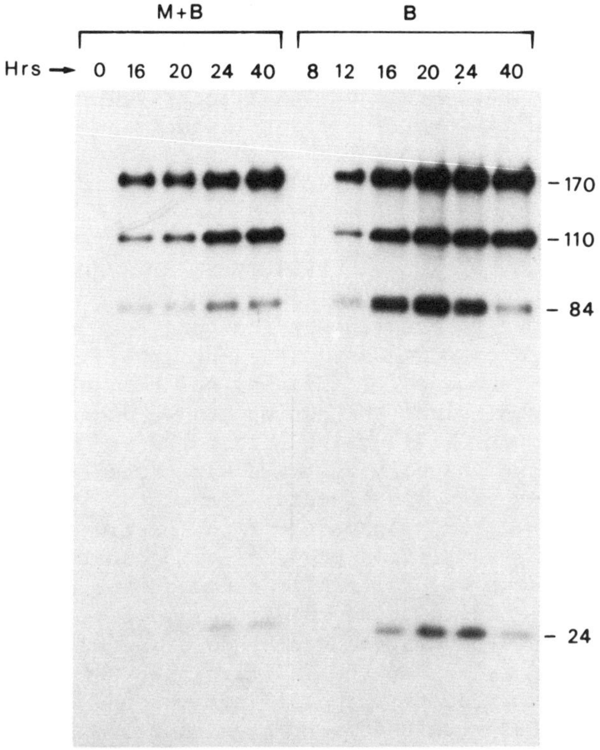

FIGURE 4. Detection of the B-RNA encoded 24K protease in cowpea protoplasts infected with CPMV (M + B) or alone B components. The figure shows a Western blot of extracts of protoplasts at different times after inoculation using peptide antibodies and ^{125}I-labeled protein A for detection. Since the sequence of the 24K protein is contained in the 170K, 110K and 84K processing intermediates these proteins are detected as well. It can be seen that in protoplasts infected with B components the B-encoded proteins are produced in larger amounts than in cells infected with the complete virus.

of the 24K protein in free form indicates that at least part of the 110K and/or 84K proteins are further processed to generate free 24K together with 87K, 58K and VPg. The time course of appearance of free 24K illustrated in Figure 4 shows that the 24K protein does not stably accumulate but turns over and the detectable amounts of 24K decrease when viral RNA replication slows down (40 hrs after infection).

DIRECT EVIDENCE FOR THE 24K PROTEIN BEARING PROTEOLYTIC ACTIVIT

In addition to previous indications that the 24K polypeptide encoded by B-RNA may represent a viral protease (3, 8) direct evidence for this has now been obtained from experiments using viral DNA copies cloned behind a phage SP6 or T7 promoter (12). Using either SP6 or T7 RNA polymerase a full-size DNA copy of B-RNA was transcribed into RNA molecules which were subsequently translated in vitro. The in vitro translation produced 200K polypeptides indistinguishable from translation products of natural B-RNA. Moreover this product was faithfully cleaved into 32K and 170K polypeptides. A deletion of 29 amino acid residues in the 24K protein sequence introduced by deleting 87 basepairs from the 24K coding region in the full-length B-DNA copy resulted in translation products which did not undergo this proteolytic processing. This result demonstrates that the 24K protein is involved in the primary cleavage at a glutamine-serine pair. In further experiments a hybrid clone was constructed, starting from full-length DNA copies of both B- and M-RNA (12), in which a HaeIII-BamH1 fragment of B-DNA containing the coding region of the 24K protein and its surrounding was inserted between two XhoI sites in the M-DNA clone. The resulting hybrid sequence contained a single uninterrupted open reading frame, which upon in vitro translation produced a 136K junction protein that by proteolytic cleavage released the small coat protein VP23 from the carboxy terminal end (see Fig. 1). This indicates that the 24K polypeptide insertion in the M polyprotein sequence is responsible for the cleaving at the glutamine-glycine site between the two capsid protein sequences. The cleavage did not occur if the proteolytic activity in the 24K polypeptide was inactivated by a deletion of 29 amino acids as described above.

CONCLUSION

The results indicate that two proteases, both encoded by B-RNA, achieve all cleavages in the polyproteins of CPMV, the 32K protein cleaving at glutamine-methionine pairs and the 24K protein at glutamine-serine and glutamine-glycine pairs. There is now good evidence that M-RNA is translated in vivo into a 95K polypeptide which is rapidly cleaved by the 32K protease into 48K and 60K products whereupon the 60K is further cleaved by the 24K protein to generate the VP37 and VP23 capsid proteins. In vitro M RNA is also translated into a 105K polypeptide overlapping the 95K polyprotein, but it is unclear if this expression of M RNA has a role in vivo.

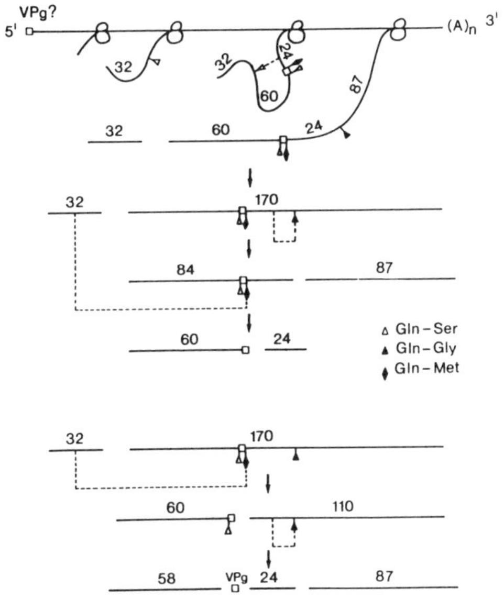

FIGURE 5. Model for the proteolytic processing of the B-RNA encoded 200K polyprotein. For explanation see the text. It is not known whether VPg is released if the RNA is used as messenger. The open and closed marks indicate the different cleavage sites.

The translation of B-RNA and the processing of the 200K polypeptides is pictured in figure 5. The processing can start before the chain of the 200K primary translation product is completed. The release of the 32K protein is started, as soon as the ribosomes have passed the 24K coding region by proteolytic activity of the 24K domain in the nascent chain. Further cleavages take place only after completion of the 170K protein. The order of these cleavages is not fixed resulting in different intermediates but the same final products: 1. The 170K is cleaved intramolecularly into 87K and 84K proteins by action of the 24K region; the 84K protein can be further processed by the 32K protein into 60K and 24K proteins. 2. Alternatively 170K is first cleaved by the 32K protease into 110K and 60K proteins followed by intramolecular cleavage of the 110K into 24K and 87K. Processing of the 60K protein into 58K and VPg has never been observed in vitro and free VPg can neither be detected in vivo. Probably this cleavage step is closely linked to the replication of virus RNA. This cleavage is carried out by the 24K protease but it remains to be established whether free 24K or one of its precursors is the active protein

ACKNOWLEDGEMENTS

The authors thank Marie-José van Neerven for typing the manuscript and Piet Madern for art work. This work was supported by the Netherlands Foundation for Chemical Research (SON) with financial aid from the Netherlands Organization for the Advancement of Pure Research (ZWO).

REFERENCES

1. Goldbach R, Van Kammen A (1985). Structure, replication and expression of the bipartite genome of cowpea mosaic virus. In Davies JW (ed): "Molecular Plant Virology" vol. II Boca Raton: CRC Press p. 86.
2. Goldbach R (1986). Comoviruses: Molecular biology and replication. In Harrison BD, Murant AF (eds) "The Plant Viruses: Viruses with bipartite genomes and isometric particles" New York, Plenum, in press.

3. Lomonossoff G, Shanks M, Evans D (1985). The structure of cowpea mosaic virus replicative form RNA. Virology 144:351.
4. Franssen H, Leunissen J, Goldbach R, Lomonossoff G, Zimmern D (1984). Homologous sequences in non-structural proteins from cowpea mosaic virus and picornaviruses. EMBO J. 3:855.
5. Toyoda H, Nicklin MJH, Murray MG, Anderson CW, Dunn JJ, Studier FW, Wimmer E (1986). A second virus-encoded proteinase involved in proteolytic processing of poliovirus polyprotein. Cell, in press.
6. Wellink J, Rezelman G, Goldbach R, Beyreuther K (1986). Determination of the sites of proteolytic processing in the polyprotein encoded by the B-RNA of cowpea mosaic virus. J. Virol., in press.
7. Zabel P, Moerman M, Lomonossoff G, Beyreuther K (1984). Cowpea mosaic virus VPg: Sequencing of radiochemically modified proteins allows mapping of the gene on B-RNA. EMBO J. 3:1629.
8. Van Wezenbeek P, Verver J, Harmsen J, Vos P, Van Kammen A, (1983). Primary structure and gene organization of the middle component RNA of cowpea mosaic virus. EMBO J. 2:941.
9. Franssen H, Moerman M, Rezelman G, Goldbach R (1984). Evidence that the 32,000-dalton protein encoded by bottom component RNA of cowpea mosaic virus is a proteolytic processing enzyme. J. Virol. 50:183.
10. Franssen H, Goldbach R, Van Kammen A (1984). Translation of bottom component RNA of cowpea mosaic virus in reticylocyte lysate: faithful proteolytic processing of the primary translation product. Virus Research 1:39.
11. Verver J, Goldbach R, Garcia J, Vos P (1986). Construction and in vitro expression of a full length DNA copy of cowpea mosaic virus B-RNA: a small deletion in the coding region of the 24-kilodalton protein abolishes proteolytic cleavage of the primary translation product, submitted.
12. Vos P. (1986). Manuscript in preparation.
13. Jaegle M. (1986). Manuscript in preparation.
14. Wellink J. (1986). Manuscript in preparation.

Positive Strand RNA Viruses, pages 209–225

REPLICATION OF ALPHAVIRUSES AND FLAVIVIRUSES: PROTEOLYTIC PROCESSING OF POLYPROTEINS[1]

James H. Strauss, Ellen G. Strauss, Chang S. Hahn, Young S. Hahn, Ricardo Galler[2], W. Reef Hardy and Charles M. Rice[3]

Division of Biology, California Institute of Technology
Pasadena CA 91125

ABSTRACT Large polyproteins are produced by the translation of the genomic RNAs of both alphaviruses and flaviviruses, or by translation of the subgenomic mRNA of alphaviruses, and subsequently processed by proteolytic cleavage. We propose that both viral enzymes (active in the cytosolic phase of the cell) and cellular proteases (active in subcellular organelles of the infected cell) are required to process these polyproteins.

INTRODUCTION

The genomes of the type alphavirus, Sindbis virus, family Togaviridae, and of the type flavivirus, yellow fever virus, family Flaviviridae, have been sequenced in their entirety (1,2) and details of the genome organizations of these viruses determined. It appears clear that in both virus families, proteolytic processing of polyprotein precursors is extensively used to produce the final

[1]This work was supported by grants AI20612 and AI10793 from NIH, by grant DMB-8316856 from the NSF, and by contract DAMD17-85-C-5223 from the U.S. Army Medical Research and Development Command.

[2]On leave from the Oswaldo Cruz Foundation of Brazil and supported by a fellowship from CNPq.

[3]Present address: Department of Microbiology and Immunology, Washington University School of Medicine, St. Louis, MO 63110.

products required for virus replication. We have proposed that these cleavage events fall into two categories (3,4). The first category includes cleavages that occur in the cytosolic phase of the infected cell. These usually include production of any nonstructural proteins required for RNA replication but in some cases structural proteins, required for virion assembly, may also be processed in the cytosol. These cleavages we hypothesize are performed by virus encoded proteases. The second category includes cleavages that occur in subcellular organelles such as the lumen of the endoplasmic reticulum or the Golgi apparatus that we propose are catalyzed by cellular proteases. These latter cleavages are restricted to viral proteins that have access to these compartments which, for the most part, are structural glycoproteins. We have probed the nature of the cleavage sites used and the enzymes responsible for the cleavages by a combination of techniques including comparative sequencing of viruses, sequencing of temperature-sensitive mutants defective in the protease at a nonpermissive temperature, and site-specific mutagenesis.

CLEAVAGE OF THE CAPSID PROTEIN OF THE ALPHAVIRUSES

The alphavirus structural proteins are produced from a subgenomic 26S messenger RNA 4,106 nucleotides in length (for Sindbis virus) that is translated to produce a polyprotein that is subsequently cleaved into the major structural proteins of the virus: the nucleocapsid protein and two major envelope glycoproteins called E1 and E2 (3,5,6). In addition, two smaller products given the names E3 and the 6 K protein are also produced (7). Cleavage of the capsid protein from the polyprotein precursor has long been thought to be an autoproteolytic event. This was first postulated from the fact that cleavage of the capsid from a polyprotein precursor was quantitative when small amounts of 26S messenger RNA were translated in extracts of rabbit reticulocytes (8). In addition, it has been shown that the defect in certain mutants of Sindbis virus, the type alphavirus, which fail to cleave the nucleocapsid protein from the polyprotein precursor at a nonpermissive temperature, can be complemented when cells are co-infected with a mutant of a different phenotype, such as wild-type virus (9). This implies that the cleavage can be catalyzed by a diffusible factor and therefore that the capsid protein can act as a diffusible protease as well as an autoprotease. Furthermore, it is known that this cleavage of the capsid protein from the polyprotein precursor can be inhibited with amino acid analogs (10). Finally, it has been suggested that the alphavirus capsid protein is a serine protease (11).

We have probed the nature of the capsid autoprotease by sequencing four temperature-sensitive mutants whose defect is a failure to cleave the capsid protein from the polyprotein precursor (12; unpublished observations). These mutants were each found to have a single amino acid substitution in the C-terminal half of the capsid protein that led to temperature-sensitive cleavage. Since these mutations were far removed from the site of the actual cleavage, this suggested that the areas changed were involved in the proteolytic activity. The changes found were lysine to leucine at position 138 of the Sindbis capsid protein, proline to serine at position 170, and proline to serine at position 218. In contemplating the location of these changes and comparing the sequences of the capsid proteins to those of serine proteases, we developed the hypothesis that the alphavirus capsid protein is a serine protease whose catalytic triad is formed by histidine-141, aspartic acid-147 and serine-215 of the Sindbis capsid protein. The location of the temperature-sensitive mutations that have been mapped are consistent with this hypothesis. The change from lysine-138 to leucine involves a residue that is lysine or arginine in the four alphaviruses sequenced to date and that is three residues removed from the histidine that we postulate to be active in proteolysis. Similarly, the change from proline to serine in position 218 is three residues removed from the serine that is hypothesized to be involved in proteolysis. The mutation at position 170 is more distant from these sites but since it affects a conserved proline residue positioned between the aspartic acid and serine residues involved in the catalytic triad, it could alter the folding of the protein and the positioning of serine with respect to the histidine and aspartic acid residues. The locations of these temperature-sensitive mutants and comparison of the sequence of alphavirus capsid proteins with that of five serine proteases is shown in Fig. 1. Note that the two proline residues that are affected are in each case conserved not only among the four alphavirus capsid proteins whose sequences are known but also among the five serine proteases that are shown. The lysine at position 138 affected in *ts*13 is not found in serine proteases. This charged residue could be of importance because of the close proximity of the histidine and aspartic acid residues thought to be involved in the alphavirus capsid protease, residues that are much further separated in serine proteases.

The suggestion that the capsid protein might be a serine autoprotease is also consistent with the fact that the cleavage event that releases the capsid protein occurs after a tryptophan residue (Fig. 1), which is characteristic of enzymes with chymotryptic-like activity. An earlier observation from Pfefferkorn's laboratory had in fact shown that cleavage of the capsid protein could be inhibited by chymotrypsin inhibitors (14). If the alphavirus capsid protein is a serine protease, it is unusual among viral encoded proteases. Many animal viruses, especially RNA viruses, are known to encode proteases (15-17), but most of these proteases appear to be sulfhydryl proteases. The only other report of a putative serine protease encoded by an animal virus of which we are aware is one dealing with protein VP1 of polyoma virus (18). It is now possible to perform site-specific mutagenesis of the capsid protein to test the importance of various residues in the cleavage event. Such experiments are under way to test the serine protease hypothesis as well as to examine the functions of other residues.

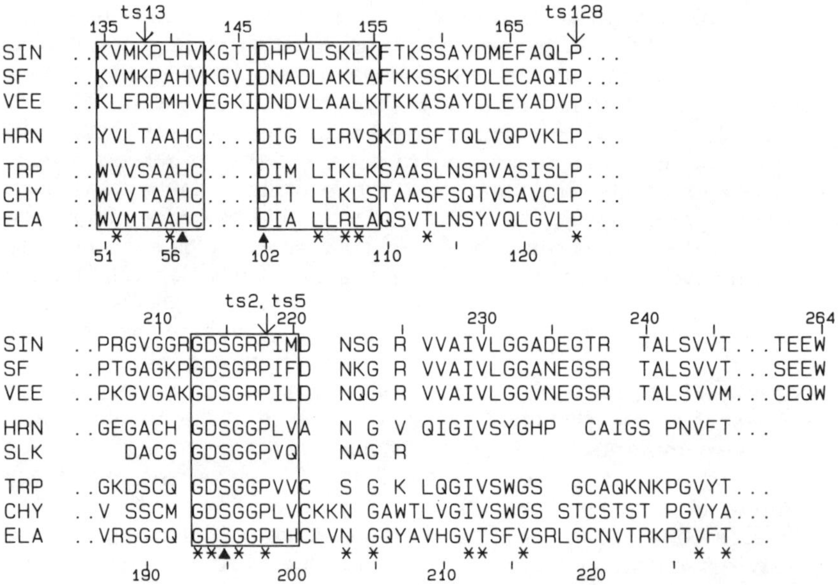

FIGURE 1. Amino acid sequences from regions of the capsid proteins of three alphaviruses (Sindbis, Semliki Forest and Venezuelan equine encephalitis viruses) are compared to regions from sequences of five serine proteases, two from insects (hornet and silkworm) and three from mammals (trypsin, chymotrypsin and elastase). The single letter amino acid code is used. The amino acids found in the catalytic triad of serine proteases, histidine, aspartic acid and serine, are marked by solid triangles. The location of four temperature-sensitive mutants of Sindbis virus are shown. (Adapted from ref. 12; data for VEE are from ref. 13; data for *ts*128 are unpublished.)

CLEAVAGE OF THE GLYCOPROTEINS OF ALPHAVIRUSES

After cleavage of the nucleocapsid protein from the structural polyprotein precursor of alphaviruses, an N-terminal signal sequence appears to function leading to the insertion of the nascent polypeptide chain into the endoplasmic reticulum (reviewed in 19). This N-terminal signal sequence is not cleaved. After insertion of the glycoprotein, a stop transfer signal near the C-terminal end of the first glycoprotein (PE2) leads to anchoring of this glycoprotein by a hydrophobic anchor near the C-terminus. Following this, an internal signal sequence is believed to function which leads to the insertion of glycoprotein E1 (the second glycoprotein produced) into the endoplasmic reticulum. Separation of the two glycoproteins occurs when a small polypeptide of approximately 60 amino acids (called 6 K) is cut from this polyprotein precursor, releasing the C-terminus of E2 and the N-terminus of E1. These cleavages occur after alanine residues and the sequence around these cleavage sites for a number of alphaviruses is shown in Fig. 2. In each case the cleavage sites are homologous to the sites preferred by the enzyme signalase, active in the lumen of the endoplasmic reticulum, and it has been proposed that these cleavage events are catalyzed by signalase

```
                       ↓          ↓
SIN    SVID....ANA ETF....VDA YEH....RR
                       ↓          ↓
SF     SVSQ....AHA ASV....ARA YEH....RR
                       ↓          ↓
RR     SVTE....ANA ASF....AKA YEH....RR
                       ↓          ↓
VEE    STEE....ARA AST....APA YEH....HN

       ←—E2—→←—6K—→←—E1—→
```

FIGURE 2. Amino acid sequences around the cleavage sites that are used to separate the 6 K protein from the glycoprotein precursor polyprotein of alphaviruses. (Data for Sindbis virus are from ref. 3, for Semliki Forest virus from ref. 6, for Ross River virus from ref. 20 and for Venezuelan equine encephalitis virus from ref. 13.)

(3,6). This would imply that the C-terminal end of E2 must be exposed to this enzyme in the lumen of the endoplasmic reticulum, at least transiently, and that, therefore, glycoprotein E2 penetrates the lipid bilayer twice. This sort of insertion mechanism to position two or more glycoproteins appears to be a common theme among a number of virus groups and as will be noted below appears to be used to produce the glycoproteins of the flaviviruses.

There is another cleavage event that occurs in the production of alphavirus glycoproteins. Glycoprotein E2 is produced in a precursor form referred to as PE2 which is subsequently cleaved, approximately 20 minutes after synthesis, to produce E3, the N-terminal 60 amino acids or so of PE2, and the mature glycoprotein E2. It is thought that this cleavage event is important for maturation of the virus in some way and is analogous to cleavage events undergone by glycoproteins of many enveloped animal viruses such as the influenza viruses, paramyxoviruses, retroviruses, and flaviviruses, as well as alphaviruses (reviewed in 21). This cleavage event occurs after two basic residues in succession, either lysine-arginine or arginine-arginine, in all the cases that have been studied, and in addition there is always one or more arginine or lysine residues upstream from this polypeptide and separated often by one uncharged residue (such that the minimal canonical site appears to be Arg-X-Arg/Lys-Arg). After the primary cleavage event, one or more of the newly produced C-terminal residues are usually removed by an exopeptidase activity. The sequences around these cleavage sites for a number different viruses are shown in Fig. 3 in which the canonical site has been boxed. Because these cleavage sites are all similar to one another and resemble sites utilized by cellular enzymes known to function either in the Golgi apparatus or in transport organelles, and that may be related to the cathepsins, these cleavage events have been hypothesized to be catalyzed by a cellular protease (3,6,20,21).

CLEAVAGE OF THE ALPHAVIRUS NONSTRUCTURAL POLYPROTEIN

The alphavirus nonstructural proteins are produced as a polyprotein by translating the genome length RNA. Four final protein products are produced which we have referred to as nonstructural proteins 1, 2, 3 and 4 (nsP1, nsP2, nsP3 and nsP4). The cleavage sites used to produce these four nonstructural proteins are shown in Fig. 4 for a number of different alphaviruses where the sequence is known. Cleavage in each case follows an amino acid with a short side chain (glycine, cysteine, or alanine) which is preceded by a glycine residue. The amino acid following the cleavage sites after nsP1 and nsP2 also has a short side chain (alanine or glycine), whereas that after the nsP3/nsP4 sites is a tyrosine. We hypothesize that these cleavage events are all catalyzed by a virus encoded protease that resides in one of the nonstructural proteins, perhaps nsP3. It is of interest that the

VIRUS	PROTEIN			CLEAVAGE ↓
ALPHAVIRUS				
SIN		CGSSG	RSKR	SVIDGF
VEE	pE2 → E3 + E2	AAVKC	RKRR	STEELF
SFV		CRNGT	RHRR	SVSQHF
MYXOVIRUS				
FPV	HA → HA1 + HA2	EPSKK	REKR	GLFGAI
PARAMYXOVIRUS				
RSV	F → F2 + F1	TLSKK	RKRR	FLGFLL
SV5		LIPTR	RRRR	FAGVVI
RETROVIRUS				
RSV	pr95 → gp85 + gp35	SRTGI	RRKR	SVSHLD
MLV	pr95 → gp70 + p15E	FERSN	RHKR	EPVSLT
FLAVIVIRUS				
YF		AGRSR	RSRR	AIDLPT
MVE	prM → M + ?	ARHSK	RSRR	SITVQT
DEN2		TGEMR	REKR	SVALVP

FIGURE 3. Amino acid sequences around the cleavage sites for a number of virus precursor glycoproteins. In each case cleavage of the precursor glycoprotein occurs after translation and during transport of the glycoprotein or during virus maturation. The precursor glycoproteins and the products produced are shown in each case. Sequences for the three alphaviruses are from the references listed in Fig. 2. Sequence data for the myxovirus, fowl plague virus (FPV), are from ref. 22; for the paramyxoviruses respiratory synctial virus (RSV) and simian virus 5 (SV5) from ref. 23 and 24, respectively, for retroviruses Rous sarcoma virus (RSV) and murine leukemia virus (MLV) from ref. 25 and 26, respectively; for flaviviruses yellow fever virus (YF) from ref. 2, Murray Valley encephalitis virus (MVE) from ref. 27, and dengue 2 virus from unpublished data from our laboratory.

penultimate glycine gives this proteolytic activity a unique twist. Virus encoded proteases in general each seem to have a specificity unique to the virus and usually different from specificities of cellular encoded proteases that have been described in the past.

In the case of alphaviruses, the production of nsP4 is regulated so that it is produced in smaller amounts. We had originally reported that in the case of Sindbis virus and Middelburg virus, modulation was effected by the presence of an opal termination codon between nsP3 and nsP4, so that the majority of time translation terminates after production of nsP3 (30). Thus most of the nsP3 produced has a C-terminus determined by the position of the stop codon rather than by the position of the cleavage site that separates nsP4 from nsP3. Readthrough is required to produce a longer polyprotein, which upon cleavage could give rise to nsP4, and for this reason very small quantities of nsP4 are found in cells infected with Sindbis virus (31). The complete nucleotide sequence of Semliki Forest virus has recently been obtained (K. Takkinen, personal communication) and it was found that Semliki Forest virus did not possess such an opal termination codon. Nonetheless, it has been known for some time that nsP4 is found in relatively small quantities in Semliki Forest virus-infected cells, although in relatively larger quantities than in Sindbis-infected cells. It has been postulated that the lower quantity of nsP4 in Semliki Forest virus-infected cells was due to rapid turnover of the protein (32) although this has not been shown conclusively. It may, in fact, be the case that there is some type of attenuation during translation of the RNA such that translation terminates prematurely in this region of the RNA. Because of the fact that different alphaviruses appear to regulate production of

SIN	NH_2-MEKPVV.....	ADIGA ↓ ALVETPR....	DGVGA ↓ APSYRT....	XLTGVGG ↓ YIFSTDT....	LYGGPK-COOH
MID	NH_2-MARPVV.....	ARAGA ↓ GVVNTPR....	HTAGC ↓ APSYRV....	XLDRAGA ↓ YIFSSDT....	LYGGPK-COOH
SFV	NH_2-MAAKV	↓ GVV???R			
RRV	NH_2-		....HTAGC ↓ APSYRV....	XLGRAGA ↓ YIFSSDT....	LYGGPK-COOH
ONN	NH_2-		....TRAGC ↓ APSYRL....	RLDRAGG ↓ YIFSSDT....	LYGGPK-COOH

←nsP1→ ←nsP2→ ←nsP3→ ←nsP4→

FIGURE 4. Amino acid sequences around the cleavage sites used to process the alphavirus nonstructural polyprotein for five alphaviruses. (Data for Sindbis virus are from ref. 1; for Semliki Forest virus from ref. 28,29; while data for Middelburg, Ross River virus and O'Nyong-nyong viruses are unpublished data from our laboratory.)

nsP4 in different ways, we recently undertook to sequence the appropriate regions of two more alphavirus genomes, that of Ross River virus and of O'Nyong-nyong virus. We found that Ross River virus possessed an opal termination codon in the appropriate place, whereas O'Nyong-nyong virus did not possess such a termination codon. Thus of five viruses examined, three alphaviruses have a termination codon to regulate nsP4 production and two do not. It would be of considerable interest to determine whether the presence or absence of a termination codon is essential for virus growth and/or whether it influences viral virulence or host range or epidemiology.

TRANSLATION OF THE FLAVIVIRUS GENOME

The complete sequence of the type flavivirus, yellow fever virus, has been determined (2) and the most remarkable feature about this genome is the presence of a very long open reading frame of 10,233 nucleotides that could encode a polyprotein of 380,763 daltons. This leaves 5' and 3' noncoding regions of 118 and 511 nucleotides, respectively. In this open reading frame the structural proteins occupy the 5' terminal one fourth of the genome and the nonstructural proteins occupy the remainder. The presence of such a long, open reading frame and the existence of a consistent set of potential cleavage sites that could explain the generation of the yellow fever polypeptides by posttranslational cleavages (to be discussed in more detail below) led us to hypothesize that the flavivirus genome is translated into one, continuous polyprotein that is processed posttranslationally. Similar to the case with alphaviruses, we postulate that the structural proteins in the amino terminal region of the polyprotein precursor are cleaved by cellular proteases, active in the lumen of the endoplasmic reticulum or within the Golgi apparatus; and that the cleavage of the nonstructural polyprotein is effected in the cytosol, catalyzed by a virus encoded protease (2).

THE FLAVIVIRUS STRUCTURAL PROTEIN CLEAVAGES

The flavivirus structural proteins are found along the genome in the order capsid protein C, glycoprotein precursor prM, and envelope glycoprotein E (2,33,34). These three proteins are followed by another glycoprotein, a nonstructural protein called NS1 not present in the mature virion but which, because it is a glycoprotein, may be involved in virus maturation or assembly rather than, for example, in viral RNA replication. Furthermore, this glycoprotein NS1 appears to be cleaved in a manner analogous to the structural glycoproteins and to the capsid protein, and will be considered together with the structural proteins.

The nucleocapsid protein, in contrast to the situation with alphaviruses, has at its carboxyterminal end a possible membrane spanning domain of uncharged amino acids. (It should be noted, however, that in the case of all of the flavivirus proteins, protein sequence data to date has been limited to amino terminal sequencing of mature proteins or to sequencing of selected tryptic peptides derived from the mature proteins. To date no carboxyterminal sequencing of mature proteins has been performed and it is unknown whether polypeptide domains are removed from between the carboxy terminus of one mature protein and the amino terminus of the following protein, whether by exo- or endopeptidase activity. For the sake of this discussion we assume that each protein terminates adjacent to the beginning of the following protein.)

The cleavage sites used to separate the capsid protein from protein prM in several flaviviruses are illustrated in Fig. 5. This cleavage site is homologous to those preferred by signalase. This, together with the fact that there is a putative membrane-spanning anchor at the C-terminal end of the capsid protein, leads to the hypothesis that this cleavage is effected by signalase and that the capsid protein produced is anchored in the membrane at its C-terminal end.

The three glycoproteins that follow, prM, E and NS1, appear to resemble the situation for the alphaviruses. Each has near its C-terminal end two hydrophobic domains that could span the lipid bilayer and that are separated from one another by charged residues. The first of these two domains could act as a stop transfer signal and hydrophobic anchor, while the second domain could act as an internal signal sequence leading to the insertion of the following glycoprotein into the endoplasmic reticulum. In any event, it seems reasonable to hypothesize that signalase is the enzyme responsible for separating the proteins from one another by proteolytic cleavage. The cleavage following NS1 could also be effected by signalase. The cleavage sites used to separate these glycoproteins in a number of different flaviviruses are also illustrated in Fig. 5 and in each case the cleavage sites are homologous to those known to be preferred by signalase.

The glycoprotein precursor prM is cleaved in a late cleavage event in a manner analogous to that undergone by alphavirus glycoprotein precursor PE2 or the precursors of several other groups of enveloped animal viruses referred to above. This cleavage follows two basic residues and is presumably catalyzed by the cellular protease active in the Golgi or in the transport vesicles. This cleavage site has been illustrated in Fig. 3.

Assembly of the flaviviruses appears to differ in detail from the assembly events involved in other enveloped viruses. Budding figures are infrequently seen in infected cells, and the assembly takes place in association with internal membranes, presumably smooth endoplasmic reticulum (reviewed in 37). The cleavage of prM, or possibly other cleavages such as in the capsid protein in a maturation event, could be involved in this assembly process in ways that we do not understand at the present time.

CLEAVAGE OF THE FLAVIVIRUS NONSTRUCTURAL PROTEINS

The flavivirus nonstructural proteins following NS1 along the genome, which we have referred to as ns2a, ns2b, NS3, ns4a, ns4b, and NS5, respectively, are hypothesized to be cleaved from the

			CLEAVAGE SITE ↓	
prM	YF	MLLMTGG		VT-LV--RKNRWLLLNV
	SLE		?	
	MVE	MLIGFAA		ALKLSTFQGKIMMTVNA
	WN	LIACAGA		VT-LSNFQGKVMMTVNA
	DEN2	TV-MAFH		LT-TRNGEPHMIVSRQE
E	YF	AVGPAYS		AHCIGITDRDFI
	SLE		?	
	MVE	LVAPAYS		FNCLGTSNRDFI
	WN	LVAPAYS		FNCLGMSNRDFL
	DEN2	AVAPSMT		MRCIGISNRDFV
NS1	YF	LSLGVGA		DQGCAINFGKRELKCGDGIFIF
	SLE	LATSVQA		DSGCAISLQRRELKCGGGIFVY
	MVE	LATNVHA		DTGCAIDITRRELKCGSGIFIH
	WN	LSVNVHA		DTGCAIDIGRQELRCGSGVFIH
	DEN2	LGVMVQA		DSGCVVSWKNKELKCGSGIFVT

FIGURE 5. Amino acid sequences around the cleavage sites that are used to separate the structural proteins and protein NS1 of flaviviruses. (Data for yellow fever virus are from ref. 2; for St. Louis encephalitis virus from ref. 35; for Murray Valley encephalitis virus from ref. 27; for West Nile virus from ref. 33, 34, 36; for dengue 2 virus, data are unpublished from our laboratory.)

polyprotein precursor in a number of events (2). Cleavage follows two basic residues in succession, either arginine-arginine (all yellow fever sites as well as a number of sites in other viruses) or lysine-arginine. These cleavage sites are illustrated in Fig. 6 for a

```
                              CLEAVAGE
                                SITE
                    -5                           10             20
                     |            ↓         |    |       |      |
NS3     YF     HVRGARR        SGDVLWDIPTPKIIEECEHLED
        SLE                      -AL V   KV  K   K
        MVE    TLKYTKR        GG-VFWDTPSPKVYPKGDTTPG
        WN     TLQYTKR        GG-VLWDTPSPKEYKKGDTTT-
        DEN2   WEVKKPR        AG-VLWDVPSPPPVGKAE-LED

                    -5                           10
                     |                      |    |    |
ns4a    YF      --EGRR        GAAEVLVVLSEL-PDF
        MVE     LGSGRR        S?--FFEVPGRM-PEH
        WN      --SGKR        YTKR---GG-VLWDTP
        DEN2    --AGRK        SLLNLI---TEMGPTF

                    -5                           10
                     |                      |    |    |
ns4b    YF      KLAQRR        VFHGVAENPVVDGNPT
        MVE     RAAQKR        TAAGIMKNAVVDGIVA
        WN      RSAQRR        TAAGIMKNVVVDGIVA
        DEN2    REAQKR        AAAGLMKNPTVDGITV

                    -5                           10             20
                     |                      |    |       |      |
NS5     YF     KMKTGRR        GSAN-GKTLGEVWKRELNLLDK
        SLE                      K- ATL  T  K
        MVE    EKPAFKR        GRAG-GRTLGEQWKEKLNAMGK
        WN     EKPGLKR        GGAK-GRTLGEVWKERLNHMTK
        DEN2   NTTNTRR        GTGNIGETLGEKWKSRLNALGK
```

FIGURE 6. Amino acid sequences around the cleavage sites utilized to produce flavivirus nonstructural proteins from the polyprotein precursor. (References for the different viruses are the same as in Fig. 5.) A number of the cleavage sites are conjectural and many have been postulated from homologies among the flavivirus polyproteins (see ref. 2, 38).

number of different flaviviruses. We hypothesize that these cleavage events are effected by a viral-encoded protease that is presumably resident within one of these nonstructural proteins. We presume that the first cleavage event is autocatalytic to release the protease and that the protease then acts as a diffusible protease to effect the remainder of the cleavages. Evidence for a polyprotein precursor that is subsequently processed has recently been obtained by G. Cleaves (39) and by C. Blair (personal communication).

The cleavage sites utilized for production of the flavivirus nonstructural proteins appear to involve not simply the two basic amino acids preceding the cleavage site, but also the amino acids on either side of this dibasic peptide. Note from Fig. 6 that most of the sites are preceded and/or followed by amino acids with short side chains: glycine, serine, alanine, or threonine. We therefore believe that these amino acids form important components of the recognition site (35) and that once again the cleavage site has a unique twist characteristic of the flaviviruses.

CONCLUDING REMARKS

It is of interest to speculate upon the origin of such virus-encoded proteases. The most reasonable hypothesis would appear to be that these enzymatic activities have been captured from the host cell in some way, at some time in the evolutionary history of the viruses. It would seem reasonable that a recombinational event could have occurred during the evolution of the viruses such that the viral genome acquired coding sequences for a protease (although convergent evolution leading to the proteases present in viruses is also a possibility). During subsequent evolution the protease could have undergone many changes, but might still retain a number of characteristics of the ancestral protease from which it was derived. Thus in the case of the postulated serine protease activity of the alphavirus capsid protein, significant amino acid homologies appear to be retained between the alphavirus capsid protein and the currently extant serine proteases. In the case of the putative alphavirus nonstructural protease, or the putative flavivirus nonstructural protease, no homologies with serine proteases are seen. Thus it seems more likely that these proteases might be cysteine based proteases. The sites recognized by the flavivirus nonstructural protease resemble those cleaved by the cellular enzyme invoked as being responsible for processing the glycoprotein precursors of enveloped viruses. Because the flavivirus nonstructural polyprotein is thought to be cleaved in the cytosol, and the glycoprotein protease appears to be localized within the Golgi or transport vesicles, it seems unlikely that the cellular enzyme could be responsible for the flavivirus nonstructural cleavages. However, the similarities in the cleavage site could conceivably have resulted if the flavivirus nonstructural protease was derived from the cellular enzyme, that is if the flaviviruses captured the Golgi enzyme at some point. To date no sequence information is available for this Golgi protease and it will be of considerable interest to compare the amino acid sequences of this enzyme or family of enzymes with that of the flavivirus nonstructural proteins when such results are available.

REFERENCES

1. Strauss EG, Rice CM, Strauss JH (1984). Complete nucleotide sequence of the genomic RNA of Sindbis virus. Virology 133:92.
2. Rice CM, Lenches EM, Eddy SR, Shin SJ, Sheets RL, Strauss JH (1985). Nucleotide sequence of the yellow fever virus: Implications for flavivirus gene expression and evolution. Science 229:726.
3. Rice CM, Strauss JH (1981). Nucleotide sequence of the 26S mRNA of Sindbis virus and deduced sequence of the encoded virus structural proteins. Proc Natl Acad Sci USA 78:2062.
4. Strauss JH, Strauss EG, Hahn CS, Rice CM (1986). The genomes of alphaviruses and flaviviruses: Organization and translation. In Rowlands DJ, Mahy BWJ, Mayo M (eds): "The Molecular Biology of the Positive Strand RNA Viruses," in press.
5. Garoff H, Frischauf AM, Simons K, Lehrach H, Delius H (1980). The capsid protein of Semliki Forest virus has clusters of basic amino acids and prolines in its amino terminal region. Proc Natl Acad Sci USA 77:6376.
6. Garoff H, Frischauf AM, Simons K, Lehrach H, Delius H (1980). Nucleotide sequence of cDNA coding for Semliki Forest membrane glycoproteins. Nature (London) 288:236.
7. Welch WJ, Sefton BM (1979). Two small virus specific polypeptides are produced during infection with Sindbis virus. Virology 29:1186.
8. Simmons DT, Strauss JH (1974). Translation of Sindbis virus 26S RNA and 49S RNA in lysates of rabbit reticulocytes. J Mol Biol 86:397.
9. Scupham RK, Jones KJ, Sagik BP, Bose HR (1977). Virus directed posttranslational cleavage in Sindbis virus infected cells. J. Virol. 22:568.
10. Aliperti G, Schlesinger M (1978). Evidence for an autoprotease activity of Sindbis virus capsid protein. Virology 90:366.

11. Boege U, Wengler G, Wengler G, Wittmann-Liebold B (1981). Primary structures of the core proteins of the alphaviruses, Semliki Forest virus and Sindbis virus. Virology 113:293.
12. Hahn CS, Strauss EG, Strauss JH (1985). Sequence analysis of three Sindbis virus mutants temperature sensitive in the capsid protein autoprotease. Proc Natl Acad Sci USA 82:4648.
13. Kinney RM, Johnson BJB, Brown VL, Trent DW (1986). Nucleotide sequence of the 26S mRNA of the virulent Trinidad donkey strain of Venezuelan equine encephalitis virus and deduced sequence of the encoded structural proteins. Virology, in press.
14. Pfefferkorn ER, Boyle MK (1972). Selective inhibition of the synthesis of Sindbis virion proteins by an inhibitor of chymotrypsin. J Virol 9:187.
15. Palmenberg AA, Pallansch MA, Rueckert RR (1979). Protease required for processing the picornaviral coat protein resides in the viral replicase gene. J Virol 32:770.
16. von der Helm K (1977). Cleavage of Rous sarcoma viral polypeptide precursor into internal structural proteins in vitro involves viral protein p15. Proc Natl Acad Sci USA 74:911.
17. Bhatti AR, Weber J (1979). Protease of adenovirus 2: Partial characterization. Virology 96:478.
18. Bowen JH, Chlumecky V, D'Obrenan P, Colter JS (1984). Evidence that polyoma polypeptide PP1 is a serine protease. Virology 135:551.
19. Strauss EG, Strauss JH (1986). Structure and replication of the alphavirus genome. In Schlesinger S, Schlesinger MJ (eds): "The Togaviridae and Flaviviridae," New York: Plenum Press, Chapter 3, p. 35.
20. Dalgarno L, Rice CM, Strauss, JH (1983). Ross River virus 26S RNA: Complete nucleotide sequence and deduced sequence of the encoded structural proteins. Virology 129:170.
21. Strauss EG, Strauss JH (1985). Assembly of enveloped animal viruses. In Casjens S (ed): "Virus Structure and Assembly," Boston: Jones and Bartlett, Chapter 6, pp. 205.
22. Bosch FX, Garten W, Klenk H-D, Rott R (1981). Proteolytic cleavage of influenza virus hemagglutinins: Primary structure of the connecting peptide between HA_1 and HA_2 determines proteolytic cleavability and pathogenicity of avian influenza viruses. Virology 113:725.
23. Collins PL, Huang YT, Wertz GW (1984). Nucleotide sequence of the gene encoding the fusion (F) glycoprotein of human respiratory syncytial virus. Proc Natl Acad Sci USA 81:7683.

24. Paterson RG, Harris TJR, Lamb RA (1984). Fusion protein of the paramyxovirus simian virus 5: Nucleotide sequence of mRNA predicts a highly hydrophobic glycoprotein. Proc Natl Acad Sci USA 81:6706.
25. Schwartz DE, Tizard R, Gilbert W (1983). Nucleotide sequence of Rous sarcoma virus. Cell 32:853.
26. Shinnick TM, Lerner RA, Sutcliffe JG (1981). Nucleotide sequence of Moloney murine leukemia virus. Nature 293:543.
27. Dalgarno L, Trent DW, Strauss JH, Rice CM (1986). Partial nucleotide sequence of the Murray Valley encephalitis virus genome: Comparison of the encoded polypeptides with yellow fever virus structural and nonstructural proteins. J Mol Biol 187:309.
28. Kalkkinen N, Laaksonen M, Söderlund H, Jörnvall H (1981). Radio-sequence analysis of in vivo multilabeled nonstructural protein ns86 of Semliki Forest virus. Virology 113:188.
29. Ou J-H, Strauss EG, Strauss JH (1983). The 5'-terminal sequences of the genomic RNAs of several alphaviruses. J Mol Biol 168:1.
30. Strauss EG, Rice CM, Strauss JH (1983). Sequence coding for the alphavirus nonstructural proteins is interrupted by an opal termination codon. Proc Natl Acad Sci USA 80:5271.
31. Lopez S, Bell JR, Strauss EG, Strauss JH (1985). The nonstructural proteins of Sindbis virus as studied with an antibody specific for the C terminus of the nonstructural readthrough polyprotein. Virology 141:235.
32. Keränen S, Ruohonen L (1983). Nonstructural proteins of Semliki Forest virus: Synthesis, processing and stability in infected cells. J Virol 47:505.
33. Castle E, Nowak T, Leidner U, Wengler G, Wengler G (1985). Sequence analysis of the viral core protein and the membrane associated proteins V1 and NV2 of the flavivirus West Nile virus and of the genome sequence for these proteins. Virology 145:227.
34. Wengler G, Castle E, Leidner U, Nowak T, Wengler G (1985). Sequence analysis of the membrane protein V3 of the flavivirus West Nile virus and of its gene. Virology 147:264.
35. Rice CM, Aebersold R, Teplow DB, Pata J, Bell JR, Vorndam AV, Trent DW, Brandriss MW, Schlesinger JJ, Strauss JH (1986). Partial N-terminal amino acid sequences of three nonstructural proteins of two flaviviruses. Virology 151: 10. press.
36. Castle E, Leidner U, Nowak T, Wengler G, Wengler G (1986). Primary structure of the West Nile flavivirus genome region coding for all nonstructural proteins. Virology 149:10.

37. Westaway EG (1980). Replication of flaviviruses. In Schlesinger RW (ed): "The Togaviruses," New York: Academic Press, Chapter 19, p. 531.

38. Rice CM, Strauss EG, Strauss JH (1986). Structure of the flavivirus genome. In Schlesinger S, Schlesinger MJ (eds): "The Togaviridae and Flaviviridae," New York: Plenum Press, Chapter 10, p. 279.

39. Cleaves GR (1985). Identification of dengue type 2 virus-specific high molecular weight proteins in virus-infected BHK cells. J Gen Virol 66:1767.

Positive Strand RNA Viruses, pages 227–237

ASSEMBLY OF MOLONEY MURINE LEUKEMIA VIRUS: REQUIREMENT FOR MYRISTYLATION SITE IN PR65^{gag}[1]

Alan Rein,[2] Melody R. McClure,[2] Nancy R. Rice,[2] Ronald B. Luftig,[3] and Alan M. Schultz[2]

LBI-Basic Research Program, NCI-Frederick Cancer Facility, Frederick, MD 21701[2]
and
Dept. of Microbiology, Louisiana State University Medical Center, New Orleans, LA 70112[3]

ABSTRACT Type C retroviruses normally assemble at the plasma membrane of the infected cell. A single viral protein, Pr65^{gag}, is sufficient for virus assembly and release. The nature of the interaction of Pr65^{gag} with the membrane is not known; however, recent studies from this laboratory have demonstrated that Pr65^{gag} is modified by the covalent attachment of myristic acid to the N-terminal glycine residue (Henderson, Krutzsch, and Oroszlan, PNAS 80:339, 1983; Schultz & Oroszlan, J. Virol. 46:355, 1983). In an effort to determine the significance of myristylation of Pr65^{gag}, we have used site-directed mutagenesis to change the codon for the N-terminal glycine in Moloney MuLV Pr65^{gag} to an alanine codon, or to delete it. Both of these mutants were found to encode Pr65^{gag} which is not myristylated. Further analysis showed that the mutant, unmyristylated Pr65^{gag} molecules (a) are predominantly soluble cytoplasmic proteins; and (b) do not assemble into

[1]This work was supported by the National Cancer Institute, DHHS, under contract No. N01-CO-23909 with Litton Bionetics, Inc.

virus particles or recognizable virus-specific structures. These results are consistent with the hypothesis that the myristate moiety plays a crucial role in the association of wild-type $Pr65^{gag}$ with the plasmic membrane, and that this association is essential for virus assembly.

INTRODUCTION

Type C retroviruses assemble at the plasma membrane of virus-producing cells. Remarkably, this assembly is apparently accomplished by a single protein, termed $Pr65^{gag}$ in the case of the murine leukemia viruses (1-4; Rein and Schultz, unpublished). Thus this single protein must be able to (i) selectively package the viral genomic RNA; (ii) interact with the plasma membrane and with other $Pr65^{gag}$ molecules so as to initiate particle formation and bud from the membrane; and (iii) incorporate viral envelope proteins, if they are present, into the surface of the particle.

The present work deals with the nature of the interaction between $Pr65^{gag}$ and the cell membrane. The protein is not glycosylated and lacks any obvious "signal" sequence. However, it was recently shown to have an unusual modification: the N-terminal glycine is covalently linked to the 14-carbon saturated fatty acid, myristic acid (5,6). It seemed possible that this moiety might play a role in the association of $Pr65^{gag}$ and the plasma membrane, and that this association might, in turn, facilitate virus assembly. The experiments described here provide strong support for both of these hypotheses.

All known myristylated proteins have glycine at their N-termini (7-11). Glycine has been shown to be necessary for myristylation of $pp60^{src}$ in avian cells (12). It therefore seemed likely that $Pr65^{gag}$ molecules lacking N-terminal glycine would not be myristylated in mammalian cells. Accordingly, our approach to analyzing the biological significance of myristate in the gag polyprotein has been to eliminate the N-terminal glycine from Moloney MuLV $Pr65^{gag}$ by site-directed mutagenesis.

RESULTS

A portion of an infectious clone of Moloney MuLV (in Charon 4A; a kind gift of D. Steffen) was subcloned into M13 mp11. This subclone was mutagenized by the two-primer method of Zoller and Smith (13). Two mutants were induced: in one the glycine codon (immediately downstream from the *gag* initiator methionine codon) was replaced by an alanine codon, and in the other the glycine codon was deleted. In each case a 176bp Pst I fragment was completely sequenced, to ensure that it carried the desired base changes but no other changes; it was then used in the reconstruction of an intact viral genome. The genome was then inserted in the plasmid vector pSV2 Neo (14). The use of this vector eliminates the need for co-transfection, since the mutant viral genome is introduced into the cells on the same molecule as the neo^r gene, carrying resistance to the antibiotic G418.

The Mutants Synthesize Unmyristylated $Pr65^{gag}$.

The mutant viral genomes (together with appropriate controls) were transfected into NIH/3T3 mouse cells and CHO hamster cells, and transfectants were selected with G418. (We have obtained very similar results in the two cell lines. Mouse cells provide higher-level expression of the MuLV genome, while hamster cells have the advantages of being completely negative for cross-reacting endogenous sequences, and of lacking receptors for ecotropic MuLV. The absence of receptors means that wild-type genomes are "defective" in these cells: they cannot give rise to virus particles capable of infecting the cells. This "defectiveness" is helpful for direct, quantitative comparisons of the properties of wild-type and mutant genomes). Initial tests showed that the transfectants released no infectious virus, but did produce "complementation plaque forming units" (particles capable of forming XC plaques in the presence of a helper virus) (15) upon superinfection with a helper virus (data not shown). These results show that the mutant clones represent replication-defective ecotropic MuLV genomes, and suggest that N-terminal

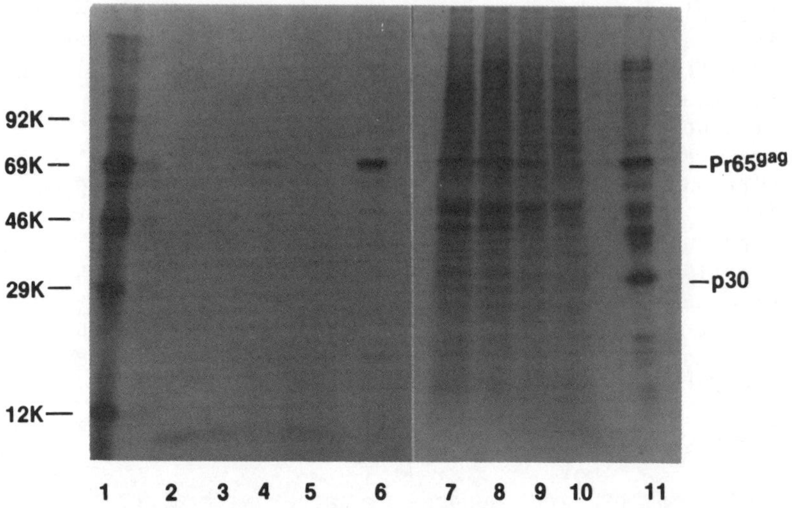

FIGURE 1. Synthesis of unmyristylated Pr65^{gag} in cells containing myristylation site mutants of MuLV. G-418 resistant hamster cells were labeled with ^{3}H-myristate (0.5 mCi/ml) (lanes 2-6) or ^{35}S-methionine (50 μCi/ml) (lanes 7-11) for 1 hr and analyzed by precipitation of cell extracts with anti-p30 serum and SDS-PAGE as described (6). The cells contained pSV2 Neo with the following inserts: lanes 2 and 7, gly → ala mutant; 3 and 8, gly deletion mutant; 4 and 9, Moloney MuLV mutant with a 288 bp deletion in pol (Rein, unpublished); 5 and 10, none; 6 and 11, wild-type Moloney MuLV.

glycine on Pr65^{gag} is necessary for the production of infectious progeny by Moloney MuLV.

We then tested for the presence of Pr65^{gag} in the transfected cells. Cells were labeled with ^{35}S-methionine, and extracts were precipitated with anti-p30 serum and analyzed by SDS-PAGE. As shown in Fig. 1, lanes 7 and 8, cells transfected with either the gly → ala mutant or the gly deletion mutant synthesize $Pr65^{gag}$, while control cells containing the pSV2Neo vector alone do not (lane 10). To determine whether the mutant $Pr65^{gag}$ molecules were myristylated, we labeled parallel cultures with ^{3}H-myristate. Lanes 2 and 3 of Fig. 1 show that these cultures did not incorporate ^{3}H label into $Pr65^{gag}$, while the label was easily detectable in transfectants containing a wild-type viral genome (lane 6) or a genome with a deletion in pol (lane 4). We conclude that if the N-terminal glycine in $Pr65^{gag}$ is replaced with alanine or is deleted, the protein is not myristylated. These results suggest that in mammalian cells, as in avian cells (12), N-terminal glycine is required for myristylation.

The Mutant $Pr65^{gag}$ Molecules do not Assemble into Virus Particles.

It was of great interest to determine whether the absence of myristic acid in $Pr65^{gag}$ would affect virus assembly. We therefore tested culture fluids from the transfected cultures in several assays for released virus particles. As is shown in Table 1, these fluids were negative for particle-associated reverse transcriptase activity. They were also negative in a sensitive immunoblotting assay for particle-associated p30-related protein (Fig. 2, lanes 2 and 3). This result is especially striking because it was obtained in hamster cells, in which the mutant and wild-type $Pr65^{gag}$ are synthesized at equal rates (data not shown). We estimate that this assay could detect 1-2% of the wild-type level of virus particle production. These negative results strongly suggest that the mutant $Pr65^{gag}$ molecules do not form virus particles which are released from the cell.

It seemed possible that myristate, the normal blocking group on the N terminus of $Pr65^{gag}$ (1,2), might be required to protect the molecule from rapid degradation in the cell. However, pulse-chase studies showed that, on the contrary, the unmyristylated $Pr65^{gag}$ molecules are metabolically very stable, with no proces-

TABLE 1
LACK OF DETECTABLE REVERSE TRANSCRIPTASE ACTIVITY IN CULTURE FLUIDS OF CELLS CONTAINING MYRISTYLATION SITE MUTANTS

Transfected DNA	Cells	RT Activity[a]
gly → ala	Mouse	< 0.004
gly deletion	Mouse	< 0.006
pSV2Neo	Mouse	< 0.004
Mo-MuLV	Mouse	5
gly → ala	Hamster	< 0.0015
gly deletion	Hamster	< 0.0015
pSV2Neo	Hamster	< 0.0015
Mo-MuLV	Hamster	0.03

[a]Reverse transcriptase activity, pmol ^{3}H-TMP incorporated per ml of culture supernatant, assayed in the presence of poly rA·oligo dT as described by Gerwin et al. (18).

sing or degradation detectable over many hours (data not shown). Thin sections of the cells were also examined in the electron microscope. 50 cell sections of mouse cells containing the mutant genomes were studied: no virus particles were seen, nor were any recognizable virus-specific structures observed in the cells. Indeed, the cells containing mutant MuLV genomes could not be distinguished from control cells in the electron microscope. In contrast, 275 particles were seen in sections of cells with wild-type virus (data not shown). Similar results were also obtained in hamster cells, except that the level of wild-type virus production is much lower than that seen in mouse cells (data not shown). It thus appears that in the absence of myristylation, $Pr65^{gag}$ not only does not assemble into released virus particles, but also does not accumulate in the cells in recognizable virus-specific structures.

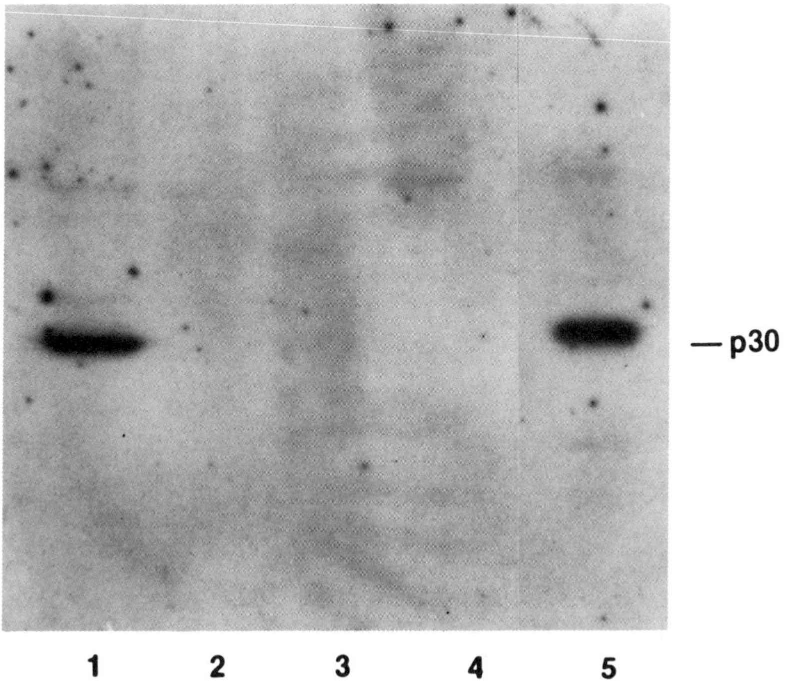

FIGURE 2. Lack of detectable virus particle production in cells containing myristylation site mutants of MuLV. Culture fluids were harvested from G-418 resistant hamster cells (lanes 1-4) or NIH/3T3 cells productively infected with Moloney MuLV (lane 5) and pelleted through 20% sucrose. The pellets were analyzed by Western blotting with anti-p30 serum as described (19). The hamster cells contained pSV2 Neo with the following inserts: lane 1, wild-type Moloney MuLV; 2, gly → ala mutant; 3, gly deletion mutant; 4, none. Lane 5 contained 1/300 of the amount of culture fluid as lanes 1-4.

Mutant $Pr65^{gag}$ is not Membrane-Associated.

We also tested the idea that the myristate group is involved in the interaction of wild-type $Pr65^{gag}$ with the cell membrane. Mouse cells containing mutant or wild-type viral genomes were suspended in hypotonic buffer and lysed by Dounce homogenization. Nuclei were removed by centrifugation, and the supernatant was made 0.3 M NaCl and centrifuged at 100,000 x g for 30 min. The amount of $Pr65^{gag}$ in each fraction (including the nuclear pellet) was then quantitated by immunoblotting. As is shown in Table 2, myristylation has a dramatic effect on the distribution of $Pr65^{gag}$ in this simple cell fractionation: it reduces the fraction of $Pr65^{gag}$ found in the post-microsomal supernatant from ~ 2/3 to ~ 1/6.

TABLE 2
FRACTION OF MUTANT AND WILD-TYPE $PR65^{gag}$
PRESENT IN CYTOSOL

Genotype	% $Pr65^{gag}$ in S_{100}
Gly → ala	73
Gly deletion	63
Wild type	15

DISCUSSION

The results presented here have several interesting implications concerning the process of assembly in the mammalian Type C retroviruses. First, it seems clear that myristate is involved in the interaction of $Pr65^{gag}$ with cellular membranes. Second, the results suggest that this interaction with membranes is a prerequisite for virus assembly.

Myristate might promote membrane association of $Pr65^{gag}$ simply by means of a hydrophobic interaction with lipid bilayers. However, extensive studies with the myristylated protein pp60^{src} have shown that myristate is necessary, but not sufficient, for association with cellular membranes (16). In addition, analysis of acylated proteins in BC_3H1 cells has demonstrated the existence of at least one major myristylated species which is not membrane-bound (17). These data suggest that the role of myristate is more subtle than the simple hydrophobic-interaction hypothesis. One intriguing possibility is that there is a specific interaction between the 14-carbon chain and a cellular component. The extreme rarity of myristate in lipoproteins is consistent with this idea.

It is also striking that the avian type C viruses, which appear quite similar to the mammalian type C viruses in their mode of assembly and overall structure, lack myristic acid in their core polyprotein, Pr76^{gag}. We have shown here that myristate performs some function in the assembly of the mammalian viruses. Presumably this function is performed by some other structural feature in the avian gag polyprotein. Perhaps further mutagenesis experiments, and/or the exchange of genetic information between different viruses, will help in identifying this hypothetical feature; in turn, this may shed further light on what the function of myristate is in the mammalian viruses.

ACKNOWLEDGEMENTS

We thank Mark Zoller and Robert Stephens for helpful discussions; Jànet Hanser and Diane Hudson for excellent technical assistance; Stephen Oroszlan for advice and support; and Cheri Rhoderick and Carolyn Phillips for preparing the manuscript.

REFERENCES

1. Bassin RH, Phillips LA, Kramer MJ, Haapala DK, Peebles PT, Nomura S, Fischinger PJ (1971) Transformation of mouse 3T3 cells by murine sarcoma virus: Release of virus-like particles in the absence of replicating murine leukemia helper virus. Proc Nat Acad Sci USA 68:1520.

2. Donoghue DJ, Sharp PA, Weinberg RA (1979) Comparative study of different isolates of murine sarcoma virus. J Virol 32:1015.
3. Yoshinaka Y, Luftig RB (1982) p65 of Gazdar murine sarcoma viruses contains antigenic determinants from all four of the murine leukemia virus (MuLV) gag polypeptides (p15, p12, p30, and p10) and can be cleaved in vitro by the MuLV proteolytic activity. Virology 118:380.
4. Scheele CM, Hanafusa H (1971) Proteins of helper-dependent RSV. Virology 45:401.
5. Henderson LE, Krutzsch HC, Oroszlan S (1983) Myristyl amino-terminal acylation of murine retrovirus proteins: an unusual post-translational protein modification. Proc Nat Acad Sci USA 80:339.
6. Schultz AM, Oroszlan S (1983) In vivo modification of retroviral gag gene-encoded polyproteins by myristic acid. J Virol 46:355.
7. Aitken A, Cohen P, Santikarn S, Williams DH, Calder AG, Smith A, Klee, CB (1982) Identification of the NH_2-terminal blocking group of calcineurin B as myristic acid. FEBS Lett 150:314.
8. Carr SA, Biemann K, Shoji S, Parmelee DC, Titani K (1982) N-Tetradecanoyl is the NH_2-terminal blocking group of the catalytic subunit of cyclic AMP-dependent protein kinase from bovine cardiac muscle. Proc Nat Acad Sci USA 79:6128.
9. Ozols J, Carr SA, Strittmatter P (1984) Identification of the NH_2-terminal blocking group of NADH-cytochrome B_5 reductase as myristic acid and the complete amino acid sequence of the membrane-binding domain. J Biol Chem 259:13349.
10. Schultz AM, Oroszlan S (1984) Myristylation of gag-onc fusion proteins in mammalian transforming retroviruses. Virology 133:431.
11. Schultz AM, Henderson LE, Oroszlan S, Garber EA, Hanafusa H (1985) Amino terminal myristylation of the protein kinase p60src, a retroviral transforming protein. Science 227:427.
12. Kamps MP, Buss JE, Sefton BM (1985) Mutation of NH_2-terminal glycine of $p60^{src}$ prevents both myristoylation and morphological transformation. Proc Nat Acad Sci USA 82:4625.

13. Zoller MJ, Smith M (1984) Oligonucleotide-directed mutagenesis: A simple method using two oligonucleotide primers and a single-stranded DNA template. DNA 3:479.
14. Southern PJ, Berg P (1982) Transformation of mammalian cells to antibiotic resistance with a bacterial gene under control of the SV40 early region promoter. J Mol Appl Genetics 1:327.
15. Rein A, Benjers BM, Gerwin BI, Bassin RH, Slocum DR (1979) Rescue and transmission of a replication-defective variant of Moloney murine leukemia virus. J Virol 29:494.
16. Garber EA, Cross FR, Hanafusa H (1985) Processing of $p60^{v\text{-}src}$ to its myristylated membrane-bound form. Mol Cell Biol 5:2781.
17. Olson EN, Spizz G (1986) Fatty acylation of cellular proteins: Temporal and subcellular differences between palmitate and myristate acylation. J Biol Chem 261:2458.
18. Gerwin BI, Rein A, Levin JG, Bassin RH, Benjers BM, Kashmiri SVS, Hopkins D, O'Neill BJ (1979) Mutant of B-tropic murine leukemia virus synthesizing an altered polymerase molecule. J Virol 31:741.
19. Schultz AM, Copeland TD, Oroszlan S (1984) The envelope proteins of bovine leukemia virus: Purification and sequence analysis. Virology 135:417.

V. GENOME REPLICATION

Positive Strand RNA Viruses, pages 241–250

REPLICATION AND PACKAGING SEQUENCES IN DEFECTIVE INTERFERING RNAS OF SINDBIS VIRUS[1]

Sondra Schlesinger, Robin Levis[2], Barbara G. Weiss, Manuel Tsiang and Henry Huang

Department of Microbiology and Immunology
Washington University School of Medicine
St. Louis, MO 63110

ABSTRACT Defective interfering (DI) genomes provide a valuable tool for identifying those sequences in a viral genome required for replication and packaging. We have analyzed a DI genome of Sindbis virus to identify such essential sequences. To achieve this goal, we cloned a cDNA copy of a complete DI genome directly downstream from the promoter for the SP6 bacteriophage DNA dependent RNA polymerase. The cDNA was transcribed into RNA which was transfected into chicken embryo fibroblasts in the presence of helper Sindbis virus. After one to two passages the DI RNA became the major viral RNA species in infected cells. Using this transfection and amplification assay we established that only 19 nucleotides at the 3' terminus and sequences in the 162 nucleotide region at the 5' terminus are absolutely required for DI RNA amplification. In addition, these DI genomes can serve as vectors for the introduction of foreign genes into cells.

1 This work was supported by grants from the National Institute of Allergy and Infectious Diseases and the Monsanto/Washington University Biomedical Research Contract.

2 R.L. has been supported by a Cellular and Molecular Biology Training Grant from the National Institutes of Health. She is presently a Stephen Morse Fellow in Microbiology and Immunology.

INTRODUCTION

Defective-interfering (DI) particles are deletion mutants of viruses that were first characterized by their ability to interfere with the replication of homologous or closely related viruses (1,2). The deleted genomes compete for proteins coded by the standard virus and replicate at the expense of the infectious particle. DI genomes provide a valuable tool for identifying sequences in a viral genome required for replication and packaging. They are less complex than the genome of the infectious virion from which they are generated, and, although they need not retain coding information, they must contain sequences required for replication and encapsidation. We have been analyzing DI genomes derived from the alphavirus, Sindbis virus, with the object of identifying the essential recognition sequences in this virus.

Sindbis virus, a plus strand RNA enveloped virus, is a member of the alphavirus genus of the Togaviridae family. The complete sequence of the Sindbis genome is known (3) as is that of one of the DI genomes (4). Furthermore, enough sequence information exists for several other alphaviruses to identify those regions of the genome that are conserved among different alphaviruses. Four regions of conservation have been identified. They are: (i) the nineteen nucleotides at the 3' terminus (5); (ii) twenty-one nucleotides spanning the start of the subgenomic 26S mRNA (6); (iii) a span of fifty-one nucleotides located from nucleotide 155 to nucleotide 205 in the Sindbis virus genome and (iv) the 5' terminal forty nucleotides (7). These regions are considered to be important as recognition sites for replication or encapsidation.

DI genomes of Sindbis virus and of the closely related Semliki Forest virus are about 15 to 20% the length of the infectious genome. Their sequence complexity is considerably less than this because several regions of the viral genome are represented as repeats in the DI genomes (4,8,9). A surprising feature of several DI RNAs of Sindbis virus is the presence of nucleotides 10 to 75 of $tRNA^{Asp}$ at the 5' terminus (10).

Our strategy for using DI genomes to identify sequences essential in the infectious virus required that we be able to modify the RNA as has been done with DNA. To this end, we cloned a cDNA copy of a complete DI genome directly downstream from the promoter for the SP6 bacteriophage DNA dependent RNA polymerase. This cDNA, or DNA in which deletions have been made, was transcribed into RNA which was then transfected into chicken embryo fibroblasts in the presence of helper Sindbis virus. After one to two passages the DI RNA became the major viral RNA species in infected cells. Using this transfection and amplification assay we were able to determine the essential sequences in this genome (11) and were also able to show that DI genomes can tolerate the insertion of foreign genes. Furthermore these foreign sequences can be translated.

RESULTS

Transcription and Transfection

The plasmid used in these experiments has been described in detail (11). The essential features are: (i) the 5' terminus of the DI cDNA positioned directly downstream of the promoter for the SP6 polymerase; (ii) the several unique restriction enzyme sites downstream of the 3' terminus of the DI cDNA at which the DNA can be linearized to permit runoff transcription and (iii) the specific restriction enzyme sites used to make deletions throughout the genome.

For transfection, monolayers of chicken embryo fibroblasts are treated with DEAE dextran and then are exposed to varying concentrations of transcribed DI RNA in the presence of infectious Sindbis virus. Essentially all of the cells receive helper virus and some fraction also receive the transcribed DI RNA. The treated cells are incubated overnight (passage 1) and then a fraction of the supernatant fluid containing virus is used to infect a new monolayer of chicken cells. The medium from this infection is harvested 8 to 16 hours later (passage 2). Cells are routinely assayed during the formation of passage 3 for the presence of DI RNA which, as shown in Fig. 1, becomes the predominant viral RNA species in those cells transfected with DI RNA. As described previously, the amplification of transfected RNA is sensitive to RNase but not to DNase and is also dependent on the presence of helper virus (11). A typical result is seen in Fig. 1 in which the concentration of transcribed RNA was varied over a 100-fold range for transfection. These data show the viral RNA species detected during the generation of passage 3. DI RNA is also readily detected during the formation of passage 2 when cells are transfected with the higher concentrations of RNA (11).

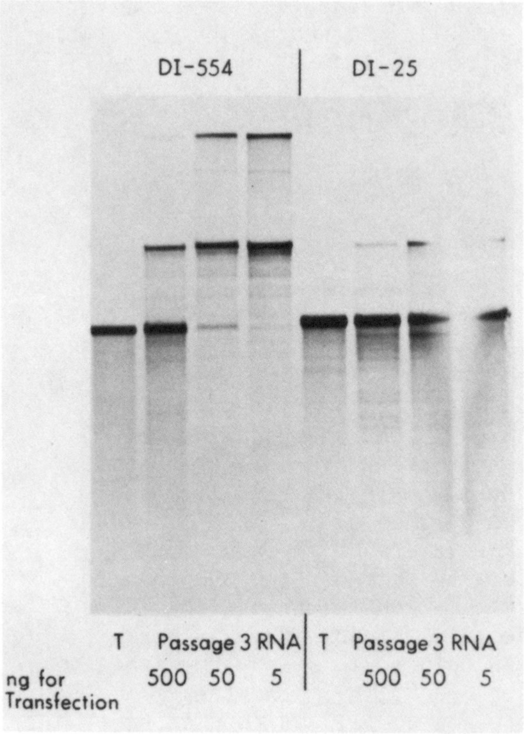

FIGURE 1. The amplification of DI RNAs of Sindbis virus in chicken embryo fibroblasts after transfection with varying concentrations of transcribed RNA. The intracellular RNA was analyzed during the formation of passage 3. T refers to the transcript used for transfection. DI-25 is derived from the complete DI cDNA (see Fig. 2). DI-554 lacks the conserved 51 nucleotide sequence (7,11).

We have generated a series of deletions extending over the entire DI genome. Each of these deletions was analyzed in the transfection and amplification assay and the results are summarized in Fig. 2. These data show that only 19 nucleotides at the 3' terminus and a maximum of 162 nucleotides at the 5' terminus are absolutely required for the biological activity of Sindbis DI RNAs. The nineteen nucleotides at the 3' terminus are those defined by Ou et al to be highly conserved among alphaviruses (5). Thus, these nucleotides appear to be necessary and may be sufficient to define the recognition site for initiation of transcription of the negative strand.

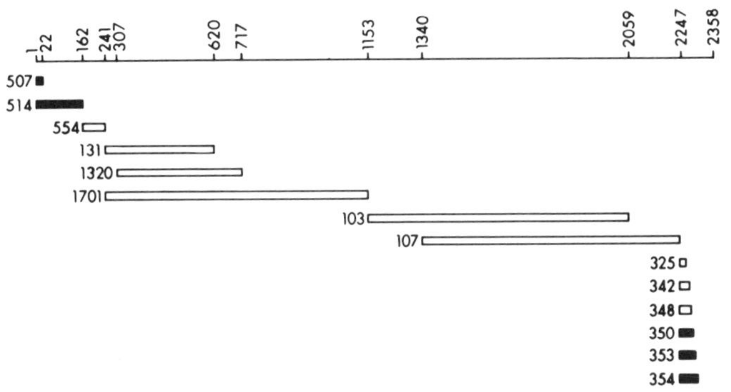

Figure 2. A deletion map of the genome of a Sindbis DI RNA. The deletions, which cover the entire DI genome are all indicated on this diagram. The open boxes represent regions which were deleted without causing a loss of biological activity. The closed boxes are regions which when deleted led to a complete loss in biological activity. Published with permission of Cell (11).

The 5' terminus of the DI RNA has two regions of interest - the extreme 5' end and the 51 conserved nucleotides located from nucleotides 155 to 205 in the virion RNA. The observations that naturally-occurring DI RNAs vary at the extreme 5' terminus and that this latter region is the least conserved among alphaviruses (7) led to the proposal that the 51 nucleotide region plays a crucial role in one or both functions of replicase binding and of encapsidation. Deletion of this region, however, does not destroy the biological activity of the DI RNA and therefore the function of these sequences remains in question (see below and Fig. 1).

Although the 51 nucleotide conserved sequence is not essential for the biological activity of DI RNA, sequences further upstream from this region are required. A cDNA with a deletion of the first 7 nucleotides and one with a deletion from nucleotides 22 to 162 produce RNA transcripts that are not viable. The 5' terminal 7 nucleotides are part of the $tRNA^{Asp}$ sequence which is not present either in several other DI RNAs or in Sindbis virion RNA and may be only one of several sequences able to be recognized at the 5' terminus. Based on their comparisons of the exact 5' terminus of several different alphaviruses, Ou et al. (7) suggest that the structural features of the termini may be more important than the actual sequence and our data are consistent with this possibility.

Deletion Analysis at the 5' Terminus

The 5' terminal sequences of DI-25 are of special interest because they are essentially identical to nucleotides 10-75 of the rat $tRNA^{Asp}$ (10). Sequence analysis proved that the 3' terminal CCA of the tRNA is joined to viral sequences at nucleotide 31 of the virion RNA. We have made several deletions in the tRNA region of the DI cDNA (Fig. 3). Deletions extending from nucleotide 18 to nucleotide 59 do not inactivate the DI genome. These deletions cover the anticodon (AC) and ψ stem and loop regions. Deletions extending either upstream to nucleotide 5 or downstream to nucleotide 104 produce RNA transcripts which are not amplified after transfection and passaging.

Insertion of Foreign Sequences into Sindbis DI Genomes.

Only a small fraction of Sindbis sequences are essential for propagation of the DI genomes. This observation suggested that the defective genomes might serve as vectors for the expression of foreign genes. To determine the feasibility of this idea we inserted two foreign genes into DI-25. The two genes are the bacterial chloramphenicol acetyltransferase (CAT) gene and the vesicular stomatitis virus (VSV) glycoprotein gene - the G gene. In both cases, the insertion was first made between the HindIII site at nucleotide 162 and a site at nucleotide 1928 deleting 1766 nucleotides of DI-25 including the 51 base highly conserved region. The total CAT gene is actually less than 700 nucleotides but to maintain the appropriate size of the DI RNA a 1.6×10^3 nucleotide fragment including bacterial and SV40 sequences as well as the CAT gene was inserted.

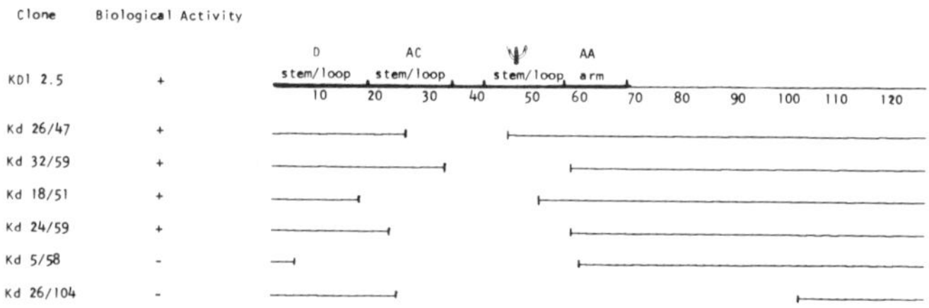

FIGURE 3. A diagram of the 5' termini of DI-25 and deletion mutants. The first 68 nucleotides include nucleotides 10-74 of $tRNA^{Asp}$ (10,11). The specific domains of the tRNA are indicated. The deleted clones are labeled to indicate the first and last nucleotides included in the deletion. Biological activity refers to the appearance of DI RNA by passage 3. The two negative samples were also negative at passages 4 and 5.

These cDNA plasmids produced RNA transcripts that were not amplified in our standard transfection and passaging assay. In both cases, however, when the foreign sequences were inserted into a plasmid in which the 51 nucleotide 5' conserved region had been retained, the transcripts were amplified. In retrospect, these results are not surprising (Fig. 1). The 51 nucleotide conserved region is clearly not essential for biological activity (11), but its presence permits the DI RNA to be biologically active at lower concentrations. That the DI RNAs isolated from infected cells actually contain the foreign sequences was demonstrated by blot analysis. In addition cells that are replicating DI CT25, the DI RNA containing the CAT gene, are also producing an enzymatically active chloramphenicol acetyltransferase (Fig. 4).

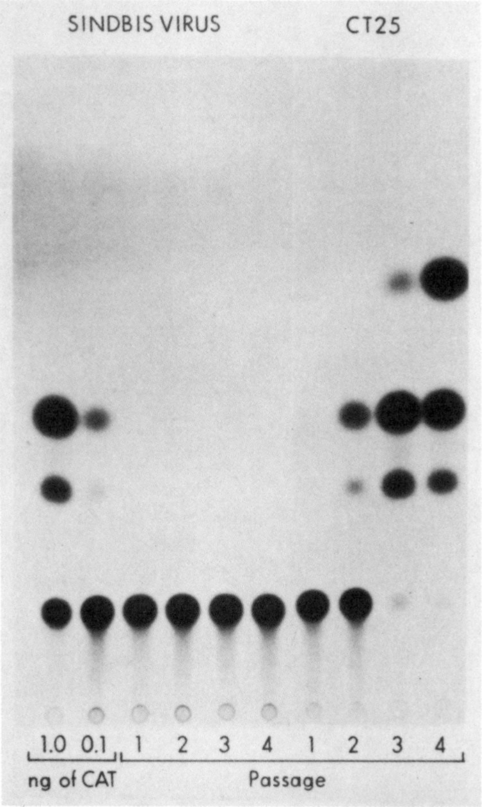

FIGURE 4. Chloramphenicol acetyltransferase (CAT) activity in cells infected with Sindbis virus and DI particles containing the CAT gene insert. The first two lanes show the CAT activity obtained with 1.0 and 0.1 ng of purified enzyme.

DISCUSSION

Our studies show by deletion mapping and by insertion of foreign sequences that only small regions at the 3' and 5' termini of the DI genome of Sindbis virus are essential for these DI RNAs to be replicated and encapsidated. As documented in Fig. 2, however, some deletions may have a strong quantitative effect and produce DI RNA molecules that are less active. Amplification of DI RNAs depends on their efficiency of replication and on their ability to be encapsidated. Furthermore, the stability of the transfected RNA may be a factor in its success in being amplified. In particular, an important but unanswered question is whether or not recognition sites for replication and for packaging are distinct or are overlapping. It should be possible to answer this question by determining the level of replication during the first passage and such experiments are in progress.

The feasibility of introducing foreign genes into cells as RNA was first demonstrated by French et al in plant protoplasts using brome mosaic viral RNA as a vector (12). Our studies with DI RNAs of Sindbis virus show that insertion of foreign sequences into these genomes is compatible with replication and encapsidation of these genomes. The DI RNA containing the CAT mRNA is translated to produce active enzyme. Although the efficiency of these RNAs has not yet been optimized our results indicate that Sindbis virus and its DI particles may prove to be useful vectors for introducing genes - in the form of RNA - into eucaryotic cells.

ACKNOWLEDGEMENTS

We thank Sherry Gee, David Kingsbury II and Rebecca Wright for their important contributions to this work.

REFERENCES

1. Holland JJ, Kennedy SIT, Semler BL, Jones CL, Roux L, Grabau EA (1980). Defective interfering RNA viruses and host-cell response. In Fraenkel-Conrat H, Wagner RR (eds): Comprehensive Virology New York: Plenum Press 16: pp 137-192.

2. Perrault J (1981). Origin and replication of defective interfering particles. Curr Top Microbiol Immunol 93: 151-207.
3. Strauss EG, Rice CM, Strauss JH (1984). Complete nucleotide sequence of the genomic RNA of Sindbis virus. Virology 133: 92-110.
4. Monroe SS, Schlesinger S (1984). Common and distinct regions of defective-interfering RNAs of Sindbis virus. J Virol 49: 865-872.
5. Ou JH, Strauss EG, Strauss JH (1981). Comparative studies of the 3'-terminal sequences of several alphavirus RNAs. Virology 109: 281-289.
6. Ou JH, Rice CM, Dalgarno L, Strauss EG, Strauss JH (1982). Sequence studies of several alphavirus genomic RNAs in the region containing the start of the subgenomic RNA. Proc Natl Acad Sci USA 79: 5235-5239.
7. Ou JH, Strauss EG, Strauss JH (1983). The 5'-terminal sequences of the genomic RNAs of several alphaviruses. J Mol Biol 168: 1-15.
8. Lehtovaara P, Soderlund H, Keranen S, Pettersson RF, Kaarianen L (1981). 18S defective-interfering RNA of Semliki Forest virus contains a triplicated linear repeat. Proc Natl Acad Sci USA 78: 5353-5357.
9. Lehtovaara P, Soderlund H, Keranen S, Pettersson RF, Kaarianen L (1982). Extreme ends of the genome are conserved and rearranged in the defective-interfering RNAs of Semliki Forest virus. J Mol Biol 156: 731-748.
10. Monroe SS, Schlesinger S (1983). RNA's from two independently isolated defective-interfering particles of Sindbis virus contain a cellular tRNA sequence at their 5' ends. Proc Natl Acad Sci USA 80: 3279-3283.
11. Levis R, Weiss BG, Tsiang M, Huang H, Schlesinger S (1986) Deletion mapping of Sindbis virus DI RNAs derived from cDNAs defines the sequences essential for replication and packaging. Cell 44: 137-145.
12. French R, Janda M, Ahlquist P (1986). Bacterial gene inserted in an engineered RNA virus: efficient expression in monocotyledonous plant cells. Science 231: 1294-1297.

Positive Strand RNA Viruses, pages 251–259

ALPHAVIRUS PLUS STRAND AND MINUS STRAND RNA SYNTHESIS[1]

Dorothea Sawicki and Stanley Sawicki

Department of Microbiology, Medical College of Ohio
Toledo, Ohio 43614

ABSTRACT During the replication cycle of alphaviruses, minus synthesis stops normally at the same time that plus strand synthesis reaches a maximal rate. We have studied alphavirus temperature sensitive mutants and their revertants to elucidate the mechanisms regulating minus strand synthesis and determining the maximal rate of plus strand synthesis. We present our model for the temporal regulation of alphavirus minus strand synthesis which involves the preferential utilization and stable association of minus strands by the replication complex.

INTRODUCTION

Togaviruses are a family of enveloped, plus-strand RNA viruses that include the alphaviruses (the best studied members of which are Sindbis virus, SIN, and Semliki Forest virus, SFV), pestiviruses and rubella virus. The successful synthesis of 49S genome RNA and subgenomic 26S mRNA is dependent on the synthesis of genome-length minus strand RNA. Alphavirus minus strand RNA synthesis occurs only early after infection

[1]This work was supported by the US-Japan Medical Sciences Program through grant AI-15123 and AI-00510 from the NIH.

and requires concomitant protein synthesis (1,2). The failure to continue minus strand synthesis late after infection might be caused by excessive production of 26S mRNA which would be translated at the expense of the 49S plus strand and result in the failure to continue synthesis of the viral nonstructural proteins, or might result from the 26S mRNA competitively inhibiting the minus strand polymerase, or from the binding of capsid to a site on the 49S plus strand that is required for replicase binding (reviewed in 3). We have recently tested several of these hypotheses in an attempt to understand the mechanisms responsible for regulating minus strand synthesis and for determining the maximal rate of plus strand synthesis. We were aided in this study by the availability of several temperature-sensitive (ts) mutants of SFV and the heat-resistant (HR) strain of SIN, and their revertants.

RESULTS

The synthesis of alphavirus minus strand RNA requires plus strand templates and viral nonstructural proteins. Minus strand synthesis is detected beginning about 1 h p.i. in SIN and SFV infected cells (1,2). After alphavirus transcription has commenced, minus strand but not plus strand synthesis is sensitive to inhibition of protein synthesis. Figure 1 shows that treatment of SFV infected cells with cycloheximide during the early period of the replication cycle resulted in the premature cessation of minus strand synthesis. We concluded from these kinds of experiments that the minus strand polymerase had a short functional half-life, whereas the plus strand polymerase activity was stable. We characterized ts mutants of SIN HR that were unable to synthesize viral RNA at nonpermissive temperature (complementation groups A,B,F and G) in order to identify cistrons in which ts mutations would affect the synthesis of minus strands. We demonstrated that in cells infected with ts11 (the sole member of the B group) minus strand synthesis, but not plus strand synthesis, was temperature sensitive (2). Our recent studies (4) indicated that in addition to ts11, ts4 of the A group probably also contains a ts mutation in the B cistron. Since ts6 of the F group also possessed ts minus strand synthesis (2),

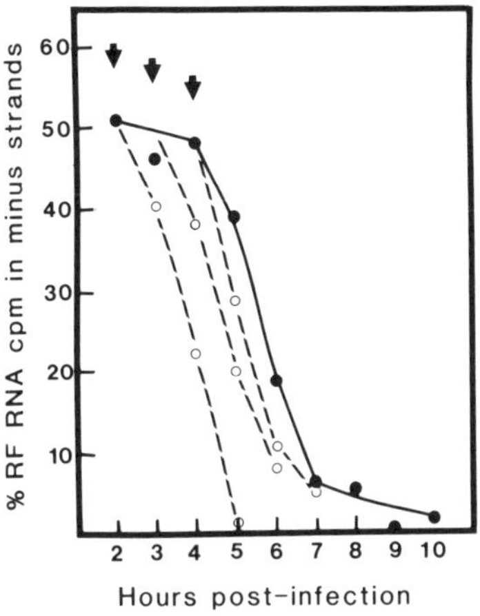

Figure 1. Alphavirus minus strand synthesis is short-lived. CEF cells were infected with SFV at 30C and treated with 100 ug/ml of cycloheximide at 2, 3 or 4 h p.i. (arrows). The treated and untreated cultures were labeled with ^{3}H-uridine (200uCi/ml) for 60 min in the presence (O) or absence (0) of cycloheximide. The % of labeled RF RNA that was in minus strand RNA was determined as described (8).

but complemented ts11 and ts4, the ts mutation in ts6 that affected minus strand synthesis was in a different cistron that the ts mutation in ts11 or ts4. We concluded that at least two cistrons function in minus strand synthesis and that the B cistron functions selectively in minus strand synthesis.

Figure 2 demonstrates that alphavirus minus strand synthesis is temporally regulated. Minus strands are synthesized early but not late after infection. Up to 50% of pulse-labeled RNA in double-stranded RNA cores (RFs) of the replicative intermediates is in minus strands when infected cells are labeled early after infection. Also shown in Figure 2 is the pattern of minus strand RNA synthesis in cells infected with a murine coronavirus. Although minus strand synthesis decreased late after infection, it was not reduced to the very low levels observed in SIN and SFV infected cells. In contrast to alphaviruses, plus strand synthesis in coronavirus

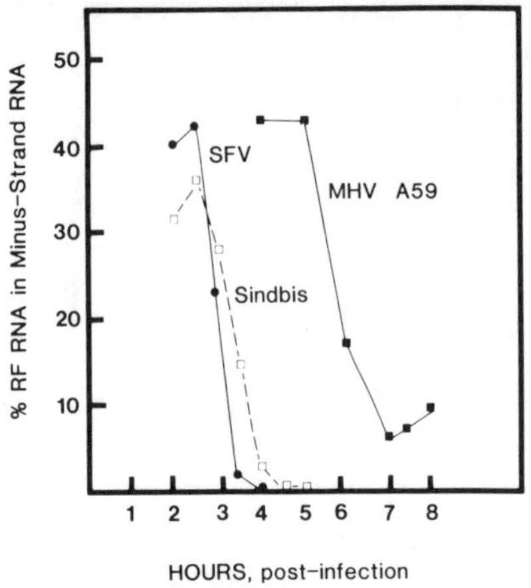

Figure 2. Comparison of minus strand synthesis by alphaviruses and coronavirus. Cells were infected at 37C with SFV, SIN HR or the A59 strain of mouse hepatitis virus, a murine coronavirus. The % of the labeled RF RNA in minus strands synthesized during 30 min periods was determined as described in Figure 1.

infected cells is inhibited by cycloheximide (5) as is poliovirus plus strand synthesis (6). Thus, alphaviruses are unique among plus-stranded RNA viruses because they regulate minus strand synthesis and form a stable plus-strand polymerase. The cessation of alphavirus minus-strand synthesis is due to the failure to initiate transcription of new minus strands rather than to the rapid turnover of newly synthesized minus strands (1). The cessation of minus strand synthesis does not appear to be caused by accumulation of capsid protein. Cells infected with SFV ts3, a mutant defective in the cleavage of the structural polyprotein (7), shut off minus strand synthesis normally at 40C (unpublished results). Also, cells infected with SIN HR ts24 and shifted from 30C to 40C late in infection when large amounts of capsid were present in the infected cells resumed minus strand synthesis (8).

Since the minus strand polymerase activity is normally short-lived and disappears quickly after inhibition of protein synthesis, the cessation of minus strand synthesis might result from the failure to synthesize viral nonstructural proteins late in infection. However, we found in ts17, ts24, and ts133 infected cells, and in cells infected with revertants of ts24, that functional minus strand polymerase was present long after synthesis of minus strands had terminated (4,8,9). Thus, proteolytic cleavage cannot be responsible for the inactivation or loss of an essential component of the minus strand polymerase. A mutant of SFV, ts1, which overproduced the viral nonstructural proteins and synthesized much less 26S mRNA and viral structural proteins at 40C than at 30C (7), synthesized the same quantity of minus strand RNA and shut off minus strand synthesis at the same time as did wild type SFV (9). On the other hand, as shown in Figure 3, cells infected with a revertant of ts24, that fails to shut off minus strand synthesis at 40C (8), continued to synthesize viral nonstructural proteins late in infection. Thus the cessation of minus strand synthesis is not due to a reduction in the synthesis of viral nonstructural proteins late in infection or to the overproduction of nonstructural proteins.

Four nonstructural proteins have been identified in SIN and SFV infected cells (11,12), but only three of these (nsp1-3) are produced in large amounts in infected cells. We found that the synthesis of nsp4 in cells infected with SFV ts1 was readily detectable after treatment with tunicamycin and that all four nonstructural proteins continued to be synthesized late in infection after minus strand synthesis had ceased, as well as early in infection when minus strands were being actively made (9). Furthermore, nsp4 did not appear to turnover during long chase periods. Therefore, minus strand synthesis does not appear to be regulated by the temporal synthesis of viral nonstructural proteins.

We have suggested that the cessation of minus strand synthesis may be regulated by a viral encoded function that is ts in ts24 (8). If one of the viral cistrons encoded a catalytic polypeptide domain that inactivated selectively the minus strand polymerase activity, then the accumulation of viral proteins would instantaneously inactivate minus strand polymerase and result in the failure to synthesize minus strands late but not early in infection. This inactivating function would be ts in ts24, which is an A group mutant. If regulation of minus strand synthesis were a function of the A cistron, then

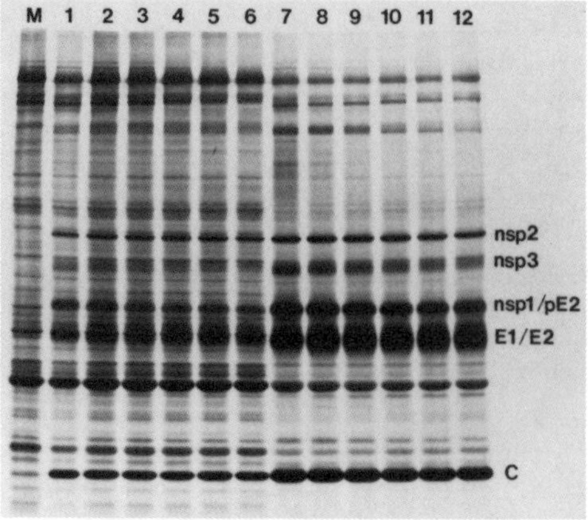

Figure 3. Alphavirus nonstructural proteins are synthesized and are stable early and late after infection. CEF cells were infected with a revertant of ts24 at 40C in the presence of actinomycin D. At 2.5 h p.i. (ln 1-6) or at 4.5 h p.i. (ln 7-12), the cells were pulse-labeled with ^{35}S-methionine (50uCi/ml) and then chased for 0 min (ln 1,7), 15 min (ln 2,8), 30 min (ln 3,9), 60 min (ln 4,10), 90 min (ln 5,11) or 120 min (ln 6,12). Mock infected cells (M) were labeled for 60 min. Proteins were displayed on 8-12% linear gradient gels (8).

other A group mutants would also possess the phenotype of ts24. Analysis of all 10 members of the A group indicated that only three (ts17, ts24, ts133) were ts in the regulation of minus strand synthesis. Furthermore, we isolated revertants of ts24 that replicated efficiently at 40C but which nevertheless retained the ts defect in the regulation of minus strand synthesis (4). Therefore, the ts defect in the regulation of minus strand synthesis is not conditionally lethal and might map outside the A cistron.

In cells infected with revertants of ts24, minus strand synthesis did not shut off at 40C but continued at a high rate. Nevertheless, the rate of plus strand synthesis plateaued at the same level and at the same time after infection as in SIN HR infected cells (9). The minus strands accumulated as double-stranded RNA and entered RIs that were engaged in 26S mRNA synthesis. Continued synthesis of viral nonstructural proteins (as demonstrated in Figure 3) coupled with continued minus strand synthesis at 40C did not result in an increase in the rate of plus strand synthesis late after infection with revertants of ts24 (9). Furthermore, viral protein synthesis was not required late after infection for newly made minus strands to function as templates.

DISCUSSION

Based on the above results, we propose the following working model for alphavirus transcription (Figure 4). Infected cell possess only a limited number of potential sites to form alphavirus replication complexes. Early in infection, viral nonstructural proteins associate with

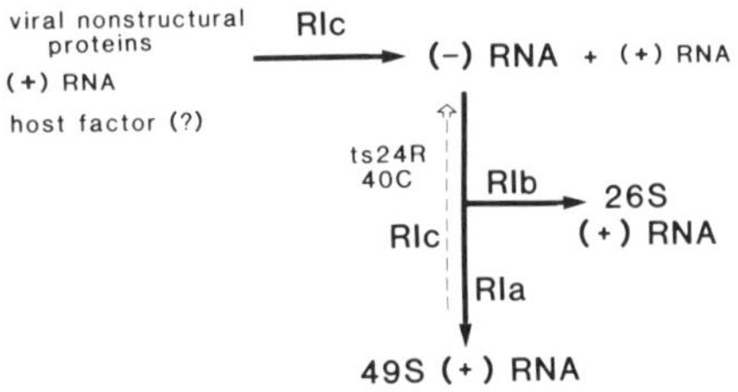

Figure 4. Preferred template model for alphavirus RNA synthesis. Early after infection nonstructural proteins associate with 49S plus strands to form a replication complex. The newly formed replication complex (RIc) synthesizes a minus strand which is the preferred template of the replication complex and is converted to a replication complex engaged in plus strand synthesis (RIa and RIb). In some Sindbis virus mutants the stable association of the minus strands with the replication complex is temperature sensitive. At 40C a newly made 49S plus strand exchanges with the minus strand template and converts an RIa to an RIc which in turn is converted back to an RIa and RIb.

these sites to form replication complexes actively engaged in viral RNA synthesis. Therefore, inhibition of protein synthesis early after infection prevents the number of replication complexes from increasing; and even if more minus strands are made, they cannot be used to increase the rate of viral plus strand synthesis (9). Initially the replication complex binds a plus strand enabling synthesis of minus strands. However, once synthesized, the minus strand becomes the preferred template because it associates stably with the replication complex. With time, therefore, all the replication complexes will be engaged in plus strand synthesis; because all of the potential sites are occupied, plus strand synthesis will have reached its maximum rate. Neither the accumulation of greater numbers of minus strands as occurs at 40C in cells infected with revertants of ts24 nor the overproduction of nonstructural proteins as occurs in SFV ts1 infected cells will cause an increase in the rate of plus strand synthesis. For minus strand synthesis to resume, the replication complex must replace the minus strand template with a plus strand template. The stable association of the minus strand with a replication complex in ts17, ts24, ts133 and revertants of ts24 would be temperature sensitive. Therefore, at 40C minus strand synthesis resumes in cells infected with these mutants but does not result in an increase in the rate of plus strand synthesis.

REFERENCES

1. Sawicki DL, Sawicki SG (1980). Short-lived polymerase for Semliki Forest virus. J Virol 34:108.
2. Sawicki DL, Sawicki SG, Keranen S, Kaariainen L (1981). A Sindbis virus coded function for minus strand synthesis. J Virol 39:348.
3. Strauss EG, Strauss JH (1983). Replication stategies of the single stranded RNA viruses of eukaryotes. Curr Topics Microbiol Immunol 105:1.
4. Sawicki DL, Sawicki SG (1985). Functional analysis of the A complementation group mutants of Sindbis HR virus. Virology 144:20.

5. Sawicki SG, Sawicki DL (1986). Coronavirus minus strand RNA synthesis and effect of cycloheximide on coronavirus RNA synthesis. J Virol 57:328.
6. Baltimore D (1968). Structure of the poliovrius replicative intermediate. J Mol Biol 32:359.
7. Keranen S, Kaariainen L (1979). Functional defects of RNA negative temperature sensitive mutants of Sindbis and Semliki Forest virus. J Virol 32:19.
8. Sawicki SG, Sawicki DL, Kaariainen L, Keranen S (1981). A Sindbis virus mutant temperature sensitive in the regulation of minus stand synthesis. Virology 115:161.
9. Sawicki DL, Sawicki SG (1986). The effect of regulation of minus strand RNA synthesis on Sindbis virus replication. Virology 151: in press.
10. Sawicki SG, Sawicki DL (1986). The effect of overproduction of nonstructural proteins on alphavirus plus strand and minus strand RNA synthesis. Virology, in press.
11. Keranen S, Ruohonen L (1983). Nonstructural proteins of Semliki Forest virus: synthesis, processing and stability in infected cells. J Virol 47:505.
12. Lopez S, Bell JR, Strauss EG, Strauss JH (1985). The nonstructural proteins of Sindbis virus as studied with an antibody for the C terminus of the nonstructural readthrough polyprotein. Virology 141:235.

Positive Strand RNA Viruses, pages 261–272

REPLICATION OF FLAVIVIRUSES[1]

Margo A. Brinton and Janet B. Grun

The Wistar Institute, Philadelphia, PA 19104

ABSTRACT We are attempting to characterize flavivirus polymerase activity and to identify the flavivirus RNA transcriptional control sequences and the proteins involved in viral RNA synthesis through the study of West Nile virus (WNV), strain E101, replication complexes.

INTRODUCTION

Flaviviruses are small (50-55 nm), enveloped plus strand viruses. After uncoating, the genome RNA, which is also the only flavivirus mRNA, must first be translated to produce viral replicase protein(s) needed for viral RNA synthesis. Flaviviruses encode 7 non-structural proteins, but little is currently known about the functions of these proteins (1). It has been suggested that the two largest, NS3 (67K) and NS5 (96K), are polymerases. NS5 contains some sequence homology with the non-structural regions of the alpha togaviruses and some plant viruses (1). However, since there are two large, viral nonstructural proteins (NS3 and NS5), it is possible that the flavivirus plus and minus strand replication complexes may differ in their composition.

During the viral replication cycle, minus strand RNA is never found as a free strand in infected cells, but is always within replicative intermediates (2). Although minus strand RNA synthesis continues throughout the infection, after the latent period plus strand synthesis is more efficient than minus strand synthesis (3).

[1]This research is supported by AI-18382 from NIAID.

RESULTS

Search for flavivirus transcriptional signal sequences.

The non-coding terminal regions of RNA genomes contain sequences which function as transcriptional control signals. Such sequences would be expected to be highly conserved among different flaviviruses and to be present within the 3'-terminal regions of both the plus and minus strand viral RNAs.

3'-terminal sequence of the plus strand RNA. We have directly sequenced the 3' termini of end-labeled WNV, strain E101, and St. Louis encephalitis (SLEV), strain 75V 14532, genome RNAs (4). The RNAs were first labeled at their 3' termini with [^{32}P]-pCp using T_4 RNA ligase (5) and then partially digested with one of five RNases (6). The five RNases were inefficient at digesting certain regions within the first 80 nucleotides (4). While stable secondary structures are generally resistant to enzymatic hydrolysis (7, 8), such regions can be sequenced with the chemical modification method of Peattie (8). The WNV and SLEV 3' sequences obtained were compared to the published sequence of yellow fever virus (YFV), strain 17D (Fig. 1; 1).

```
          1         2                              3            4
    3'
WNV HO-UCAUAGG...ACACAAGAGCGUGGUGGUCGGUGGU....AACGAGCCGC.GUGACGACGGCACCACCGACC
YFV HO-UCACCAAA..ACACAAACAGUAGGUUUCCAGAC..GAAUAA.GAACUCGUUUGACACGAGUGUCUGGAGAC
SLE HO-UCAGAACGG.ACACAGXXGUUUGXGCGGGXAGCA.GA.GAA.GC.XXXXGGAGACX---------------

             5          6          7
       ACCAUGCUCUCGUCUUCUA...GAGGACCAGACAGGGUCCACAG.UAUACGACAAACGACUCACAAACACU
       CUCCUUU..UCGUCUCUUGG.UGAGG.CCAGAAAGGGACCGCAGUUAUACCAGGGUGAUAAACAAGGGACC
       ---------------------XGAGGAC.AGAAAGGGACCUCAGAUAUACG--------------------

       GGUGUGUGUAGUA 5'
       UCCCAGAGGAGAU
       -------------
```

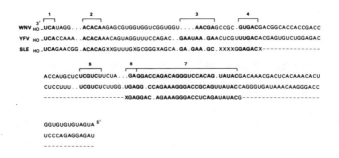

Figure 1. Comparison of the 3'-terminal sequences of the genomes of three flaviviruses. Gaps introduced to align homologous sequences are indicated by dots. Unidentified bases are indicated by x's or dashes. Regions of conserved sequence are indicated by numbered brackets. Reproduced by permission from Brinton et al., 1986.

Using the RNA5 computer program (9), a very stable stem and loop structure was predicted for the WNV 3'-terminal sequence (Fig. 2). The RNase digestion pattern predicted a similar structure (4). Among the three flaviviruses compared, sequence conservation was observed in 7 regions as indicated by brackets in Fig. 1, but only regions 1, 2, and 5 contained highly conserved sequences. These 7 regions were all located in loops, or, in the case of region 7, outside the terminal secondary structure (Fig. 2). Although the WNV, YFV and SLEV 3'-terminal sequences can be folded into secondary structures of similar size and shape (1, 4), there appears to be no conservation of the sequences which constitute the stems of these structures. This suggests that the form of the structure serves an important function for the virus. The conservation of sequences located within loops implies that these sequences may be conserved by the specificity of their interactions with viral or cellular proteins which may be involved in RNA replication.

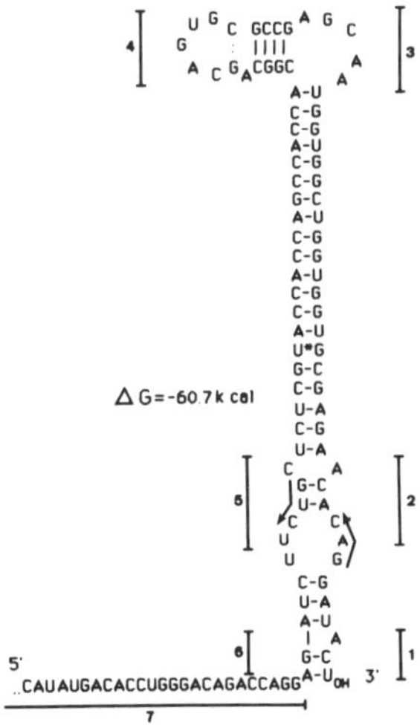

Figure 2. 3'-terminal sequence of WNV genome RNA folded into a stable secondary structure. Brackets indicate regions containing conserved sequences. Reproduced by permission from Brinton et al., 1986.

<u>5'-terminal sequence of the plus strand RNA</u>. The 5'-terminal sequence of the plus strand RNA was analyzed in order to deduce the complementary 3'-terminal sequence of the minus strand RNA. The entire non-coding regions of the WNV and SLEV RNAs were sequenced by the dideoxy method using a synthetic DNA primer. Some sequence conservation was observed in 8 regions (Fig. 3) as indicated by bold letters. The initial three nucleotides and the region 2 nucleotides (Fig. 1) were found to be present at the 3' terminus of both the plus and minus strand RNAs (4, 10). However, the region 2 sequence, while completely conserved in the plus strand, shows only partial conservation in the minus strand (Fig. 3).

```
WNV   5' m7GXXXAG·UUCGCCUGUGUGAG················
SLEV     m7GXXXAUGUUCGCGUCGGUGAGCGGAGAG·········
YFV      m7GAGUAAAUCC···UGUGUGCUAAUUGAGGUGCAUUGG

         ·CUGACAAAC····UUAGUAG·············UUUG
         ······GAAACAGAUUUCUUU·············UUUG
         UCUG·CAAAUCGAGUUGCUAGGCAAUAAACACAUUUG

         UGAGGA··UUAACAACAAUUAACACGGUGC·GAGCUCU
         GA·GGA···UAACAAC··UUAA················
         GA······UUAAUU····UUAAUCGUUCGUUGAGCG··
                                                 Primer
                                                  ↓     3'
         UUCUUAGCAC···G·AAGAUCXXX···············AUG...
         ··CUU·G·ACUGCG·AACAGUUUUUUAGCAGGGAAUXXXAUG...
         ··AUUAGCAGA··G·AACUGACCAGAAC···········AUG...
```

Figure 3. Comparison of 5'-terminal sequences of WNV, YFV, and SLEV. The bold letters indicate regions of conservation. Gaps introduced to align homologous sequences are indicated by dots. (Brinton and Dispoto, submitted).

The 5'-terminal sequences were folded using the RNA5 program. The resulting secondary structures had a large side loop of variable size (Fig. 4). The 5'-terminal, type I cap is not shown. The 5' structure is less stable than the 3'-terminal structure. The energy shown for the WNV structure is the combined energy for both stems. The energy for the longer stem alone is -21.4 kcal. In the 5'-structure the region 2 conserved sequence was usually located within a base-paired stem region, rather than in a loop region as it is in the 3'-terminal structure.

The 3'- and 5'-terminal sequences of a non-temperature sensitive, replication-efficient mutant (11, 12) of WNV have also been compared with those of the parental virus RNA. No sequence changes were observed between the two genomes in the first 150 nucleotides of the 3' terminus or in the 5'-non-coding region. This indicates that the relevant change may be located in one of the viral non-structural proteins or in an internal replication signal sequence.

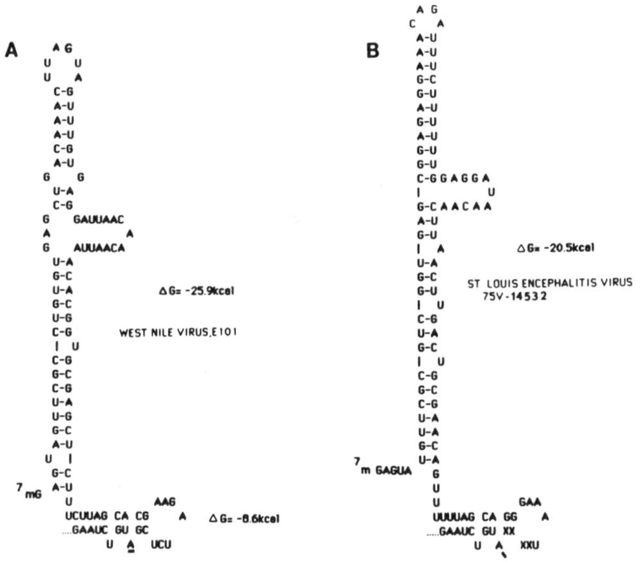

Figure 4. Secondary structures formed by 5'-terminal sequences of WNV and SLEV. The A of the AUG translation initiation codon is underlined. (Brinton and Dispoto, submitted).

Viral proteins associated with in vitro viral polymerase activity.

As a first step in characterizing WNV replication complexes, we developed an in vitro transcription assay using pelleted cytoplasmic membrane fragments obtained from infected BHK cells (13). Cells are first swollen in hypotonic buffer and then disrupted in a Dounce homogenizer. The nuclei were pelleted and the supernatant was subjected to high speed centrifugation. The viral-specific polymerase activity was associated with these pellets. All of the viral proteins are associated with cytoplasmic membranes, especially those in the perinuclear region of the cell. Perinuclear membranes are the site of viral macromolecular synthesis and the recovery of viral specific proteins and polymerase activity could be doubled by passing the pelleted nuclei through syringes to strip off more of the perinuclear membranes (13). However, even after this treatment much of the viral protein remained associated with the nuclei.

Through further purification of the WNV replication complexes, we have separated the majority of the NS5 (96K) protein from the other viral proteins, including the NS3 (67K) protein (data not shown). Ninety percent of the viral polymerase activity remained associated with the 67K protein fraction, while about 8% of the activity was associated with the 96K protein. These preliminary data indicate that the 67K protein may be responsible for the majority of the polymerase activity measured in vitro. The majority of the RNA synthesized in these reactions is plus strand. Therefore, it is reasonable to hypothesize that different complexes synthesize plus and minus strand RNAs, with the NS3 (67K) protein copying plus strands from minus strands and the NS5 (96K) protein copying minus strands from plus strands.

Analysis of the product RNAs from in vitro transcription.

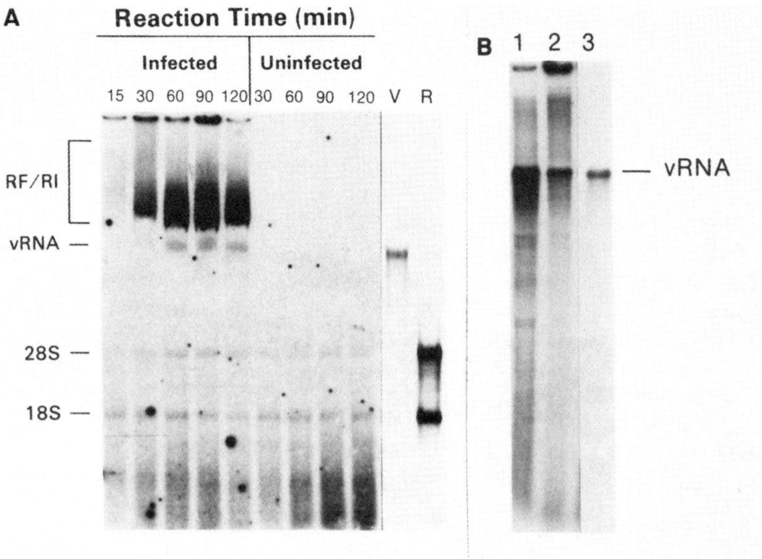

Figure 5. Time course of WNV in vitro polymerase activity. A. [α-^{32}P] UTP labeled RNA products from reactions containing either uninfected or infected cell cytoplasmic pellets were electrophoresed under non-denaturing conditions on 0.8% agarose gels. Markers were genome RNA (V) and cell ribosomal RNA (R). B. The RNA in the RF/RI band was separated by LiCl and the resulting RI (lane 1) and RF (lane 2) RNAs were electrophoresed under denaturing conditions. Lane 3 - marker genome RNA.

Cytoplasmic pellets from infected cells were used as the source of WNV replication complexes and reaction conditions were optimized for WNV activity (13). Incorporated radioactivity was first detectable by 15 minutes in a broad virus-specific band migrating near the top of the gel (Fig. 5A). By 30 minutes, radioactivity was also detected in a band which comigrated with genome RNA. Cellular RNAs were labeled in reactions containing extracts from both infected and uninfected cells. The RNA in the upper virus-specific band could be separated into LiCl-insoluble and -soluble fractions (13), indicating that both replicative intermediate (RI) and replicative form (RF) RNAs were labeled during the reaction. When the RI RNA was denatured and then electrophoresed on a denaturing gel, it was found to contain both genome-sized RNA and heterogeneous smaller RNAs (Fig. 5B). The RF RNA contained mostly genome-sized RNA. When the in vitro labeled, virus-specific RNA contained in the lower band (Fig. 5A) was compared to orthophosphate-labeled genome RNA by oligonucleotide fingerprinting on 2-dimensional mini gels, similar patterns were observed (data not shown). This same RNA also co-migrated with genome RNA under denaturing conditions and reacted with a cDNA probe which was complementary to the genome RNA.

Possible involvement of host proteins in WNV replication.

Host cell proteins have been shown to play specific roles in the replication of Qβ phage (14, 15) and picornaviruses (16). The poliovirus host factor apparently has uridylyl transferase activity (17). The cytoplasmic pellets used for our in vitro WNV transcription assays were found to contain both cellular terminal uridylyl and adenylyl transferase activities. When [α-^{32}P]-GTP (Fig. 6) or [α-^{32}P]-CTP (data not shown) were used, only viral-specific RNAs were labeled by extracts from infected cells and no labeled products were observed with extracts from uninfected cells. In contrast, if [α-^{32}P]-UTP or [α-^{32}P]-ATP were used, both cellular and viral RNAs were labeled in reactions with infected cell extracts and cellular RNAs were labeled with uninfected cell extracts. WNV infection had no effect on the relative activities of these cellular enzymes (Fig. 6).

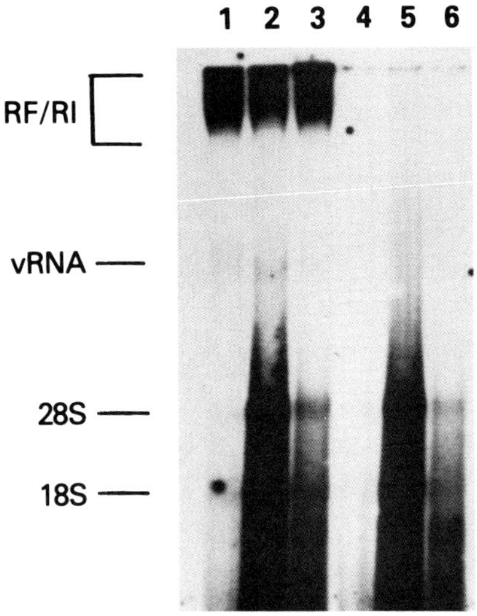

Figure 6. Viral polymerase and cellular transferase activities. RNAs were labeled during *in vitro* reactions with either [α-^{32}P]GTP (lanes 1 and 4), [α ^{32}P]UTP (lanes 2 and 5) or [α-^{32}P]ATP (lanes 3 and 6). RNAs were electrophoresed on 0.8% agarose gels. Infected cell extracts (lanes 1 through 3); uninfected cell extracts (lanes 4 through 6).

Further studies showed that the adenylyl transferase preferentially labeled cellular mRNA, while the uridylyl transferase labeled all types of cellular RNAs (13). Although a cellular uridylyl transferase is thought to be responsible for initiating polio RNA transcription (17), we could not demonstrate terminal addition to WNV RNA by either cellular transferase. This is not surprising, since the 3' terminus of the WNV genome RNA is base paired within a stable secondary structure (4; Fig. 2).

Another method for detecting host factors is to search for cellular proteins which can bind to the viral RNA. Pelleted cytoplasmic extracts from infected and uninfected cells were run on SDS-polyacrylamide gels and the proteins were then transferred to nitrocellulose and probed with orthophosphate-labeled WNV genome RNA by the method of Bowen et al (18) modified for detection of RNA binding proteins. Five cellular proteins bound the WNV RNA after washes containing 200 mM salt (data not shown). Two of these proteins, 44 and 45K, were enriched in the perinuclear membranes and showed the strongest binding activity. The binding of VSV RNA and rRNA, was also investigated. The 44 and 45K proteins bound the rRNA, but not the VSV RNA. These cellular proteins may be associated with ribosomes and may be able to distinguish mRNAs from minus-strand RNAs. Whether these proteins function in flavivirus RNA replication still needs to be determined.

DISCUSSION

Interactions between viral template RNAs and replication complex proteins are presumed to be mediated by signal sequences and secondary structures present on the viral RNA. Only the first three nucleotides and the region 2 sequence, 3'-A C A C A - 5', are conserved in the 3' termini of both the flavivirus plus and minus strand RNAs. However, which protein recognizes these signals is not known. Sequence conservation was also observed in 5 to 6 additional regions in the terminal sequences of flavivirus RNAs. Different secondary structures were predicted for the 3'-terminal sequences of the plus and minus strands. Although the sequences forming the stems of these structures were not conserved, the size and shape of the structures was conserved among the 3 flaviviruses compared. This indicates that the forms of the flavivirus secondary structures are subject to functional constraints.

Preliminary data indicate that the composition of the replication complexes for the viral plus and minus strands may differ; the NS5 protein may copy the plus strand, while the NS3 protein may copy the minus strand. Host protein(s) have been reported to be involved in the replication of other positive strand RNA viruses (15, 16). The existence of a flavivirus-specific host resistance gene suggests that host proteins do play a role in flavivirus replication (2). Flavivirus resistance is inherited as a single Mendelian dominant allele which is not linked to the major histocompatibility locus. Cells from congenic resistant and susceptible mice are equally infectible, but resistant animals and cell cultures made from them produce less virus. In resistant cells, viral RNA synthesis is less efficient and defective-interfering viral RNAs are preferentially amplified (19). These cell-specific differences could be explained by the existence of two isotypes of a cellular protein which is involved in flavivirus RNA synthesis. The two isotypes would differ in their ability to provide a function required by the virus. Much still remains to be learned about the control of flavivirus RNA synthesis and the composition of the viral replication complexes.

ACKNOWLEDGEMENTS

We thank Janice Dispoto and Edith Gavin for expert technical assistance and Cheryl McFadden for typing the manuscript.

REFERENCES

1. Rice CM, Lenches EM, Eddy SR, Shin SJ, Sheets RL, and Strauss JH (1985). Nucleotide sequence of yellow fever virus: Implications for flavivirus gene expression and evolution. Science 229:726.
2. Brinton MA (1986). Replication of Flaviviruses In Schlesinger S and Schlesinger M (eds): "The Togaviridae and Flaviviridae", New York: Plenum Publishing Co., p327.
3. Stollar V, Schlesinger RW, and Stevens TM (1967). Studies on the nature of dengue viruses. III. RNA synthesis in cells infected with type 2 dengue virus. Virology 33:650.
4. Brinton MA, Fernandez AV, and Dispoto JH (1986). The 3'-nucleotides of flavivirus genomic RNA form a conserved secondary structure. Virology, in press.
5. Keene JD, Schubert M, Lazzarini RA, and Rosenberg M (1978). Nucleotide sequence homology at the 3'-termini of RNA from vesicular stomatitis virus and its defective interfering particles. Proc Nat Acad Sci USA 75:3225.
6. Donis-Keller H, Maxam AM, and Gilbert W (1977). Mapping adenines, guanines, and pyrimidines in RNA. Nucleic Acid Res 4:2527.
7. Stanley J and Vassilenko S (1978). A different approach to RNA sequencing. Nature (London) 274:87.

8. Peattie DA (1979). A direct chemical method for sequencing RNA. Proc Nat Acad Sci USA 76:1760.
9. Zuker M and Sankoff D (1984). RNA secondary structures and their prediction. Bull Mathemat Biol 46:591.
10. Wengler G, and Wengler G (1981). Terminal sequences of the genome and replicative form RNA of the flavivirus West Nile virus: Absence of poly (A) and possible role in RNA replication. Virology 113:544.
11. Brinton MA (1981). Isolation of a replication efficient mutant of West Nile virus from a persistently infected genetically resistant mouse cell culture. J Virol 39:413.
12. Brinton MA, and Fernandez AV (1983). A replication-efficient mutant of West Nile virus is insensitive to DI particle interference. Virology 129:107.
13. Grun J, and Brinton MA (1986). Characterization of West Nile viral RNA-dependent RNA polymerase and cellular terminal adenylyl and uridylyl transferases in cell-free extracts. J. Virol, in press.
14. Landers TA, Blumenthal T, and Weber K (1974). Function and structure in ribonucleic acid phage Qβ ribonucleic acid replicase. J Biol Chem 249:5801.
15. Blumenthal J (1979). Qβ RNA replicase and protein synthesis elongation factors EF-Tu and EF-Ts. Methods in Enzymology 60:628.
16. Dasgupta A, Zabel P, and Baltimore D (1980). Dependence of the activity of the poliovirus replicase on a host cell protein. Cell 19:423.
17. Andrews NC, Levin D, and Baltimore D (1985). Poliovirus replicase stimulation by terminal uridylyl transferase. J Biol Chem 260:7628.
18. Bowen B, Steinberg J, Laemmli UK, and Weintraub H (1980). The detection of DNA-binding proteins by protein blotting. Nuc Acid Res 8:1.
19. Brinton MA (1983). Analysis of extracellular West Nile virus particles produced by cell cultures from genetically resistant and susceptible mice indicates enhanced amplification of defective interfering particles by resistant cultures. J Virol 46:860.

Positive Strand RNA Viruses, pages 273–284

MECHANISM OF RNA REPLICATION BY THE POLIOVIRUS RNA POLYMERASE, HeLa CELL HOST FACTOR, AND VPg.[1]

J. Bert Flanegan, Dorothy C. Young, Gregory J. Tobin, Mary Merchant Stokes, Carol D. Murphy, and Steven M. Oberste

Department of Immunology and Medical Microbiology
University of Florida, College of Medicine
Gainesville, Florida 32610

ABSTRACT The poliovirus polymerase is a strict RNA dependent RNA polymerase that copies a wide variety of homopolymeric and heteropolymeric RNA templates. The error frequency for the in vitro polymerase reaction was found to be very high and ranged from 0.7 x 10^{-3} to 5.0 x 10^{-3}. This is consistent with the high mutation frequency observed with picornaviruses and other RNA viruses. The purified polymerase requires an oligo(U) primer or a HeLa cell host factor to initiate RNA synthesis in vitro. In the presence of the host factor, the polymerase appears to use a template priming mechanism to initiate RNA synthesis and this results in the production of dimer sized product RNA. The synthesis of monomer sized product RNA by the polymerase purified on poly(U) sepharose appears to result from contaminating oligo(U) in the enzyme. In addition, the host factor was shown to greatly enhance priming by oligo(U). The immunoprecipitation of product RNA by anti-VPg antibody was shown to be mediated by VPg on the template RNA. These results indicate that VPg was not required for the initiation of RNA synthesis in vitro. The addition of synthetic VPg to the dimer sized product RNA resulted in the linkage of VPg to the labeled product RNA. The addition of VPg occurred after the product was synthesized. The synthetic VPg was linked to the poly(U) sequence at the 5' end of the negative strand product RNA.

1.This work was supported by PHS grant AI15539.

INTRODUCTION

Poliovirus replicates in the cytoplasm of infected cells and has a single stranded RNA genome of positive polarity. Poliovirion RNA (Mr = 2.5×10^6) contains a 3' terminal poly(A) sequence and a small virus specific protein (VPg) that is covalently linked to the 5' terminal nucleotide by a phosphodiester bond to the single residue of tyrosine in VPg (1,2). Replication of poliovirus RNA requires a virus-specific RNA dependent RNA polymerase that is not present in uninfected cells. The primary site of RNA replication in infected cells is the membrane bound replication complex which is composed of one complete negative strand RNA, several nascent chains of positive strand RNA, and the viral RNA polymerase. VPg is linked to the 5' ends of both the positive and negative strand RNAs in the replication complex (3). Purification of the replication complex showed that only one virus coded protein ($3D^{pol}$, also P3-4b, p63, and NCVP4 in previous publications) was required for elongating polymerase activity (4,5).

A soluble and template dependent form of the poliovirus polymerase was isolated from the cytoplasm of infected cells by using a poly(A):oligo(U) template:primer (6). Preparation of highly purified forms of the polymerase showed that $3D^{pol}$ was again the only virus-specific protein that copurified with the soluble polymerase (7). The highly purified polymerase synthesizes full sized copies of poliovirion RNA and various other polyadenylated RNAs, but only when oligo(U) is added as a primer to the in vitro reaction (8-10). The requirement for the oligo(U) primer can be eliminated by the addition of a cellular protein or host factor that was first reported by Dasgupta et al. (11). The host factor can be purified from uninfected cells and is known to copurify with the polymerase during the early steps in its purification (12,13). This explains why partially purified forms of the polymerase can initiate RNA synthesis in the absence of added oligo(U) or purified host factor. Thus, it appears that the host factor plays an important role in the initiation of RNA synthesis but we are now only beginning to understand its mechanism of action.

RESULTS AND DISCUSSION

Polymerase Elongation Activity.

Template specificity. The purified poliovirus polymerase shows very little specificity for poliovirus RNA. In a previous study the polymerase was active on all homopolymeric RNA template:primers tested and was active on a wide variety of polyadenylated and nonpolyadenylated heteropolymeric RNAs (14). In more recent work we have examined the activity of the polymerase using DNA templates, primers, and substrates. The results showed that the polymerase was not active on DNA templates and would not use deoxyribonucleoside triphosphates as substrates on either DNA or RNA templates. Thus, the polymerase acted as a strict RNA dependent RNA polymerase. The enzyme would, however, efficiently use a DNA primer in place of an RNA primer and this may prove to be of some practical use in view of the ease in preparing synthetic oligodeoxynucleotides.

Fidelity of RNA replication. The fidelity of RNA replication in vitro was examined by measuring polymerase error frequencies for product RNA synthesized on synthetic homopolymeric RNA templates. The frequency at which incorrect nucleotides were incorporated by the polymerase was measured by using a ^{3}H-labeled nucleotide as the correct (complementary) substrate and a ^{32}P-labeled nucleotide as the incorrect (noncomplementary) substrate. The base substitution frequency that was observed was very high, ranging from 0.7×10^{-3} to 5.0×10^{-3}. These values have been confirmed by a direct determination of the base composition of the product RNA synthesized in the presence of ^{32}P-labeled complementary and noncomplementary ribonucleoside triphosphates. The error frequency was found to increase with changes in the reaction conditions that are known to increase the elongation rate of the polymerase reaction. Thus, there appears to be a direct correlation between the polymerase error frequency and the elongation rate of the polymerase reaction. The high error frequencies observed in our studies are consistent with the very high mutation frequencies observed for poliovirus, other picornaviruses, and for RNA viruses in general (15). Sobrino et al. (16) have compared multiple clones of foot-and-mouth disease virus by oligonucleotide map comparisons. They have estimated that 2 to 8 base differences exist per viral genome which would translate to an error frequency of 0.3×10^{-3} to 1.1×10^{-3}. These values are in accord with the error frequencies obtained by Steinhauer and Holland (15) for vesicular stomatitis virus RNA polymerase (0.1×10^{-3} to 0.4×10^{-3}) and with the 10^{-3} to 10^{-4} mutation frequencies estimated for RNA retrovirus reverse transcriptase (17,18).

Initiation of RNA Synthesis.

The mechanism of initiating poliovirus RNA synthesis in vitro has been studied in several laboratories using either the membrane-bound replication complex or the soluble purified polymerase and host factor. A uridylylated form of VPg (i.e., VPg-pUpU) is synthesized in infected cells (19) and in vitro by the membrane-bound replication complex (20). The mechanism involved in the formation of VPg-pUpU is not known, but it has been postulated that VPg-pUpU may act as a primer for the polymerase to initiate RNA synthesis. Studies in Dr. Eckard Wimmer's laboratory with partially purified forms of the membrane replication complex are in progress to characterize the ability of pulse-labeled VPg-pUpU to be chased into plus-strand product RNA. This finding would suggest that VPg-pUpU can either act as a primer or can be linked to newly synthesized plus strand RNA.

We and others have used the soluble purified polymerase to investigate the mechanism of initiating RNA synthesis. Questions concerning the role of the polymerase, host factor and VPg in initiating RNA synthesis have been addressed in these studies. Product RNA synthesized by the purified polymerase and host factor has been characterized for its size, polarity, and for the presence of covalently linked VPg.

Host Factor Purification.

Several different protocols have been developed for purifying the host factor from infected cells. Regardless of the protocol used, however, the addition of the host factor stimulates the initiation of RNA synthesis on a wide variety of RNA templates by highly purified forms of the poliovirus polymerase (12). As discussed below, there is now considerable evidence to indicate that the polymerase uses a template-priming mechanism to initiate RNA synthesis in the presence of the host factor.

Protein kinase activity. A protein kinase acivity has been reported by Morrow *et al*. (21) to copurify with the host factor that was isolated from a ribosomal salt wash fraction. Because of its molecular weight (Mr = 67,000), its stimulation by dsRNA, and its ability to phosphorylate itself and the β-subunit of eIF2, it was suggested that the host factor- associated protein kinase may be the same protein kinase induced by interferon. Results from our laboratory indicate that a self phosphorylating protein kinase activity (Mr = 67,000) is present in host factor preparations purified from the ribosomal salt wash fraction. Additional studies are now required to determine if the protein kinase and host factor activities are on the same protein and the function of the kinase in RNA replication.

Terminal uridylyl transferase. Andrews and Baltimore (22) reported that a terminal uridylyl transferase (TUT) activity copurifies with the host factor isolated from the soluble fraction of the cytoplasm. This activity adds a short poly(U) sequence to the 3' end of various RNAs including poly(A). We have also found that TUT activity copurifies with the host factor and that a terminal adenylyltransferase (TAT) is also present. We have not yet determined if the TUT and TAT activities can be separated from each other and the host factor after more extensive purification. The addition of a short poly(U) sequence on the 3' end of the poly(A) sequence in poliovirion RNA may allow the formation of a 3' terminal hairpin that could act as a primer for the polymerase.

Polymerase Purification.

Several approaches have been used to purify the polymerase. The primary objectives have been the removal of endogenous host factor and contaminating ribonuclease activity. At early stages of purification the polymerase will initiate RNA synthesis due to the presence of endogenous host factor (12). The polymerase and host factor are known to physically interact with each other in infected cell extracts and when mixed together in vitro. Gradient elution chromatography on hydroxylapatite (12) and poly(U) sepharose (11) have been used to separate the polymerase from the host factor.

Size of Product RNA.

Monomer length product RNA was synthesized on poliovirion RNA by the poly(U) sepharose purified polymerase in the presence of purified host factor. This contrasts with the dimer-sized product RNA that is synthesized by the polymerase purified on hydroxylapatite (12) or sephacryl S-200 (23). The dimer-sized product RNA was also observed when globin mRNA and several non-polyadenylated RNAs were used as templates (23,24). This difference in the size of product RNA has been confusing and was at first difficult to understand. We have now shown, however, that the polymerase purified by published protocols using poly(U) sepharose chromatography contains some oligo(U) which is derived from a partial digestion of the poly(U) on the column by contaminating ribonuclease in the polymerase load. We have also shown that the host factor greatly enhances the priming of product RNA synthesis by oligo(U). In reactions containing polymerase, host factor, and oligo(U), most of the product RNA is monomer length. The addition of host factor to reactions containing hydroxylapatite purified polymerase and oligo(U) can increase the amount of monomer length product RNA by five fold or more. Thus, in reactions containing oligo(U) the host factor greatly enhances the synthesis of oligo(U) primed product RNA and most of the resulting product RNA is monomer length. In reactions containing only polymerase and host factor, the polymerase appears to use a template-priming mechanism to synthesize dimer-sized product RNA.

Presence of VPg on Product RNA.

The immunoprecipitation of labeled product RNA with anti-VPg antibody has been used to detect VPg linked product RNAs. Although VPg has not been detected in any of the polymerase preparations purified by the various published protocols, the product RNA synthesized on poliovirion RNA was efficiently immunprecipitated. In cases where monomer length product RNA was synthesized, it was suggested that trace amounts of a VPg related protein in the polymerase may have acted as a primer. We and others have now shown, however, that when VPg is first removed from the template RNA by proteinase K treatment that no labeled product RNA will immunoprecipitate (24,25). This is true for the poly(U) sepharose purified polymerase that synthesizes monomer length product RNA as well as the hydroxylapatite purified polymerase (24). Thus, VPg on the poliovirion RNA template and not VPg on the product RNA mediates the immunoprecipitation of the labeled product RNA recovered from these reactions. These findings argue that VPg or a VPg related protein is not required by the polymerase and host factor to initiate RNA synthesis in vitro.

Linkage of Synthetic VPg to Product RNA.

The addition of synthetic VPg (22 amino acids) to the in vitro reaction results in the linkage of VPg to the product RNA. The presence of VPg linked product RNA has been measured by the immunoprecipitation of product RNA synthesized on poliovirion RNAs pretreated with proteinase K. The linkage appears to be covalent in that it is resistant to SDS treatment and phenol extraction. The linkage of VPg to the product RNA appears to occur after the product RNA is synthesized since the same amount of VPg linked labeled product RNA is recovered when VPg is added to the product RNA after its synthesis is complete. The linkage reaction appears to be specific for product RNA synthesized by the polymerase and host factor on poliovirion RNA template. The linkage of VPg to the product RNA does not occur with monomer sized product RNA synthesized in the presence of oligo(U) or with dimer sized or monomer sized product RNA synthesized on a globin mRNA template. Control experiments with a truncated peptide of VPg (14 amino acids) that does not contain tyrosine gave negative results as expected. Studies to date indicate that the linkage reaction does not require the polymerase, host factor, ATP or other ribonucleoside triphosphates. The reaction does require magnesium with a fairly sharp optimum at 7 mM. Complete digestion of the VPg linked product RNA with ribonuclease releases labeled VPg-pUp that can be immunoprecipitated with anti-VPg antibody and identified by polyacrylamide gel electrophoresis and high voltage paper electrophoresis. Digestion of the VPg linked product RNA with ribonuclease T1 results in the recovery of what appears to be VPg linked poly(U). Thus our results indicate that synthetic VPg is linked to the poly(U) sequence that is known to be present at the 5' end of the labeled negative strand product RNA.

Conclusions and RNA Replication Model.

The poliovirus polymerase is a strict RNA dependent RNA polymerase that is required for the replication of poliovirus RNA. The polymerase ($3D^{pol}$) is the only known viral or cellular protein that is required for the elongation reaction. The polymerase shows a very high error frequency similar to that observed for other RNA dependent RNA polymerases. This is consistent with the high mutation frequency that is observed for poliovirus and other picornaviruses.

The poliovirus polymerase is a primer dependent enzyme that can efficiently initiate RNA synthesis when host factor is added to the in vitro reaction. In the presence of the host factor, the polymerase appears to use a template priming mechanism to initiate RNA synthesis (Figure 1). The presence of terminal uridylyl transferase activity in the host factor may allow a poly(A):oligo(U) hairpin to form at the 3' end of polyadenylated RNA templates. It appears likely that this structure would be stabilized by the host factor itself. This possibility is consistent with our finding that the host factor significantly enhances the priming by exogenously added oligo(U) presumably by stabilizing the interaction of the oligo(U) primer with the poly(A) sequence in the template. This model (Figure 1) predicts that initiation would occur at the 3' end of the poly(A) sequence in the template RNA and this is supported by our ability to isolate labeled poly(U) from product RNA synthesized by the polymerase and host factor. It is clear, however, that the polymerase can also use a template priming mechanism to initiate synthesis on non-polyadenylated RNAs and on subgenomic fragments of poliovirion RNA (24,25). The exact mechanism involved in these reactions will require additional study and characterization.

The generation of unit length product RNA linked to VPg may involve a concerted cleavage and linkage reaction (Figure 1). The results suggest that this reaction is specific for product RNA synthesized on poliovirus RNA by the polymerase and host factor. Because the host factor and polymerase are not required for the linkage of VPg to the product RNA and because of the catalytic processing activities associated with other RNAs, it is interesting to speculate that specific sequences and/or structures in poliovirus RNA may play an important role in catalyzing this reaction.

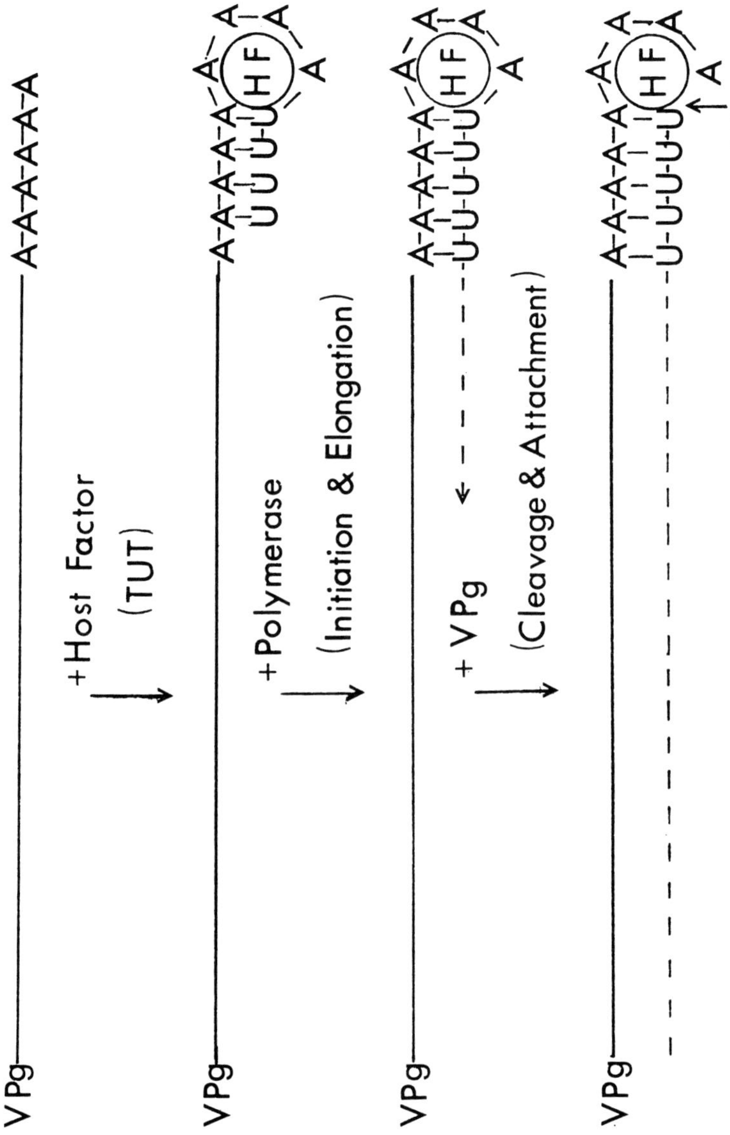

Figure 1. Model for negative strand RNA synthesis.

The relevance of these findings to the synthesis of positive strand RNA is not yet known. It has been shown by Kaplan et al. (26) and in our laboratory that the polymerase and host factor can synthesize infectious plus strand RNA in vitro using poliovirus minus strand RNA templates. The detailed characterization of this reaction is now in progress. This work in combination with the studies on the membrane bound replication complex and genetic studies on RNA replication in vivo, should give a detailed understanding of the molecular mechanisms involved in poliovirus RNA replication.

REFERENCES

1. Flanegan JB, Pettersson RF, Ambros V, Hewlett MJ, Baltimore (1977). Covalent linkage of a protein to a defined nucleotide sequence at the 5'terminus of the virion and replicative intermediate RNAs of poliovirus. Proc. Natl. Acad. Sci. USA 74:961.

2. Lee YF, Nomoto A, Detjen BM, Wimmer E (1977). A protein covalently linked to poliovirus genome RNA. Proc. Natl. Acad. Sci. USA 74:59.

3. Pettersson RF, Ambros V, Baltimore D (1978). Identification of a protein linked to nascent poliovirus RNA and to the polyuridylic acid of negative-strand RNA. J Virol 27:357

4. Flanegan JB, Baltimore D (1979). Poliovirus polyuridylic acid polymerase and RNA replicase have the same viral polypeptide. J Virol 29:352.

5. Lundquist RE, Ehrenfeld E, Maizel JV (1974). Isolation of a viral polypeptide associated with poliovirus RNA polymerase. Proc Natl Acad Sci USA 71:4773.

6. Flanegan JB, Baltimore D (1977). Poliovirus-specific primer-dependent RNA polymerase able to copy Poly(A). Proc Natl Acad Sci USA 74:3677.

7. Van Dyke TA, Flanegan JB, (1980). Identification of poliovirus polypeptide P63 as a soluble RNA-dependent RNA polymerase. J Virol 35:732.

8. Dasgupta A, Baron MH, Baltimore D (1979). Poliovirus replicase: a soluble enzyme able to initiate copying of poliovirus RNA. Proc Natl Acad Sci USA 76:2679.

9. Van Dyke TA, Rickles RJ, Flanegan JB (1982). Genome-length copies of poliovirion RNA are synthesized in vitro by the poliovirus-RNA-dependent RNA polymerase. J Biol Chem 257:4610.

10. Baron MH, baltimore D (1982). In vitro copying of viral positive-strand RNA by poliovirus replicase:characterization of the reaction and its products. J Biol Chem 257:12359.

11. Dasgupta A, Zabel P, Baltimore D (1980). Dependence of the activity of the poliovirus replicase on a host cell protein. Cell 19:423.

12. Young DC, Tuschall DN, Flanegan JB (1985). Poliovirus RNA-dependent RNA polymerase and host cell protein synthesize product RNA twice the size of poliovirion RNA in vitro. J Virol 54:256.

13. Dasgupta A, Hollingshead P, Baltimore D (1982). Antibody to a host protein prevents initiation by the poliovirus replicase. J Virol 42:1114.

14. Tuschall DM, Hiebert E, Flanegan JB (1982). Poliovirus RNA dependent RNA polymerase synthesizes full length copies of poliovirion RNA, cellular mRNA, and several plant virus RNAs in vitro. J Virol 44:209.

15. Steinhauer DA, Holland JJ (1986). Direct method for quantitation of extreme polymerase error prequencies at selected single base sites in viral RNA. J Virol 57:219.

16. Sobrino F, Davila M, Ortin J, Domingo E (1983). Multiple genetic variants arise in the course of replication of foot-and-mouth disease virus in cell culture. Virology 123:310.

17. Clark SP, Mak TW (1984) Fluidity of a retrovirus genome. J Virol 50:759.

18. Darlix JL, Spahr PF (1983). High spontaneous mutation rate of Rous sarcoma virus demonstrated by direct sequencing of the RNA genome. Nucleic Acids Res 11:5953.

19. Crawford NM, Baltimore D (1983). Genome linked protein VPg of poliovirus is present as free VPg and VPg-pUpU in poliovirus infected cells. Proc Natl Acad Sci USA 80:7452.

20. Takegami Y, Kuhn RJ, Anderson CW, wimmer E (1983). Membrane dependent uridylation of the genome-linked protein VPg of poliovirus. Proc Natl Acad Sci USA 80:7447.

21. Morrow CD, Gibbons GF, Dasgupta A (1985). The host protein required for in vitro replication of poliovirus RNA is a protein kinase that phosphorylates eukaryotic initiation factor-2. Cell 40:913.

22. Andrews NC, Baltimore D (1986). Purification of a terminal uridylyltransferase that acts as host factor in the in vitro poliovirus replicase reaction. Proc Natl Acad Sci USA 83:221.

23. Hey TD, Richards OC, Ehrenfeld E (1986). Synthesis of plus and minus strand RNA from poliovirion RNA template in vitro. J Virol (in press).

24. Young DC, Dunn BM, Tobin GJ, Flanegan JB (1986). Ant-VPg antibody precipitation of product RNA synthesized in vitro by the poliovirus polymerase and host factor is mediated by VPg on the template RNA. J Virol (in press).

25. Andrews NC, Baltimore D (1986) Lack of evidence for Vpg-priming of poliovirus RNA synthesis in the host factor dependent in vitro replicase reaction. J Virol 58:212.

26. Kaplan GJ, Lubinski J, Dasgupta A, Racaniello VR (1985). In vitro Synthesis of infectious poliovirus RNA. Proc Natl Acad Sci USA 82:8424.

Positive Strand RNA Viruses, pages 285–297

LEADER RNA-PRIMED TRANSCRIPTION AND RNA RECOMBINATION OF MURINE CORONAVIRUSES

Michael M.C. Lai, Shinji Makino, Ralph S. Baric, Lisa Soe, Chien-Kou Shieh, James G. Keck and Stephen A. Stohlman

Departments of Microbiology and Neurology, University of Southern California School of Medicine, Los Angeles, CA 90033

ABSTRACT Murine coronaviruses contain a nonsegmented RNA genome and encode six subgenomic mRNAs with a common leader sequence. Previous studies have supported a model of leader RNA-primed transcription. In this report, we present two additional pieces of evidence for this model: (1) Leader sequences can be exchanged between the mRNAs of two co-infecting viruses; and (2) The expression of anti-sense leader RNA inhibits the transcription of coronavirus RNA. We also describe the isolation of multiple RNA recombinants which suggest that RNA replication may proceed by a discontinuous and nonprocessive mechanism, thus generating segmented RNA intermediates. In addition, we have determined the leader sequence of the genomic RNA, which reveals the detailed mechanism of leader-primed transcription.

INTRODUCTION

Murine coronaviruses, or mouse hepatitis viruses (MHVs), are a group of enveloped viruses with a positive-stranded RNA genome of 5.4×10^6 molecular weight (1), which replicate in the cytoplasm of several established mouse cell lines. Upon entry into the cells, the virion RNA encodes an RNA polymerase which transcribes the RNA genome into a full-length negative-stranded RNA (2,3). The latter is then transcribed by a different virus-specific RNA polymerase into a positive-sensed genomic RNA and six subgenomic mRNA species (4). These mRNAs have a nested-set structure, containing sequences from the 3'-end of the genomic RNA that extend for various distances toward the 5'-end (4). Each mRNA and genomic RNA also contain an identical leader sequence of approximately 72 nucleotides at the 5'-end (5,6). UV transcriptional mapping studiessuggested that the coronavirus subgenomic mRNAs are not derived by cleavage of larger precursor RNAs (7). Thus, the joining of the leader sequences to coronavirus mRNAs does not utilize conventional eukaryotic RNA splicing mechanisms. Studies of replicative-intermediate RNA and double-stranded replicative form RNA further suggest that the leader RNA is joined to the mRNAs during transcription, but not post-transcriptionally, and does not involve "looping out" of intervening sequences in the RNA template (8). We thus proposed that MHV synthesizes a free leader RNA which is utilized as a primer for transcription of subgenomic mRNAs, thus generating mRNAs containing the leader sequence (8). This model has recently been supported by the detection of free leader RNA species in the cytoplasm of MHV-infected cells and also by the isolation of a temperature-sensitive mutant which synthesizes only leader RNA but not mRNAs at nonpermissive temperature (9).

In addition to this novel mechanism of mRNA transcription, MHVs also display an unusual biological property, i.e. they can undergo RNA recombination at a very high frequency (10,11). This phenomenon of RNA recombination has previously been observed in only one RNA virus, the picornaviruses (12). Furthermore, the high frequency of RNA recombination almost matches the frequency of gene reassortment between viruses with segmented RNA genomes, such as reoviruses (13). It is, thus, likely that segmented RNA intermediates are involved in the RNA replication of coronaviruses.

In this report, we will present additional studies in support of the model of leader-primed RNA transcription. We also describe a series of coronavirus RNA recombinants which, together with the biochemical studies of RNA intermediates, suggest that coronavirus RNA replication proceeds in a discontinuous and nonprocessive manner (10). These studies establish that the mechanisms of RNA transcription and replication of coronaviruses are unique.

MATERIALS AND METHODS

RNA preparation, polyacrylamide gel electrophoresis and oligonucleotide fingerprinting: were as described elsewhere (10).
DNA transfection: followed the calcium phosphate procedure (14).

Superinfection with MHV was performed 24 hours after DNA transfection.
Molecular cloning: cDNA cloning followed the method of Gubler and Hoffman (15). DNA sequencing was performed by Sanger's dideoxynucleotide chain termination method (16).

RESULTS

THE LEADER RNA CAN BE FREELY EXCHANGED BETWEEN THE mRNAs OF DIFFERENT MHVS DURING MIXED INFECTION

The leader-primed transcription model proposes that free leader RNA species are utilized as primer for the transcription of subgenomic mRNAs. The detection of such leader RNA species in MHV-infected cells (9) fulfills this criterion. However, the data can not rule out the possibility that these leader RNA species might be abortive transcriptional products rather than the true intermediates of transcription. To demonstrate that MHV transcription indeed utilizes a free leader RNA species as primer, we examined the structure of individual mRNA species in cells co-infected with two different viruses. It was predicted that the free leader RNA from the different viruses would be mixed and randomly exchangeable between the mRNAs of co-infecting viruses. For this purpose, we used two MHVs, B1 and CA21 (11), with similar growth kinetics and rates of RNA synthesis. To separate individual mRNAs, we synthesized two oligomers complementary to a region of mRNA7 where B1 and CA21 are divergent. After one of the oligomers was hybridized to the mRNA from the co-infected cells, the RNA·DNA hybrid was digested by RNase H, leaving intact the mRNA of only one of the viruses. This approach allowed the separation of virus-specific RNAs of two parental viruses. Thus, it was possible to examine the individual mRNA from the mix-infected cells and determine whether the leader sequences had been exchanged during mixed infection. Fig. 1 shows such a result. The B1-specific mRNA7 isolated from the mix-infected cells contains the leader sequences of both parental viruses. The relative ratio of these sequences from the two viruses is roughly equal, suggesting that almost half of the B1-specific mRNAs contain the CA21-specific leader sequences. The reciprocal study examining the CA21-specific mRNAs also revealed the presence of B1-specific leader sequences, although the amount of such mixing was less. Similar results have been obtained with mRNA 6 (Data not shown). These results strongly suggest that the leader RNA can be freely exchanged between two co-infecting viruses as if the leader RNA exists as a free and separate transcriptional unit. Thus, the free leader RNA species can act in trans, and are true intermediates of RNA transcription.

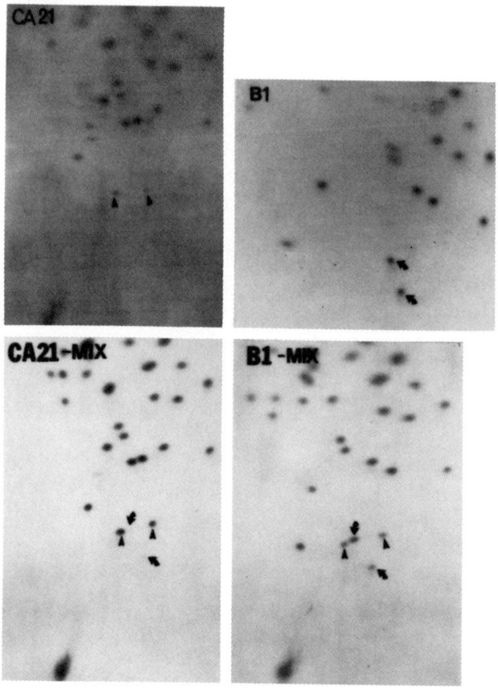

FIGURE 1 Oligonucleotide fingerprints of B1 and CA21-specific mRNA7 isolated from single infection and mixed infected cells. The arrows denote the leader-specific oligonucleotides of these two viruses respectively.

ANTI-SENSE LEADER RNA CAN INHIBIT VIRAL RNA TRANSCRIPTION

If the free leader RNA indeed serves as a primer for transcription, the removal of the free leader RNA can be expected to result in the inhibition of viral RNA transcription. On the other hand, if the leader RNA is added to mRNAs after initiation of transcription, such a treatment will not have an effect on transcription. We approached this problem by examining the effects of anti-sense leader RNA expression on the virus-specific RNA transcription of MHV. The 72-nucleotide leader-specific cDNA was synthesized chemically, attached with a Bam HI linker and cloned into an expression vector, which contains an LTR of feline leukemia virus (17) (Fig. 2).

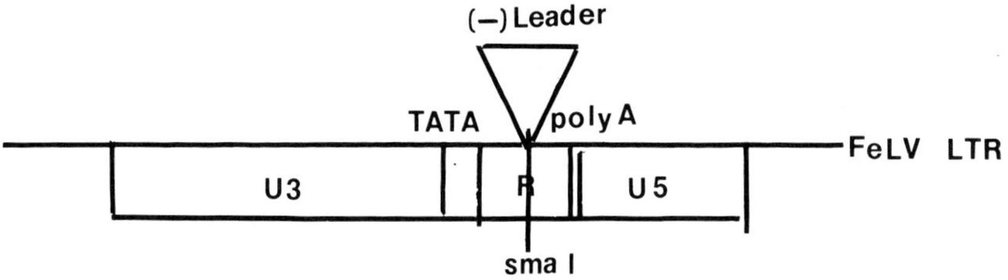

FIGURE 2. The construction of plasmid expressing anti-sense leader RNA

Approximately 20 clones were isolated and used to separately transfect L2 cells. The cytoplasmic RNA of the transfected cells were probed with the ^{32}P-labeled cDNAs representative of the negative-and positive-sensed leader sequences of MHV. The clones which express the negative-sensed leader RNA were selected (Data not shown). One of these clones, pFL4-7, was used for the following experiments. The L-2 cells were transfected with this clone expressing anti-sense leader RNA. Twenty-four hours later, the cells were superinfected with MHV-A59, and the kinetics of virus production and RNA synthesis were examined at hourly intervals. Table 1 shows that, at earlytime points post-infection, there was a 60% inhibition of virus-specific RNA synthesis in the cells expressing anti-sense leader RNA. This inhibition disappeared after 10 hours post-infection. This result suggests that anti-sense leader RNA causes a delay of RNA transcription. Thus free leader RNA appears to be required for MHV RNA transcription. A similar degree of inhibition of virus yield in the cells expressing anti-sense leader RNA was also observed (Table 2).

TABLE 1
The effect of anti-sense leader on the kinetics of viral RNA synthesis

Time (hrs p.i.)	Trichloroacetic acid precipitable counts (cpm)		%Inhibition
	(-)-sense leader	carrier DNA	
6	0	0	0
7	6,132	14,824	59
8	14,276	35,617	60
9	8,170	22,519	64
10	44,337	43,914	0
11	44,439	44,822	0

TABLE 2
The effect of anti-sense leader on virus titer

Time (hrs p.i.)	Virus titer (p.f.u./ml x 10^3)		% inhibition
	(-)-sense leader	carrier DNA	
4	4.0	6.3	37
6	6.1	14.5	58
8	30	220	86
10	350	1,100	68
12	850	1,500	43
24	6,600	6,300	0

HIGH-FREQUENCY AND MULTIPLE RNA RECOMBINATION EVENTS BETWEEN CORONAVIRUS RNAs

We have previously shown that MHV RNAs can undergo RNA recombination at a very high frequency (10), which is reminiscent of gene segment reassortment between RNA viruses with segmented RNA genomes.

This finding suggests that segmented RNA intermediates might be involved in the replication of coronaviruses. We have now obtained additional RNA recombinants which further demonstrate that RNA recombination can occur at multiple sites. Two different previously described strategies for obtaining recombinants were employed: one selecting recombinant wild-type viruses from genetic crosses involving two temperature-sensitive mutants (11), and the other screening progeny viruses from crosses between a temperature-sensitive mutant of A59 and a wild-type JHM (10). The latter procedure appeared to select recombinants which contain the 5'-end sequences of A59 (10). We have obtained several unusual recombinants from the cross between a JHM ts mutant and an A59 ts mutant. The first recombinant, A1-1, contains roughly 50 nucleotides at the 5'-end of the genome identical to the corresponding region of JHM, while the rest of the genome were derived from A59 (Fig 3).

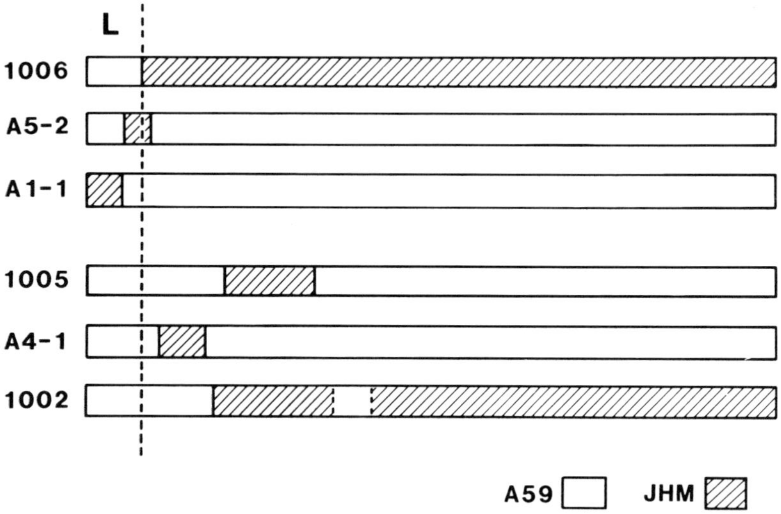

Figure 3. The genetic map of MHV recombinants

The second recombinant, A5-2, contains roughly 50 nucleotides at the 5'-end which is identical to that of A59, while the rest of the leader sequences are identical to that of JHM. The remaining genomic sequences were derived from A59. Thus, it appears to have two cross-over sites. A third recombinant, A4-1, also appears to have two cross-over sites, with the majority of the genome sequence derived from A59, while a portion of the gene A was derived from JHM.

The second class of recombinants were derived by crossing wild-type JHM with an A59 ts mutant, LA10, which has been shown to synthesize only the leader RNA at the nonpermissive temperature (9). We have characterized three of these recombinants. One, clone 1006, appears to have recombined at the leader-body junction site, although the precise point of cross-over could not be determined. The second recombinant, 1005, appears to have two cross-over sites, while the third recombinant, 1002, may have as many as three cross-over sites. All the recombination sites are clustered close to the 5'-end. The isolation of these new classes of recombinants suggests that RNA recombination occurs at multiple sites along the genome.

The occurrence of RNA recombination within the leader region is reminiscent of the multiple leader-containing RNA intermediates detected in the MHV-infected cells (9). To determine whether these RNA species could be the precursor to RNA recombination, we have further characterized the leader-containing RNA species.

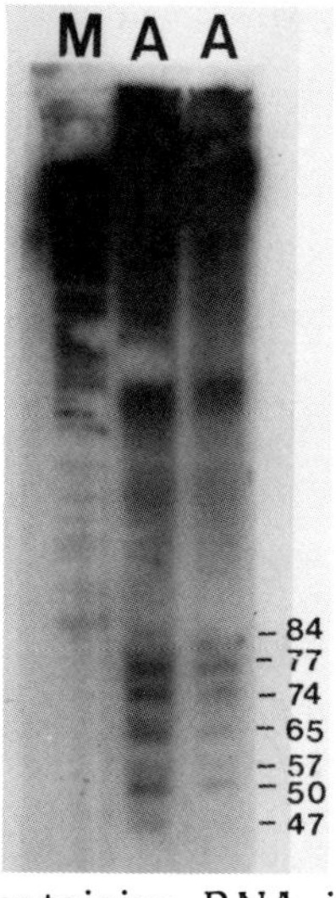

FIGURE 4. The leader-containing RNA intermediates in MHV-infected cells. The lane M is the marker DNA representing T tract of dideoxy sequencing of M13 DNA. The two lanes A represent two independently isolated A59 intracellular RNA.

The intracellular RNA from the MHV-infected cells was separatedby polyacrylamide gel electrophoresis, transferred to membrane filters and probed with the leader-specific DNA. Fig. 4 shows that three RNA species smaller than the known MHV leader RNA (72 nucleotides) (5) were detected in the MHV-infected cells. The size of these leader-containing species were 47, 50 and 57 nucleotides respectively, corresponding to RNA termination at three separate sites around a possible loop within the leader RNA region (Fig. 6). These termination points all fall within the possible cross-over sites in recombinant viruses A1-1 and A5-2. Thus, they could potentially be involved in RNA recombination. RNA termination around the secondary structure is consistent with the model of discontinuous, nonprocessive RNA transcription.

MOLECULAR CLONING OF THE 5'-END SEQUENCES OF THE MHV GENOME AND THE GENE ENCODING RNA POLYMERASES

The leader-primed transcription mechanism and the occurrence of RNA recombination in MHV suggest that the RNA polymerases of MHV have unique properties. To understand the structure and properties of these polymerases, and also to understand the precise mechanism of leader-priming, we cloned most of the sequences encoding the RNA polymerase, i.e., the approximately 8 kilobases (kb) of sequences at the 5'-end of the JHM genome. For cDNA cloning, we sequenced two RNase T_1-resistant oligonucleotides which had previously been mapped close to the 5'-end of the genome, one located within the polymerase gene and the other slightly downstream from the 3'-end boundary of the gene (18). We then synthesized two oligodeoxyribonucleotides complementary to these two T1-oligonucleotides and used them as primers for reverse transcription. We have obtained cDNA clones ranging in size from 0.5 to 4.5 kb. The 5'-ends of these cDNA clones were sequenced and additional cDNA clones were made by using primers complementary to these sequences. Clones containing the leader sequences were identified using a leader-specific probe. Fig 5 shows the map of some of the clones obtained. The leader sequence at the 5'-end of the genome was determined from these cDNA clones. The sequences are exactly identical to the leader sequences present at the 5'-end of the subgenomic mRNAs (Data not shown) (5).

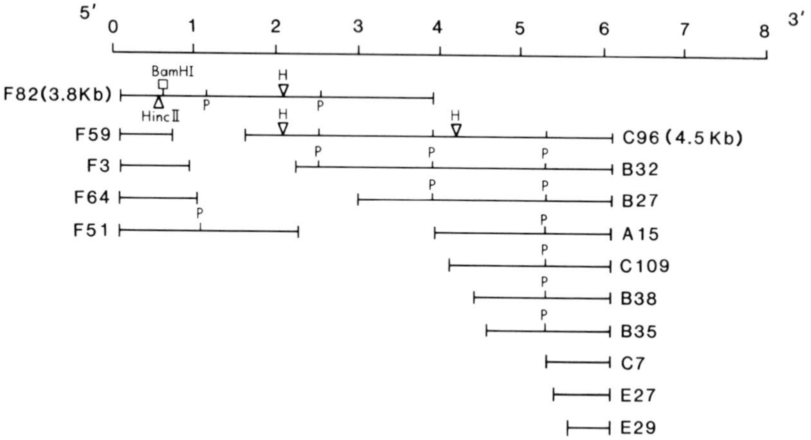

FIGURE 5 The genetic map of cDNA clones representing the gene encoding the RNA polymerase of JHM

At the 3'-end of the leader region, there are 4 repeats of UCUAA. The leader-body junction occurs at the 3'-end of the third repeat (Fig. 6). Thus the total leader sequence is 74 nucleotides for the JHM strain. There is a possible hairpin loop structure at 75-88 nucleotides from the 5'-end, which could provide termination points for the leader RNA synthesis.

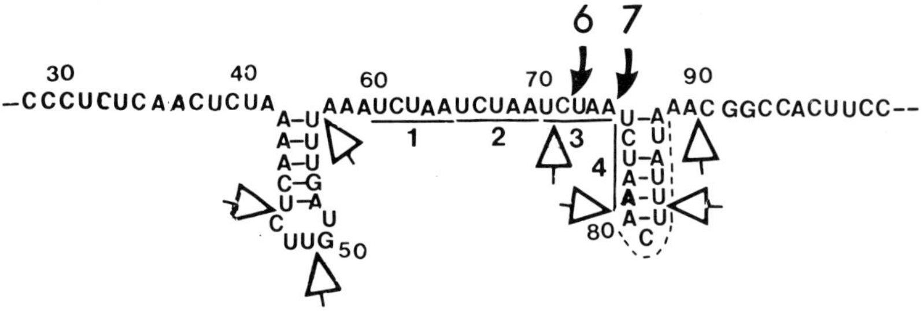

FIGURE 6 The leader sequence of JHM. The arrows represent the possible leader-body junction sites of mRNAs 6 and 7. The triangles denote the transcription "pausing" sites.

Indeed, four RNA intermediates terminate around and within this loop structure (Fig 4). Some of these intermediates may act as primers for mRNA transcription.

DISCUSSION

Our laboratory and others have obtained evidence which support the mechanism of leader RNA-primed transcription of coronaviruses (8). These supporting data include the following: (1) Free leader RNA species, which are dissociated from the RNA template, have been detected in MHV-infected cells (9); (2) A temperature-sensitive mutant, which synthesizes only the leader RNA but not mRNAs at nonpermissive temperature, has been isolated (9), suggesting that the leader RNA and mRNAs are synthesized independently; (3) During a mixed infection, the leader RNA sequences of the mRNAs from one virus could be attached to the mRNAs of the co-infecting virus at a very high frequency, as if the leader RNA functions as a separate transcription unit; (4) The expression of anti-sense leader RNAs could inhibit the transcription of MHV, suggesting that the leader RNA is required for transcription; (5) Sequence analysis of the leader RNA and the intergenic regions of various mRNAs have revealed sequence homology of 6-10 nucleotides between the 3'-end of the leader RNA and the initiation sites of various subgenomic mRNAs, thus providing a possible mechanism for the leader RNA to bind to the initiation sites of various subgenomic mRNAs (19). The presence of this homology has also been noted in another coronavirus, avian infectious bronchitis virus (20). From the sequence data, it appears that the free leader RNA species might be longer than the leader RNA in the subgenomic mRNAs. Therefore, the 3'-end of the leader RNA must be cleaved before serving as a primer. This mechanism of leader-primed transcription represents a very unique mechanism of RNA transcription and an alternative mechanism of RNA splicing. A similar phenomenon probably occurs in African trypanosomes (21), although the mechanism of transcription in this system has not yet been clearly established. The leader RNA thus provides a potential regulatory element for RNA transcription and other biological properties of coronaviruses.

The observation of multiple and high-frequency RNA recombination between coronaviruses provides an important tool for studying the pathogenicity of the viruses since various recombinants between the neuropathogenic JHM and nonpathogenic A59 strains, or recombinants of other MHV strains are readily available. These recombinants also provide insights into the mechanism of coronavirus RNA replication. The occurrence of high-frequency RNA recombination suggests that segmented RNA intermediates may be generated during RNA replication. This interpretation is supported by the isolation of RNA recombinants with cross-over sites which correspond to the sizes of leader-containing RNA intermediates detected in MHV-infected cells. Furthermore, these leader RNA termination points appear to cluster around the secondary structure of the template RNA. This observation is reminiscent of the RNA transcriptional pausing in QB and T7 (22,23). Whether such RNA intermediates are indeed utilized for transcription would require the use of an in vitro transcription system.

The different modes of leader-primed transcription and discontinuous RNA replication of coronaviruses suggest that different RNA polymerases might be generated during the growth cycle of coronaviruses. The previous studies on the RNA polymerase activities of MHV suggested that at least two different RNA polymerases might be involved (2). In vitro translation of the MHV genomic RNA yields three large proteins, which may correspond to the primary translation products of the polymerase gene (24). Whether these proteins are indeed different forms of the polymerase and whether they are involved in different transcription and replication functions are unknown at the present time. The availability of the antibodies against RNA polymerases would make such studies feasible.

REFERENCES

1. Lai MMC, Stohlman SA (1982). J Virol 26:235.
2. Brayton, PR, Ganges RG, Stohlman SA (1982). J Virol 42:847.
3. Lai MMC, Patton CD, Stohlman SA (1982). J Virol 44:487.
4. Lai MMC, Brayton PR, Armen RC, Patton CD, Pugh C, Stohlman SA (1981). J Virol 39:823.
5. Lai MMC, Baric RS, Brayton PR, Stohlman, SA (1984). Proc Natl Acad Sci USA 81:3626.
6. Spaan W, Delius H, Skinner M, Armstrong J, Rottier P, Smeekens S, van der Zeijst BAM, Siddell SG (1983). EMBO J 2:1939.
7. Jacobs L, Spann JM, Horzinek MC, van der Zeijst BAM (1981). J Virol 39:401.
8. Baric RS, Stohlman SA, Lai MMC (1983). J Virol 48:633.
9. Baric RS, Stohlman SA, Razavi MK, Lai MMC (1985). Virus Res 3:19.

10. Makino S, Keck JG, Stohlman SA, Lai MMC (1986). J Virol 57:729.
11. Lai MMC, Baric RS, Makino S, Keck JG, Egbert J, Leibowitz JL, Stohlman SA (1985). J Virol 56:449.
12. King AM, McCahon QD, Slade WR, Newman JWI (1982). Cell 29:921.
13. Field BN (1981). Curr Top Microbiol Immunol 91:1
14. Graham FL, van der Eb, AJ (1973). Virology 52:456.
15. Gubler U, Hoffman BJ (1983). Gene 25:263.
16. Sanger F, Nicklen S, Coulson AR (1977). Proc Natl Acad Sci USA 74:5463.
17. Wong TC, Goodenow RS, Sher BT, Davidson N (1985). Gene 34:27.
18. Makino SF, Taguchi N, Fujiwara K (1984). Virology 133:9
19. Budzilowicz CJ, Wilczynski SP, Weiss, SR (1985). J Virol 53:834.
20. Brown TDK, Boursnell MEG, Binns MM, Tomley FM (1986). J Gen Virol.
21. Campbell DA, Thornton DA, Boothroyd JC (1984). Nature 311:350.
22. Bills DR, Dabkin C, Kramer FR (1978). Cell 15:541.
23. Kassavetis GA, Chamberlin MJ (1981). J Biol Chem 256:2777.
24. Leibowitz JL, Devries JR, Haspel WV (1982). J Viol 42:1080.

Positive Strand RNA Viruses, pages 299–316

THE SPATIAL FOLDING OF THE 3' NONCODING REGION OF AMINOACYLATABLE PLANT VIRAL RNAs

Cornelis W.A. Pleij, Jan Pieter Abrahams, Alex van Belkum, Krijn Rietveld[1] and Leendert Bosch

Department of Biochemistry, University of Leiden, The Netherlands

ABSTRACT An essential feature in the construction of the aminoacyl acceptor arm of all tRNA-like structures of plant viral RNAs studied so far, is the presence of a so-called pseudoknot. In this paper we show that a number of tandemly located pseudoknots are found in the 3' noncoding region of aminoacylatable plant viral RNAs, upstream of the tRNA-like structure. These consecutive pseudoknots can give rise to the formation of a long quasi-continuous double helix or stalk-like region. Some features of the pseudoknotted structures appear to be strongly conserved among different virus groups. The possible function of these tandemly arranged pseudoknots is discussed.

INTRODUCTION

The 3' noncoding region of plus stranded viral RNAs plays a crucial role in the viral replication cycle. This region harbours recognition signals for the viral replicase in order to guarantee a proper initiation of the minus strand synthesis (1). However, no detailed information is as yet available how this recognition and the binding of the replicase to the viral RNA comes about. A full understanding of this process certainly requires insight into the three-dimensional folding of the 3' terminus of the viral RNA.

[1]Present address: Gist Brocades NV P.O. Box 1 2600 MA Delft

Despite an increasing number of RNA molecules sequenced so far, no detailed three-dimensional models for viral 3' noncoding regions have been described.
One group of plant viruses lends itself very well for such studies. Plant viral RNAs, like tobacco mosaic virus (TMV), turnip yellow mosaic virus (TYMV) or brome mosaic virus (BMV) RNA can esterify a specific amino acid to their 3' end in a manner analogous to tRNA (2). This implies that the folding of these 3' termini somehow must resemble that of canonical tRNA, the only natural RNA molecule with a 3D structure known to date (3). Secondary structures with cloverleaf features were therefore proposed as soon as the nucleotide sequences of these plant viral 3' termini were determined (4-6). We recently examined the secondary structure and spatial folding of several tRNA-like structures by means of chemical modification, enzymatic digestion and sequence comparisons (7-9). Interestingly, the aminoacyl acceptor domain was found to display a characteristic type of RNA folding, the so-called pseudoknot (9,10). Pseudoknotting was found to be present in all plant viral tRNA-like structures known to date. Analysis of the region of TMV RNA, located upstream of the tRNA-like structure, showed that pseudoknot formation is basic to the higher order folding of the 3' terminal last 200 nucleotides of TMV RNA (11). In this paper we report that pseudoknots are a common feature in the structure of a number of other plant viral RNA 3' termini. Some of the pseudoknots found have intriguing well conserved properties, pointing to a possible important function in the life cycle of these plus stranded RNA plant viruses.

tRNA-LIKE STRUCTURES IN PLANT VIRAL RNAs

Up till now, three different classes of tRNA-like structures in plant viral RNAs can be discerned, depending on whether they can be aminoacylated with valine, histidine or tyrosine. The tRNA-like structure of TYMV RNA, consisting of about 85 nucleotides is the most simple one, clearly resembling standard $tRNA^{Val}$ (7,12). The secondary and tertiary structure proposed for the tyrosine accepting 3' terminus of the bromoviral RNAs like BMV RNA is somewhat more complicated which is also reflected by the larger number of nucleotides (135) in the tRNA-like structure (8,13).
In the present paper we restrict ourselves to a discussion of the secondary and tertiary structure of the tRNA-like structure of the histidine-accepting TMV RNA. The spatial

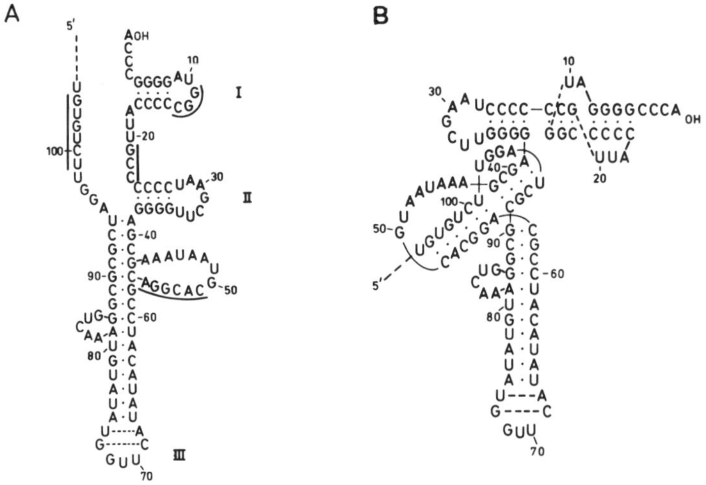

FIGURE 1. Model of the secondary (A) and tertiary (B) structure of the 3′ terminus of TMV RNA (vulgare strain). Hairpins are indicated with Roman numbers. Solid lines in (A) indicate complementary nucleotide sequences. Numbering is from the 3′ end.

folding previously proposed by us is fully supported by experimental data from chemical modification and enzymatic digestion studies and from sequence comparisons (9,11). Furthermore, it clearly illustrates the important role of pseudoknotting in the folding of the 3′ noncoding regions of plant viral RNAs and can therefore serve as a reference structure for a discussion of other plant viral 3′ noncoding regions.

Fig. 1A shows the secondary structure of the TMV RNA fragment comprising the 105 3′ terminal nucleotides. This structure, which harbours all the tRNA-like properties (14), is strongly supported by a comparison with the sequences of two other histidine-accepting tobamoviral RNAs from the tomato strain and the watermelon strain of TMV (9). This model basically consists of three hairpins (I-III), two relatively short ones (I and II) and a longer one containing two bulge loops (III). A remarkable feature of this secondary structure is the lack of a classical cloverleaf as in standard tRNA, in which the 3′ and 5′ end come together to form the aminoacyl acceptor stem of 7 base pairs. Instead, an anomolous hairpin I was found, which raises the question how the aminoacyl acceptor arm is formed. This problem could be solved by proposing a Watson-Crick base pairing of the sequence CGG in the loop of hairpin I with the complementary region CCG flanking hairpin II. A similar folding was proposed earlier in the tRNA-like structure of TYMV RNA (7).

The advantage of such a tertiary interaction becomes evident when the resulting double helical segment of three base pairs is coaxially stacked between the stems of hairpin I and II (Fig. 1B). A stack of 11 G•C base pairs in a so-called quasi-continuous double helix is formed with the sequence $ACCA_{OH}$ at one end and a seven-membered loop with the sequence UUCG (comparable with TΨCG in standard tRNA) at the other. In this way this domain strongly resembles the aminoacyl acceptor arm of normal tRNA.

When hairpin III is positioned perpendicular to the aminoacyl acceptor arm, the molecule becomes L shaped characteristic for standard tRNAs. (Fig. 1B). We obtained evidence for a second tertiary interaction, involving the large bulge loop of hairpin III and the 5'terminal single stranded region U_{99}-G_{104}. This long-range interaction, which also meets the definition of a so-called pseudoknot (see below) so far appears to be specific for tobamoviral RNAs and was not found in the tyrosine-accepting (e.g. BMV RNA) or valine-accepting (e.g. TYMV RNA) tRNA-like structures.

THE PRINCIPLE OF PSEUDOKNOT FORMATION

The type of folding, proposed here for the construction of the aminoacyl acceptor arm of TMV RNA, was earlier introduced on theoretical grounds as a possible feature of RNA folding (15). In general terms, it involves Watson-Crick base pairing of regions in hairpin, bulge or interior loops with a complementary single stranded stretch elsewhere in the RNA chain. If the latter region itself belongs to a loop structure, this comes down to a loop-loop interaction. Such base pairings give rise to a so-called pseudoknotted structure or shortly pseudoknot. In the tRNA-like structure of TMV RNA two examples are shown (Fig. 1). One in which a hairpin loop (hairpin I) and a second in which a bulge loop (hairpin III) participate in pseudoknot formation. We note here that such interactions are not considered in current computer programs predicting secondary structures of RNA, partially because no examples in nature had been observed (16). We here will focus on a special type of pseudoknotting because it appears to play a dominant role in the folding of plant viral 3' noncoding RNA regions. This particular

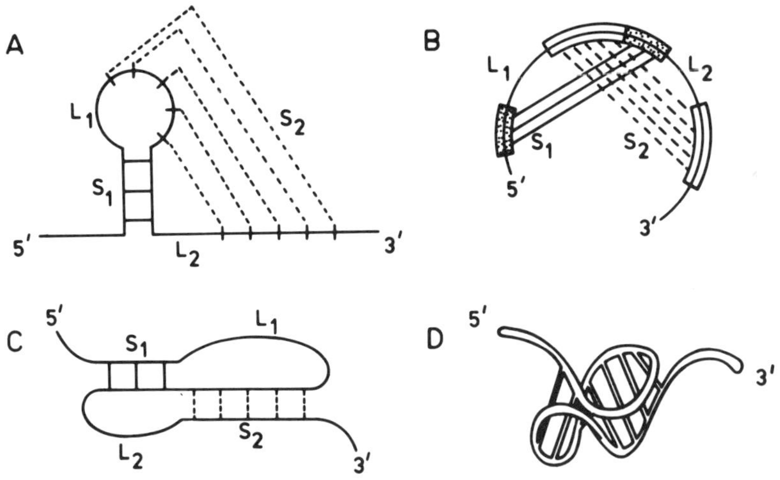

FIGURE 2. Schematic presentation of the formation of a pseudoknot of a special type. A nucleotide stretch from a hairpin loop, bordering the stem region, base pairs with a complementary single stranded region elsewhere in the chain. S_1 and S_2 indicate normal Watson-Crick base paired stem structures. L_1 and L_2 are single stranded RNA regions connecting the double helical segments. The principle of pseudoknotting is given in a conventional (A), circular (B), "2½D" (C) and a 3D (D) presentation.

pseudoknot is illustrated in Fig. 2. It shows various schematic representations of the interaction of a nucleotide sequence from a hairpin loop and directly bordering the stem with a complementary stretch elsewhere in the RNA chain. The result is a quasi-continuous double helix formed by coaxial stacking of the two stem segments S_1 and S_2. In fact this coaxial stacking is an assumption rather than an experimentally demonstrated phenomenon and was originally put forward to explain the folding of the aminoacyl acceptor arm in various tRNA-like structures. As a consequence of the stacking of the two helical segments two single stranded stretches connect opposite strands either bridging the shallow groove (L_2) or the deep groove (L_1), respectively. These two single stranded regions or connecting loops are not equivalent because of the geometric properties of the RNA-A double helix. We have analysed the minimal length requirements of these two different loops (10). It turned out that L_2 is minimally three nucleotides long when the shallow groove is bridged. In that case one to three base pairs are spanned, whereas L_1 only needs two nucleotides upon crossing the deep groove over five to seven base pairs. These calculations were based on a completely regular RNA-A duplex. Below we will present data suggesting that the deep groove can be bridged by one nucleotide only. For further details of the properties of this special type of pseudoknot see Pley et al. (10).

A further extension of the quasi-continuous double helix as depicted in Fig. 2D is conceivable if the pseudoknot is immediately flanking another stem segment. This is realised for instance in the aminoacyl acceptor arm of TMV RNA (Fig. 1B) where the pseudoknot of seven base pairs borders the stem of hairpin II. It is just this coaxial stacking of three stem segments which makes this domain to resemble one of the two limbs of the tRNA molecule. It will be shown below that in principle such a quasi-continuous double helix can be built of much more than three separate stem segments as a result of two or more pseudoknots located in tandem. This was first demonstrated to occur in the 3' noncoding region of TMV RNA.

THE 3' NONCODING REGION OF TMV RNA UPSTREAM OF THE tRNA-LIKE STRUCTURE

Experimental analysis of the 3' noncoding region upstream of the tRNA-like structure of TMV RNA revealed the presence of three more pseudoknots, all built according to the same principle as illustrated in Fig. 2. As can be seen in Fig. 3 these pseudoknots appear to be located in tandem, and form a long quasi-continuous duplex structure consisting of six stem segments. Apart from the experimental evidence obtained from chemical modification and enzymatic digestion, even more convincing evidence for this model came from sequence comparisons among four related RNAs from the tobamovirus group (11). Three similar consecutive pseudoknots could be proposed in exactly the same region of the RNA from the cowpea, watermelon and tomato strain of TMV. Even in the case of the cowpea strain of TMV (C_cTMV), which has a valine-accepting tRNA-like structure instead of the usual histidine-accepting one, this stalk-like region was conserved, despite strong differences with the vulgare strain at the level of primary structure.

A few interesting aspects of this RNA domain are noteworthy in view of the discussion below of the other plant viral 3' noncoding regions. The 3' proximal pseudoknot

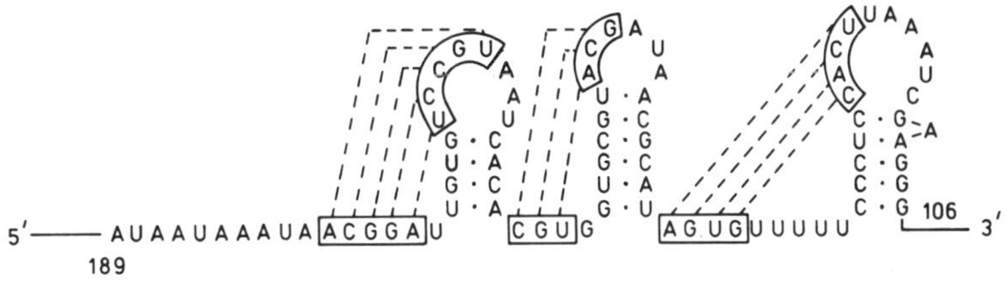

FIGURE 3. Secondary structure of part of the 3′ noncoding region of TMV RNA (vulgare strain). The region shown is located immediately upstream of the tRNA-like structure (cf Fig. 1). The stopcodon of the coat protein cistron is at position 205-207. Numbering is from the 3′ end. Interactions between complementary (boxed) regions which are responsible for pseudoknot formation are indicated with dashed lines.

in Fig. 3 has a bulged A residue, positioned in a strictly conserved eight nucleotide stretch. This pseudoknot in its turn might be stacked on the one formed by the long range interaction involving the large bulge loop in hairpin III (see Fig. 1). (Our experimental data do not support the existence of the base pair $G_{50} \cdot U_{105}$, compare Fig. 4A). We also note that the pseudoknot with the bulged A residue and the the pseudoknot in the middle have the same characteristics in all four TMV strains analysed. This is in contrast to the 3′ distal pseudoknot which appears to be more variable in terms of stem or connecting loop size. Especially the middle pseudoknot is structurally very well conserved. It always contains five or six base pairs in one and three base pairs in the other stem region, while the deep groove is bridged by one single G residue and the shallow groove by an AUA or AAA stretch. It will be demonstrated below that just this particular pseudoknot with the single G residue in the deep groove appears to be a conserved feature in a number of other plant viral RNAs. The finding of the three consecutive pseudoknots in the 5′ half of the 3′ noncoding region, together with our proposal of the tRNA-like structure as shown in Fig. 1 has enabled us to propose a model of the spatial folding of the entire 3′ noncoding region of TMV RNA (see Fig. 4). Essentially it consists of three major domains. The aminoacyl acceptor arm harbouring a pseudoknot as described above, is the first one. The second domain consists of one hairpin (III, Fig. 1)

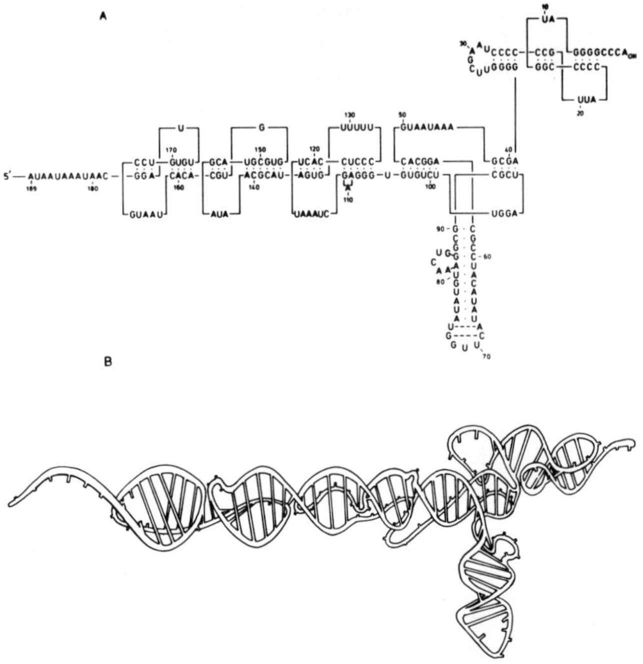

FIGURE 4. Spatial folding of the 3′ noncoding region of TMV RNA (vulgare strain). A. Schematic lettered presentation. See also Fig. 1 for the tRNA-like structure. Numbering is from the 3′ end. B. Artist's view of the three-dimensional structure. For further details see text.

which is assumed to function as the anticodon arm. Its orientation perpendicular to the aminoacyl acceptor arm yields the well-known L shape of standard tRNA. The three consecutive pseudoknots farther upstream (see Fig. 4A) form a third domain, which together with the longe range interaction of the bulge loop in hairpin III has a length of about 100 Å.

The entire 3′ noncoding region of TMV RNA contains five pseudoknots, four of which are of the special type as described in Fig. 2. It should be emphasized that all double stranded regions present in the model can be accounted for by experimental data and sequence comparisons (11). The exact orientation of the several domains, the precise conformation or the demonstration of other tertiary interactions has to await further analysis, mainly with

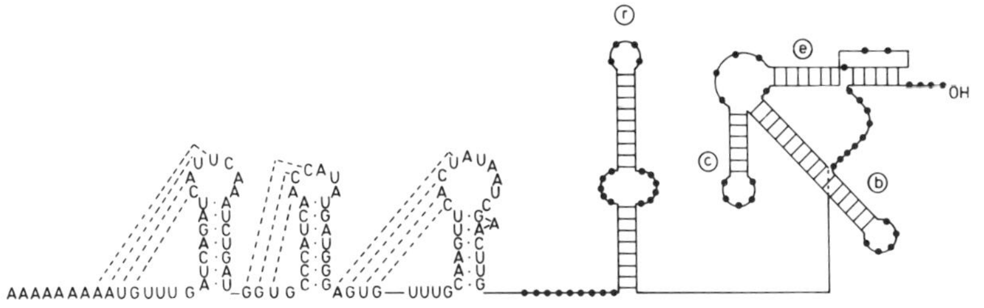

FIGURE 5. The secondary structure of the 3′ noncoding region of barley stripe mosaic virus (BSMV) RNA2. The nucleotide sequence was determined by Kozlov et al. (17), who also proposed hairpin b, c, e and r, given here in a schematic way. Dashed lines represent Watson-Crick base pairings.

biophysical techniques. Nevertheless, the model presented in Fig. 4 is the most detailed structure described so far for RNA molecules or fragments of this size.

PSEUDOKNOTS IN THE 3′ NONCODING REGION OF OTHER PLANT VIRAL RNAs

The question can be raised whether the presence of pseudoknotted structures is a special feature of tRNA-like structures and of the 3′ noncoding region of TMV RNA. A search for (conserved) pseudoknots in the coding region of TMV RNA did not give any positive result so far. It therefore seems reasonable to look at the 3′ noncoding regions of other aminoacylatable plant viral RNAs. The RNAs from the bromoviruses (e.g. BMV) and the hordeiviruses (e.g. BSMV) contain sufficient nucleotides (200-250) in their 3′ noncoding region for harbouring one or more pseudoknots upstream of the tRNA-like structure. In this respect the publication of the 3′ terminal sequence of barley stripe mosaic virus (BSMV) RNA2 by Kozlov et al. (17) was revealing. The genomic RNAs of this tricornavirus can be aminoacylated with tyrosine and the secondary structure proposed for the 3′ terminus (17) indeed has many features in common with that of the tyrosylatable bromoviral RNAs as proposed by us (8) and others (13,18). (see Fig. 5). An important difference is the relative short putative anticodon arm and the absence of hairpin d (using the

nomenclature of Ahlquist et al. (18)) like in the broad bean mottle virus (BBMV) RNAs (18). A hairpin designated r, immediately upstream of the tRNA-like structure was proposed as being characteristic for BSMV RNA (17). Between hairpin r and the internal poly A stretch farther upstream, three other hairpins were proposed, designated f, g and h respectively. However, a closer look at the latter three hairpins revealed, to our surprise, that there were many similarities with the pseudoknotted stalk-like region in TMV RNA. By disrupting hairpin f and maintaining hairpins g and h, proposed by Kozlov et al. (17), we found that the RNA can be folded into three pseudoknots as illustrated in Fig. 5. The two 3' proximal potential pseudoknots are very similar to those found in TMV RNA (see also Fig. 3). Note the position of the bulged A residue and the conserved hairpin loop sequence in the 3' proximal pseudoknot of both models. The middle pseudoknot contains seven base pairs in one of the two stem regions involved, instead of six or five as found in the tobamoviral RNAs. These seven base pairs are again spanned by a single G residue, which is not in contradiction with the geometric properties of an RNA A helix as outlined above and described earlier (10). Furthermore, the hairpin loop again consists of six nucleotides, the characteristic AUA sequence of which spans the shallow groove of this pseudoknot.
All together this means that the secondary and tertiary structure of BSMV RNA2 between the poly A stretch, which separates the coding from the noncoding region (19), and hairpin r, is essentially the same as that of the stalk-like region of the histidine- or valine-accepting tobamoviral RNAs (see Fig. 4).

Encouraged by this finding for BSMV RNA, we surveyed all published 3' terminal nucleotide sequences of plant viral RNAs, both polyadenylated or nonpolyadenylated. Our conclusion is that in almost all capped nonpolyadenylated plant viral RNAs one or more potential pseudoknots are found in the 3' noncoding region close to the stopcodon (20). We shall describe briefly two other interesting examples from the bromovirus and the tobravirus groups.

Fig. 6 shows the proposals for the secondary structures of the 3' noncoding regions of the three genomic RNAs of brome mosaic virus (BMV). These 3' termini contain a number of hairpins (b-g) which form the tRNA-like structure and have identical base sequences in RNA1, 2 and 3, except for a few substitutions near stem b (8,13,18). The variability in the three sequences starts upstream of hairpin g, just at the point where we have found a series of consecutive pseudoknots. The pseudoknots proposed here are strongly

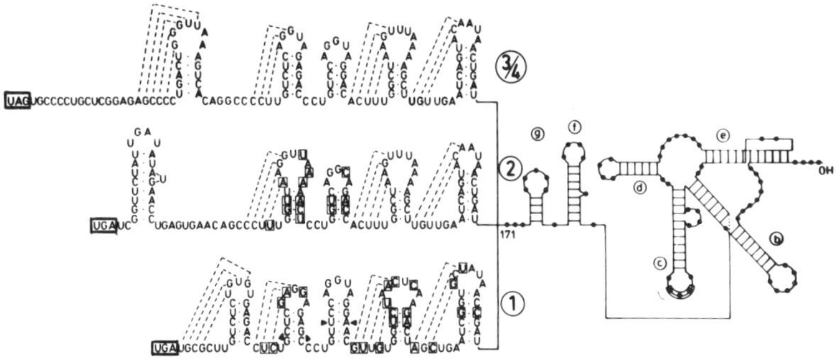

FIGURE 6. The secondary structure of the 3′ noncoding region of genomic RNAs 1, 2 and 3 of brome mosaic virus (BMV). The subgenomic RNA4 is derived from RNA3. Hairpin b-g which are involved in the formation of the tRNA-like structure are discussed elsewhere (8,13,18). Dashed lines indicate Watson-Crick base pairings giving rise to pseudoknotted structures. Boxed residues and arrows in RNA1 and 2 indicate substitutions and deletions or insertions as compared with RNA3 and 4. The position of the stopcodon at the 5′ end is indicated.

supported by their conservation among the three genomic RNAs. Each stem region contains at least one compensating base change in one of the two other BMV RNAs, including the tertiary interactions which are responsible for the pseudoknot formation. There is one single, but interesting, exception in the case of the hairpin in the middle of the pseudoknot region. A possible base pairing of the AGG sequence in the loop with the CCU sequence adjacent to the stem is not supported by such a natural second site revertant. If this pseudoknot would exist, a long quasi-continuous double helix of four or five pseudoknots could be formed. However, such a pseudoknot would require that the wide groove is spanned by no residue at all which is hardly conceivable at the moment. Some of the hairpins, which in our model are assumed to be involved in pseudoknot formation, were proposed earlier, but the possibility of pseudoknot formation was not considered (18,21).

Note also that the pseudoknot of the 3′ proximal hairpin closely resembles the conserved one in BSMV and TMV RNA (see above). It contains two double helical segments of six and three base pairs, respectively and its shallow groove is bridged by the characteristic AUA sequence. The only difference is that here three nucleotides bridge the deep groove instead of a single G residue. Such a single G containing connecting loop was found again in one of the three genomic RNAs of the related cowpea chlorotic mottle virus (CCMV) at exactly the same position (not shown). A full comparison can not be made with these latter RNAs because no complete sequences are available, which is also true for BBMV (18).

So far we were not able to find conserved pseudoknots in the cucumber mosaic virus (CMV) RNAs. The 3' noncoding regions of these cucumoviral RNAs, which can also be aminoacylated with tyrosine, are relatively long and show a complicated organisation (22).

The 3' termini of certain RNAs from the tobravirus group are another interesting example of the presence of pseudoknotted structures. Tobraviruses, like tobacco rattle virus (TRV), have a bipartite genome. RNA1 has a relatively constant size of about 6500 nucleotides no matter the strain or serotype, whereas RNA2 is of variable length from 2000 to as much as about 4000 nucleotides (23). Recent sequencing studies of tobraviral RNAs have shed new light on this length heterogeneity of RNA2. On the one hand it was the presence of an extra cistron, as observed in RNA2 of the TCM strain, while on the other hand the sequence at the 3' end of RNA2 turned out to be 100% homologous with RNA1 over a variable length for the different strains (23-26).

What is of interest here is that we have found a set of potential pseudoknots in RNA2, just upstream of the region completely homologous to RNA1. These pseudoknots again show strong resemblances with those described for TMV RNA and other plant viral RNAs mentioned above. In Fig. 7 we propose for three different strains of TRV the secondary structure of the RNA stretch located between the coding region of RNA2 at its 5' side and the region homologous to RNA1 at its 3' side. This model strongly suggests that a duplication of the stalk-like structure as present in TMV and BSMV RNA has taken place in the case of the PSG and TCM strains. An extra hairpin is present in the PSG strain, separating the two sets of pseudoknots. In all three TRV RNA2 molecules there is one particular pseudoknot which again is very similar to the one found in TMV, BSMV or CCMV RNA. This pseudoknot also has the single G residue bridging the deep groove and the AUA stretch bridging the shallow groove. Interestingly, the upstream located copy of this particular pseudoknot contains imperfections like mismatches or bulge loops in just the two strains showing the duplication. One might wonder whether this pseudoknot represents a copy which is not functional

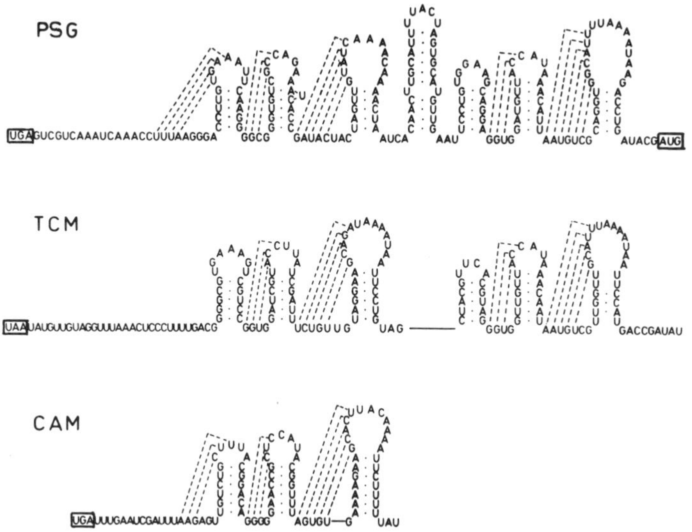

FIGURE 7. The secondary structure of TRV RNA2 between coding region and the region homologous to RNA1. Nucleotide sequences and genome organisation of TRV RNA2 from the PSG, TCM and CAM strain were published by others (23,25,26). Stop and initiation codons are boxed. Dashed lines indicate Watson-Crick base pairings typical for pseudoknot formation.

anymore. The occurrence of such a "pseudo-pseudoknot" might be related somehow to the duplication itself. Similar potentially pseudoknotted structures were not found in RNA1 to date, neither in the 3′ noncoding region nor in the intercistronic regions. We also want to emphasize here again that the probability of finding two or three consecutive pseudoknots giving rise to a quasi-continuous double helix of 20-25 base pairs is very low (11), let alone when they occur in noncoding regions only or when the number of pseudoknots increases to five or six like in the PSG strain of TRV. Furthermore, the sequence homology between the two series of pseudoknots in either the PSG or TCM strain is surprisingly low and is of the order of 50%, which is similar to the homology found between these regions in the different TRV strains.

DISCUSSION

In this paper we have shown that so-called pseudoknots are major structural elements in the higher order folding of 3' noncoding regions of several plant viral RNAs. All tRNA-like structures in plant viral RNAs harbour at least one such pseudoknot in the aminoacyl acceptor arm (7-10). Most pseudoknots in the 3' noncoding regions were discovered farther upstream in the vicinity of the stopcodon. In the case of TMV RNA experimental evidence and support from sequence comparisons was obtained for the presence of three pseudoknots in tandem, giving rise to a so-called stalk-like region just adjacent to the tRNA-like structure (11). It is striking that the latter series of pseudoknots appear at similar positions in a number of other plant viral RNAs. The viruses involved belong to groups as different as mono-, bi- and tripartite viruses. In structural terms, this stalk region seems to be even better conserved than the tRNA-like structures. In this respect we recall that the RNA from BSMV, TMV and the cowpea strain of TMV (C_CTMV) is aminoacylated with tyrosine, histidine and valine, respectively and have different three-dimensional structures at their 3' terminus whereas they share very similar pseudoknots farther upstream. This means that the amino acid accepting specificity of the 3' terminus may not be very important as long a tRNA-like structure is present and that the latter can be exchanged relatively easy by recombinational events as suggested earlier (11) (see also below).

Analysis of published sequences of plant viral RNAs indicates that one or more potential pseudoknots occur near the stopcodon in almost all capped nonpolyadenylated plant viral RNAs. No (conserved) pseudoknots could be detected in animal viral RNAs so far. The 3' termini of picornaviral RNAs for instance may have other structural features in common (27). This does not mean that the folding of RNA into pseudoknotted structures is only restricted to the 3' termini of plant viral RNAs. We have shown previously that pseudoknots play a role in ribosome functioning and in autocatalytic selfsplicing of RNA precursor molecules (10).

A major question which emerges from these studies is that of the possible function of these pseudoknotted stalk-like regions. To date no clear answer is available. To the best of our knowledge no experimental data are available about the function of the 5' proximal part of these 3' noncoding regions. Nevertheless, we can at least exclude a few possibilities. One of the most obvious possible functions would be a role in the replication of the plus strand. In view of the results, recently obtained for BMV, this seems less likely. It was shown that 3′ terminal fragments of BMV RNA as short as 134 nucleotides have high template activity for virus specific RNA dependent RNA polymerase (28). From this it can be inferred that the series of pseudoknots upstream of the tRNA-like structure of BMV RNA do not seem essential for replication. Another possibility might be that this stalk-like region of pseudoknots plays a role in encapsidation. Apart from the fact that these pseudoknots occur in RNAs from various, structurally widely differing, helical and icosohedral viruses, detailed studies on the assembly of TMV have shown that the origin of assembly and also the so-called pseudo-origin of assembly are located in the region coding for the coat protein and in some strains of TMV even farther upstream (24).

The presence of pseudoknots between the coding region and the tRNA-like structure, might help in keeping apart or regulating the two processes of translation and replication. If the function of these pseudoknots would be to bridge only two functionally different RNA domains, then, in our view, the strong conservation of some structural features of the pseudoknots involved would be less well understandable. Because of the peculiar situation in TRV, where the same stalk-like region is found in RNA2 only, just upstream of the sequence which is completely homologous to RNA1, we suggest that these regions might be involved in recombination events in RNA or in some sort of "communication" (25) between RNA molecules. We also want to mention here again the apparent exchangeability of tRNA-like structures as suggested by the structural organisation of various plant viral RNAs (see above). Recent work of Bujarski and Kaesberg (30) who obtained evidence for recombination on the RNA level in the 3′ noncoding region of BMV RNA3, further fosters speculations along these lines. Future work, especially using cloned cDNA transcripts may shed light on a possible function of these pseudoknots which appear as strongly conserved elements in the 3′ noncoding region of plant viral RNAs.

REFERENCES

1. Strauss EG, Strauss JH (1983). Replication strategies of the single stranded RNA viruses of eukaryotes. In: "Current Topics in Microbiology and Immunology", New York: Springer-Verlag Vol 105, p 1.
2. Haenni AL, Joshi S, Chapeville F (1982). tRNA-like

structures in the genomes of RNA viruses. In Cohn WE (ed): "Progress in Nucleic Acids Research and Molecular Biology" New York: Academic Press Vol 27, p 85.

3. Kim SH, Suddath FL, Quigley GJ, Mc Pherson A, Sussman JL, Wang AHJ, Seeman NC, Rich A (1974). Three-dimensional tertiary structure of yeast phenylalanine transfer RNA. Science 185:435.
4. Briand JP, Jonard G, Guilley H, Richards K, Hirth L (1977). Nucleotide sequence (n = 159) of the amino-acid-accepting 3' OH extremity of turnip yellow mosaic virus RNA and the last portion of its coat protein cistron. Eur J Biochem 72:453.
5. Silberklang M, Prochiantz A, Haenni AL, RajBhandary UL (1977). Studies on the sequence of the 3' terminal region of turnip yellow mosaic virus RNA. Eur J Biochem 72:465.
6. Guilley H, Jonard G, Kukla B, Richards KE (1979). Sequence of 1000 nucleotides at the 3' end of tobacco mosaic virus RNA. Nucleic Acids Res 6:1287.
7. Rietveld K, Van Poelgeest R, Pley CWA, Van Boom JH, Bosch L (1982). The tRNA-like structure at the 3' terminus of turnip yellow mosaic virus RNA. Differences and similarities with canonical tRNA. Nucleic Acids Res 10:1929.
8. Rietveld K, Pley CWA, Bosch L (1983). Three-dimensional models of the tRNA-like 3' termini of some plant viral RNAs. EMBO J 2:1079.
9. Rietveld K, Linschooten K, Pley CWA, Bosch L (1984). The three-dimensional folding of the tRNA-like structure of tobacco mosaic virus RNA. A new building principle applied twice. EMBO J 3:2613.
10. Pley CWA, Rietveld K, Bosch L (1985). A new principle of RNA folding based on pseudoknotting. Nucleic Acids Res 13:1717.
11. Van Belkum A, Abrahams JP, Pleij CWA, Bosch L (1985). Five pseudoknots are present at the 204 nucleotides long 3' noncoding region of tobacco mosaic virus RNA. Nucleic Acids Res 13:7673.
12. Van Belkum A, Jiang BK, Rietveld K, Pleij CWA, Bosch L (1986). Submitted for publication.
13. Joshi RL, Joshi S, Chapeville F, Haenni AL (1983) tRNA-like structures of plant viral RNAs: conformational requirements for adenylation and aminoacylation. EMBO J 2:1123.
14. Joshi RL, Chapeville F, Haenni AL (1985). Conformational requirements of tobacco mosaic virus RNA for aminoacylation and adenylation. Nucleic Acids Res 13:347.

15. Studnicka GM, Rahn GM, Cummings IW, Salser WA (1978). Computer method for predicting the secondary structure of single stranded RNA. Nucleic Acids Res 5:3365.
16. Zuker M, Stiegler T (1981). Optimal computer folding of large RNA sequences using thermodynamics and auxiliary information. Nucleic Acids Res 9:133.
17. Kozlov YV, Rupasov VV, Adyshev BM, Belgelarskaya SN, Agranovsky AA, Mankin AS, Morozov SY, Dolja VV, Atabekov JG (1984). Nucleotide sequence of the 3′ terminal tRNA-like structure in barley stripe mosaic virus genome. Nucleic Acids Res 12:4001.
18. Ahlquist P, Dasgupta R, Kaesberg P (1981). Near identity of 3′ RNA secondary structure in bromoviruses and cucumber mosaic virus. Cell 23:183.
19. Stanley J, Hanau R, Jackson AO (1984). Sequence comparison of the 3′ ends of a subgenomic RNA and the genomic RNAs of barley stripe mosaic virus. Virology 139:375.
20. Pley CWA (1986). In preparation.
21. Ahlquist P, Dasgupta R, Kaesberg P (1984). Nucleotide sequence of the brome mosaic virus genome and its implications for viral replication. J Mol Biol 172:369.
22. Symons RH (1979). Extensive sequence homology at the 3′ termini of the four RNAs of cucumber mosaic virus. Nucleic Acids Res 7:825.
23. Angenent GC, Linthorst HJM, Van Belkum AF, Cornelissen BJC, Bol JF (1986). RNA2 of tobacco rattle virus strain TCM encodes an unexpected gene. Nucleic Acids Res 14:4673.
24. Boccara M, Hamilton WDO, Baulcombe DC (1986). The organisation and interviral homologies of genes at the 3′ end of tobacco rattle virus RNA1. EMBO J 5:223.
25. Bergh ST, Koziel MG, Huang S-C, Thomas RA, Gilley DP, Siegel A (1985). The nucleotide sequence of tobacco rattle virus RNA2. Nucleic Acids Res 13:8507.
26. Cornelissen BJC, Linthorst HJM, Brederode FTh, Bol JF (1986). Analysis of the genome structure of tobacco rattle virus strain PSG. Nucleic Acids Res 14:2157.
27. Pley CWA (1986). Structural similarities in the 3′ noncoding regions of poly(A) containing RNA viruses. J Cell Biochemistry 10D:290. Abstract Q61.
28. Miller WA, Bujarski JJ, Dreher TW, Hall TC (1986). Minus-strand initiation by brome mosaic virus replicase within the 3′ tRNA-like structure of native and modified RNA templates. J Mol Biol 187:537.
29. Lebeurier G, Nicolaieff A, Richards KE (1977). Inside-out model for self-assembly of tobacco mosaic virus. Proc Natl Acad Sci USA 74:149.

30. Bujarski JJ, Kaesberg P (1986). Genetic recombination between RNA components of a multipartite plant virus. Nature 321:528.

Positive Strand RNA Viruses, pages 317–325

MUTATIONAL ANALYSIS OF THE FUNCTIONS OF THE tRNA-LIKE REGION OF BROME MOSAIC VIRUS RNA[1]

Theo W. Dreher and Timothy C. Hall

Biology Department, Texas A & M University
College Station, Texas 77843-3258

ABSTRACT Each of the three genomic and the single subgenomic RNAs of brome mosaic virus (BMV) is highly conserved in sequence and secondary structure near the 3′ end. The so-called tRNA-like structure imparts a number of tRNA-associated activities on the viral RNA, notably specific aminoacylation, and also functions as the promoter used by BMV replicase in copying virion RNAs. We have created a comprehensive set of mutant RNAs with sequence alterations within this conserved 3′ region, and have studied the effects these mutations have on aminoacylation and replicase template activities *in vitro*. These mutants have given an indication of the way in which the aminoacyl tRNA synthetase and replicase interact with the viral RNA, and are a valuable resource for further studies. Many of the mutations within the 134 nt-long 3′ tRNA-like structure of BMV RNA result in a loss of activity as template for BMV replicase, indicating that the replicase interacts with most of the regions of the structure. The most critical sequence-specific interaction between replicase and BMV RNA involves the loop of arm C, with which synthetase does not appear to interact. Synthetase activity is far less sensitive to sequence changes than the replicase; nevertheless, mutants with selectively decreased aminoacylation, which will be useful for *in vivo* studies, have been made.

---

[1]This work was supported by NIH grants AI 11572 and 22354 and by Texas A&M University.

INTRODUCTION

This laboratory has been studying the RNA-dependent RNA polymerase (replicase) that is induced in barley plants by infection with brome mosaic virus (BMV) (1-3). Replicase preparations made from BMV-infected leaves are able to initiate and complete synthesis of (-) strands using virion RNAs as templates (4). The marked specificity in copying BMV RNAs, and the absence of activity in uninfected plants, suggests that the activity studied <u>in vitro</u> represents the enzyme used in viral replication. The signals responsible for template recognition and selection of the site for complementary strand initiation reside within the c. 200 nt-long homologous region present at the 3′ ends of all 3 genomic and single subgenomic virion RNAs (5). The replicase "promoter" has been further localized to the 134 nt-long 3′ region (4), which is folded into a structure that in some aspects resembles tRNA (ref. 6; refer Fig. 1). This tRNA-like structure is also the core structural feature responsible for a number of tRNA-like properties of the viral RNA: aminoacylation (specifically with tyrosine), formation of a ternary complex of the charged RNA with GTP and elongation factor, and the ability to act as substrate for ATP:CTP tRNA nucleotidyl transferase (7).

We are interested in the relationship between the tRNA-like properties and the replicase promoter function of the 3′ homologous region of BMV RNA. The multiple functions of this region may be indicative of an involvement of the tRNA-associated activities with the assembly of an active replicase complex (8), in a way analogous to the inclusion of host elongation factors in the replicase of the RNA bacteriophage Qβ (9). At least one tRNA-like property, that of substrate for nucleotidyl transferase, is likely to serve an independent role in post-replicational adenylation and in maintaining an intact 3′ terminus (4). The tRNA-associated activities may also be important in regulating the relative usage of the viral RNAs in translation and replication. Finally, an understanding of the function of the 3′ region and its interaction with the tRNA-associated proteins could help decide whether this part of the viral RNA has origins as a host tRNA, or has evolved convergently towards tRNA function.

Production <u>in vitro</u> of wild type and mutant 3′ BMV RNA fragments.

In order to study the relationships between the

multiple functions of the 3′ region of the BMV RNAs, we have constructed a comprehensive set of mutants with sequence alterations focussed within the tRNA-like structure (Fig. 1). Such a set of mutants is a valuable pool of variants that can be subjected to analysis for the various tRNA-associated activities and for replicase promoter function in vitro, as well as for the effects of the sequence changes on infectivity in vivo. This paper describes the mutations that have been made as an extension of our earlier work (10,11), and their activities in vitro as replicase templates and in aminoacylation assays.

Natural or mutagen-induced mutations mapping to the 3′ homologous region of BMV RNA have not been reported; the multiple functions of this part of the RNA suggests that many mutations within this region are likely to be lethal. In order to overcome this limitation and to enable the construction of any desired mutant, we engineered a cDNA clone of the 3′ region of BMV RNA in such a way that a consistent proportion of run-off transcripts produced in vitro by SP6 RNA polymerase has the correct -CCA 3′ terminus of native virion RNAs (10). These transcripts show appropriate aminoacylation and replicase template activities in vitro. Large amounts of transcript can be produced from such constructs placed behind the SP6 promoter in either pUC-derived plasmids or in M13. The latter constructs have been used recently to facilitate site-directed mutagenesis.

RESULTS

Description of mutations within the 3′ region of BMV RNA

The locations of mutations that have been constructed in the BMV RNA are shown in Fig. 1, and the sequence alterations are detailed in Table I. All mutations were designed to keep changes to a minimum, in order to maximize the chance of producing mutants specifically inhibiting a single activity, and to avoid transmitting significant steric perturbations to other parts of the overall structure as far as possible. Two considerations were applied in deciding which mutations to produce. Firstly, it was desired to probe the effect of changes in each of the structural parts of the tRNA-like region. This would enable a compilation of a structure-function map for each activity. Secondly, we were interested in producing mutants capable of replication but deficient in aminoacylation, for future in

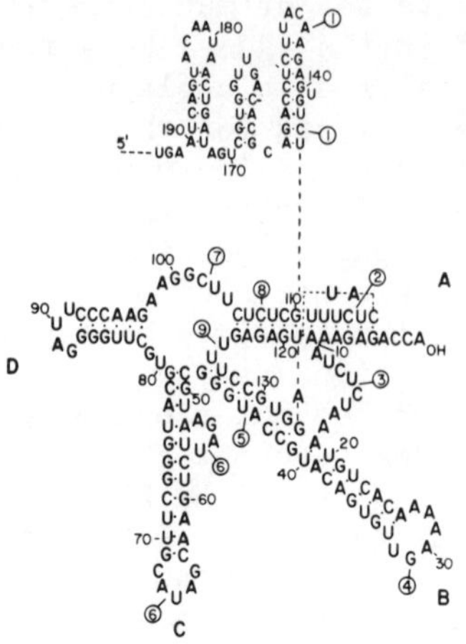

Fig. 1. Representation of secondary structure of 3' homologous region of BMV RNA 3 (based on ref. 6). Nucleotides are numbered from the 3' end. Letters A→D identify the principle arms of the tRNA-like structure; circled numbers indicate loci at which mutations were made.

vivo experiments designed to test the relevance of tyrosylation to replication. Attempts were made to predict the locations in the BMV RNA of bases analogous to those of tRNAs that are thought to interact with synthetases, including some of the invariant bases. However, the complexity of the BMV RNA structure relative to that of tRNAs has made this difficult. All mutants reported here have been constructed by deoxyoligonucleotide-directed mutagenesis, and were transcribed and assayed for tyrosylation and replicase template activities as described previously (10). Despite the stability of the stems in the RNA structure of this region, no difficulties were encountered in mutagenizing the single-stranded cDNA of M13 clones. This work will be described in more detail elsewhere.

Effect of mutations on ability to aminoacylate in vitro

Mutations in several regions of the tRNA-like structure have no detectable effect on the tyrosylation characteristics of the BMV RNA (Fig. 2, Table I). Thus, either the synthetase does not interact with these regions of the RNA, or if there are contacts, there is enough flexibility in the interaction to accommodate certain sequence changes. Such loci include arm C, the so-called anticodon stem. Neither

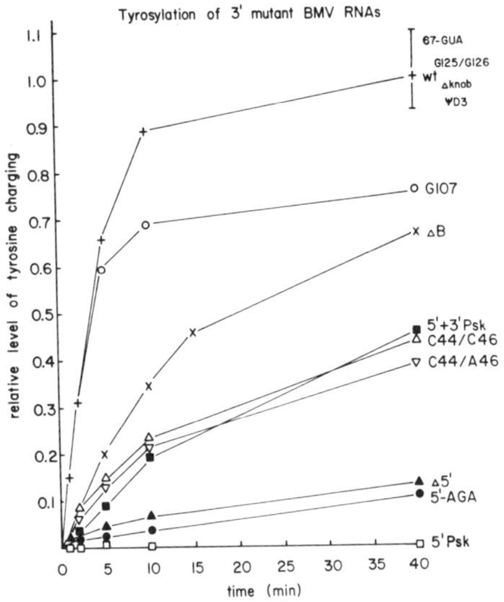

Fig. 2. Relative aminoacylation rates of wild type (wt) and various mutant BMV RNAs.

subtle nor drastic mutations in this stem and its loop affect tyrosine charging. The same results are obtained with purified yeast tyrosyl tRNA synthetase, an enzyme known to aminoacylate less efficiently its cognate tRNA carrying substitutions in the anticodon (12). We conclude that synthetase does not intimately interact with stem C, and that the presence in the loop of an -AUA- tyrosine anticodon is not related to the aminoacylation. The results of arm B mutations (Table I) demonstrate that this stem does interact with synthetase, but it is not clear whether this is in a fashion analogous to the interaction of synthetase with the anticodon stem of tRNA. It is possible that no clear analogue to the anticodon stem or bases of tRNA is present in the BMV RNA.

Other loci that have been changed without affecting tyrosylation are in arm D (11), and at loci 7 and 9, which are very conserved in the RNAs among viruses closely related to BMV.

Mutations in three separate loci have resulted in drastic loss of tyrosylation, in two cases (loci 1 and 2) without a large loss of replicase template activity (Table I). Changes in the base of arm B affected both <u>in vitro</u> activities. A very informative pair of mutants (5′PsK and 5′+3′PsK) was made in order to test the dependence of the <u>in vitro</u> activities on the presence of a tRNA-like aminoacyl acceptor stem (arm A, Fig. 1), by first substituting bases which would prevent stable "pseudoknot" (6) formation, and then substituting complementary bases in the opposite strand, so that a pseudoknot of equal thermodynamic stability but inverted sequence could form. Aminoacylation was dependent on the presence of an intact aminoacyl accceptor stem. The activity of mutant 5'+3'PsK was low, however, and it is possible that this RNA may charge more efficiently with an amino acid other than tyrosine.

Affect of mutations on replicase template activity in vitro.

Mutations in most regions of the tRNA-like structure have resulted in less efficient replicase template activity (Fig. 3, Table I). The mutations selectively affecting aminoacylation (mentioned above) are the exceptions: the region 5' to the tRNA-like structure (locus 1), and the 3' end of the aminoacyl acceptor stem (locus 2). We have also previously shown that deletion of arm D actually enhances in vitro replication (11). Mutations causing the greatest loss of template activity are those in arm C (locus 6), studied previously (10) and extended here, and that preventing the formation of an aminoacyl acceptor stem (5' PsK). The interaction of replicase with the loop residues of arm C is stringently sequence-specific (10; Table I); this loop is

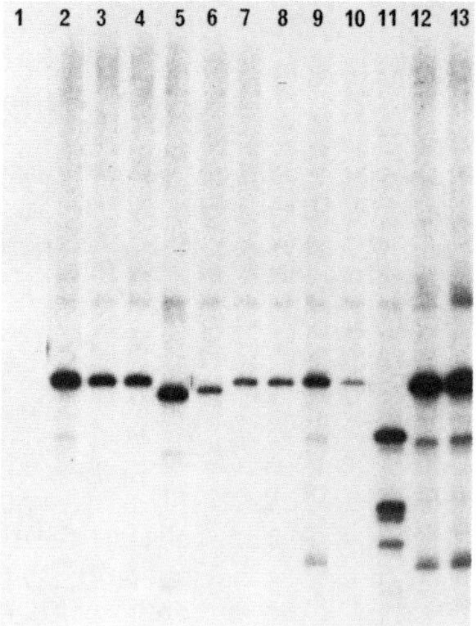

Fig. 3. Replicase products analyzed on 6% sequencing PAGE. RNA transcripts used as templates were: lanes 1, no RNA; 2, ψD3; 3, C44/C46; 4, C44/A46; 5, ΔB; 6, Δknob; 7, 67-GAA; 8, 67-GUA; 9, 5'+3' PsK; 10, 5'PsK; 11, Δ5'; 12&13, wt.

Table I.

Relative replicase template and aminoacylation activities of BMV 3′ mutants.

Locus	Mutant	Sequence Alteration	Relative Aminoacylation	Relative Replicase Template Activity
	wt		1.00	1.00
1.	Δ5′	Δ 156-135	0.19	0.90
1.	5′-AGA	137-UCU→AGA	0.17	0.59
2.	5′PsK	115-UCU→AGA	0	0.08
2.	5′+3′PsK	151-UCU→AGA & 8-AGA→UCU	0.54	0.71
3.	SSA_{12}	16-UCUCU→A_5	0.95	0.39
4.	ΔB	Δ32-27	0.75	0.34
5.	C44/C46	G46→C & A44→C	0.48	0.17
5.	C44/A46	G46→A & A44→C	0.30	0.20
6.	-GAA-	67-AUA→GAA	1.08	0.07
6.	-GUA-	67-AUA→GUA	1.04	0.06
6.	Δknob	Δ56-53	0.98	0.06
7.	ψD3	103-UC→AG	0.96	0.48
8.	G107	C107→G	0.78	0.38
9.	G125/G126	126-UU→GG	1.00	0.19

RNA Sequence positions are numbered from the 3′ end (see Fig. 1), which also details the mutated loci.
Aminoacylation data is expressed as final level of tyrosylation relative to wt transcripts.

the most critical recognition determinant on the RNA that we have observed. Single or double base substitutions as well as structural alterations in this arm result in equally drastic loss of in vitro replication activity. Studies aimed at understanding the nature of the interaction of replicase with this region will surely provide answers to the mode of specific template selection by BMV replicase.

Mutations at loci 3,4,7,8 and 9 all cause significant loss of template activity, exemplifying the widespread contacts that the replicase has with the RNA. Multiple contact points presumably help to ensure stringent template specificity. Locus 7 comprises the analogous sequence to the conserved TψCG(A) in the pseudouridine arm of tRNAs. This sequence is part of the internal promoter region for RNA polymerase III in tRNA genes (13). It is intriguing that a partial promoter sequence for polymerase III may also function as such in BMV RNA for the replicase. It remains to be seen whether there is any mechanistic similarity between RNA polymerase III transcription and the use by BMV replicase of its internal promoter, the tRNA-like structure, to direct initiation.

CONCLUSIONS

The ability to synthesize correctly terminated partial BMV RNAs by transcription in vitro has greatly facilitated our analysis of the properties of the 3′ tRNA-like structure as both replicase promoter and tRNA mimic. BMV replicase clearly has a complex interaction with many parts of the structure, but especially, in a sequence-specific manner, with arm C (Fig. 1). Tyrosyl tRNA synthetase has far more limited essential contacts with the RNA, but nucleotides upstream of the replicase promoter (nt 137-134) are important for efficient charging. Extension of the analysis of the comprehensive set of mutants described here to include other tRNA-associated activities should reveal whether segments of the tRNA-like structure not essential for aminoacylation, such as arms C or D, lack functional homology to tRNAs, but perhaps exhibit properties of other cellular RNAs, such as ribosomal RNAs. Experiments in which those mutants described here with specific loss of aminoacylation are tested in vivo for their effect on viral replication should shed light on the role of aminoacylation during infection processes.

REFERENCES

1. Hardy SF, German TL, Loesch-Fries LS, Hall TC (1979). Highly active template-specific RNA-dependent RNA polymerase from barley leaves infected with brome mosaic virus. PNAS (USA) 76:4956.
2. Bujarski JJ, Hardy SF, Miller WA, Hall TC (1982). Use of dodecyl-β-D-maltoside in the purification and stabilization of RNA polymerase from brome mosaic virus-infected barley. Virology 119:465.
3. Miller WA, Hall TC (1984). Use of micrococcal nuclease in the purification of highly template dependent RNA-dependent RNA polymerase from brome mosaic virus-infected barley.
4. Miller WA, Bujarski JJ, Dreher TW, Hall TC (1986). Minus-strand initiation by brome mosaic virus replicase within the 3′ tRNA-like structure of native and modified RNA templates. J. molec. Biol. 187:537.
5. Ahlquist P, Dasgupta R, Kaesberg P (1981). Near identity of 3′ RNA secondary structure in bromoviruses and cucumber mosaic virus. Cell 23:183.
6. Rietveld K, Pleij WA, Bosch L (1983). Three-dimensional models of the tRNA-like 3′ termini of some plant viral RNAs. EMBO J. 2:1079.
7. Joshi S, Haenni A-L (1984). Plant RNA viruses: strategies of expression and regulatin of viral genes. FEBS Letters 177:163.
8. Hall TC (1979). Transfer RNA-like structures in viral genomes. Intl. Rev. Cytology 60:1.
9. Blumenthal T, Carmichael GG (1979). RNA replication: function and structure of Qβ replicase. Ann. Rev. Biochem. 48:525.
10 Dreher TW, Bujarski JJ, Hall TC (1984). Mutant viral RNAs synthesized <u>in vitro</u> show altered aminoacylation and replicase template activities. Nature 311:171.
11. Bujarski JJ, Dreher TW, Hall TC (1985). Deletions in the 3′ terminal tRNA-like structure of viral RNA differentially affect aminoacylation and replication <u>in vitro</u>. PNAS (USA) 82:5636.
12. Bare L, Uhlenbeck OC (1985). Aminoacylation of anticodon loop substituted yeast tyrosine transfer RNA. Biochemistry 24:2354.
13. Ciliberto G, Castagnoli L, Cortese R (1983). Transcription by RNA polymerase III. Curr. Top. Devel. Biol. 18:59.

Positive Strand RNA Viruses, pages 327–336

MUTATIONAL ANALYSIS OF THE INTERNAL PROMOTER FOR TRANSCRIPTION OF THE SUBGENOMIC RNA4 OF BMV[1]

Loren E. Marsh, Theo W. Dreher and Timothy C. Hall

Department of Biology, Texas A&M University
College Station, Texas 77843-3258

ABSTRACT. Brome mosaic virus, a (+) strand RNA virus, is comprised of three genomic RNAs, designated RNAs 1, 2 and 3. A (-)-sense RNA3, produced by in vitro transcription with SP6 or T7 polymerase from cloned RNA3 cDNA, was shown to function in vitro as a template for replication of the subgenomic (+) RNA4 (Miller et al., Nature 313:68, 1985) that serves as mRNA for coat protein production. Analysis of deletion mutants has shown that the core of the subgenomic promoter lies within a region of some thirty bases located primarily upstream of the subgenomic RNA initiation site. Considerable sequence homology with the probable subgenomic promoters of alphaviruses and of some other plant (+) strand RNA viruses has been found for this core sequence. The postulated internal subgenomic promoters of the plant viruses generally possess A and U-rich regions immediately upstream and downstream of the promoters. These regions probably serve mechanical roles in facilitating transcription and translation.

INTRODUCTION

Brome mosaic virus, BMV, has a tripartite genome (RNAs 1, 2, 3). A fourth component, RNA4, is also encapsidated, but serves as the subgenomic mRNA for the synthesis of coat protein (1). The entire sequence of RNA4 is colinear with and contained within the 3′ end of the dicistronic RNA3 (2); however, only the 5′ reading frame, encoding protein

1This work was supported by NIH Grant AI 22354 and by Texas A&M University.

3a, is translated from RNA3 (2,3). Furthermore, in mixed infections, progeny RNA4 reflects the RNA3 sequence of the inoculum (4), implying that RNA4 is generated from RNA3.

Many other (+) strand RNA viruses possess polycistronic genomic RNAs from which the internal cistrons are expressed via subgenomic mRNAs. Experiments with temperature sensitive mutants of alphaviruses have strongly suggested internal initiation by the replicase and subsequent transcription as the mechanism for generation of the 26S subgenomic RNA from the 42S genomic RNA (5). Alternative mechanisms have been proposed (6,7) for generation of subgenomic RNAs for plant viruses. However, in vitro experiments show that BMV replicase can initiate internally on (-) strand RNA3 template and synthesize (+) strand RNA4 product (8), implying that the former mechanism functions in both (+) strand plant RNA viruses and alphaviruses.

Progress in understanding the origin of subgenomic BMV RNA4 results in large measure from three technical advances: (a) utilization of bacteriophage DNA-dependent RNA polymerases (such as SP6 RNA polymerase) for generation in vitro of relatively large amounts of specific RNA transcripts (9); (b) isolation and characterization of cDNA clones of BMV RNAs (10); and (c) development of an in vitro BMV replicase system (11-13). The BMV replicase used is an extract from infected barley leaves which, when treated with micrococcal nuclease, is dependent on exogenous template and shows high specificity for BMV RNA. In the present study, we have taken advantage of the in vitro replicase system to further characterize, by deletion analysis, the promoter internal to BMV RNA3 that generates the (+) strand subgenomic RNA4.

METHODS

Three transcriptional vectors were used for construction of mutants. One, pSP4HE5, has been described previously (10); the other two are derivatives of the same BMV RNA3 cDNA clone (see Fig. 1 for details). Deletions were constructed by cutting the parental vectors with restriction enzymes followed by digestion with mung bean nuclease and subsequent relegation. Additional deletions were made by oligonucleotide-directed site specific mutagenesis (14). The conditions for in vitro transcription of BMV (-)RNA3 template from the vectors and

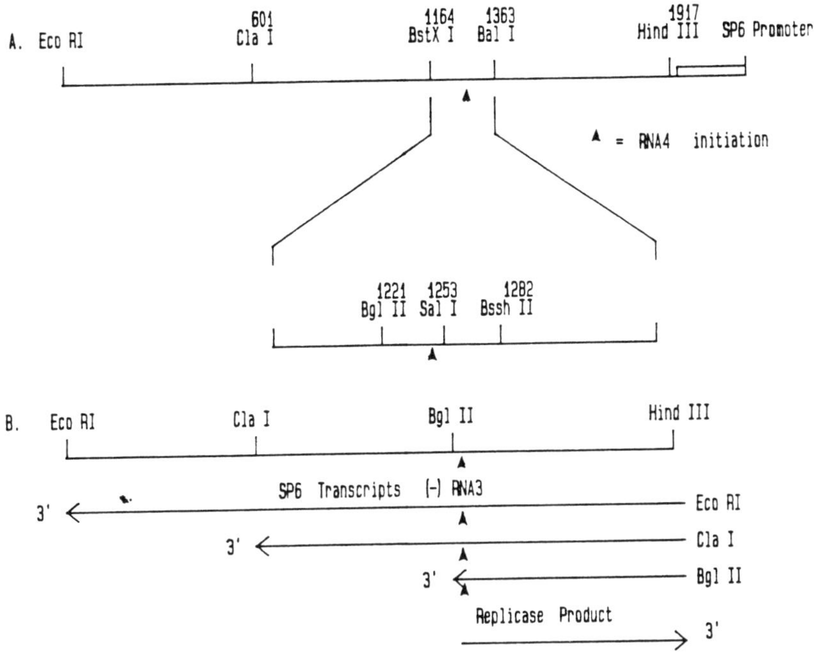

Figure 1. Diagram of transcriptional plasmids and phage used to define the subgenomic promoter, with the expected transcripts and replicase products.

A proportional restriction map of the relevant region of the cDNA of BMV RNA3 present in the transcriptional plasmids is shown in **A**. The area of interest is expanded below. The numbering of the bases given over the restriction sites corresponds to (+)RNA3. The cDNA clone is identical in all of the wild type transcriptional plasmids to that of pSP4HE5 (10). Two other transcriptional vectors were also used in this study; mp3SP6HE5 and pH2T7HE5. The former is an M13 phage construct having the relevant region of pSP4HE5 cloned into mp18. The latter vector is derived from pIBI76 and differs in that the (-)RNA3 is transcribed from a T7 polymerase promoter. These two vectors facilitate sequencing in that single stranded DNA can be isolated. SP6 or T7 transcripts which function as truncated replicase templates *in vitro* are shown in **B**. They result from linearization of the vectors described above with different restriction enzymes. The resulting replicase product is shown below the transcripts.

the conditions for the in vitro replicase reaction have been described previously (8,10). Replicase reaction products were treated with RNase in high salt (2x SSC) (11) and analyzed by electrophoresis on 3-4% polyacrylamide gels (15).

RESULTS

Prior to mutagenesis, sequences in RNA3 close to the site of initiation of RNA4 were analyzed for features reminiscent of promoters because it was assumed that they might predict the nature of the internal promoter. Homology was observed at and downstream from the site of RNA4 initiation to the consensus sequences implicated as promoters for Pol III genes, such as tRNAs, where the promoters are internal to the transcripts (16,17). Also, there is significant homology upstream of the RNA4 initiation site to the consensus sequences implicated in the transcription of the 26S subgenomic RNAs of the alphaviruses (18) (Table 1). An AU-rich region containing a variable poly(A) stretch of as many as 22 A's is found 21 bases upstream of the initiation site for RNA4 (2).

In order to extend observations that the replicase preparation could initiate and faithfully synthesize (+) strand subgenomic RNA4 from (-) template in vitro (10) we concentrated on delineating the internal promoter sequence by two approaches. (a) Truncated (-)RNA3 templates, generated in vitro by run-off transcription of linearized cDNA sequences and described in the legends to Figs. 1 and 2, were used for the in vitro replicase reactions. (b) The deletion mutants described in Fig. 2 were constructed in the cloned BMV3 cDNA sequence, transcribed in vitro and tested for replicase template activity.

The truncated (-)RNA3 templates yielded transcripts shorter than the wild type RNAs at the 3′ end. Such truncated templates are functionally analogous to deletions upstream from the subgenomic initiation site. Tests using this approach have shown that the core sequence of the promoter region is downstream from the Bgl II site (9) (Figs. 1 and 2). Transcripts shorter than those produced from vectors linearized at the Bgl II site are inactive as replicase templates. Mutant (-)RNA3 transcripts obtained by approach (b) were functional templates for the replicase if the deletions were downstream (with respect to the subgenomic initiation site) from the Sal I site in the

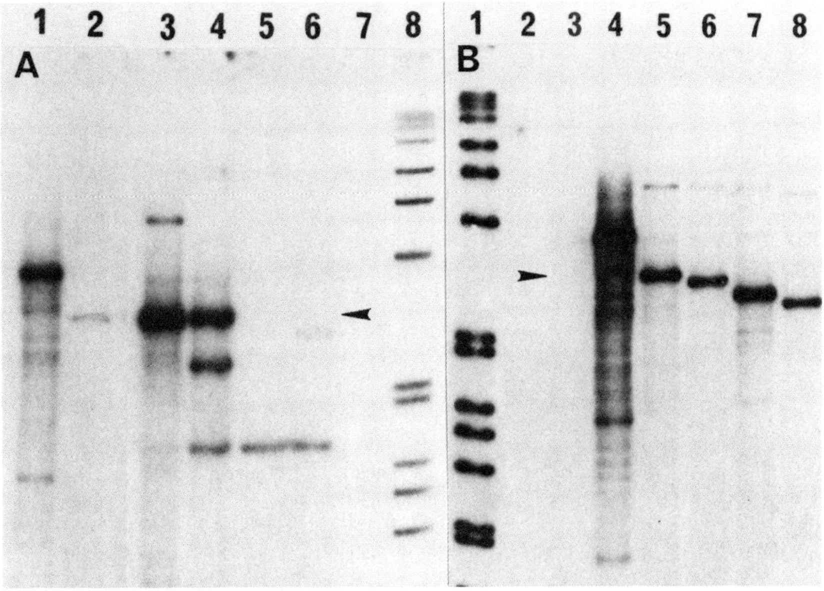

Figure 2. Replicase products from truncated templates and deletions downstream from the subgenomic initiation site.

In **A** the transcriptional plasmid was linearized at various restriction sites near the subgenomic initiation site prior to SP6 transcription. Replicase assays were done with identical amounts of transcript RNAs (2 μg) and the products were electrophoresed on a 3% acrylamide gel: (+)RNA4 (lane 1), 0 RNA (lane 2), Cla I product (lane 3), Bgl II product (lane 4), Sal I product (lane 5) and Bssh II product (lane 6). Lane 7 was not loaded and lane 8 contained molecular weight standards. The (→) indicates the wild type replicase product. The replicase products were 200 bases shorter than RNA4 because the cDNA was inserted in the vectors as an Eco RI-Hind III fragment of the RNA3 cDNA.

In **B** the transcriptional plasmids were linearized with Cla I prior to SP6 transcription. Replicase reactions were performed as above and the samples were then electrophoresed on a 3.5% gel: wild type (lane 5); deletions Sal I to Bssh II (lane 6), Bssh II to Bal I (lane 7) and Sal I to Bal I (lane 8). Lanes 3 and 4 were, respectively, the products from 0 RNA and (+)RNA4. Lane 1 contained molecular weight standards and lane 2 was not loaded.

cDNA. Mutant (-)RNA3 transcripts with sequences deleted between the Bgl II and Sal I sites were not functional replicase templates. Therefore, the promoter core sequence appears to be located in the 32 bases between these two restriction sites.

Further support for the concept that the region between the Bgl II and Sal I sites contains the functional core promoter sequence is provided by the comparisons shown in Table 1 (abbreviations used: AMV, alfalfa mosaic virus; BMV, brome mosaic virus; CMV, cucumber mosaic virus; MBV, Middleburg virus; RRV, Ross River virus; SFF, Semiliki Forest virus; SIN, Sindbis virus; TMV, tobacco mosaic virus and TRV, tobacco rattle virus). In section **A** of Table 1 a consensus sequence is derived from alphavirus putative subgenomic promoters (18). Homology within the Bgl II-Sal I region of BMV RNA3 to this consensus is found within a 25 base sequence, underlined in section **B** of Table 1. The three blocks which contain 16 bases of homology out of the total 25 bases are shown in section **C** of Table 1. Similar blocks of homology can be found upstream of the subgenomic initiation sites of other plant viruses (section B of table 1) providing additional support for the notion that these sequences are intrinsic to the promotion of subgenomic RNAs.

Although the transcript truncated at the Bgl II site is functional, mutants with sequences deleted at the Bgl II site are inactive. Additionally, a mutant in which the AU-rich region upstream of the promoter (including the poly(A) region of RNA3) was deleted had greatly reduced activity. Also, although deletions downstream from the RNA4 initiation site are, in general, active replicase templates, deletion of the short AU-rich stretch immediately downstream from the subgenomic initiation site inactivates the template.

DISCUSSION

Of the features observed in the region of the subgenomic promoter, the homology to the putative alphavirus promoters between the Bgl II and Sal I sites appears to be most significant (see Table 1). This concept is supported by the observation that deletions are tolerated at the Sal I site, which is downstream from the RNA4 initiation site and these regions of homology. In contrast, deletions are not tolerated at the Bgl II site

TABLE 1

COMPARISON OF ALPHAVIRUS PROMOTERS TO THOSE OF PLANT RNA VIRUSES

A. Putative alphavirus promoters (19).

SIN AUCUCUACGGUGGUCCUAA^AUAGUCAGCAUA
MBV ACCUCUACGGCGGUCCUAA^AUAGUUGCGUGA
SFV ACCUCUACGGCGGUCCUAA^AUUGGUGCGUUA
RRV ACCUCUACGGCGGUCCUAA^AUAGAUGCAGAGA
ACCUCUACGGCGGUCCUAA^AUAGUUGCG(A OR U)(U OR G)A -consensus

B. Postulated plant viral subgenomic promoters.

BMV Bgl II Sal I
AGUUAUUAUUAAAAAAAAAAAAAAAGAUCUAUGUCCUAAUUCAGC^GUAUUAAUAAUGUCGAC

CMV (19).
AAGGUUCAAUUCAAUUUGCAUCCCUGUUAGGCAAGGCCUUACUUUCUAUGGAUGCUUCUCCGCG-
AGUUAGC^GUUUAGUUGUUCACCUGAGUCGUGUUUUCUUUGUUUGCGUCUCAGUGUGCCUAUG

TRV (20 and personal communication, B. Cornelissen).
AUUAUUGUCAAGUGAAGAUGUUAAGAGAGCGUCUAAUAAGAAAAACUCGUCUUAAUGC^AUAAA-
GAAAUUUAUUGUCAAUAUG

AMV (21).
GUUCUCCAAAGGGUCUUGGAGUUCCGAAAGGGUUUACAUAUGAAAGUUUUAUUAAAGAUG-
AAAUAUUACCUGAUCAUUGAUCGGUAAUGGGCC^GUUUUAUUUUUAAUUUUCUUUCAAAUACUUCCAUCAUG

TMV (22).
UUUAAAAAGAAUAAUUUAAUCGAUGAUUCGGAGGCUACUGUCGCCGAAUCGGAUUC^GUUUAAAUAUG

C. Comparison of homologous sequence blocks

Alphaviruses -	UCUA	CGGCGGUCCUAA	UGCG(U OR A)
BMV -	UCUA	GUCCUAA	AGCGUA
TRV -	UCUA	CGUCUUAA	UGCAU
CMV -	UCUA	CCGCGAGUUA	AGCGUU

See text for details. The ^ precedes the initial base of the subgenomics.

which contains a portion of the homologous region. Although not part of the promoter core sequence, the AU-rich region upstream of the initiation site also seems to be significant in that its deletion reduces template activity. This AU-rich region upstream of the subgenomic promoter probably serves a mechanical role in making the promoter accessible to the replicase because computer modelling shows this region to be virtually free of base pairing. The short AU-rich stretch downstream from the RNA4 initiation site may serve a similar function. The homology to the tRNA promoters is probably not as significant in that deletions extending downstream from the Sal I site delete a significant amount of the homology to the tRNA promoters and yet the derived transcripts remain active as replicase templates.

Postulated subgenomic promoters of other (+) strand RNA plant viruses also show homology to the subgenomic promoter of BMV and those of the alphaviruses. This is shown in section B of Table 1 where significant sequences, such as the postulated promoter core sequences and the AUG beginning the reading frame, are underlined. The probable promoter core sequences of the plant viruses show more divergence than do those of the alphaviruses. This divergence probably reflects the greater evolutionary distance between the individual plant viruses. The postulated core sequences of BMV and TRV begin with AAGA. Although the promoter core sequences of other plant viruses do not always begin with AAGA, they often have this sequence slightly upstream of the core sequence. This is seen in the postulated promoter core sequence of AMV, which begins with ACC (as do those of the alphaviruses) but has AAGA only 10 bases upstream. We have also observed that the AAGA motif can also be found just upstream of the alphavirus putative promoters. The significance of this sequence is not obvious, although it may facilitate replicase recognition of the core sequence. Other plant viruses also have AU-rich regions upstream and downstream of the putative subgenomic promoters; BMV RNA3 appears to be unusual in having a poly(A) stretch in the upstream region.

Although deletions between the Sal I site and the SP6 and T7 promoters produced functional templates for the replicase, some templates functioned better than others (Fig. 2). The differences observed may be due to conformational changes in the tertiary structure of the mutant template RNAs. Therefore, an interesting area for future research is the role that sequence context plays in the modulation of the core sequence of the subgenomic promoter. Additionally, the functionality of the postulated subgenomic promoter sequence remains to be unequivocally demonstrated.

REFERENCES

1. Dasgupta R, Kaesberg P (1982). Complete nucleotide sequences of the coat protein messenger RNAs of brome mosaic virus and cowpea chlorotic mottle virus. Nucl. Acids Res. 10:703.
2. Ahlquist P, Luckow V, Kaesberg P (1981). Complete nucleotide sequence of brome mosaic virus RNA3. J. Mol. Biol. 153:23.
3. Shih DS, Kaesberg P (1973). Translation of brome mosaic virus ribonucleic acid in a cell-free system derived from wheat germ embryos. PNAS (USA) 70:1799.
4. Lane LC, Kaesberg P (1971). Multiple genetic components in bromegrass mosaic virus. Nature New Biol. 232:40.
5. Keranen S, and Kaariainen L (1979). Functional defects of RNA-negative temperature-sensitive mutants of sindbis and semliki forest viruses. J. Virol. 32:19.
6. Goelet P, Karn J (1982). Tobacco mosaic virus induces the synthesis of a family of 3′ coterminal messenger RNAs and their complements. J. mol. Biol. 154:541.
7. Gonda TJ, Symons, RH (1979). Cucumber mosaic virus replication in cowpea protoplasts: time course of virus, coat protein and RNA synthesis. J. Gen. Virology 45:723.
8. Melton DA, Krieg PA, Rebagliati MR, Maniatis T, Zinn K, Green MR (1984). Efficient in vitro synthesis of biologically active RNA and RNA hybridization probes from plasmids containing a bacteriophage SP6 promoter. Nucl. Acids Res. 12:7035.
9. Miller WA, Dreher TW, Hall TC (1985). Synthesis of brome mosaic virus subgenomic RNA in vitro by internal initiation on (-)-sense genomic RNA. Nature 313:68.
10. Ahlquist P, French R, Janda M, Loesch-Fries S (1984) Multicomponent plant viral infection derived from cloned viral cDNA. PNAS (USA) 81:7066.

11. Hardy SF, German, TL, Loesch-Fries LS, Hall TC (1979). Highly active template-specific RNA dependent RNA polymerase from barley leaves infected with brome osaic virus. PNAS (USA) 76:4956.
12. Bujarski JJ, Hardy SF, Miller, WA, Hall TC (1982). Use of dodecyl-β-D-maltoside in the purification and stabilization of RNA polymerase from brome mosaic virus-infected barley. Virology 119:465
13. Miller WA, Hall TC (1984). Use of micrococcal nuclease in the purification of highly template dependent RNA-dependent RNA polymerase from brome mosaic virus-infected barley. Virology 125:236.
14. Zoller MJ, Smith M (1984). Oligonucleotide-directed mutagenesis: a simple method using two oligonucleotide primers and a single-stranded DNA template. DNA 3:479.
15. Peacock AC, Dingman CW (1968). Molecular weight estimation and separation of ribonucleic acid by electrophoresis in agarose-acrylamide composite gels. Biochemistry 7:668.
16. Koski RA, Clarkson SG, Kurjan J, Hall BD, Smith M (1980). Mutations of the yeast SUP4 $tRNA^{Tyr}$ locus: transcription of the mutant genes in vitro. Cell 22:415.
17. Hall BD, Clarkson SG, Tocchini-Valentini G (1982). Transcription initiation of eucaryotic transfer RNA genes. Cell 29:3.
18. Ou JH, Rice CM, Dalgarno L, Strauss L, Strauss JH (1982). Sequence studies of several alphavirus genomic RNAs in the region containing the start of the subgenomic RNA. PNAS (USA) 79:5235.
19. Gould AR, Symons RH (1982). Cucumber mosaic virus RNA3: determination of the nucleotide sequence provides the amino acid sequence of protein 3a and viral coat protein. Eur. J. Biochem. 126:217.
20. Boccara M, Hamilton WDO, Baulcombe DC, (1986). The organization and interviral homologies of genes at the 3′ end of tobacco rattle virus RNA1. EMBO J. 5:223.
21. Barker RF, Jarvis NP, Thompson DV, Loesch-Fries LS, Hall TC (1983). Complete nucleotide sequence of alfalfa mosaic virus RNA3. Nucl. Acids Res. 11:2881.
22. Goelet P, Lomonossoff GP, Butler PJG, Akam ME, Gait MJ, Karn J (1982). Nucleotide sequence of tobacco mosaic virus RNA. PNAS (USA) 79:5818.

VI. TRANSLATION, PROTEIN MODIFICATION, AND ASSEMBLY

Positive Strand RNA Viruses, pages 339–349

PROCESSING OF CORONAVIRUS PROTEINS AND ASSEMBLY OF VIRIONS[1]

K.V. Holmes[2], J.F. Boyle[2], R.K. Williams[2], C.B. Stephensen[2], S.G. Robbins[3], E.C. Bauer[2], C.S. Duchala[2], M.F. Frana[2], D.G. Weismiller[2], S. Compton[3], J.J. McGowan[3], and L.S. Sturman[4]

[2]Departments of Pathology and [3]Microbiology, Uniformed Services University of the Health Sciences, Bethesda, MD 20814 and [4]New York State Department of Health, Albany, NY 12201

ABSTRACT Coronaviruses exhibit marked species and tissue specificity, and host genes can determine susceptibility to coronavirus infection. New approaches to analysis of coronavirus protein processing and assembly and virus-receptor interactions are providing insight into the molecular mechanisms of host dependence and virus strain dependence of coronavirus maturation.

INTRODUCTION

The fascinating events in the replication and transcription of coronavirus RNA are described elsewhere in this volume. The paper will analyze the complex interactions between viral structural components which occur during the processing, intracellular transport and assembly of coronaviruses, and interaction of these viruses with receptors on the plasma membrane of target cells. These details of virus assembly and receptor binding may determine whether coronavirus infection will produce new infectious virions and/or induce cytopathic effects in different cell types. Since coronaviruses show stringent host species and tissue specificity, host cell factors which affect virus maturation can determine the outcome of infection with coronaviruses.

The three structural proteins of mouse hepatitis virus strain A59 (MHV-A59) have been described in detail

[1]This work was supported by NIH grants AI 18997 and GM 31698, USUHS grant RO7403 and AID/SCI grant 2H-13.

and the general scheme for their synthesis, intracellular transport, assembly into virions and release from cells has been determined (1). Coronavirus nucleocapsids assemble in the cytoplasm from single, plus-strand genomic RNA and many copies of the phosphorylated nucleocapsid protein, N. The budding site of MHV-A59 virions on Golgi-associated vesicles appears to be determined by the accumulation at this site of the membrane glycoprotein, E1. The other virus glycoprotein, E2, which forms the spikes or peplomers, is also transported from the RER to the Golgi assembly site, but excess E2 can be transported to the plasma membrane.

The molecular interactions of virus structural components which are required for coronavirus assembly and infection are summarized in Table 1. In this paper, we will discuss a series of experiments on interactions between structural elements of MHV or between virions and host cell receptors which provide new insight into coronavirus assembly and infection.

TABLE 1
INTERACTIONS IN CORONAVIRUS ASSEMBLY AND ATTACHMENT

Interacting molecules	Associated assembly function
N-RNA and N-N	*Assembly of nucleocapsid (NC)
E1-membrane	Insertion into internal membranes
E1-E1	Determination of budding site
NC-E1	*Binding nucleocapsid at budding site
E2-E2 and E2-E1(?)	Assembly and orientation of spike
Cell protease-E2	*Cleavage of E2, activates cell fusion
E2-cell receptor	*Virus binding to membrane

*Topics to be discussed in this paper

MOLECULAR INTERACTIONS OF NUCLEOCAPSID PROTEIN, N

The N protein of MHV is the only protein in the helical nucleocapsid. It is a phosphorylated protein which varies in molecular weight from about 46 to 55K in virions of different strains of MHV (1,2). Although there is only a single molecular weight species of N in virions from each MHV strain, infected cell lysates analyzed by immunoprecipitation or immunoblotting frequently demonstrate one or more faster migrating species which we call N' and N''. The origin and functions of these forms of N are unclear. Pulse chase experiments show that they are derived from N, and that they are made in increasing amounts relative to N during the course of MHV infection. They may result from premature termination of N synthesis, altered phosphorylation or specific protease cleavages in the infected cell. We have compared the N, N' and N'' proteins of several strains of MHV in different cell lines, and have found that both the host cell and the virus strain determine the size and number of faster migrating N species in infected cells. Apparently, amino acid sequence differences make N proteins from various strains differentially susceptible to host cell processing events. By comparing immunoblots of MHV-A59 or MHV-JHM infected cell extracts done with antibody to full length N with antibody to a synthetic peptide of the carboxy terminus of N, we found that the faster migrating N species in infected cells lacked the carboxy terminal antigenic domain (Figure 1). This could result either from premature termination or from a specific protease cleavage.

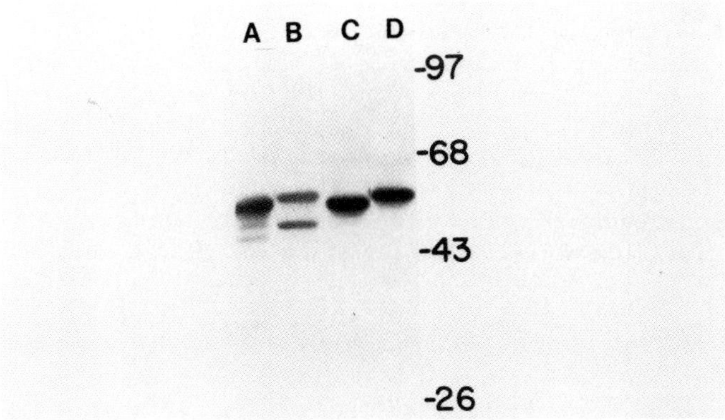

FIGURE 1. Nucleocapsid proteins in extracts of cells infected with MHV-A59 (lanes A and C) or MHV-JHM (lanes B and D). Western blots were done with antisera to the complete N protein of MHV-A59 (lanes A and B) or to a synthetic peptide near its carboxy terminus (lanes C and D).

To assemble the helical nucleocapsid of coronaviruses, one or a few molecules of N probably interact with an organizing site on the viral genome, and then additional molecules of N must be recruited, similar to the assembly of helical plant viruses. We have studied the interaction of coronavirus N protein to RNA by an RNA-overlay protein blot assay (ROPBA; 3). The N protein has such a high affinity for RNA that we have not yet been able to identify the postulated nucleotide sequence specific RNA binding activity which would initiate encapsidation in vivo. Instead, we find that isolated N binds to MHV genomic RNA in a nucleotide sequence independent manner, and even binds to RNAs from cells or other viruses, and to DNA. The ROPBA assay should allow us to identify the domain of N associated with binding to the genomic RNA, and may help to explain why N' and N'' do not assemble into virions even when they are present in large amounts in the infected cell.

At first, we expected that binding of N to host nucleic acids might just be an artifact of the in vitro binding assay, but recently we have found that N can also bind to other nucleic acids in vivo. Studies on the intracellular location of N in infected 17 Cl 1 cells showed that 2 to 5 hours post inoculation (p.i.) N antigen was found in small aggregates in the perinuclear area. Later, N antigens was found in larger, more numerous cytoplasmic aggregates and in virions attached to the plasma membrane. Since MHV infected 17 Cl-1 cells, unlike many other cell lines, are not rapidly killed by fusion, we were able to study cells up to 24 hours p.i. Multinucleate cells showed large amounts of N antigen concentrated in some but not all nuclei (Figure 2). The N antigen in nuclei co-localized with DNA detected by anti-DNA antibodies. The mechanism for transport of this cytoplasmic protein to the nucleus late in the infectious cycle is not yet clear, but it appears not to be due to loss of the nuclear envelope during cell division. The effects on host cell metabolism of large amounts of N in the nucleus must be determined. Possibly this could facilitate development of persistent infection.

A collection of temperature sensitive, RNA positive mutants of MHV-A59 has recently been developed. Several of these mutants have alterations in the N protein, and study of them should help to identify functions of N in virus replication and assembly. Two of these mutants show thermolability of transport of N to the nucleus, and several others show altered patterns of distribution of N antigen in the cytoplasm or nucleus. One mutant which should be particularly useful for studies of virus structure is called Alb-4. It is temperature sensitive, producing minute plaques at 39 C and normal, larger plaques at 32 C. The N protein of Alb-4 migrates faster than that of wild type MHV-A59 at both temperatures, suggesting that the mutation affects the N gene and that the function of the altered protein is temperature sensitive. This conclusion is supported by the observation that virions of Alb-4 are much more thermolabile than wild type virus. Possibly high temperature induces allosteric changes in the mutant N protein, which then interacts abnormally with N, RNA and/or E1.

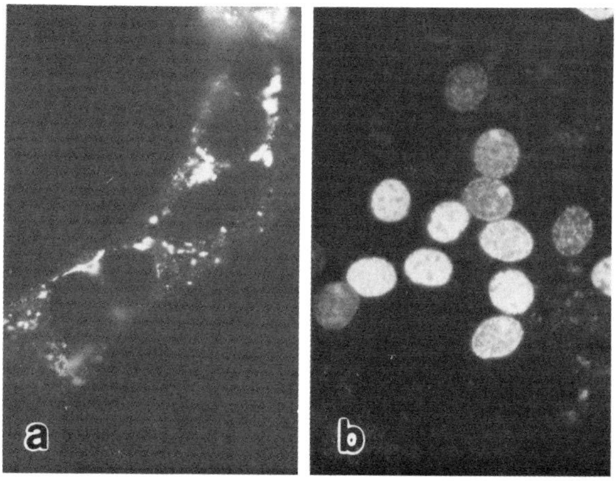

FIGURE 2. Intracellular locations of the N protein of MHV-A59 at 7 hours (a) and 16 hours (b) after virus inoculation. After 14 hr p.i., N protein becomes concentrated in nuclei of 17 Cl-1 cells. Immunofluorescence with monospecific anti-N antibody on acetone fixed cells.

MOLECULAR INTERACTIONS OF MEMBRANE GLYCOPROTEIN, E1

The E1 glycoprotein of coronaviruses is an unusual and very interesting molecule (4-8). Its structure and synthesis will be briefly summarized here. In MHV-A59 virions, E1 has three domains: a small external amino terminal domain with short oligosaccharides linked through O-glycosidic linkages to serine and threonine residues; a central domain which has three segments of hydrophobic amino acids each long enough to traverse the lipid bilayer; and a cytoplasmic domain which may interact with nucleocapsids. We showed that E1 accumulates in the Golgi apparatus and suggested that this might be the reason for budding of coronaviruses from this site (5). We also found that, in the presence of tunicamycin, virions can form normally without incorporating any E2, suggesting that E1 plays the central role in interacting with nucleocapsids and causing virus to bud from the membranes (6).

We have recently used a monoclonal antibody to perturb the spatial arrangement of E1 in intracellular membranes and to examine its effect on virus budding. Hybridoma cells producing monoclonal antibodies to the external domains of MHV-A59 E1 or E2 were inoculated with MHV-A59. The infected cells produced monoclonal antibody, inserted or secreted it into the RER and transported it to the Golgi and the plasma membrane. In the lumen of RER and Golgi vesicles, antibody could interact with external domains of E1 or E2 glycoproteins made in the infected cell. We found that virions were made and released normally in hybridomas producing antibody to the external domain of E2, but in a hybridoma producing monoclonal antibody to the external domain of E1, no virions were produced and viral nucleocapsids accumulated in the cytoplasm. Thus, virus budding was prevented by antibody-induced perturbation of the insertion or orientation of the external domain of E1 in the membranes where virus budding should occur. This illustrates the central role of E1 in the budding process of coronaviruses.

Binding of E1 to nucleocapsids can be studied in an in vitro system. Several years ago we observed that nucleocapsids released from NP40-disrupted virions quantitatively bound solubilized E1 but not E2 if incubated at 37 C (9). This in vitro system may be useful for identifying domains of E1 and N which interact.

MOLECULAR INTERACTIONS OF THE SPIKE GLYCOPROTEIN, E2

The E2 glycoprotein of MHV-A59 is a very large (180K) multifunctional molecule which forms the spikes or peplomers of the viral envelope. It is responsible for attachment to cell surface receptors, for coronavirus-induced cell fusion and for induction of neutralizing antibodies. As E2 is transported through the Golgi, some of it is incorporated into budding virions which are then released from the cell, and the remainder of the E2 goes to the plasma membrane. During this transport, some of the E2 molecules are cleaved in about the center of the molecule by a host cell protease (10). The two halves of the molecule, termed 90A and 90B, can be separated after detergent solubilization or by treatment of virions at pH 8 (11), but normally 90A and 90B remain attached together in a peplomer composed of two or more E2 molecules. Little is known about the assembly of the peplomer or its incorporation into the viral envelope.

The cleavage of E2 depends upon the host cell, since the ratio of 180/90K E2 varies considerably in virions released from different cell lines and since the size of the E2 cleavage products also varies (10). Several different enzymes can cleave E2 at approximately the same site, including trypsin and thermolysin suggesting that a cluster of protease susceptible sites is available near the center of the E2 molecule (12). Trypsin treatment of virions activates cell fusing activity (12) and may be necessary for viral infectivity. The new amino terminal region on 90A generated by trypsin cleavage of E2 of MHV-A59 was sequenced at the Rockefeller University protein sequencing facility: SerValSerThrGlyTyrArgLeuThr ThrPheGlu. Thus, trypsin treatment of coronavirus E2 does not reveal a hydrophobic domain at the new amino terminus such as those found at the cleavage sites of the fusion glycoproteins of orthomyxo- and paramyxoviruses (13). The pH optimum for coronavirus fusion is alkaline, rather than acidic like those of orthomyxo- and alphaviruses. These observations suggest that the mechanism of cell fusion induced by a coronavirus may differ from that induced by some other cell fusing viruses.

CORONAVIRUS RECEPTOR INTERACTIONS

One of the functions of the E2 glycoprotein is to bind to plasma membrane receptors. We have analyzed the interaction of MHV-A59 with cells susceptible to or resistant to infection. Coronavirus infection tends to be quite species specific. For example, MHV-A59 will not grow in hamster (BHK), human (HeLa), monkey (BSC1) or amphibian (XTC-2) cell lines. In addition, infection is quite tissue specific. For example, different strains of MHV-A59 show markedly different tropisms for intestinal epithelium, hepatocytes, oligodendrocytes, neurons and other cell types (14). Susceptibility to MHV-A59 infection is dominant and controlled by a host gene (Mhv-1;15). Mouse strains have been classified as to whether they are susceptible (BALB/c), semi-susceptible (C3H) or resistant (SJL) to MHV infection. To determine whether susceptibility to MHV infection was due to availability of cell surface receptors which could interact with the E2 glycoprotein, we studied the interaction of MHV virions with purified plasma membranes from normal target tissues in genetically susceptible and resistant mouse strains.

We developed a solid phase virus binding assay to analyze coronavirus receptors. Purified plasma membranes bound to nitrocellulose were exposed to MHV, and bound virus was detected by radioimmunolabelling (Figure 3). This simple assay showed that in susceptible Balb/c mice, MHV binds to plasma membrane receptors on enterocytes and hepatocytes, whereas in resistant SJL mice, no MHV receptors are present on either of these target tissues. Thus, profound resistance to MHV appears to be due to absence of virus receptors on the membranes. C3H Mice which are semi-susceptible to MHV do have receptors on enterocytes and hepatocytes and do permit some virus replication. Therefore, their partial resistance to MHV is probably determined by a later step in virus replication.

The MHV receptor on murine enterocytes and hepatocytes could be characterized as a 95 to 110K glycoprotein, because it retained activity after treatment with SDS. A single band with virus binding activity was detected in SDS-PAGE gels of membrane proteins by a virus-overlay protein blot assay (VOPBA; Figure 4). This receptor protein was partially purified by anion exchange and gel filtration FPLC. It was inactivated by proteases but not by neuraminidase or several other glycosidases. Further studies of the characteristics of the MHV receptor and its interaction with E2 are in progress.

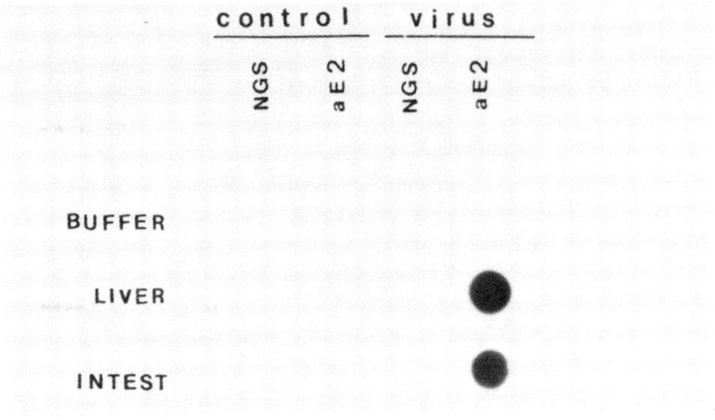

FIGURE 3. Solid phase virus receptor assay. Purified plasma membranes from BALB/c liver (LIVER), enterocytes (INTEST) or buffer alone (BUFFER) was dotted onto nitrocellulose and probed for virus receptor activity with clarified culture supernatant from 17 Cl-1 cells inoculated with MHV-A59 (VIRUS) or medium (CONTROL). Bound virus was detected with goat antibody to the E2 glycoprotein of MHV-A59 (aE2) or normal goat serum (NGS) followed by ^{125}I labeled Staphylococcal protein A.

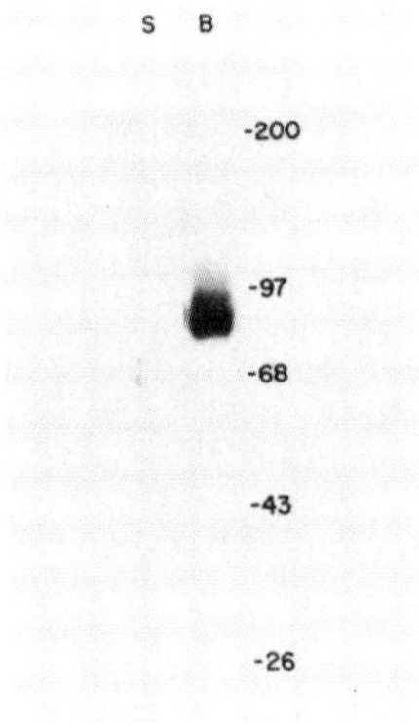

FIGURE 4. Virus overlay protein blot assay (VOPBA) for MHV receptor activity in proteins from hepatocyte plasma membranes. Equal amounts of proteins from purified plasma membranes from MHV susceptible BALB/c(B) or genetically resistant SJL(S) mice were separated by SDS-PAGE, electroblotted onto nitrocellulose and probed for virus binding activity using the solid phase receptor method.

These studies suggest that the susceptibility of different mouse strains to MHV is determined at least in part by the ability of the E2 glycoprotein to recognize a specific receptor protein on the membrane of target cells. Similar studies are in progress to determine whether the specificity of E2 for receptors can also account for species specificity and tissue tropism of coronavirus infection.

ACKNOWLEDGEMENTS

The opinions expressed are the private views of the authors and should not be construed as official or necessarily reflecting the views of the Uniformed Services University or the Department of Defense.

REFERENCES

1. Sturman LS, Holmes KV (1983). The molecular biology of coronaviruses. Adv Virus Research 28:35.
2. Siddell S, Wege H, ter Meulen V (1982). The structure and replication of coronaviruses. Curr Top Microbiol Immunol 99:131.
3. Robbins SG, Frana MF, McGowan JJ, Boyle JF, Holmes KV (1986). RNA-binding proteins of coronavirus MHV: Detection of monomeric N protein with an RNA overlay-protein blot assay. Virol 150:402.
4. Sturman LS, Holmes KV (1977). Characterization of a coronavirus. II. Glycoproteins of the viral envelope: Tryptic peptide analysis. Virol 77:650.
5. Holmes KV, Frana MF, Robbins SG, Sturman LS (1984). Coronavirus maturation. Adv Exptl Med Biol 173:37.
6. Holmes KV, Doller EW, Sturman LS (1981). Tunicamycin resistant glycosylation of a coronavirus glycoprotein: Demonstration of a novel type of viral glycoprotein. Virol 115:334.
7. Niemann H, Boschek B, Evans D, Rosing M, Tamura T, Klenk H-D (1982). Post-translational glycosylation of coronavirus glycoprotein E1: Inhibition by monensin. EMBO J 1:1499.
8. Armstrong J, Niemann H, Smeekens S, Rottier P, Warren G (1984). Sequence and topology of a model intracellular membrane protein, E1 glycoprotein, from a coronavirus. Nature 308:751.

9. Sturman LS, Holmes KV and Behnke JN (1980). Isolation of coronavirus envelope glycoproteins and interaction with the viral nucleocapsid. J Virol 33:449.
10. Frana MF, Behnke JN, Sturman LS, Holmes KV (1985). Proteolytic cleavage of the E2 glycoprotein of murine coronavirus: Host-dependent differences in proteolytic cleavage and cell fusion. J Virol 56:912.
11. Ricard CS, Sturman LS (1985). Isolation of the subunits of the coronavirus envelope glycoprotein E2 by hydroxyapatite high performance liquid chromatography. J Chromatog 326:191.
12. Sturman LS, Ricard CS, Holmes KV 91985). Proteolytic cleavage of the E2 glycoprotein of murine coronavirus. I. Activation of cell fusing activity of virions by trypsin treatment and separation of two different 90K cleavage fragments. J Virol 56:904.
13. Klenk H-D, Rott R (1981). Cotranslational and posttranslational processing of viral glycoproteins. Curr Top Microbiol Immunol 90:19.
14. Barthold SW, Smith AL (1984). Mouse hepatitis virus strain-related patterns of tissue tropism in suckling mice. Arch Virol 81:103.
15. Smith MS, Click RE, Plagemann PG (1984). Control of mouse hepatitis virus replication in macrophages by a recessive gene on chromosome 7. J. Immunol 133:428.

Positive Strand RNA Viruses, pages 351–363

BIOSYNTHESIS OF THE STRUCTURAL PROTEINS OF SFV - A RECOMBINANT DNA APPROACH

Henrik Garoff, Paul Melançon, Daniel Cutler
Laurie Roman, Jim Hare, Marino Zerial
and Danny Huylebroeck

European Molecular Biology Laboratory
Postfach 10.2209
D-6900 Heidelberg, F.R.G.

ABSTRACT We have used cDNA mutagenesis and expression analyses to study the various proteolytic cleavages and topogenic signals which are required for correct biosynthesis of the SFV structural proteins (capsid, p62 and E1). Our results confirm that the capsid (C) protein is an autoprotease and show that it is only involved in the C-p62 cleavage but not in any of the other proteolytic events which are needed to generate the individual chains from the SFV polyprotein. The signal peptide for p62 has been localized to a stretch of 33 amino acids at the N-terminus of the chain. The translocation signal for E1 is positioned within the C-terminal one third of the 6kD peptide. (The 6kD peptide is interspaced between the p62 and E1 on the polyprotein sequence). Both signals require SRP for function. The properties of the membrane binding region of the p62 protein has also been analyzed in detail. The typical consensus features for spanning peptide segments were shown to be important for a stable association of the chain in the membrane but all of them were not required for the generation of a transmembrane topology during synthesis.

INTRODUCTION

Simple enveloped viruses, like Semliki Forest virus (SFV), have been studied intensively not only because of a genuine virological interest, but also as a model system to gain insight into several more general cell biological questions (1,2). These include protein-RNA interactions (as represented by the viral nucleocapsid formation), synthesis and assembly of spanning membrane glycoproteins in the endoplasmic reticulum (ER) (biosynthesis of the p62 and E1 spike subunits), membrane protein oligomerization (formation of a p62-E1 complex), cell surface transport (routing of spikes to the plasma membrane, PM), interactions between cytoplasmic and transmembrane proteins (nucleocapsid-spike interactions during virus budding), membrane protein sorting (selective inclusion of spikes into viral membranes during budding), receptor-mediated endocytosis (virus entry) and membrane glycoprotein catalyzed membrane fusion (virus fusion in endosomes).

We have recently applied recombinant DNA technology to study in detail the generation of the virus structural proteins from the subgenomic 26S mRNA molecule. During infection these are translated as a polyprotein using a single initiation site (2). The capsid (C) protein is made first and this is followed by the synthesis of two membrane glycoproteins, p62 and E1. A small nonstructural peptide, the 6kD peptide, separates the membrane glycoproteins from each other on the polyprotein sequence. The correct synthesis of the SFV proteins thus involves a series of proteolytic cleavage events and in addition the expression of various topogenic signals to assure correct assembly of the glycoproteins in the ER membrane. We have here followed these events during synthesis. These will be discussed in view of the new data obtained through SFV cDNA mutagenesis followed by the in vivo and in vitro expression of viral structural genes.

Synthesis of the C Protein. The C-p62 Cleavage.

The C protein is formed by cleavage between Trp (267) and Ser (268) in the polyprotein sequence (3,4). This cleavage occurs efficiently in vitro using human, rabbit or wheat germ lysates programmed with 26S mRNA (5-7). Therefore, it must be a cytoplasmic event. As proteolytic cleavages, especially with Trp-Ser specificity, are unlikely to be catalyzed by a cytoplasmic protease, it has been suggested that the C-p62 cleavage is brought about by the polyprotein itself as in the case of poliovirus. However, in contrast to the polio polyprotein which harbours the protease at its C-terminal region (24), the N-terminal C protein appears to be responsible for its own release from the SFV polyprotein.

This is supported by instant cleavage of C protein when the corresponding chain length has been translated in vitro (7, 8). Several pieces of evidence support autoprotease activity contained within the C protein. Firstly, Aliperti and Schlesinger (9) showed in in vitro translation, that the p62-C cleavage was inhibited by the incorporation of amino acid analogues into the polyprotein chain and thereby resulted in the accumulation of large ^{35}S-methionine labelled precursor molecules. Upon a subsequent chase with an excess of unlabelled amino acids, labelled precursors were found to be efficiently cleaved into C protein. This suggested that the newly synthesized (and unlabelled) SFV polyprotein, probably the C protein, carried a protease activity which could act in trans. Secondly, Boege and coworkers (10) noticed that the C protein of both SFV and Sindbis virus contain the sequence -Gly-Asp-Ser-Gly- which is characteristic for the catalytic center of serine proteases like chymotrypsin. Thirdly, Hahn and Strauss (11) sequenced the C region of the 26S mRNA of three p62-C-cleavage defective Sindbis mutants and found that each of them contained a mutation in the C gene that was changed back to wild type in revertants. Two of these mutations involved Pro (218) just adjacent to the Ser (215) of the putative catalytic center.

We have studied the C-p62 cleavage in an in vitro translation system programmed with mRNA transcribed from wild type and mutagenized SFV cDNA (Melancon and Garoff, unpublished). The mutagenesis (Nae I cleavage, Bal 31 exonuclease, linker ligation) was designed to change the proposed active center sequence of the C protein. Nucleotide sequence studies showed that one of our mutants coded for -Gly(217)-Asp-Arg-Ser-Thr- instead of the wild type "active center peptide" -Gly(217)-Asp-Ser-Gly-: one basic amino acid residue (Arg) has been inserted, and one residue (Gly) has been changed (to Thr) (Fig.1). When this cDNA was transcribed into RNA and the latter used for translation in vitro, no C protein cleavage took place but the complete 130kD polyprotein of SFV accumulated (Fig.1). In contrast, the wild type RNA was translated into C and a 97kD protein encompassing the sequences for the membrane glycoproteins. This result lends further support for the autoprotease function of the C protein.

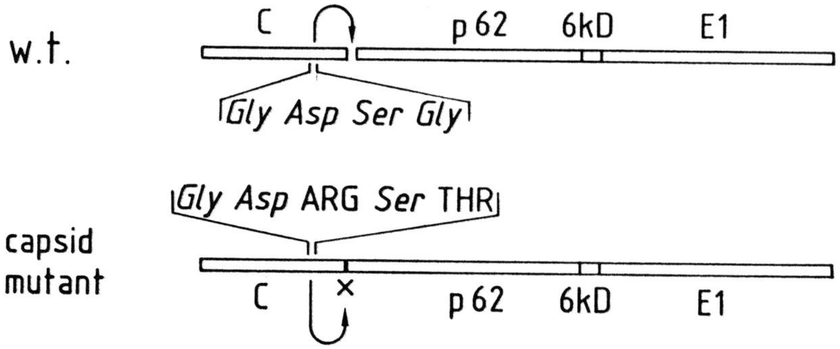

FIGURE 1. Cleavage phenotype of SFV structural polyproteins translated in vitro from mRNA which has been transcribed from wild type cDNA and from capsid mutant cDNA.

Translocation of the Lumenal Domain of p62 across the ER Membrane.

After the C protein has been cleaved off from the growing polyprotein chain, the translating ribosome becomes membrane associated to allow insertion of the first membrane glycoprotein (p62). This has been shown to occur in a SRP dependent and a cotranslational fashion, suggesting that a signal analogous to the cleavable signal peptide is responsible for this event (7,8). Mapping experiments, using either SRP or microsome additions at different times after starting a synchronized 26S mRNA translation have tentatively localized the translocation signal of the p62 protein to its N-terminal one third (7,8). A more precise localization of this signal has been made by D. Huylebroeck (unpublished) in our laboratory. We have inserted the genes of two reporter molecules, chimpanzee α-globin and mouse dihydrofolate reductase (dhfr) in frame at a site 33 codons away from the 5' end of the p62 coding region (Fig.2). Translations of RNA transcribed from these constructs in the presence of membranes yield authentic C protein and translocated p62(N-term)-DHFR hybrids. Both fusion proteins remain uncleaved and are glycosylated. The sugar unit is presumably added to Asn (14) of the p62 piece as it is the only potential site present in either hybrid and it is normally glycosylated (12,13). We conclude that the N-terminal 33 residues of p62 contains a translocation signal.

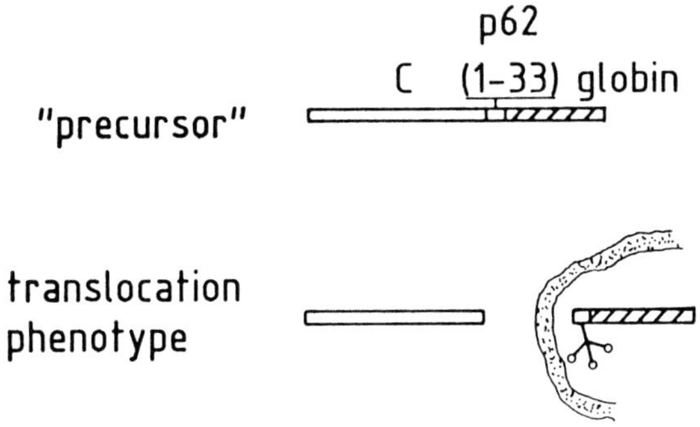

FIGURE 2. Translocation phenotype of a C-p62(1-33)-globin hybrid.

Analyses of the published amino acid sequence of the p62 protein show that it contains a stretch of 18 nonpolar residues at its N-terminus (3). The same feature is also observed in the p62 protein of the closely related Sindbis virus (14). As a stretch of apolar (and hydrophobic) amino acid residues appear to be the most critical property of signal peptides (15), we believe that the first 18 residues of p62 comprise its signal peptide. It should, however, be pointed out that the p62 signal peptide shows some unusual features too. It does not possess an amino acid with a positively charged side chain at its N-terminal region as do most cleavable signal peptides (3,15). Even the positively charged α-NH_2 group at the end of the chain has been modified by acetylation (13). Furthermore, the p62 signal is glycosylated at Asn(14) as already mentioned. Perhaps, most remarkable is that the signal peptide itself becomes completely translocated across the ER membrane to become a part of the lumenal domain of the p62 protein (2). Although the exact fate of cleavable signal peptides is not known, these are generally believed to remain associated with the ER membrane after cleavage. According to one popular hypothesis, the signal-peptide loops into the membrane with its hydrophobic segment leaving the N-terminus (with the positive charge) on the cytoplasmic surface and bringing the distal chain across (16). It is conceavable that the pressure for a most economical use of the viral genome has resulted in a signal peptide that can be used for an other purpose after first initiating translocation. One might postulate that the acetylation and glycosylation reactions occur after the p62 signal has fulfilled its function in translocation. As these modifications most likely will result in a less stable association with the ER membrane, removal of the peptide from the membrane will thus be effected first after initiation of translocation. In the future, it would be interesting to test whether the p62(N-term)-globin or -DHFR hybrids, which were translocated in vitro, would remain associated with the microsomal membrane via the p62 signal if the glycosylation event is prevented.

The Stop-Translocation Signal of p62

The p62 protein is known to have about 360 amino acid residues in its lumenal domain, a hydrophobic membrane spanning segment comprising about 30 residues and a C-terminal cytoplasmic domain consisting of 31 amino acid residues (3). During synthesis the spanning segment is thought to arrest the translocation process and thus generate a transmembrane polypeptide. This we tested directly by analyzing whether a secretory protein, chicken lysozyme, can be converted into a transmembrane protein by fusing it to the transmembrane region of the p62 protein (Melançon and Garoff, unpublished). For this purpose the lysozyme coding sequence was joined to a truncated form of the SFV cDNA at a point immediately in front of the coding sequence for the spanning segment of p62 (Fig.3). When the hybrid protein was translated in vitro in the presence of microsomal membranes, lysozyme-p62(C-term) fusion polypeptides were found to be inserted across the membrane in the expected orientation. In contrast, control lysozyme was not accessible to protease and therefore completely sequestered within the microsomes. We conclude that the transmembrane segment of p62 functions as a stop-translocation signal during synthesis.

Next, we analyzed what features of the C-terminal, membrane binding region of the p62 protein, were important for its function as a stop translocation and membrane binding signal. A comparison of all known signal spanning (bitopic) membrane proteins shows that this region displays two consensus features (15). These are: (1) a stretch of about 20 apolar (mostly hydrophobic) amino acid residues and (2) one or more residues with a basic charge flanking the spanning segment on its cytoplasmic side.

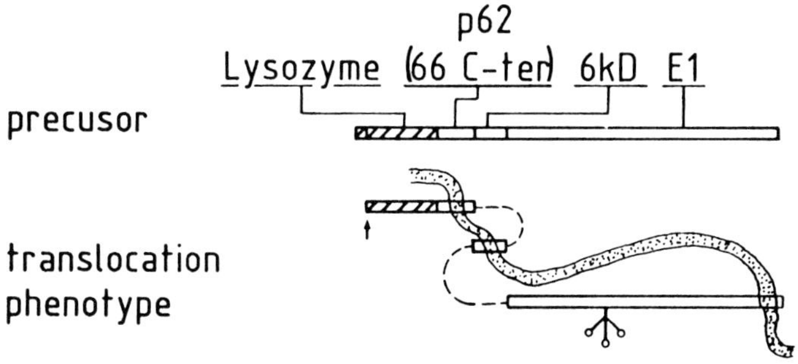

FIGURE 3. Translocation phenotype of a lysozyme-p62 (66C-term)-6kD-E1 hybrid.

We have used site directed mutagenesis to test whether changing these features in the p62 protein leads to altered topology of this protein in the membrane (17,18). Altogether, three SFV cDNA mutants were made (Fig.4). The first two were designed to change the basic cluster flanking the spanning segment into neutral and acid residues, respectively. The third one was designed to change a hydrophobic residue into a charged one in the middle of the spanning region.

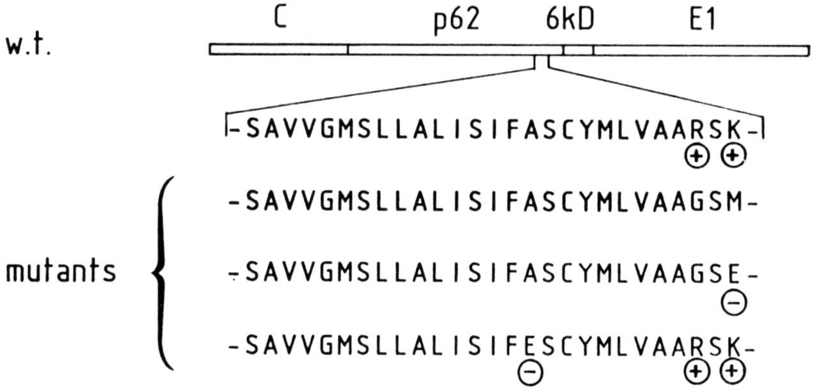

FIGURE 4. Mutants of the membrane binding region of p62.

Expression analyses made in vivo showed that all of the altered p62 proteins exhibited a normal transmembrane topology and that they could be transported to the cell surface in a functionally active form. However, when the membrane binding properties were assayed by extraction at high pH, all mutants were shown to be less stably associated with the lipid membrane than the wild form of p62. Similar results have been obtained with transmembrane mutants of other bitopic membrane proteins (discussed in 18). This suggests that the observed consensus features of the membrane binding region of single spanning membrane proteins are required for optimal stability in the membrane, but may not all be required for correct assembly in the membrane during synthesis. Apparently only a certain threshold level of hydrophobicity is required in the membrane binding polypeptide region to assure a spanning configuration. Although not true for the SFV proteins, this "relaxed" requirement might be crucial for the synthesis of proteins, like the T cell receptor complex (19). In this oligomer, the individual subunits are p62-like bitopic proteins, however, with charged residues in their spanning segments. Perhaps these subunits are synthesized as transmembrane proteins which are less stably associated with the ER membrane (as our p62 protein mutants) until they form an oligomeric complex in which the charged residues can be neutralized by subunit interactions.

Reinitiation of Polypeptide Chain Translocation. The Synthesis of the E1 Glycoprotein.

After the ribosome has synthesized the membrane binding region of the p62 protein and this has arrested the chain translocation process the cytoplasmic domain of p62 is translated. Soon thereafter, chain translocation has to be reinitiated to account for the insertion of the E1 polypeptide across the ER membrane. So far, no translocation signal has been identified for this process. We argued that the E1 signal might show some interesting features in comparison with that of the p62 protein since ribosomes which start to translate the E1 chain have already been targeted to the ER membrane by the signal of p62. For instance, if the ribosome-membrane interaction persists during the synthesis of the cytoplasmic portions of the SFV glycoprotein, there does not seem to be any obvious need for a second SRP-dependent "docking" event on the ER membrane. On the other hand, pretargeting of the ribosome to the ER membrane through the p62 signal peptide might represent an absolute requirement for E1 translocation. These questions were tested by in vitro translation of the cDNA constructions shown in Figure 5. In the first construction, the p62 coding sequence has been deleted. Expression of this polyprotein deletion mutant yielded authentic E1 which was glycosylated and inserted across the membranes of the added microsomes. We conclude that the E1 protein has a translocation signal which can function independently of the p62 signal. Furthermore, the experiment suggested that the E1 signal is located within the 6kD peptide or in the E1 sequence itself. A precise location of the signal was made through the expression analyses of a series of cDNA deletion mutants.

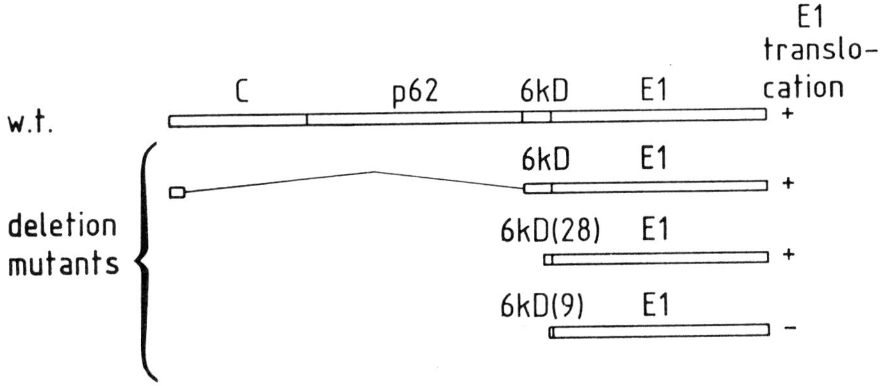

FIGURE 5. E1 translocation phenotype of C+p62+6kd(N-term) deletion mutants.

In these all of the coding regions for the C and p62 proteins were removed together with progressively larger parts of the 6kD peptide region from its 5' end. In these constructions the natural initiation codon was replaced with a synthetic one as provided by an oligonucleotide linker molecule. The analyses of the expressed protein products showed that E1 translocation occurred when the precursor contained 28 C-terminal amino acid residues of the 6kD peptide but failed to take place if only 9 residues were left (see Fig.5).

This suggests that the E1 signal is located within the C-terminal one third of the 6kD peptide. An examination of the amino acid sequence of this part of the 6kD peptide shows features which match those of cleavable signal peptides (basic residues followed by a stretch of hydrophobic residues and a series of amino acids with sequential, small, large and then small side chains, before the cleavage site). The activity of the proposed E1 signal peptide was also tested in a positive way, by fusing it to a cytoplasmic reporter protein and analyzing the hybrid for translocation in vitro.

The construction is shown in Figure 6. A segment of the SFV cDNA encoding the 28 C-terminal amino acids of the 6kD peptide plus the 4 first residues of E1 were fused to the α-globin sequence. The translation of the hybrid protein in the presence of membranes yielded a product smaller than the precursor made in the absence of membranes but somewhat larger than the reporter molecule. Obviously, the E1 signal peptide in the 6kD portion is responsible for the chain translocation event and is subsequently cleaved at the 6kD peptide-E1 junction which was left intact in the construction.

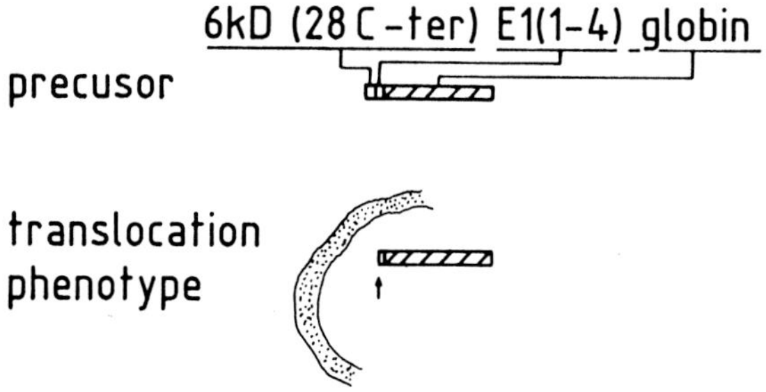

FIGURE 6. Translocation phenotype of 6kD(C-term)-globin hybrid.

Further experiments using the same construction demonstrated that the E1 translocation signal was SRP-dependent. We conclude that this signal is very similar to the signal peptides of secretory proteins. Most likely, the reinitiation of chain translocation that accounts for the insertion of the E1 glycoprotein during 26S mRNA translation requires a second round of SRP and docking protein interactions.

The sequential biosynthesis of the p62 and the E1 glycoproteins in the ER membrane is reminiscent of the way by which multispanning membrane proteins are assembled in the membrane. This is thought to occur through the co-translational expression of several intermittent signal peptides and stop translocation signals. In the cases of the bovine rhodopsin (20) and the human glucose transporter molecule (21), internal signal peptides have been demonstrated and shown to be SRP-dependent.

Cleavages of the 6kD Peptide.

The small 6kD peptide located in the SFV polyprotein between the p62 and E1 parts is probably released by co-translational cleavage events (22,23). This is suggested by the fact that precursors corresponding to p62 + 6kD or 6kD + E1 products have never been observed during *in vivo* or *in vitro* studies (e.g. 7). The 6kD-E1 cleavage is probably catalyzed by the signal peptidase (see above) and is therefore taking place at the lumenal membrane surface. On the other hand, the nature of the p62-6kD cleavage and its topological location are unclear. If the cleavage were to occur on the cytoplasmic side of the membrane, one might argue that also this could be catalyzed by the C protease. However, using the lyoszyme-SFV polyprotein construct shown in Figure 3, we were able to exclude the involvement of the C protein in the cleavages at the 6kD peptide. When this construct was expressed *in vitro* in the presence of microsomal membranes, we could observe the formation of the 6kD peptide in addition to lysozyme-p62(C-term) hybrids and E1 glycoprotein.

Membrane Binding of the E1 Protein.

The polyprotein synthesis ends with the translation of two Arg residues immediately after the synthesis of a long stretch of hydrophobic amino acid residues. This region, most likely represents the stop translocation signal, i.e. the membrane binding region of the E1 polypeptide. We have not tested the properties of this region. These are probably similar to those of the corresponding region of p62.

Altogether, the use of the cDNA mutagenesis and the in vitro/in vivo expression approach has in a most convincing way shown how the SFV structural proteins are synthesized from the polyprotein. Several earlier postulations have now obtained experimental support. The major open questions in this system are related to the nature of the p62-6kD cleavage and the actual topology of the 6kD peptide itself.

REFERENCES

1. Simons K, Warren G (1984). Semliki Forest virus: A probe for membrane traffic in the animal cell. Adv Protein Chemistry 36:79-132.
2. Garoff H, Kondor-Koch C, Riedel H (1982). Structure and assembly of alphaviruses. In "Current Topics in Microbiology and Immunology" 99:1-50.
3. Garoff H, Frischauf AM, Simons K, Lehrach H, Delius H (1980). Nucleotide sequence of cDNA coding for Semliki Forest virus membrane glycoproteins. Nature 288:236-241.
4. Kalkkinen N (1980). Carboxyl-terminal sequence analysis of the four structural proteins of Semliki Forest virus. FEBS Letters 115:163-166.
5. Bonatti S, Canceda R, Blobel G (1979). Membrane biogenesis: in vitro cleavage, core glycosylation and integration into microsomal membranes of Sindbis virus glycoproteins. J Cell Biol 80:219-224.
6. Glanville N, Morser J, Uomala P, Kääriäinen L (1976). Simultaneous translation of structural and nonstructural proteins from Semliki Forest virus RNA in two eukaryotic systems in vitro. Eur J Biochem 64:167-175.
7. Garoff H, Simons K, Dobberstein B (1978). Assembly of the Semliki Forest virus membrane glycoproteins in the membrane of the endoplasmic reticulum in vitro. J Mol Biol 124:587-600.
8. Bonatti S, Migliaccio G, Blobel G, Walter P (1984). Role of signal recognition particle in the membrane assembly of Sindbis viral glycoproteins. Eur J Biochem. 140:499-502.
9. Aliperti G, Schlesinger MJ (1978). Evidence for an autoprotease activity of Sindbis virus capsid protein. Virology 90:366-369.
10. Boege U, Wengler G, Wengler G, Wittmann-Liebold B (1981). Primary structuresof the core proteins of the alphaviruses Semliki Forest virus and Sindbis virus. Virology 113:293-303.

11. Hahn CS, Strauss EG, Strauss JH (1985). Sequence analysis of three Sindbis virus mutants temperature-sensitive in the capsid protein autoprotease. Proc Natl Acad Sci USA 82:4648-4652.
12. Kalkkinen N, Jörnvall H, Kääriäinen L (1981). Polyprotein processing of alphaviruses: N-terminal structural analysis of Semliki Forest virus proteins p62, E3 and ns70. FEBS Letters 126:33-37.
13. Bell JR, Rice CM, Hunkapiller MW, Strauss JH (1982). The N-terminus of PE2 in Sindbis virus-infected cells. Virology 119:255-267.
14. Rice CM, Strauss JH (1981). Nucleotide sequence of the 26S mRNA of Sindbis virus and deduced sequence of the encoded virus structural proteins. Proc Natl Acad Sci USA 78:2062-2066.
15. von Heijne G (1985). Structural and thermodynamic aspects of the transfer of proteins into and across membranes. Currents Topics in Membranes and Transport 24:151-179.
16. Inouye M, Halegoua S (1979). Secretion and membrane localization of proteins in Escherichia coli. CRC Crit Rev Biochem 7:339-371.
17. Cutler DF, Garoff H (1986). Mutants of the membrane-binding region of Semliki Forest virus E2 protein. I. Cell surface transport and fusogenic activity. J Cell Biol 102: 889-901.
18. Cutler DF, Melançon P, Garoff H (1986). Mutants of the membrane-binding region of Semliki Forest virus E2 protein. II. Topology and membrane binding. J Cell Biol 102: 902-910.
19. Saito H, Kranz DM, Takagaki Y, Hayday AC, Eisen HN, Tonegawa S (1984). Complete primary structure of a heterodimeric T-cell receptor deduced from cDNA sequences. Nature 309:757
20. Friedlander M, Blobel G (1985). Bovine opsin has more than one signal sequence. Nature 318:338-343.
21. Mueckler M, Lodish HF (1986). The human glucose transporter can insert posttranslationally into microsomes. Cell 44:629-637.
22. Welch WJ, Sefton BM (1979). Two small virus-specific polypeptides are produced during infection with Sindbis virus. J Virol 29:1186-1195.
23. Welch WJ, Sefton BM (1980). Characterization of a small, nonstructural viral polypeptide present late during infection of BHK cells by Semliki Forest virus. J Virol 33: 230-237.
24. Palmenberg AC, Pallansch MA, Rueckert RR (1979). J Virol 32:770-778.

Positive Strand RNA Viruses, pages 365–378

STRUCTURE-FUNCTION RELATIONSHIPS IN THE GLYCOPROTEINS OF ALPHAVIRUSES[1]

Ellen G. Strauss
Division of Biology, California Institute of Technology
Pasadena, CA 91125

Alan L. Schmaljohn
Department of Microbiology, University of Maryland
School of Medicine, Baltimore, MD 21201

Diane E. Griffin
Department of Medicine and Neurology
Johns Hopkins University, Baltimore, MD 21205

James H. Strauss
Division of Biology, California Institute of Technology
Pasadena, CA 91125

ABSTRACT The sequences of the region of the genome encoding the glycoproteins of alphavirus Sindbis have been determined for strain AR339 and its heat resistant variant HR, as well as for a number of Sindbis *ts* mutants and other types of variants derived from these strains. We have located the nucleotide changes responsible for the temperature sensitive phenotype of mutants *ts*10 and *ts*23 of complementation group D and of *ts*20, the only member of the group E in our mutant catalog. The group D mutant changes mapped to glycoprotein E1 and the group E mutant change mapped to glycoprotein E2. We have also determined the sequence in the region of E2 for a number of antigenic variants of Sindbis which were selected for altered reactivity to neutralizing monoclonal antibodies. In this way we

[1]EGS and JHS were supported by grants AI 10793 and AI 20612 from the National Institutes of Health and grant DMB 8316856 from the National Science Foundation. DEG was supported by grant NS 18596 from the National Institutes of Health. ALS was supported by a contract from the U.S. Army Research and Development Command.

determined that a particular domain of glycoprotein E2 contains a number of neutralization epitopes. This domain is highly charged and contains an attachment site for a polysaccharide chain, and may be important for the interaction of the virus with the vertebrate immune system. Currently, we are extending these studies to examine a number of Sindbis variants which are altered in their neurovirulence for mice, in order to identify the residues on the glycoprotein E2 which are responsible for the ability of certain strains to cause fatal encephalitis in these animals.

INTRODUCTION

Sindbis virus is the type virus of the genus Alphavirus in the family Togaviridae. It consists of an icosahedral nucleocapsid surrounded by a lipid bilayer in which are embedded two integral membrane glycoproteins, called E1 and E2. The genome of Sindbis virus is a single-stranded RNA molecule of plus polarity which is 11,703 nucleotides in length, capped at the 5' terminus and polyadenylated at the 3' end. The genomic RNA is complexed with approximately 280 molecules of the capsid protein (C) to form the nucleocapsid or core, which self assembles in the cytoplasm of the infected cell. Core particles migrate to the cell surface and bud out through the cell membrane in regions of modified plasmalemma containing exclusively the virus glycoproteins E1 and E2, thereby releasing the mature enveloped virus into the extracellular space. The structure and replication of alphaviruses are reviewed in ref. 1.

The nonstructural proteins of the virus are involved in RNA replication and are translated from the genome length 49S RNA. The structural proteins found in mature virions, i.e., C, E2, and E1, are translated as a polyprotein from a subgenomic messenger known as 26S RNA in the order C, E3, E2, 6K and E1. 26S RNA constitutes the 3' one-third of the genome and is present in the cytoplasm of infected cells in roughly three-fold molar excess over genomic RNA. The capsid protein C is cleaved from the remainder of the polyprotein while nascent by an autoproteolytic mechanism (2) which will be discussed in another chapter of this volume. The small glycoprotein E3 is present as the N terminus of the precursor to E2, known as PE2, and serves in part as an uncleaved signal sequence to enable the glycoprotein to be inserted into the rough endoplasmic reticulum. The 6K protein, which is located between the C terminus of E2 and the N terminus of E1, apparently serves as an internal signal sequence for the insertion of E1 (3). Both of these proteins are anchored in the membrane by a conventional C-terminal hydrophobic root or anchor. E1 and E2 are modified by core glycosylation in the rough endoplasmic reticulum and then pass into the Golgi apparatus where several maturation events occur. These include the trimming of the high-mannose simple oligosaccharides, addition of galactose, glucosamine, fucose and sialic acid to form the complex oligosaccharides, and the addition of palmitic acid to the hydrophobic root. Upon the completion of these modifications, E1 and E2 are transported to the cell surface and inserted into the plasmalemma. These modifications appear to be essential for proper transport since nonglycosylated forms of the envelope proteins E2 and E1 are inserted into intracellular membranes normally, but the envelope proteins never reach the cell surface and no budding viruses are produced (4,5). This observation and the fact that in certain mutants which are temperature sensitive for transport of glycoproteins the mutations occur in widely separated areas of the polypeptide (6) suggest that the correct three-dimensional conformation is essential for proper transport to the cell plasmalemma (reviewed in ref. 7).

It has been known for a number of years and shown in a variety of ways that glycoproteins E1 and E2 form a heterodimer and are found closely associated soon after their synthesis (8, 9). The two proteins are transported as a heterodimer to the cell surface and incorporated together into the budding virion. Therefore, when trying to delineate the functions of E2 and E1, we must take into account this association, since *in vivo* each protein affects the conformation and the availability of sites on the other.

Some of the functional characteristics of E2 and E1 have been elucidated by the use of antibodies. Both polyclonal and monoclonal antisera have been prepared which are specific for each of the Sindbis glycoproteins and used in their characterization (reviewed in ref. 10). In general the antibodies to glycoprotein E1 are cross-reactive and nonneutralizing while antibodies to E2 are type specific and neutralizing. Monoclonal antibodies (primarily reacting with E2) which neutralize infectivity have been used to isolate the variants which we have characterized (see Results). Isolated E1 glycoproteins, but not E2, of Sindbis (11), Semliki Forest virus (12), Western equine encephalitis (13) and Chikungunya (14) have been shown to hemagglutinate. However, monoclonal antibodies reactive with either E1 or E2 may inhibit hemagglutination (15, 16, 17, 18) depending upon the virus examined. This suggests that the E1/E2 heterodimer, rather than either protein alone, is the target of hemagglutination inhibiting antibodies and forms the native viral hemagglutinin.

We have conducted a number of studies to delineate the functional domains of the Sindbis glycoproteins using temperature sensitive mutants (6, 19) as well as antigenic variants and altered virus strains, which will be discussed below.

MATERIALS AND METHODS

Sindbis Virus Strains and Variants

The isolation and characterization of the temperature sensitive mutants of Sindbis used in these studies have been previously described (20, 21). The isolation of the monoclonal antibodies specific for the E1 and E2 glycoproteins of Sindbis are given in ref. 16 and 22 and the antigenic variants isolated for resistance to neutralization by these monoclonal antibodies as well as sensitive revertants of these variants were described by Stec. et al. (23). The isolation and characterization of the neurovirulent strains have been described (24, 25).

Sequencing Methods

The sequencing of *ts*10 and *ts*23 and their revertants has been described (6). Briefly, single stranded (ss) cDNA was synthesized, cleaved with selected restriction endonucleases which cleave ss cDNA (such as HaeIII, TaqI, and RsaI), and the resulting fragments labelled at the 5' ends and sequenced using chemical methods as previously described (26).

For *ts*20 and a number of other variants, cloned cDNA copies were used. The cloning procedure involved linearizing a pBR322 derivative by cleavage in its polylinker, tailing it with poly T, and removing one T-tail. This modified vector was used as a primer for first strand cDNA synthesis with 49S Sindbis poly(A)-containing genomic RNA as a template. After second strand cDNA synthesis, the DNA was cut with HindIII to generate specific sticky ends and closed with E. coli ligase (19). A drawback of this method is that more than one clone must be sequenced to rule out cloning artifacts.

For the neurovirulent mutants and the antigenic variants, we have used direct sequencing of the RNA by means of the chain termination protocols using dideoxynucleotide triphosphates and reverse transcriptase (27, 28). For most purposes it has not proven necessary to purify the RNA; intracellular RNA isolated at approximately 7 hr postinfection can be used directly. The presence of ribosomal RNA in the preparation does not interfere unduly with the ladders obtained.

RESULTS

General Characteristics of the Glycoproteins

In order to present an overview of the region of the glycoproteins that we are considering, we have compared glycoproteins from two divergent alphaviruses in different ways which are shown in Fig. 1. The deduced protein sequences from the nucleic acid sequences determined for Sindbis and VEE have been aligned to give the greatest possible homology using the smallest number of gaps. Throughout the proteins, the homology varies from zero in the N-terminal part of E2 to 90% in two domains of E1 (using a window of 20 amino acids). The overall homology in the E3 region

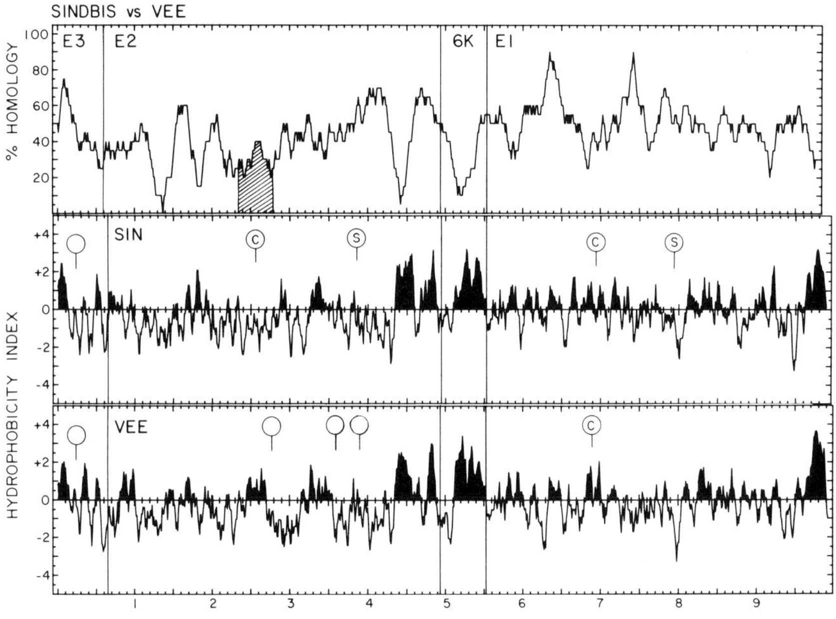

FIGURE 1. Homology and hydrophobicity profiles for the glycoproteins of Venezuelan equine encephalitis (VEE) virus and Sindbis (SIN) virus. Top panel: % homology between the proteins using a string length of 20. Shaded area of E2 is the same as that shaded in Fig. 4. Lower panels: Individual hydrophobicity plots of the SIN and VEE proteins according to the algorithm of Kyte and Doolittle (29). Symbols indicate locations of carbohydrate: C= complex oligosaccharide, S = simple oligosaccharide, blank symbol = type unknown. Data for SIN from ref. 30, VEE from 31.

is 42%, for E2 it is 39%, for 6K 28%, and for E1 50%. Among alphaviruses E2 is more variable than E1. The lower panels show hydrophobicity profiles for the individual viruses. Hydrophobic (filled) domains are above the midline and hydrophilic domains are below the line. Note that the anchors at the C-termini of E2 and E1 are quite hydrophobic. In addition both the 6K protein and the N terminus of E3 are hydrophobic, as one would expect for signal sequences. The glycosylation sites are indicated and in each protein the location of the site closest to the N-terminus is conserved between the two viruses and bears a complex oligosaccaride chain. The second glycosylation site is in general not conserved among alphaviruses, and when present usually has a simple oligosaccharide chain attached.

Mutations in *ts*10 and *ts*23 Which Affect Glycoprotein Transport.

There are two well characterized *ts* mutants of Sindbis, *ts*23 and *ts*10, which fail to transport viral glycoproteins to the cell surface at the nonpermissive temperature (32). Both of these mutants belong to complementation group D, and therefore were presumed to have defects in glycoprotein E1. We have sequenced the E1 and E2 regions of these mutants, and by comparing these sequences with those of ts^{+} revertants (to rule out adventitious changes unrelated to phenotype), we determined the changes responsible for the failure to transport glycoproteins (6). We found that *ts*23 was a double mutant: Ala-106 in HR strain was Thr in *ts*23 and Arg-267 became Gln in *ts*23. The defect in *ts*10, whose phenotype is very similar to that of *ts*23, was Lys-176 being replaced with Gly. This was also a double mutant in that two nucleotides in a single codon were affected (6). In all the cases we have examined, revertants have turned out to be same site revertants, thus simplifying the analysis. Moreover, all but one revert to the parental amino acid; for *ts*10, revertants contained Arg-176 rather than the parental Lys. This change requires only a single nucleotide substitution in the Gly-176 codon of *ts*10.

If a single contiguous linear protein domain were solely responsible for proper transport of glycoproteins to the cell surface, we would have expected to find changes that result in the loss of transportability grouped in a small region. However, these widely spaced mutations throughout the protein indicate that the overall three dimensional conformation of the protein is essential for proper transport through the Golgi apparatus to the cell plasmalemma.

A variability plot for glycoprotein E1 for four alphaviruses is shown in Figure 2. Also shown are a number of other landmarks in E1, including the locations of the three *ts* lesions mapped. Note that these mutations are found in domains that demonstrate low variability (and although not shown, the three residues mutated are invariant among the four viruses) By their very nature, *ts* lesions must affect key residues or important domains in the protein.

Antigenic Variants of Sindbis Virus

Sets of monoclonal antibodies (MAbs) which are specific for either E1 or E2 of Sindbis virus have been isolated (16, 22). A number of the anti-E2 MAbs and one anti-E1 MAb neutralized infectivity. Using these monoclonal antibodies, Stec *et al.* (23) isolated a series of antigenic variants from Sindbis AR339 which were resistant to neutralization by these antibodies; concomittantly these variants lost their ability to bind to the antibodies in an ELISA test. The isolation of resistant variants was straightforward, but it was more difficult to isolate revertants which were once more susceptible to neutralization. Revertants were obtained by an enrichment technique on a solid support (23) and by back selection on a macrophage host cell line (unpublished). (In the latter case the macrophages lacked SIN receptors, but had Fc receptors and therefore could only be infected by a soluble virus-antibody complex.) We have sequenced a number of these antigenic variants by the dideoxy technique, using a series of synthetic oligonucleotide primers and reverse transcriptase.

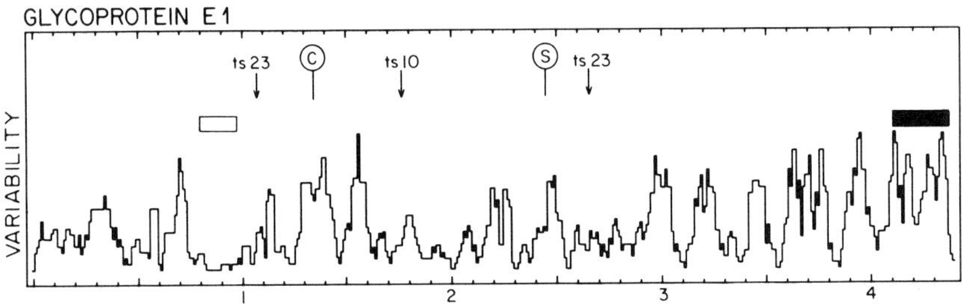

FIGURE 2. Variability plot of glycoprotein E1 for alphaviruses. The aligned protein sequences for SIN, VEE, Ross River (33) and Semliki Forest (4) viruses have been plotted to give the extent of variability at each residue, using a smoothing function of 5. Open box is the putative fusion sequence, solid box the hydrophobic anchor. Carbohydrate attachment sites are as in Fig. 1. The location of the mutation in *ts*10 and the two mutations in *ts*23 are shown. Regions of least conservation (greatest variability) appear as peaks in this graphic analysis.

Figure 3 shows a map of some of the antigenic changes which were found. The amino acid sequence shown is for AR339, the parental virus from which the variants were isolated. (AR339 is the original wild type Sindbis from which the widely studied HR strain was isolated. HR in turn was the parental strain for the *ts* mutants discussed above.) Residues 190 and 216 were the locations of the changes seen in the E2 variants shown in Figure 3. Other variants selected with different MAbs have also shown changes at residues 181, 184 and 214 (data not shown). Figure 3 illustrates two independent isolates of strains resistant to MAb50: v50 and v33/50. These both are changed at Lys-190, in one case to Met, in the other to Asn. Similarly, two isolates resistant to MAb23 are shown: v50/23 and v33/50/23. Both are changed at Lys-216, one to Glu and one to Asn. Thus appropriate changes at Lys-190 render E2 unable to interact with MAb 50 and appropriate changes at Lys-216 render it unable to interact with MAb23.

The two step variant v50/23 was isolated when v50 was selected for resistance to neutralization by MAb 23. In addition to retaining the Lys→Met change at position 190 (that rendered the variant resistant to MAb 50), this strain has a second change at position 216 from the Lys to Glu (resulting in resistance to MAb 23

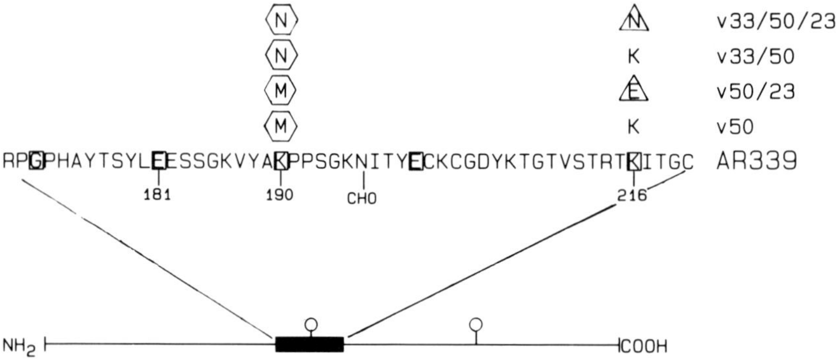

FIGURE 3. Location of selected antigenic variants of Sindbis virus. The bottom line is a map of the entire E2 glycoprotein; the solid box corresponds to the expanded line of amino acid sequence above. Boxed residues are sites that vary among the several strains or variants of Sindbis sequenced to date. Changed residues are shown for a number of variants. Strains are named for the MAb to which they are resistant. Two or three numbers indicate sequential selections. The single letter amino acid code is used as well). Residue 216 appears to be crucial for the interaction of a

number of the other MAbs as well as MAb 23 (not shown). Another interesting case is the triply resistant strain v33/50/23. MAb 33 reacts with E1 and the E1 region of this variant has not yet been sequenced. V33 was then selected with MAb 50, giving v33/50. In v33/50 a change occurs at residue 190 like that in v50 but the change is from Lys to Asn rather than Met. Finally v33/50 was subjected to a third round of selection using MAb 23. In this case, Lys-216 in AR339 is replaced by Asn. Asn-216 creates a new glycosylation site of the form Asn-X-Thr/Ser and it has been shown that this site is indeed glycosylated *in vivo* (or at least that this E2 migrates more slowly in acrylamide gels). Presumably the presence of the carbohydrate prevents the MAb from reacting with residue 216.

A number of other variants have also been examined that affect Lys-216 (data not shown), and we have found that all three nucleotides of the original Lys codon (AAA) at position 216 have been changed in one or another variant.

In Fig. 4, the locations of these two antigenic hot spots are mapped on a variability plot of glycoprotein E2. Sites of carbohydrate chain attachment are shown as well. Note that the complex oligosaccharide is found in a region of high variability and thus, although the site is retained, the amino acids around the site can vary extensively. Both v50 and v23 are in areas where somewhat less variability can be tolerated. By comparing Figure 2 and Figure 4, it is also clear that the overall variability of E2 is greater than that for E1.

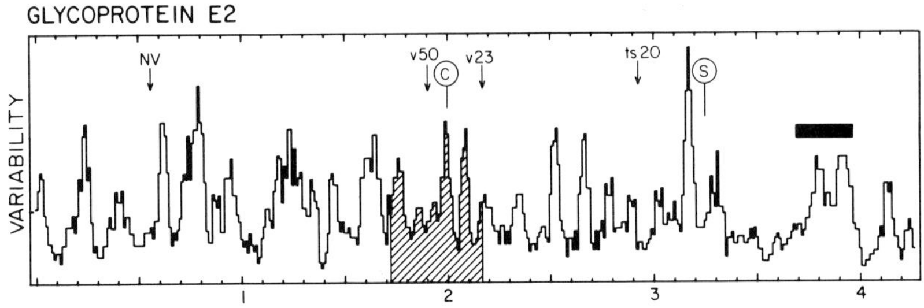

FIGURE 4. Variability plot of glycoprotein E2. Sequence sources, viruses used and conventions are the same as in Fig. 2. The shaded area represents the region of expanded sequence in E2 shown in Figure 3. The arrow (v50) is at residue 190, the arrow (v23) is at residue 216.

Analysis of *ts*20 in Complementation Group E

Also shown in Fig. 4 is the location of the *ts*20 mutation (19). *ts*20 is a mutant in Group E in which glycoproteins are transported to the surface of the cell at the nonpermissive temperature but no cleavage of PE2 to E2 plus E3 occurs and no budding virus is observed. This mutation maps to residue 291 of glycoprotein E2: His-291 in HR is replaced by Leu in *ts*20. In the revertants Leu-291 reverts to the parental amino acid, His. Note that the *ts*20 lesion occures in a domain of low variability.

Neurovirulence Mutants

A final series of experiments has used the same approach to study variants which have been selected for altered neurovirulence in mice. Wild type Sindbis, when injected intracerebrally, causes encephalitis with age-dependent mortality. Newborn mice die, adults always recover. A strain (NSV) which is neurovirulent for adult mice has been isolated (24). This NSV strain can be discriminated from wild type Sindbis by MAbs reacting with E2 (25). (These are *not* the same panel of MAbs used to select the variants in the previous section.) We are sequencing the region of the genome encoding E2 in these variants. Although the results are still preliminary, we have only been able to locate one change in NSV compared to wild type. Its location is indicated by the arrow (NV) on Fig. 4: Gln of HR is replaced by His in NSV at residue 55. This was a somewhat surprising result considering the complicated pattern of altered reactivity with MAbs. It may be that this single change has a profound effect on the tertiary configuration of the E2-E1 dimer, revealing some cryptic sites while masking other epitopes. Further study is required before firm conclusions can be drawn, however.

DISCUSSION

We can now at least in the case of protein E2 begin to draw a functional map. In the N-terminal 120 amino acids of the protein, there appear to be at least two residues involved in virulence: the one cited above at residue 55 and another site at residue 114 (S Gidwitz, NL Davis, DC Fly, FJ Fuller and RE Johnston, personal communication, and ref. 34). It is noteworthy that a naturally occurring deletion mutant of Ross River virus is missing 7 amino acids corresponding to residues 52 to 58 of Sindbis virus. This mutant strain (RRVdE2) had a greatly reduced virulence for mice (35). Presumably these changes alter the tropism for the central nervous system of the neurovirulent variants leading to fatal encephalitis. In the center of the protein is a region between residues 176 and 216 which is quite variable among alphaviruses and in which a number of the neutralization epitopes appear to reside. Up to five residues in this region have been found to be altered, leading to changes in the neutralization pattern with specific monoclonal antibodies. The residue responsible for the *ts*20 phenotype, i.e., the temperature sensitive phenotype in which the precursor to E2 is not cleaved and virus does not bud, is located downstream from this area. Many of these mutations have been confirmed by examining revertants at the same position. We hope to directly establish the validity of the sequence results by cloning the mutants or variants as cDNA copies and swapping the DNA encoding particular domains in a cassette fashion into an infectious cDNA clone of Sindbis virus. We can then examine the virus produced and show that the phenotype has been transferred with a particular genomic fragment. Furthermore, having identified the key residues for particular interactions we can engineer new mutations by site-specific mutagenesis at these locations. Eventually, from an examination of all these various types of mutants and from approaches which combine sequencing with biological studies, we hope to be able to draw a functional map of the glycoproteins of alphaviruses and to determine the functions of particular glycoprotein domains and how they interact with one another and with the host.

REFERENCES

1. Strauss EG, Strauss JH (1986). Structure and replication of the alphavirus genome. In Schlesinger, S and Schlesinger, M (eds): "The Togaviridae and Flaviviridae." Chapt. 3, Plenum Press, p. 35.
2. Hahn CS, Strauss EG, Strauss JH (1985). Sequence analysis of three Sindbis virus mutants temperature-sensitive in the capsid protein autoprotease. Proc Natl Acad Sci USA 82: 4648.
3. Hashimoto K, Erdei S, Keränen S, Saraste J, Kääriäinen L (1981). Evidence for a separate signal sequence for the carboxy-terminal envelope glycoprotein E1 of Semliki Forest virus. J Virol 38: 34.
4. Garoff H, Frischauf A-M, Simons K, Lehrach H, Delius H (1980). Nucleotide sequence of cDNA coding for Semliki Forest virus membrane glycoproteins. Nature (London) 288: 236.

5. Leavitt R, Schlesinger S, Kornfeld S (1977). Tunicamycin inhibits glycosylation and multiplication of Sindbis and vesicular stomatitis virus. J Virol 21: 375.
6. Arias C, Bell JR, Lenches EM, Strauss EG, Strauss JH (1983). Sequence analysis of two mutants of Sindbis virus defective in the intracellular transport of their glycoproteins. J Mol Biol 168: 87.
7. Strauss EG, Strauss JH (1985). Assembly of enveloped animal viruses. In Casjens, S (ed.): "Virus Structure and Assembly." Portola Valley, California: Jones and Bartlett, p. 205.
8. Bracha M, Schlesinger MJ (1976). Defects in the RNA$^+$ temperature-sensitive mutants of Sindbis virus and evidence for a complex of PE2-E1 viral glycoproteins. Virology 74: 441.
9. Rice CM, Strauss JH (1982). Association of Sindbis virion glycoproteins and their precursors. J Mol Biol 154: 325.
10. Strauss JH, Strauss E G (1985). Antigenic structure of Togaviruses. In van Regenmortel, MHV and Neurath, AR (eds): "Immunochemistry of Viruses: The Basis for Serodiagnosis and Vaccines." Amsterdam: Elsevier, p. 407.
11. Dalrymple JM, Schlesinger S, Russell PK (1976). Antigenic characterization of two Sindbis envelope glycoproteins separated by isoelectric focussing. Virology 69: 93.
12. Helenius A, Fries E, Garoff H, Simons K (1976). Solubilization of the Semliki forest virus membrane with sodium deoxycholate. Biochem Biophys Acta 436:319.
13. Yamamoto K, Suzuki K, Simizu B (1981). Hemolytic activity of the envelope glycoproteins of Western equine encephalitis virus in reconstitution experiments. Virology 109: 452.
14. Simizu B, Yamamoto K, Hashimoto K, Ogata T (1984). Structural proteins of Chikungunya virus. J Virol 51:254.
15. Chanas AC, Gould EA, Clegg JCS, Varma MGR (1982). Monoclonal antibodies to Sindbis virus glycoprotein E1 can neutralize, enhance infectivity and independently inhibit haemagglutination or haemolysis. J Gen Virol 58: 37.
16. Schmaljohn AL, Kokubun KM, Cole GA (1983). Protective monoclonal antibodies define maturational and pH-dependent antigenic changes in Sindbis virus E1 glycoprotein. Virology 130: 144.
17. Roehrig JT, Mathews JH (1985). The neutralization site on the E2 glycoprotein of Venezuelan equine encephalitis (TC-83) virus is composed of multiple conformationally stable epitopes. Virology 142: 347.
18. Boere WAM, Harmsen T, Vinjé J, Benaissa-Trouw BJ, Kraaikeveld CA, Snippe H (1984). Identification of distinct antigenic determinants on Semliki Forest virus by using

monoclonal antibodies with different antiviral activities. J. Virol. 52: 575.

19. Lindqvist BH, DiSalvo J, Rice CM, Strauss JH, Strauss EG (1986). Sindbis virus mutant *ts*20 of complementation group E contains a lesion in glycoprotein E2. Virology 151: 10.
20. Burge BW, Pfefferkorn ER (1966). Isolation and characterization of conditional-lethal mutants of Sindbis virus. Virology 30: 204.
21. Strauss EG, Lenches EM, Strauss JH (1976). Mutants of Sindbis virus I. Isolation and partial characterization of 89 temperature-sensitive mutants. Virology 74: 154.
22. Schmaljohn AL, Johnson ED, Dalrymple JM, Cole GA (1982). Non-neutralizing monoclonal antibodies can prevent lethal alphavirus encephalitis. Nature 297: 70.
23. Stec DS, Waddell A, Schmaljohn CS, Cole GA, Schmaljohn AL (1986). Antibody-selected variation and reversion in Sindbis virus neutralization epitopes. J Virol 57: 715.
24. Griffin DE, Johnson RT (1977). Role of the immune response in recovery from Sindbis virus encephalitis in mice. J Immunol 118: 1070.
25. Stanley J, Cooper SJ, Griffin DE (1985). Alphavirus neurovirulence: Monoclonal antibodies discriminating wild-type from neuroadapted Sindbis virus. J Virol 56: 110.
26. Rice CM, Strauss JH (1981). Synthesis, cleavage, and sequence analysis of DNA complementary to the 26S messenger RNA of Sindbis virus. J Mol Biol 150: 315.
27. Ou J-H, Trent DW, Strauss JH (1982). The 3' noncoding regions of alphavirus RNAs contain repeating sequences. J Mol Biol 156: 719.
28. Zimmern D, Kaesberg P (1978). 3' terminal nucleotide sequence of encephalomyocarditis virus RNA determined by reverse transcriptase and chain-terminating inhibitors. Proc Natl Acad Sci USA 75: 4257.
29. Kyte J, Doolittle RF (1982). A simple method for displaying the hydropathic character of a protein. J Mol Biol 157: 105.
30. Rice CM, Strauss JH (1981). Nucleotide sequence of the 26S mRNA of Sindbis virus and deduced sequence of the encoded virus structural proteins. Proc. Natl. Acad. Sci. USA 78: 2062.
31. Kinney RM, Johnson BJB, Brown VL, Trent DW (1986). Nucleotide sequence of the 26S mRNA of the virulent Trinidad Donkey strain of Venezuelan equine encephalitis virus and deduced sequence of the encoded structural protein. Virology, in press.
32. Saraste J, von Bonsdorff C-H, Hashimoto, K, Kääriäinen L,

Keränen S (1980). Semliki Forest virus mutants with temperature-sensitive transport defect of envelope proteins. Virology 100: 229.

33. Dalgarno L, Rice CM, Strauss JH (1983). Ross River virus 26S RNA: Complete nucleotide sequence and deduced sequence of the encoded structural proteins. Virology 129: 170.

34. Olmsted RA, Meyer WJ, Johnston RE (1986). Characterization of Sindbis virus epitopes important for penetration in cell culture and pathogenesis in animals. Virology 148: 245.

35. Vrati S, Faragher SG, Weir RC, Dalgarno L (1986). Ross River virus mutant with a deletion in the E2 gene: Properties of the virion, virus-specific macromolecule synthesis and attenuation of virulence for mice. Virology 151:222.

Positive Strand RNA Viruses, pages 379–395

MECHANISM OF RNA VIRUS ASSEMBLY AND DISASSEMBLY[1]

S. C. Harrison, P. K. Sorger[2], P. G. Stockley[3], J. Hogle[4], R. Altman[5], R. K. Strong

Department of Biochemistry and Molecular Biology
Harvard University, Cambridge, Massachusetts 02138

ABSTRACT This paper describes a series of experiments on assembly and structural rearrangements of isometric RNA virus particles <u>in vitro</u>, performed with a view to answering some questions about assembly mechanism, RNA specificity, requirements of the packaging process, and possible steps in disassembly. The experiments have been carried out primarily with turnip crinkle virus (TCV).

[1]This work was supported by PHS Grant CA-13202, awarded by the National Cancer Institute, DHHS, and by the National Science Foundation Grant PCM 82-02821.

[2]Present address: MRC Laboratory for Molecular Biology, Hills Road, Cambridge CB2 2QH, England.

[3]Present address: The Biotechnology Unit, Department of Genetics, University of Leeds, Leeds LS2 9JT, England.

[4]Present address: Scripps Clinic and Research Foundation, 10666 North Torrey Pines Road, La Jolla, California 92037

[5]Present address: School of Medicine, Stanford University, Stanford, California 94305

TCV STRUCTURE

We first review the structure of TCV, using the description to introduce what we see as the important questions concerning assembly and disassembly.

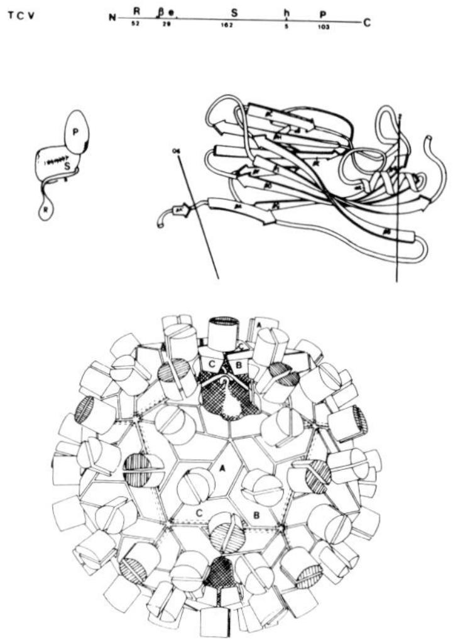

FIGURE 1. Organization of TCV. The 351-residue polypeptide chain is organized in four regions: an internal, positively charged domain (R), a connecting arm (a), a β-barrel domain that forms the shell (S), and a projecting domain (P). The fold of the chain in the S domain is shown in the middle right. The packing in the virion is shown at the bottom. There are three non-symmetry related environments: A, B and C, with 60 subunits in each type of position. The connecting arms of C subunits extend along the inner edge of the S domain (dashed lines), and they interdigitate around three-fold symmetry axes to form a β-annulus. The corresponding chain segments are marked e and β in the top part of the figure. The hinge (h) between S and P domains has one conformation in A and B subunits and a different one in C subunits.

The data in Table 1 and the overview in Fig. 1 summarize what we know about TCV from chemical and crystallographic studies. The three-dimensional structure was determined by using the atomic coordinates of tomato bushy stunt virus (TBSV), which provide a suitable first approximation for phase refinement (Hogle et al., 1986). As in TBSV, the polypeptide chain can be considered to comprise four segments: an N-terminal RNA binding region (R), an arm involved in conformational switching (a), a domain that forms the major part of the viral shell (S), and a projecting C-terminal domain (P). The S domain, shown in the secondary structural diagram in Fig. 1, is homologous to the S domains of other plant viruses such as southern bean mosaic virus (SBMV: Abad Zapatero et al., 1980), as well as to the principal domains of VP1, VP2 and VP3 in the picornaviruses (Rossmann et al., 1985; Hogle et al., 1985). Thus, it appears that the shells of many of the small, positive-stranded RNA viruses have a similar architecture, with 180 S domains (here, all identical; in picornaviruses, of three different types). The S domains of different viruses vary principally in their N-terminal extensions, in loops formed by insertions between elements of secondary structure (especially after βC, βE and βG), and in C terminal extensions (most elaborately in these plant viruses, where the C-terminal extension forms a complete P domain).

TABLE I

Composition of TCV and TBSV[+]

	Protein:	Mr	Copies/virion	RNA
TCV	coat	39K	178	4.0 kb
	p80	78K	1	
TBSV	coat	42K	178	4.5 kb
	p80	84K	1	

p80 is a covalently linked dimer of the coat protein in both cases.

[+] See Ziegler et al. (1974), Hopper et al. (1984), Morris and Carrington (1986), Stockley et al. (1986).

The structure can be considered as an assembly of subunit dimers - 120 in one conformation (A/B) and 60 in a second (C/C). The key conformational differences involve a hinge between S and P domains and the configuration of the arms. Only very subtle differences occur within S domains themselves, probably in response to packing differences in the capsid, which exert somewhat different forces on domains in A, B and C locations (see Harrison, 1984, for a description of these changes in TBSV). The C/C dimers have arms folded in an ordered way along the inner edge of the S domain, interdigitating around particle threefold axes so as to form a coherent inner scaffold. The arms of A/B dimers project in a less ordered way into the interior of the particle.

RNA packaging in these particles appears to have evolved to accommodate irregularities and changes in the nucleic acid. The R segments are not uniformly ordered, and since their principal characteristic is a high concentration of lysine and arginine residues, they serve to clamp onto the RNA. Moreover, there are no defined binding sites for RNA on the inner face of the S domain, although a number of basic residues project inward and provide further positive charge to neutralize the phosphates. Thus most of the interaction with RNA is with parts of the polypeptide chain that are flexibly linked to the S domain shell, and no highly ordered RNA is seen crystallographically. The nucleic acid is nonetheless very tightly packed - as tightly as tRNA in crystals - and there appears to be little motion (Munowitz et al., 1980). The internal disorder is therefore a reflection of the irregularity of the nonrepeating RNA structure and of the way the R segments accommodate to it.

The R segment of TCV can be prepared by chymotryptic cleavage of dissociated virus (data not shown). A strong cleavage occurs at tyr 66, as in expanded virions. Studies of the purified peptide, which corresponds to residues 1-66 (as confirmed by amino acid analysis), show little evidence for a rigidly folded structure. In particular, urea gradient gels do not give evidence for a cooperative unfolding transition, and the peptide is equally sensitive to digestion by thermolysin at 0 and 25 ^{o}C (data not shown). The properties of the isolated R region are thus consistent with its role in the virion as an accommodating clamp.

As indicated in Table 1, the minor species p80, present in one copy per virion (Ziegler et al., 1974), is a covalent dimer of the coat protein (Stockley et al., 1986). The covalent linkage is at or near the N terminus, but its chemistry is not yet known. Since this part of the subunit is not rigidly folded, there are no new constraints on the contacts, and p80 can fit into the shell just as a 'normal', unlinked dimer. We believe that the structure is thus composed of 178 copies of p40 (89 noncovalent dimers) and 1 copy of p80. A number of other plant viruses with T=3 icosahedral shells appear to contain about one copy of species having twice the M_r of the coat protein (Rice, 1974), suggesting that these dimers have a conserved function.

EXPANSION OF TBSV AND TCV

A number of T=3 plant viruses have been shown to undergo an expansion at neutral to alkaline pH, when divalent cations are sequestered (Incardona and Kaesberg, 1974; Robinson and Harrison, 1982; Kruse et al., 1982). The expanded form of TBSV is a relatively well-ordered particle, and its crystal structure is known (Fig. 2: Robinson and Harrison, 1982). There are Ca^{++} binding sites at the subunit interfaces indicated in Fig. 2, and removal of the ions leaves apposed aspartic acid groups titrating at about pH 7. The instability of negatively charged groups in close vicinity causes the subunit interfaces to dissociate and the rearrangements shown in Fig. 2 to occur. The principle changes can be described as rigid-body motions of domains with respect to each other. TCV undergoes a similar transition. It has not been crystallized, but small-angle X-ray scattering shows a similar increase in radius in both viruses, and we can reasonably suppose that the expanded TCV is essentially similar in structure to expanded TBSV.

An additional rearrangement accompanying expansion of TBSV and TCV, not evident from the crystallographic results, is exposure of the subunit arms. Native TBSV is completely resistant to digestion by various proteases, but proteolysis of expanded virions leads to cleavage. With chymotrypsin, one third of the subunits are cleaved to 35 kD fragments, and the remaining two-thirds, to 30 kD fragments (Fig. 3). With V-8 protease, only two-thirds of the subunits are sensitive to degradation, to fragments of about 38 kD. Evidence that these cleavages occur in the

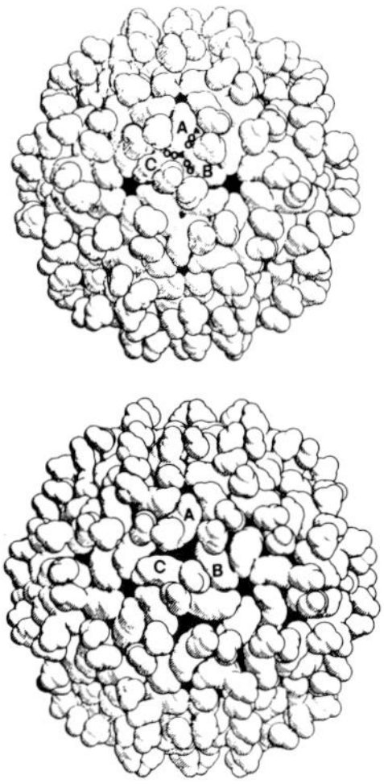

FIGURE 2. The compact and expanded structures of TBSV (from Robinson and Harrison, 1982). Some of the Ca^{++} sites that regulate the expansion are shown as circles in the upper figure. All positions symmetry-related to these also have cation sites. The corresponding interfaces move apart in the expanded structure.

amino-terminal part of the subunit is described by Golden and Harrison (1982). The position of V-8 cleavage can be uniquely assigned to glu 36, the only negatively charged residue in the first 100. The 30 kD chymotryptic fragment had probably been cleaved at phe 95, and the 35 kD chymotryptic fragment at leu 48 or leu 53. All these cleavages lie in the R segment or in the arm. The two-thirds/one-third ratio suggests differential behavior of A, B and C subunits. Expanded, cleaved and recontracted TBSV can be crystallized, and a 2.9 Å difference map shows no significant changes in ordered density (Altman, unpublished). In particular, the region around phe 95 in the C subunit arm is undisturbed. Thus, C subunits have been cleaved near residue 50 A and B subunits at residue 95.

Similar chymotryptic cleavage is observed in expanded TCV (Fig. 4), but here the extent of digestion is less extensive (one-third of the subunits) and ionic-strength dependent. The cleavage point is at tyr 66, as determined by direct sequence analysis (see Hogle et al., 1986). Sedimentation profiles show that the cleaved TCV is a homogeneous species - that is, cleavage has occurred on one-third of the subunits of all virions, not on all subunits of one-third of the virions.

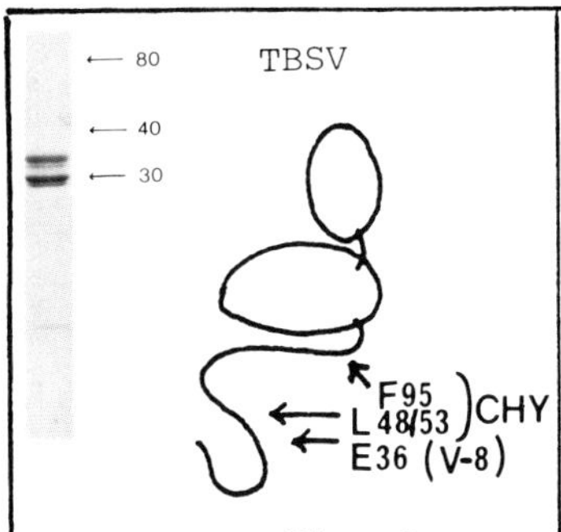

Fig. 3

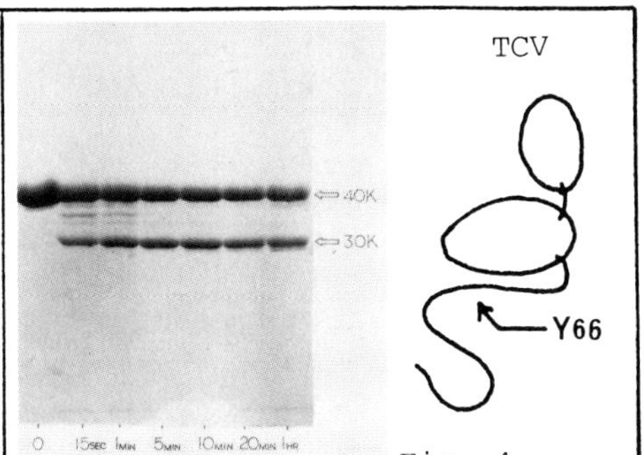

Fig. 4

FIGURE 3. Proteolytic cleavage of expanded TBSV. (a) Gel lane shows endpoint of chymotryptic cleavage of expanded virus (3 hours, room temperature, 1:100 w/w enzyme/virus). One third of the subunits have been cleaved to an approximately 35 kD species and two thirds to an approximately 30 kD species. (b) Diagram showing sites of chymotrypsin and V-8 cleavage using schematic representation of folded subunit in C conformation (see Fig. 1). The cleavages at positions 36 and 48/53 probably occur only on A/B subunits.

FIGURE 4. (a) Cleavage of expanded TCV by chymotrypsin at low ionic strength (50 mM Tris, 5 mM EDTA, pH 8.5). Note that one third of subunits are cleaved to 30 kD species. (b) Position of chymotryptic cleavage (tyr 66).

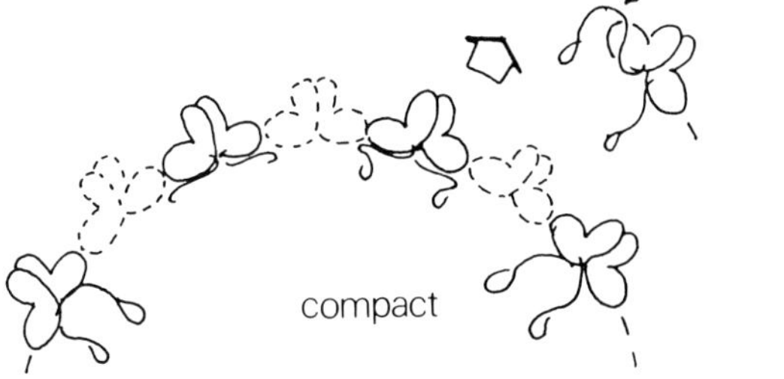

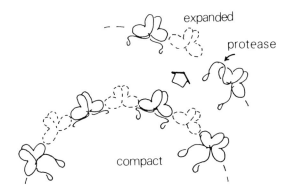

FIGURE 5. Model for looping out of arms when TCV and TBSV shells expand. In order to be exposed to chymotrypsin, the arms must move through a distance of at least 20 Å.

The proteolytic accessibility of arms and R segments is unexpected, since gaps in the expanded structure are not wide enough to admit molecules as large as chymotrypsin or V-8 protease. Our observations may be explained either by assuming that the expanded particle 'breathes' extensively, permitting entry of protease, or by postulating extension of the arm itself. Several arguments indicate that breathing alone cannot account for the cleavages. (1) Proteolytic cleavage of arms in rapidly recontracted virus suggests that accessible arms have been trapped in an exposed configuration (Sorger, unpublished). (2) RNase does not gain access to the particle interior, as judged by failure to observe RNA degradation when expanded TBSV is exposed to the enzyme (Kruse et al., 1982; Sorger and Harrison, unpublished). RNase A is substantially smaller than chymotrypsin. (3) TCV arms can be cleaved at low but not at moderate ionic strength. The radius of the expanded particle is identical in the two cases, and we therefore have no reason to believe that S domain arrangements are different. The most straightforward explanation of the observed cleavage of 1/3 of the subunits is a salt-dependent conformational change in one of the three classes of arms (probably A or B, see below). (4) V-8 protease cleaves 2/3, but not all, of the TBSV R domains. Such differential accessibility is better explained by specific conformational changes of the arms than by a general breathing of the particle. Each of these arguments is not complete by itself, but taken as a whole, the observations are difficult to reconcile with a picture other than emergence of a particular class of arms under particular expansion conditions.

A diagram with our interpretation of these results appears in Fig. 5. The tight packing of RNA and R segments in the viral interior is sufficiently relaxed on expansion that arms can loop out through the gaps between subunits shown in Fig. 2. There is good evidence from NMR studies for increased internal mobility in expanded TBSV (Munowitz et al., 1980). The similarity between plant and picornavirus structures suggests a parallel between the capacity of the subunit arm to extrude in this way and the loss of VP4 when picornaviruses bind to cells (Rueckert, 1985; Hogle et al., 1985; Rossmann et al., 1985). Some expansion of the latter structures must occur, in order to permit exit of VP4.

TCV ASSEMBLY

The mechanism of *in vitro* assembly can conveniently be studied, since disassembly of TCV can be carried out under relatively mild, non-denaturing conditions. Dissociation occurs at pH 8, 1mM EDTA (i.e., expansion conditions), if the salt concentration is raised to 0.5M (or higher); it proceeds far more rapidly at $0^{o}C$ than at room temperature. Efficient reassembly can be achieved by two-step dialysis (first to pH 6-7, then to 0.1 M NaCl) or by dilution to pH 7, 0.1M NaCl (Sorger et al., 1986).

The rp-complex.

The products of TCV disassembly are *dimers* of the coat protein subunit p40 (Golden and Harrison, 1982), and a complex of TCV RNA with six subunits of p40 and the unique copy of p80 (Fig. 6: Sorger et al., 1986). The complex of RNA and protein, which we term 'rp-complex,' is stable and resistant to high salt concentrations (1.5M). Several lines of evidence show that the protein is bound to specific sites or structures on the RNA. (1) When lightly labelled by reductive methylation with ^{3}H- formaldehyde, rp-complex is competent to reassemble. The label is retained on RNA as rp-complex if the reassembled particles, made with unlabelled coat protein dimers, are disassembled in the usual way. (2) Particles, reassembled with protein-free RNA (phenol extracted) and normal p40 dimers, can be disassembled to yield rp-complex (and free coat dimers). This experiment also shows that p80 is not required for assembly (see below). (3) Extensive digestion of rp-complex with T1 ribonuclease yields a set of seven protected fragments about 30-50 bases in length. The sequences of these fragments have been determined; three correspond to segments of the 3' one-third of the genome for which sequence is known (Carrington et al., submitted to J. Mol. Biol.). These sequences and their location are shown in Fig. 7.

Reassembly.

The properties of the *in vitro* reassembly reaction can be summarized as follows: (1) It proceeds from coat-protein dimers and rp-complex, or from coat-protein dimers and free RNA. In the latter case, the product lacks p80, which is

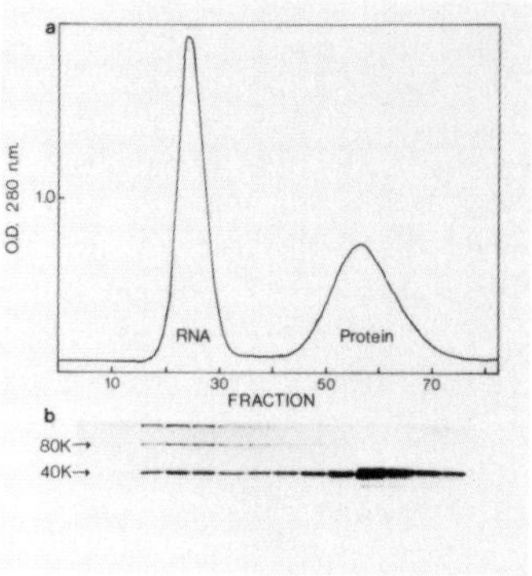

FIGURE 6. (a) Elution profile from Sephadex G-150 of TCV dissociated as described (Sorger et al, 1986). (b) SDS-polyacrylamide gel pattern of fractions from the G-150 column. Loadings from the 'protein' peak were only 1/20 as great as those from the 'RNA' peak. Protein in the 'RNA' peak shows that it is in fact an rp-complex (six subunits per RNA molecule).

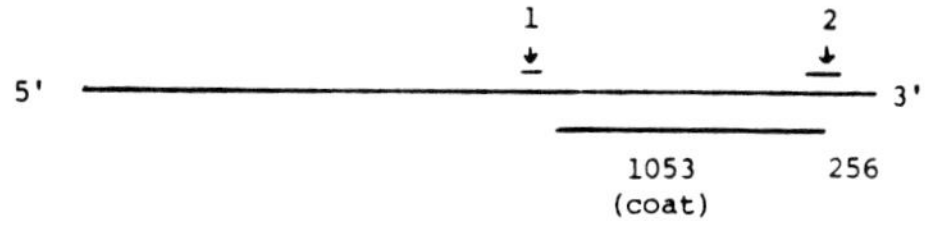

(1) G A U U C A C A C A U C C U A C A A C A C A C A C G A C U C A U C

(2) G A A A A U G G U C A C A A C U G A G G A G C A G C C A A A G G
G U A A A U G G C A A G C A C U C A G A A U U U A G U A C G G U A A

FIGURE 7. Sequences of three RNA fragments found in ribonuclease T1 limit digest of rp-complex. Fragments were isolated by filter binding of complex after digestion. Sequence 2 corresponds to two contiguous fragments. Four other T1 fragments were reproducibly isolated, but their sequences are not found in the 3' one-third of the genome for which nucleotide sequence is presently available (Carrington et al., submitted to J. Mol. Biol.).

therefore not required for morphologically correct assembly. (2) The conditions of optimal assembly are approximately physiological (pH 7, 0.1M NaCl). (3) Assembly requires RNA and intact protein subunits. No defined structures are obtained in the absence of RNA; T=1 RNA-free, 60 -subunit particles are obtained from subunits that have been proteolytically cleaved (Fig. 8). The T=1 particles have all subunits in the A/B conformation - the arm is required for conformational switching, as the character of the internal framework clearly implies. (4) The viral shell grows by addition of subunit dimers to an initiating structure. There is no evidence for defined intermediates, other than rp-complex and coat-protein dimers. Moreover, the curvature of the shell is correctly determined as it forms. That is, the selection of A/B or C/C conformation occurs as a dimer adds - subsequent adjustment or rearrangement does not occur (Fig. 9). Multiple nucleation on a single RNA molecule can lead to non-coalescing partial shells, since independently nucleated arm frameworks cannot in general merge with each other, but even these partial shells have correct local morphology (Fig. 9). (5) Assembly is selective for viral RNA. Although heterologous RNA can be packaged - 18S or 28S chick ribosomal RNA is a particularly efficient substrate - competition experiments show a strong preference for TCV RNA (Sorger et al., 1986). Free RNA is as effective as rp-complex, since under the conditions we have studies, rp-complex formation is not rate limiting. That is, when free TCV RNA is used, rp-complex is rapidly reconstituted, and free RNA thus competes as effectively as preformed complex when challenged with rRNA (Sorger et al., 1986). It is clear from the formation of 'monsters' under conditions of multiple nucleation (Fig. 9) that unique initiation is physico-chemically as well as biologically significant, and the nucleating structure can account for both accuracy and specificity in assembly.

Mechanism.

The reassembly reaction can be represented as follows:

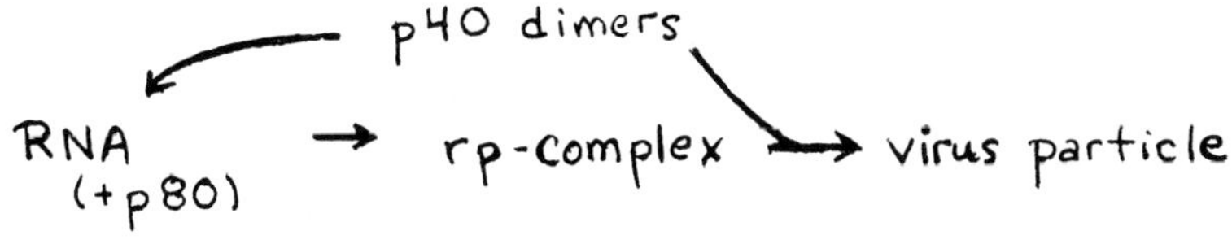

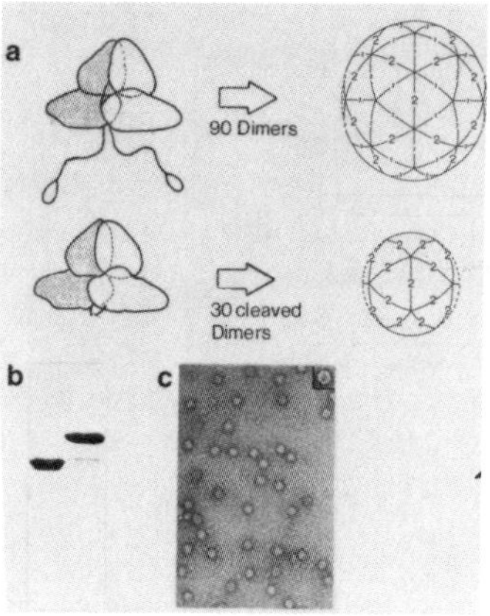

Fig. 8

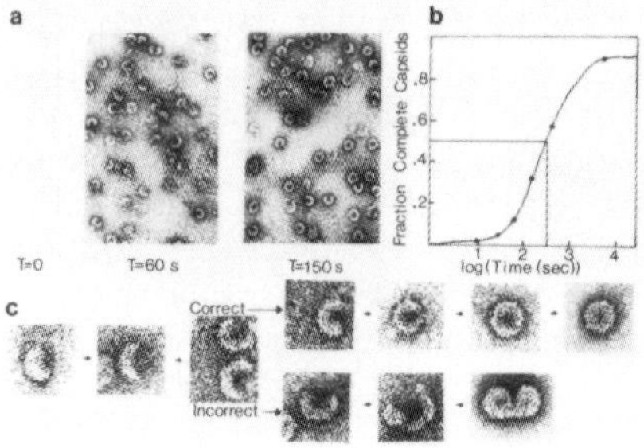

Fig. 9

FIGURE 8. (a) Intact TCV subunit dimers can reassemble to form normal T=3 capsids in the presence of RNA. Cleaved (30K) subunits assemble into RNA-free, 60-subunit, T=1 particles. (b) SDS-polyacrylamide gel of cleaved (left) and intact (right) subunits. (c) Reassembly from cleaved subunits yields T=1 particles shown in this micrograph. A T=3 particle is shown for comparison in the inset.(Sorger et al, 1986).

FIGURE 9. Dissociated TCV was reassembled by dilution to pH 7. Samples were withdrawn at various times, applied to carbon-coated EM grids, and negatively stained with uranyl acetate. (a) Typical micrographs at indicated times. (b) Fraction of particles assembled, expressed as a ratio of all visible fragments, plotted versus time. (c) A synthetic 'time course' of assembly for both native and aberrant products. The 'monster' particles appear to represent multiple nucleation on one RNA.(Sorger et al, 1986).

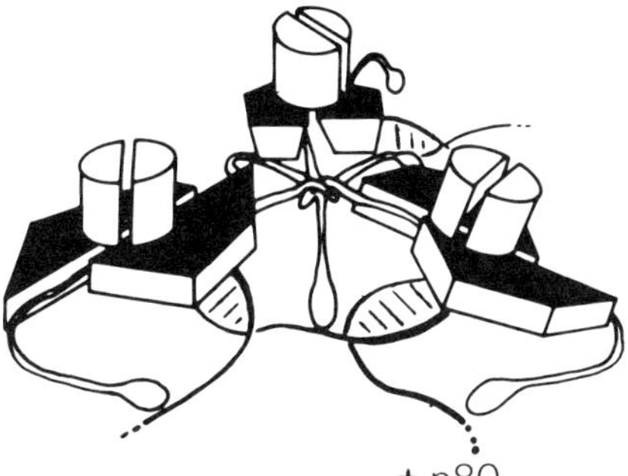

FIGURE 10. Proposed structure for the rp-complex: a trimer of dimers, connected by a β-annulus, bound to a defined structure on the viral RNA. The binding of p80, also a part of rp complex from normal virions, is probably at a different site on the RNA. For further discussion, see text.

It is clear that understanding the structure of the rp-complex is critical for visualizing the mechanism. The structure we have proposed for the complex is shown in Fig. 10. The evidence for this model comes largely from the experiments summarized above. Since rp-complex can be produced by assembly in protein-free RNA, since it survives a round of reassembly/disassembly, and since it appears as a product of gentle dissociation, it is likely to be a substructure of the virion itself. That is, the interactions among its subunits and between these subunits and RNA are likely to persist in the assembly process and in the completed particle. Moreover, since assembly can occur on heterologous RNA, we suggest that bonding properties of the subunits, rather than tertiary structure of an RNA scaffold, determine the organization of the initiating complex. There are two types of trimer interactions in the structure - the set of three C subunits linked by a β-annulus and the 'ABC' trimer linked by Ca^{++}-site interfaces (Hogle et al. 1983). The Ca^{++}-mediated, local threefold contacts are selectively destabilized in the conditions in which we prepare rp-complex, making the 'ABC' structure an unlikely model. The set of C/C dimers joined by interdigitating arms is, by contrast, an attractive model, since it is easy to see how binding of three arms or R domains to a particular RNA structure could stabilize a β-annulus. A set of three dimers forming a structure of this type is also a uniquely effective starting point in a model for accurate assembly, illustrated in Fig 11. Note that in the structure depicted in Fig 10, we show the arms in the tight conformation they adopt in a C/C dimer. In the free rp-complex, the linkage between β-annulus and S-domain may be relatively flexible, with addition of further dimers (Fig 11) required to lock it into the precise conformation shown.

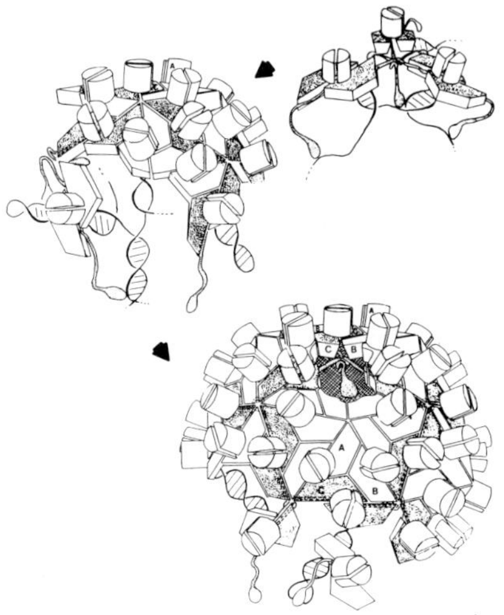

FIGURE 11. Proposed mechanism for TCV assembly. (Sorger et al, 1986).

The assembly model in Fig 11 is based on the postulated rp-complex structure, on the assumption that subsequent dimer incorporation will occur at sites of extensive interaction with the existing assembly, and on the cooperativity of the conformational switch in the coat protein dimer. The first additions of dimers to the rp-complex thus occur around the initial β-annulus, and these dimers adapt the A/B conformation because the existing structure prevents folding of their arms. The formation of B-C and C-B interfaces also locks in the arms of the rp-complex around the β-annulus. As addition continues (Fig 11b), new positions for β-annular interaction are determined by the arms of the 'other subunit' in each of the initial three dimers. That is, the two polypetide chains in C/C dimers interact closely in the region where arms fold back into S domains, and locking in of the folded arms in the first β-annulus also stabilizes the folded arm and the C conformation of the second member of each dimer. Propagation of the structure thus consists of β-annulus formation by dimers that consequently have the C/C conformation, in alternation with 'filling-out' steps by dimers that retain the A/B conformation. Recruitment of dimers to the assembly is probably accelerated by non-specific interaction of R domains with RNA (Fig 11b), and this interaction ensures that RNA is gathered into the shell as it forms.

Correct closure is guaranteed by the mechanism depicted in Fig 11, provided that multiple nucleation has not occured on the same RNA molecule (see 'monster' in Fig 9). A possible position for p80, the unique, covalently-linked coat-protein dimer, is as the last unit to enter the shell (Fig 12). If it binds near one end of the RNA, this termination step can ensure complete packaging by preventing a protruding end or loop from inhibiting closure (Fig 12). A more important function might then be in the initiation of disassembly. Wilson and co-workers have observed that SBMV will direct _in vitro_ synthesis of viral protein from intact virions and suggest a 'weak capsid element' that is displaced to allow ribosome binding (Brisco et al, 1985).

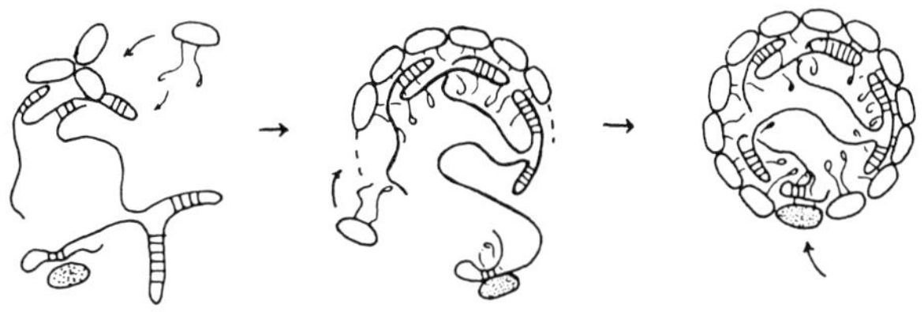

FIGURE 12. Diagram indicating one proposal for the function of p80, as a 'last in, first out' element of the viral shell. The shaded dimer represents the covalently linked p80.

SBMV and many other plant viruses appear to contain a species similar to p80 - that is, a component with M_r equal to twice that of the coat protein (Rice, 1974). Like TCV, these viruses assemble from dimers. Thus, there appears to be some importance to whatever function is served by the covalently linked species.

An important feature of the mechanism shown in Fig 11 is that it involves purely local interactions in the determination of long-range structural characteristics (curvature and closure). The 'decision' to adopt an A/B or C/C conformation is completely determined by interactions possible when a dimer adds. Moreover, these interactions do not need to be sensitive to signals transmitted across a domain or to small conformational distortions of a folded structure. When a dimer 'docks' into a shell there are two possible outcomes. If a β-annular interaction can form with an arm already folded and set to nucleate it, this interaction will occur, and a C/C dimer will result. Otherwise, the arms will remain unfolded, and the dimer will have the A/B conformation. A mechanism of this kind is robust to small changes in the conditions under which assembly occurs and to mutations that do not directly destabilize the critical interactions.

REFERENCES

Abad-Zapatero, C., Abdel-Meguid, S.S., Johnson, J.E., Leslie, A.G.W., Rayment, I., Rossmann, M.G., Suck, D., and Tsukihara,T. (1980). Nature 286, 33-39.

Brisco, M.J., Hull, R.C. and Wilson, T.M.A. Submitted to Virology.

Golden, J.S. and Harrison, S.C. (1982). Biochemistry 21, 3862-3866.

Hogle, J., Kirchhausen, T., and Harrison, S.C, (1983). J. Mol. Biol. 171, 95-100.

Hogle, J.M., Chow, M. and Filman, D.J. (1985). Science 229, 1359.

Hogle, J., Maeda, A. and Harrison, S.C. (1986). J. Mol. Biol., in press.

Hopper, P., Harrison, S.C., and Sauer, R. J. Mol. Biol. 177, 701-713 (1984).

Incardona, N.L. and Kaesberg, P. (1974). Biophys. J. 4, 11.

Kruse, J., Kruse, K.M., Witz, J., Chauvin, C., Jacrot, B., and Tardieu, A. (1982). J. Mol. Biol. 162, 393-417.

Morris, T.J. and Carrington, J.C. (1986). In Isometric Plant Viruses with Monopartate RNA Genomes, ed. R. Koenig (Plenum, New York).

Munowitz, M.G., Dobson, C.M., Griffin, R.G., and Harrison, S.C. (1980). J. Mol. Biol. 141, 327-333.

Rice, R.H. (1974). Virology 61, 249-255.

Robinson, I.K. and Harrison, S.C. (1982). Nature 279, 563-568.

Rossmann, M.G., et al. (1985). Nature 317, 145.

Rueckert, R.R. (1985). In Virology, ed. B. Fields (Pergamon, New York).

Sorger, P.K., Stockley, P.G. and Harrison, S.C. (1986). J. Mol. Biol., in press.

Ziegler, A., Harrison, S.C., and Leberman, R. (1974). Virology 59, 509-515.

Positive Strand RNA Viruses, pages 397–405

TMV ENCAPSIDATION INITIATION SITES ON NON-VIRION RNA SPECIES[1]

Albert Siegel, Chintamani Atreya,[2]
Fumihiro Terami and D'Ann Rochon[3]

Department of Biological Sciences
Wayne State University
Detroit, MI 48202

ABSTRACT Assembly of the TMV rod is initiated by reaction of a capsid protein oligomer with a specific internal region of virion RNA called the encapsidation initiation site (ei). The rod is then completed by addition of protein in an elongation reaction which is largely nucleotide sequence independent. In addition to the ei responsible for virion formation, others have been detected both on non-virion RNA species and as secondary sites on virion RNAs. These generally have a lower affinity for capsid protein than the sites involved in virion formation. Chloroplast DNA transcripts and cytoplasmic 18S rRNA are nonvirion RNA species that have eis. The chloroplast DNA transripts are encapsidated *in vivo* to form pseudovirions and can also be encapsidated *in vitro*, whereas the 18S rRNA is not encapsidated *in vivo* but is *in vitro*. The 18S RNA ei has been found to lie within a 68 nucleotide region near, but not at, the 5' terminus. The sequence of this region has been determined and

[1]This work was supported by NIH grant GM 32608, NSF grant PCM 8104436 and Agriculture Canada.
[2]Present address: Faculty of Medicine, Memorial University, St. John's, Newfoundland, Canada A1B 3V6.
[3]Present address: Agriculture Canada, 6660 NW Marine Drive, Vancouver, BC, Canada V6T 1X2.

compared with the sequence of the virion ei site. This site has been been conserved during evolution as evidenced by its presence on bovine as well as plant 18S rRNA. Native 18S RNA has an elongation block about 3/4 of the distance from its 5' terminus which may result from the presence of a modified base or bases. The block is not present on synthetic 18S RNA generated by in vitro transcription of a cloned plant 18S RNA coding sequence.

INTRODUCTION

Viral components are synthesized independently in host cells and are then assembled into virions. A well studied example of the assembly process is that of tobacco mosaic virus (TMV) rod formation. The rod is 300 nm long, has a diameter of 18 nm and consists of a single 6395 nucleotide RNA molecule incorporated into and protected by a helical array of ca. 2130 identical protein subunits with 161/3 subunits per turn and a pitch of 2.3 nm (reviewed in ref 1). Thirty years ago it was demonstrated that isolated RNA and protein components reassemble under appropriate conditions to form infectious particles (2) and the mechanism of this self-assembly has been a subject of study ever since. The current picture is that assembly occurs in at least two distinct phases: initiation and elongation. Assembly is initiated when a capsid protien oligomer reacts with a specific internal site of virion RNA, designated the encapsidation initiation site (ei). The location of the ei is strain dependent; in most strains it is located ca. 1/7 of the distance from the RNA 3' terminus and in others about 1/15 of this distance. Once initiation has been established, protein is added to both ends of the initiating oligomer so that elongation of the rod occurs bidirectionally with completion at the 5' terminus before that at the 3' terminus. It has been postulated either that addition of protein toward the 5' terminus occurs at a faster rate than towards the 3' terminus or, what seems more likely on the basis of current evidence, that 3' elongation is as rapid as 5' elongation but that elongation towards the 3' terminus does not initiate until 5' elongation is complete (1,3).

The capsid protein oligomer which initiates TMV

assembly is a 20S species which was thought until recently to be a 34 subunit, 2 layered squat cylinder called a double disc. However, new evidence indicates that the reactive protein oligomer is most probably somewhat larger than a two turn helical aggregate containing 37 to 41 subunits (4,5).

The reaction of the capsid protein oligomer with the ei, the initiation reaction, is the specific part of encapsidation and is rapid in comparison to the elongation phase under the usual conditions used for study of *in vitro* virion assembly. Once the reaction has been initiated, elongation is relatively non-specific and, with few exceptions, RNA is encapsidated without regard to nucleotide sequence. The end result of the process, the nucleoprotein rod, is a stable structure in which the RNA is protected from degradation by ribonucleases and other environmental influences.

The region encompassing the ei has been identified and characterized for four TMV strains (6-8). Three of these are closely related and include the common strain. Their eis are located at the same position in the genome, *ca.* 850-950 nucleotides in from the 3' terminus and they differ from each other by at most 20% of their nucleotide positions. The fourth strain, Cc, is distantly related to the others and has its ei in a different position, 370-460 nucleotides from the 3' terminus. It shares a greater than random nucleotide sequence homology with the others but is clearly more different from them than they are to each other. All four eis can be drawn in a stem-loop configuration which is postulated to be functionally active in the initiation reaction. Despite their differences, the eis react rapidly and specifically with their own and with each other's protein oligomers to engender encapsidation.

NON-FUNCTIONAL EIS ON VIRION RNAS

In addition to the virion eis which are efficiently recognized by capsid protein oligomers and which play a role in virus particle assembly, other RNA sites have been identified which can react with capsid protein to initiate encapsidation. These occur on non-virion RNA molecules and as secondary sites on virion RNAs. One such region located on the TMV common strain, designated SERF, reacts with capsid protein but much less efficiently than the "authentic" ei (9). It does not appear to function at all *in vivo*, nor *in vitro* when it is on the same RNA strand as the biologically functional ei, with which, nevertheless, it shares considerable homology. It is of interest that SERF is located on the common strain RNA at about the same position as the Cc ei is located on its RNA. The Cc RNA also contains an apparent "low efficiency" ei at the position where the common strain ei is located on its RNA (7) and, consequently, there has been speculation that the low efficiency eis are remnants of or precursors to the biologically active eis.

EIS ON CHLOROPLAST DNA TRANSCRIPTS

Eis have also been found on some non-virion RNA species. One class is made apparent by the presence of pseudovirions as a minor component of TMV preparations (10). These are particles that contain host rather than virus RNA. Preparations of all tested TMV strains contain pseudovirions, with one strain, U2, containing considerably more than the others. The encapsidated host RNA proves to be composed of transcripts from all regions of chloroplast DNA except that very little, if any, ribosomal RNA is present (11). The indications are that chloroplast messenger RNAs may have a structure or sequence in common which resembles the virion ei. We speculate that capsid protein has the capacity to enter chloroplasts where it reacts with and encapsidates chloroplast DNA transcripts. Non-infectious rods, shorter than virions have been observed in chloroplasts of infected plants and these may be the pseudovirions that have been identified in TMV preparations (12).

THE EI ON 18S RIBOSOMAL RNA

The conclusion that chloroplast transcripts have an ei is reinforced by the observation that they are encapsidated *in vitro* when leaf RNA is reacted with capsid protein oligomers under conditions favorable for reconstitution of virus particles. The major host RNA species encapsidated *in vitro*, however, is 18S ribosomal RNA, a species not encapsidated *in vivo* and, consequently, not found in the pseudovirion RNA population (13). We speculate that 18S rRNA fails to get encapsidated *in vivo* despite the presence of an ei because of lack of opportunity; that is, capsid protein probably does not come into contact with naked ribosomal RNA inside the cell. The *in vitro* reaction with 18S RNA is specific in the sense that it and not 25S rRNA participates and becomes encapsidated. An odd feature of the 18S RNA reaction is that not all of it becomes encapsidated and protected from ribonuclease action. There appears to be a block to further elongation after the 5' 3/4 of the 18S RNA is encapsidated (14). Only after a long period of incubation is there evidence for encapsidation of complete 18S RNA, and then only in very small amount.

TMV capsid protein not only reacts with and encapsidates a truncated version of plant 18S ribosomal RNA, it also reacts with and encapsidates a truncated version of bovine 18S RNA, indicating that both the 18S RNA ei and the elongation block are evolutionarily conserved (14). The nature of the elongation block is currently unknown but we speculate that it may consist of one or a series of modified bases. We have concluded that it is not due to secondary structure configuration because all, rather than just a portion, of *in vitro* synthesized 18S RNA is encapsidated. Synthetic RNA is generated in this case by transcription of a cloned plant (*Cucurbita maxima*) 18S RNA coding sequence placed just downstream from an SP6 promoter to form plasmid pRK18S. The ei present on native as well as on *in vitro* synthesized 18S RNA differs from the ei on virion RNA in being less efficiently recognized by the capsid protein oligomer. The initiation reaction with virion ei is rapid compared with the elongation reaction, whereas the opposite is true for the reaction with the 18S RNA ei. In this case, initiation is considerably slower than elongation so that an increasing amount of full length 18S RNA proves to be encapsidated in samples taken at increasing times after initiation of an *in vitro* reaction with very little, or no, partially encapsidated RNA being apparent.

The 18S RNA ei was located by linearizing pRK18S and other derived plasmids by cleavage at different positions in the 18S RNA coding sequence. Run-off transcripts, generated from these, were incubated with capsid protein to determine whether they would become encapsidated and made resistant to ribonuclease digestion. Only those transcripts became encapsidated which were generated from templates which contained a small XbaI - SalI restriction fragment located near, but not at, the 5' terminus of the

18S RNA coding sequence. Transcripts generated from templates which lacked this fragment failed to become encapsidated even though they contained most of the rest of the 18S RNA coding sequence. Transcripts also failed to be encapsidated if they were generated from a template containing most of the _C. maxima_ 25S RNA coding sequence. Thus, the ei responsible for encapsidation of the _C. maxima_ 18S RNA lies within the described XbaI - SalI fragment. The sequence of this fragment was determined (figure 1) and it was found to contain 68 nucleotides which could be assigned as nucleotides 157 to 224, inclusively, by comparison with published sequences for other 18S RNAs (15) and by its known position near the 5' terminus.

FIGURE 1

```
         _________________
C. max   UCUAGAGCUAAUACGUGxCAACAAACCCCGACxxxxxUUCxxx
soybean  -----------------x--------------xxxxx---xxx
maize    -----------------x--------------xxxxx---xxx
rat      --------------A--C--G-GGG-G-U---CCCCC---CCG

                      _________________________
C. max   UGGAAGGGAxxxxUGCAUUUAUUAGAUxAAAAxxxxxxxCGUC
soybean  ---------xxxx--------------x----xxxxxxx----
maize    C--G----GxxxxC-------------x----xxxxxxx--CU
rat      ---GG----CGCG----------C----C----CCAACCC----

C. max   GxxxxxxxxxxxAC   (18)
soybean  Axxxxxxxxxxx--   (15)
maize    -xxxxxxxxxxx--   (16)
rat      AGCCCCCUCCGG-    (17)
```

The observation that bovine as well as plant (tobacco and _C. maxima_) RNAs become encapsidated when reacted with TMV capsid protein indicates that the segment of _C. maxima_ 18S RNA, identified to contain the ei, might be similar to that in a number of other eucaryotes and also that this segment might have at least some features in common with known virion eis. Upon comparison of the 68 base _C. maxima_ region with available sequences of the same region of some other organisms (fig 1), it is found that the region is well conserved among dicotyledonous plants; the soybean sequence differs by only a single nucleotide from that of _C. maxima_. It is not one of the more rigidly conserved 18S RNA regions, however, because the *C. maxima* sequence differs from that of a monocot, maize, at 6 nucleotide positions and from a mammal, rat, at 13 comparable positions. Moreover, the rat region contains 31 additional nucleotides interspersed among the basic 68 present in the plant region. The rat sequence is taken as being representative for mammals because to our knowledge, bovine data are not available and the rat sequence is more like that of another mammal, rabbit (20), than that of other organisms. There are only two segments (overlined in figure 1) that are almost the same in the plant and rat regions and these are also conserved in the lower eucaryote, yeast, and in an amphibian, Xenopus (15). It is questionable, however, whether one or both of these segments imbues this region with its property of being an ei because there is no obvious sequence similarity between the segments in question and any segment of known virion ei regions. Perhaps some other feature is responsible for the ei property of this region but it is not immediately obvious what this might be. Although the *C. maxima* region can be folded to form a stem loop structure, the rat region folds into quite a different minimum energy secondary structure.

PROSPECTS

The essential features that imbue a segment of RNA with the capacity to react with a TMV capsid protein oligomer to initiate virion assembly remain largely unknown although this problem has been the subject of considerable speculation (e.g. 18). Fortunately, tools are now at hand that permit a rational experimental approach to the problem and answers should soon be forthcoming.

The biological significance, if any, of the presence of ei sites on non-virion RNA species remains speculative. It is possible that regions of RNA that are interactive with particular proteins or protein complexes in a specific manner have certain features in common and that, perhaps, the virion ei is derived from one such region. One can also imagine that capsid proteins may have structural features in common with one or more small sub-unit ribosomal proteins so that both can recognize the same RNA site to initiate coaggregation.

Because pseudovirions are present in all TMV preparations and these contain chloroplast DNA transcripts, a reasonable hypothesis can be generated that the in vivo encapsidation of these transcripts in young leaves might contribute to the classical light green-dark green mosaic symptom for which TMV is named. This does not appear to be so, however, because preparations of the masked strain (so named because it invades plants systematically without obvious symptom formation) have about the same pseudovirion content as preparations of the closely related, symptom inducing, common strain (14).

REFERENCES

1. Lomonossoff G, Wilson TMA (1985). Structure and in vitro assembly of tobacco mosaic virus. In Davis JW (ed): "Molecular Plant Virology, vol 1," Boca Raton, FL: CRC Press, p43.
2. Fraenkel-Conrat H, Williams R (1955). Reconstitution of active tobacco mosaic virus from its inactive protein and nucleic acid components. Proc Natl Acad Sci 41:630.
3. Fukuda M, Okada Y (1985). Elongation in the major direction of tobacco mosaic virus assembly. Proc Natl Acad 82:3631.
4. Correia JJ, Shire S, Yphantis DA, Schuster TM (1985). Sedimentation equilibrium measurements of the intermediate size tobacco mosaic virus protein polymers. Biochemistry 24:3292.
5. Namba K, Stubbs G (1986). Structure of tobacco mosaic virus at 3.6 A resolution: implication for assembly. Science 231:1401.
6. Zimmern D (1977). The nucleotide sequence at the origin for assembly on tobacco mosaic virus RNA. Cell 11:463.
7. Meshi T, Ohno T, Iba H, Okada Y (1981). Nucleotide sequence of a cloned cDNA copy of TMV (cowpea strain) including the assembly origin, the coat protein cistron and the 3' non-coding region. Mol Gen Genetics 184:20.
8. Takamatsu N, Ohno T, Meshi T, Okada Y (1983). Molecular cloning and nucleotide sequence of the 30K and the coat protein cistron of TMV (tomato strain) genome. Nucleic Acids Res 11:3767.

9. Guilley H, Jonard G, Richard KE, Hirth L (1975). Sequence of a specifically encapsidated RNA fragment originating from tobacco mosaic virus coat protein cistron. Eur J Biochem 54:135.
10. Siegel A (1971). Pseudovirions of tobacco mosaic virus. Virology 46:50.
11. Rochon D, Siegel A (1984). Chloroplast DNA transcripts are encapsidated by tobacco mosaic virus coat protein. Proc Natl Acad Sci 81:1719.
12. Shalla TA, Petersen LJ, Giunchedi L (1975). Partial characterization of virus-like particles in chloroplasts of plants infected with the U5 strain of TMV. Virology 66:94.
13. Rochon D, Siegel A (1985). TMV coat protein encapsidates specific species of host RNA both in vivo and in vitro. In Key JL, Kosuge T (eds): "Cellular and Molecular Biology of Plant Stress," New York: Alan R Liss, p 435.
14. Rochon D, Kelly R, Siegel A (1986). Encapsidation of 18S rRNA by tobacco mosaic virus coat protein. Virology 150:140.
15. Eckenrode VK, Arnold J, Meagher R (1985). Comparison of the nucleotide sequence of soybean 18S rRNA with the sequences of other small-subunit rRNAs. J Mol Evol 21:253.
16. Messing J, Carlson J, Hagen G, Rubenstein I, Oleson A (1984). Cloning and sequencing of the ribosomal RNA genes in maize: the 17S region. DNA 3:31.
17. Torczynsdki R, Bollon AP, Fuke M (1983). The complete nucleotide sequence of the rat 18S ribosomal RNA gene and its comparison with the respective yeast and frog genes. Nucleic Acid Res 11:4873.
18. Atreya CD, Rochon D, Terami F, Siegel A (unpublished data).
19. Zimmern D (1983). An extended secondary structure model for the tobacco mosaic virus assembly origin and its correlation with protection studies and an assembly defective mutant. EMBO J 11:1901.
20. Connaughton JF, Rairkar A, Lockard RE, Kumar A (1984). Primary structure of rabbit 18S ribosomal RNA determined by direct RNA sequence analysis. Nucleic Acids Res 12:4731.

VII. PATHOGENESIS AND VIRULENCE

Positive Strand RNA Viruses, pages 409–422

MOLECULAR DETERMINANTS OF CNS VIRULENCE OF THE CORONAVIRUS MOUSE HEPATITIS VIRUS-4[1]

Michael J. Buchmeier, Robert G. Dalziel, Marck J.M. Koolen and Peter W. Lampert[2]

Scripps Clinic and Research Foundation
La Jolla, California 92037

Mice infected intracerebrally with wild type MHV-4 normally succumb within 7 days due to an acute encephalomyelitis. In the small fraction of animals surviving this initial acute infection demyelination is evident in the brain and spinal cord white matter. Our laboratory has attempted to define the factor(s) which determine neurotropism and limit the spread of infection in vivo in the CNS. We have used monoclonal antibodies (MAb) of defined specificity in a passive transfer model to define viral proteins which are important determinants of neurovirulence. Using this approach we have shown that MAb against two topographically distinct sites on the E2 peplomer glycoprotein of MHV-4 are able to block development of fatal encephalitis following ic infection with wild type virus. Although neurons were spared, oligodendrocytes were infected and demyelination resulted. Having shown that sites on E2 were involved, we selected variant strains of MHV-4 which contained mutations in E2 and were resistant to E2 directed neutralizing MAb. These mutant strains of virus were attenuated by a factor of 2-3 logs in LD_{50}, but continued to induce chronic demyelinating disease characterized by mild inflammation, primary demyelination, remyelination and recurrent demyelination over a prolonged observation period. Thus we have directly demonstrated that the E2 glycoprotein is an important determinant of neurovirulence in MHV-4 infection. Ongoing studies are directed toward identifying the nature of the lesions in the variant viruses and their effect on tropism and virulence.

[1]This work was supported by USPHS grants NS12428 and AI16102. M.J.M. Koolen is the recipient of a long-term fellowship from EMBO.
[2]University of California, San Diego, CA 92093

INTRODUCTION

It is well established that viruses are etiologically responsible for at least two chronic progressive CNS diseases in man. Subacute sclerosing panencephalitis (SSPE) is a progressive degenerative disorder caused by persistent infection of neurons by measles virus (1). SSPE is a rare, late complication of childhood measles, and the time span between the initial attack and the onset of SSPE may be a decade or longer. Details of the host virus interaction which lead to SSPE and which occur during the long silent phase of infection are largely unknown.

Progressive multifocal leukoencephalopathy (PML) is a second example of a virus induced chronic progressive CNS disease (2). The agent of PML is a papovavirus similar to the well described SV-40. Infection with the PML agent is common, but leads only in some individuals to chronic progressive gray matter disease which, like SSPE, is invariably fatal. As in SSPE, the factors which distinguish the self-limiting infection (which occurs in most instances) from the lethal disease are poorly understood.

A number of factors implicate viruses as possible etiologic agents of human demyelinating diseases. Among these was the observation by Norrby and colleagues (3) of intrathecally produced oligoclonal IgG specific for measles and subsequently other virus in the CSF of MS patients. Epidemiologic association of MS occurrence with outbreaks of both measles and the related canine distemper viruses (4) further implicated these agents, but definitive etiology has not been established. Difficulties in establishing causal relationships between viruses and human demyelinating disease as well as the need to precisely define the factor(s) which determine the course of CNS infections have led us and others to focus on well established animal models of CNS disease (5). Using such models we are able to control factors such as the age, genetic background, and immune status of the host as well as the strain, dose and route of the infecting virus. Detailed study of the function of specific macromolecular components and genes of viruses and their interactions with host leading to the ultimate production of disease is yielding insight about the molecular basis of viral CNS pathogenesis.

Murine hepatitis virus-4 (MHV-4 = JHMV) is a neurotropic member of the Coronaviridae; a group which also includes a number of human common cold viruses (6,7). In susceptible, non-immune mice the virus causes a fatal encephalitis accompanied by demyelination (8). Intracerebral inoculation of MHV-4 results in widespread destruction of CNS neurons, and resistance to this lethal disease is controlled by a single autosomal gene on chromosome 7 which is expressed at the level of neurons and macrophages (9).

Mice which survive the initial acute encephalitis undergo chronic demyelination. Attenuation of the initial encephalitis has been achieved using ts mutants and spontaneously arising variants (10,11,12,13) of MHV-4. In contrast to the wild type infection, infection with these attenuated viruses results in a high incidence of demyelinating disease without encephalitis. The best characterized of these mutants is the ts 8 mutant of Haspel et al. (10). In ts 8 infection demyelination is a direct result of infection of the oligodendrocyte, the cell responsible for elaboration and maintenance of CNS myelin, and is thus referred to as a primary demyelination. Similar evidence of a shift from acute fatal disease to subacute demyelinating disease has been reported for an MHV-4 variant termed ds (11) and for the ts 342 mutant of MHV strain A59 (13). In the latter model, revertants of the ts 342 virus exhibiting wild type pathogenic properties were isolated indicating that the same mutation responsible for ts phenotype was also responsible for reduced pathogenicity.

MHV-4 contains 3 major classes of structural proteins including a 56-60 kd nucleocapsid protein, N, and two glycoproteins, E1 and E2 (14). Glycoprotein E1 is deeply embedded in the membrane and exists as both a non-glycosylated 23 kd and an o-glycosylated 25 kd form (15). The 180 kd E2 glycoprotein forms the peplomer, or surface spike, of the virion and is composed of two nonidentical 90 kd subunits in a dimer (16). E2 also is responsible for the biological activities of neutralization by antibody, cell attachment and cell-cell fusion (17). In this laboratory we have raised a library of monoclonal antibodies against the structural proteins of MHV-4 in order to study the function of these molecules, and the influences of specific immune responses against them on development and course of acute and chronic CNS disease (18). Using this approach we have shown that antibody responses to specific determinants on the E2 glycoprotein were sufficient to change the course of fatal encephalitis to one of chronic demyelination.

RESULTS AND DISCUSSION

Monoclonal antibodies (MAb) mapping to three topographically distinct neutralizing determinants on MHV-4 E2 have been identified (19). Two of these, MAb 5B19.2 (epitope E2A) and 5A13.5 (E2B) have been shown previously to protect mice, in a passive transfer regimen, from lethal encephalitis (18). The third MAb, 4B11.6 (E2C) neutralized virus in vitro but did not protect in vivo. These properties are summarized in Table 1.

TABLE 1

Properties of Anti-E2 Monoclonal Antibodies

Monoclonal Antibody	Epitope	Neutralization in vitro[1]	Protection in vivo[2]
5B19.2	A	+	+
5B170.3	A	+	+
5A13.5	B	+	+
4B11.6	C	+	-
5B21.5	D	-	-
5B93.9	D	-	-
5B207.7	E	-	-
5B216.8	E	-	-

[1] PRD_{50} titer + = >1/1000 - = <1/20

[2] Mice were given 150 ul of ascites fluid containing the indicated antibody then challenged 1 day later with 100 LD_{50} of MHV-4. Protected mice (+) survived indefinitely while unprotected mice (-) died within 7 days after virus challenge.

Variants of MHV-4 were selected from 3x cloned virus stocks by incubating with an excess of neutralizing MAb for 30 min at 37°, then plaquing in the presence of the antibody. Plaques escaping neutralization were picked and subjected to a second round of neutralization; stocks were then grown and re-checked for resistance to antibody. We observed a frequency of true variants in the initial population of $10^{-4.3}$ to $10^{-4.6}$, a rate consistent with other RNA virus systems (20,21,22). Six variants selected using MAb 5A13.5 and 4B11.6 were chosen for detailed study. Surprising was the observation that variants selected with either MAb resisted neutralization by both (Table 2). Other studies (23) have shown that both epitopes E2B and E2C are conformational in nature, hence it appears that selection of mutations affecting either epitope was reflected by conformational changes at both.

TABLE 2

Neutralization and Virulence of Antibody Resistent Variants of MHV-4

	PRD_{50} Titer of Antibody[1]		
Virus	5B19.2	5A13.5	4B11.6
wt MHV-4	7,943	31,600	15,800
V 5A13.1	12,600	251	36
V 5A13.2	17,800	200	90
V 5A13.3	11,200	400	80
V 4B11.1	6,300	125	36
V 4B11.2	7,943	158	40
V 4B11.3	7,943	141	<20

[1] PRD_{50} endpoint titer of antibody measured against specific wild type or variant virus.

Having selected these variants we were interested in determining whether the antigenic changes resulted in altered virulence. We first determined LD_{50} values for representative variants (Table 3) and observed a 200 to >4000 fold decrease in neurovirulence relative to the wild type parental virus.

TABLE 3

LD_{50} of Wild Type and Variant MHV-4 Strains

Virus	LD50 (PFU)	Attenuation Factor
wild type	<0.45	1
V 5A13.1	>1800	>4000
V 4B11.3	95	211

Mice were sacrificed by cardiac perfusion for histopathologic examination at intervals ranging from 4 to 65 days after infection with the variants in order to determine their in vivo pathogenic properties. Mice infected by either V5A13.1 or V4B11.3 had moderate inflammatory lesions in the white matter of their spinal cords and brains early in the infection (days 4-15) and developed extensive demyelinated lesions (Figure 1). Evidence of re-myelination, indicative of repair, was observed (Figure 2) but demyelination apparently continued since lesions in day 50 and 65 spinal cords were severe and showed evidence of both recent and old foci (Figures 3,4). No significant evidence of involvement of neurons was found, indicating that the E2 variants, like the ts 8 virus (10), had lost the property of neurovirulence.

Virus could be re-isolated by cocultivation from the brains of these animals only as late as 15 days after infection (Table 4), but the recovered virus retained the variant phenotype. Thus the system presents an interesting paradox: That virus-induced pathology persisted in the absence of demonstrable infectious virus. The likelihood exists that stimulation of specific cell mediated immune responses against viral antigen and/or host components of myelin such as myelin basic protein (MBP) are triggered early in the infectious process and contribute to the chronic demyelinating disease. Alternatively, virus may persist in limited cell populations such as the basal ganglia (24) and continually re-initiate the demyelinating process at the primary level. We have addressed the first of these alternatives. In recent studies with the ts 342 strain of MHV-A59 (13) we have observed significant levels of proliferation, measured as ^{3}H thymidine incorporation, when lymphocytes from virus infected mice were cultured in the presence of either viral antigen or MBP.

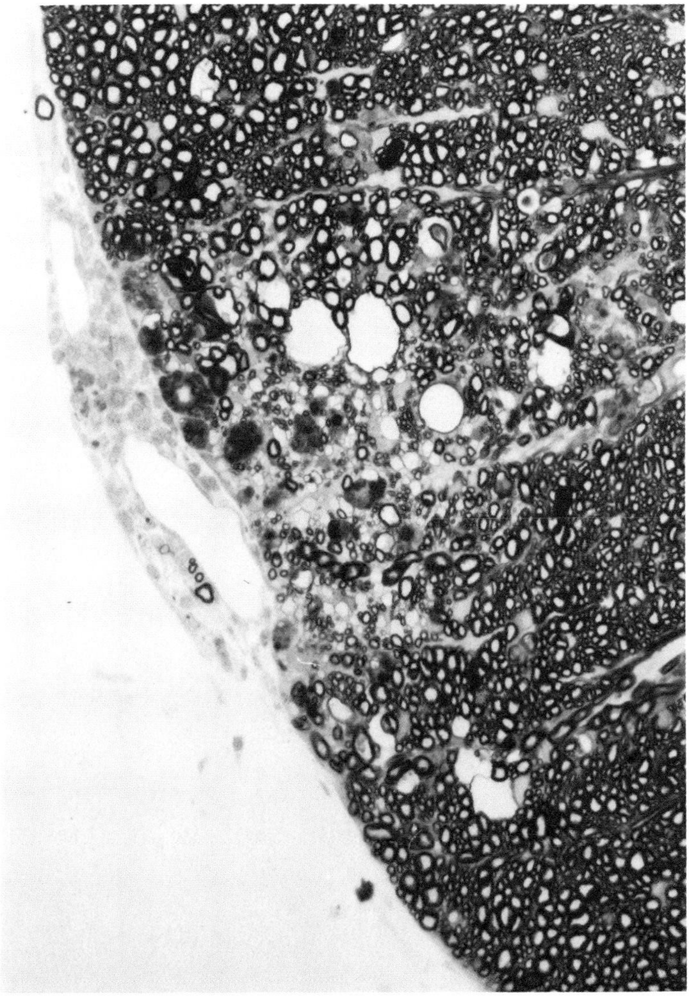

FIGURE 1. Focal area of demyelination adjacent to an inflammatory infiltrate in the spinal cord of a mouse infected 15 days earlier with variant V 5A13.1. Resin embedded, paraphenylenediamine stain. x500.

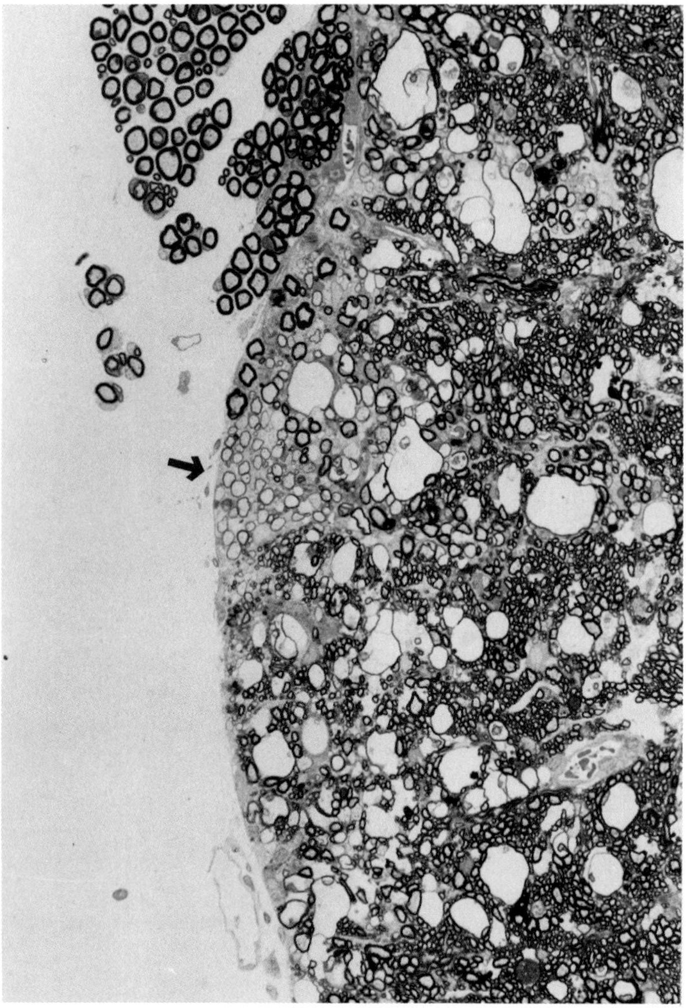

FIGURE 2. Remyelinated axons (arrow) next to myelinated axons showing vacuolated myelin in the spinal cord of a mouse 32 days after infection. x500.

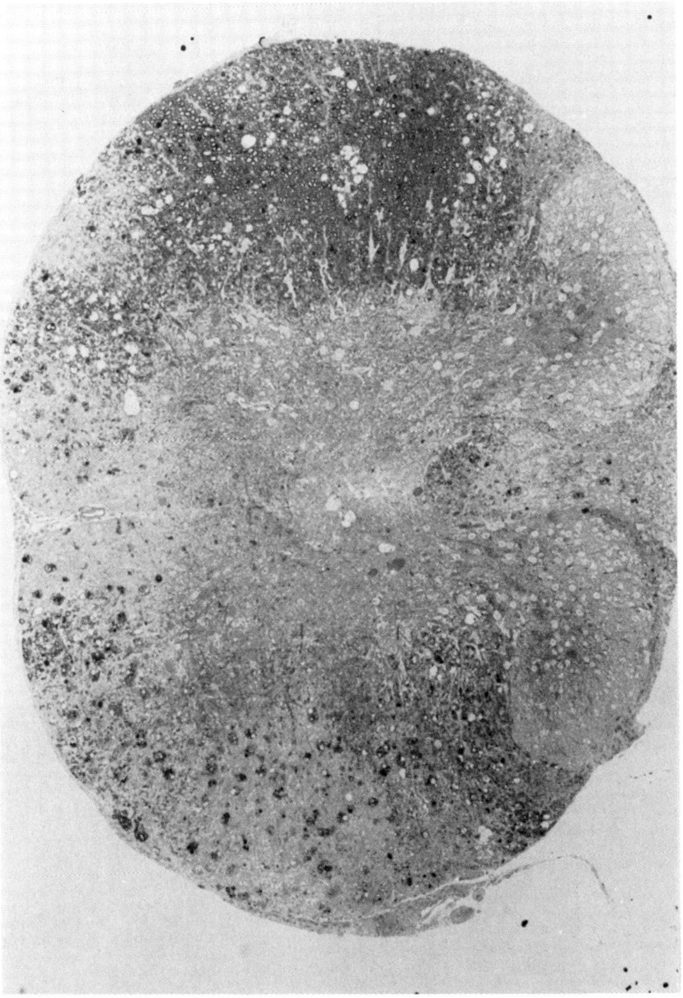

FIGURE 3. Extensive plaques of demyelination and remyelination in the spinal cord of a mouse 50 days after infection. x50.

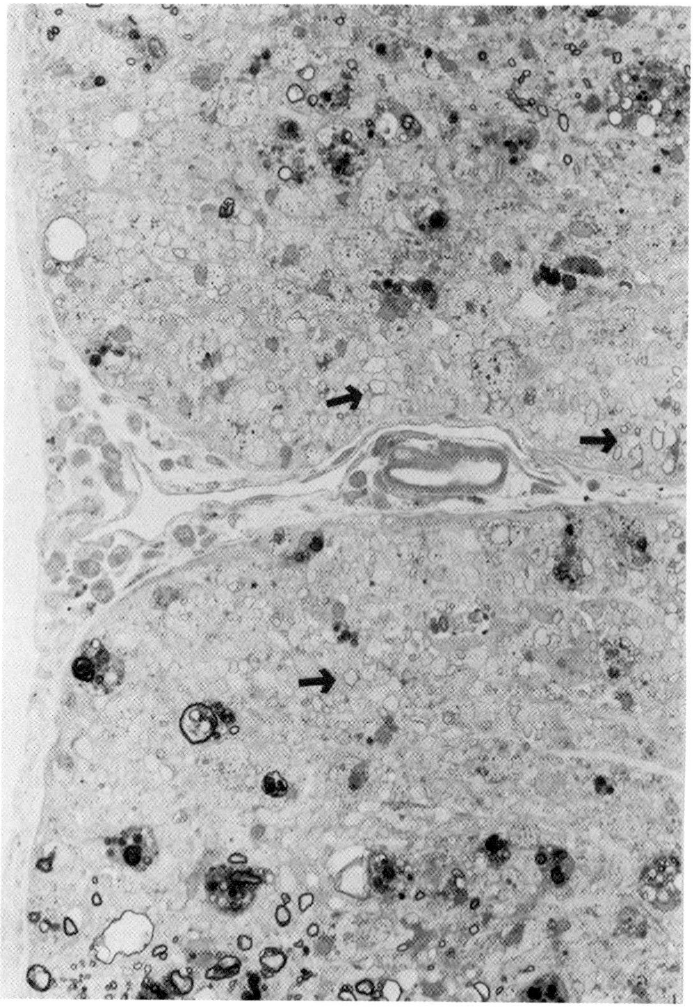

FIGURE 4. Higher magnification of an area from Figure 3 showing abundant demyelinated and a few remyelinated axons. x500.

The responder cells were shown to be T-lymphocytes, and depletion of the L3T4 population reduced in vivo the proliferative response to MBP to baseline levels. Further studies are now in progress to characterize these responses at the cellular level, to clone the responding cells, and to determine the fate and expression of our attenuated variants of MHV-4 in the CNS during acute and chronic demyelinating infections.

TABLE 4

Virus Recovery from Tissues Following Wild Type and Variant Infections

Day	Wild Type		V 4B11.3		V 5A13.1	
	Brain	Liver	Brain	Liver	Brain	Liver
2	$5x10^7$	$5x10^2$	$2.5x10^3$	<40[2]	$3.3x10^4$	<40
			$2.5x10^3$	$2x10^2$	$3x10^4$	$4x10^2$
4	$1x10^6$	$1x10^4$	$3.5x10^4$	<40	$2.5x10^4$	$1x10^5$
			$3x10^3$	<40	$4x10^3$	<40
7	$4x10^5$	$3x10^4$	<40	<40	<40	<40
			<40	<40	<40	<40
15	NA[3]	NA	+cocult			

[1] Titer of virus in PFU/gram of tissue.

[2] Limit of sensitivity.

[3] Not available, wild type infected mice normally die prior to this time point.

These studies collectively illustrate that mutational changes in a specific viral glycoprotein, E2, are reflected in substantially altered pathogenesis in vivo. It is hoped that definition of such changes at the molecular level and their effect on cellular tropism and replication in the CNS will contribute to our understanding of the pathogenesis of CNS disease in man.

ACKNOWLEDGEMENTS

We thank Kaleo Wooddell, Leslie Igarashi, and Hanna Lewicki for expert technical assistance and Jim Johnston for manuscript preparation. This is publication number 4390-IMM from the Department of Immunology, Scripps Clinic and Research Foundation.

REFERENCES

1. ter Meulen, V. and Hall, W. (1978). Slow virus infections of the nervous system: Virological, immunological and pathogenic considerations. J. Gen. Virol. 41:1-25.

2. Weiner, L.P., Herndon, R., Narayan, O., Johnson, R., Shah, K., Rubinstein, L., Prezoisi, T. and Conley, F. (1972). Isolation of a virus related to SV-40 from patients with progressive multifocal leukoencephalopathy. N. Engl. J. Med. 286:385-390.

3. Norrby, E. (1978). Viral antibodies in multiple sclerosis. Prog. in Medical Virol. 24:1-39.

4. Stroop, W. and Baringer, J. (1982) Persistent, slow and latent viral infections. Prog. in Medical Virol. 28:1-43.

5. Koolen, M. and Buchmeier, M. (1986). Experimental models of virus attenuation and demyelinating disease. Microbiol. Sciences 3:68-73.

6. Weiner, L. (1973). Pathogenesis of demyelination induced by a mouse hepatitis virus (JHM virus). Arch. of Neurol. 28:298-303.

7. Sturman, L. and Holmes, K. The molecular biology of coronaviruses. Adv. in Virus Res. 28:33-112.

8. Lampert, P., Simms, J. and Kniazeff, A. (1973). Mechanism of demyelination in JHM virus encephalomyelitis. Electron Microscopic Studies. Acta. Neuropathol. 24:76-85.

9. Knobler, R., Haspel, M. and Oldstone, M. (1981). Mouse hepatitis virus type-4 (JHM strain) - Induced fatal central nervous system disease. I. Genetic control and the murine neuron as the susceptible site of disease. J. Exp. Med. 153:832-843.

10. Haspel, M., Lampert, P. and Oldstone, M. (1978). Temperature

sensitive mutants of mouse hepatitis virus produce a high incidence of demyelination. Proc. Nat. Acad. Sci. USA 75:4033-4036.

11. Stohlman, S., Brayton, P., Fleming, J., Weiner, L. and Lai, M. (1982) Murine coronaviruses: Isolation and characterization of two plaque morphology variants of JHM neurotropic strain. J. Gen. Virol. 63:265-275.

12. Wege, H., Koga, M., Wege, H., and ter Meulen, V. (1981). JHM infection in rats as a model for acute and subacute demyelinating disease. In Biochemistry and Biology of Coronaviruses, V. ter Meulen, S. Siddell, and H. Wege, eds. Plenum, NY, p. 327-340.

13. Koolen, M., Osterhaus, A., van Steenis, G., Horzinek, M., and van der Zeijst, B. (1983). Temperature sensitive mutants of mouse hepatitis virus strain A59: Isolation, characterization and neuropathogenic properties. Virology 125:393-402.

14. Sturman, L., Holmes, K., and Behnke, J. (1980). Isolation of coronavirus envelope glycoproteins and interaction with the viral nucleocapsid. J. Virol. 33:449-462.

15. Niemann, H. and Klenk, H. (1981). Coronavirus glycoprotein E1, a new type of viral glycoprotein. J. Mol. Biol. 153:993-1010.

16. Sturman, L., Ricard, C. and Holmes, K. (1985). Proteolytic cleavage of the E2 glycoprotein of murine coronavirus: Activation of cell fusing activity of virions by trypsin and separation of two different 90k cleavage fragments. J. Virol. 56:904-911.

17. Collins, A., Knobler, R., Powell, H. and Buchmeier, M. (1982). Monoclonal antibodies to murine hepatitis virus-4 (strain JHM) define the viral glycoprotein responsible for attachment and cell-cell fusion. Virology 119:358-371.

18. Buchmeier, M., Lewicki, H., Talbot, P. and Knobler, R. (1984). Murine hepatitis virus-4 (strain JHM) induced neurologic disease is modulated in vivo by monoclonal antibody. Virology 132:261-270.

19. Talbot, P. Salmi, A., Knobler, R. and Buchmeier, M. (1984). Topographical mapping of epitopes on the glycoproteins of murine hepatitis virus-4 (strain JHM): Correlation with biological activities. Virology 132:250-260.

20. Laver, W., Air, G., Webster, R., Gerhard, W., Ward, C. and Dopheide, T. (1979) Antigenic drift in type A influenza virus: Sequence differences in the hemagglutinin of Hong Kong (H3N2) variants selected with monoclonal hybridoma antibodies. Virology 98:226-237.

21. Dietzschold, B., Wimmer, W., Wiktor, T., Lopes, A., Lafon, M., Smith, C. and Koprowski, H. (1983). Characterization of an antigenic determinant of the glycoprotein that correlates with pathogenicity of rabies virus. Proc. Natl. Acad. Sci. USA 80:70-74.

22. Minor, P., Schild, G., Bootman, J., Evans, D., Ferguson, M., Reeve, P., Spitz, M., Stanway, G., Cann, A., Hauptmann, R., Clarke, L., Mountford, R. and Almond, J. (1983). Location and primary structure of a major antigenic site for poliovirus neutralization. Nature 301:674-679.

23. Talbot, P., Knobler, R. and Buchmeier, M. (1984). Western and dot immunoblotting analysis of viral antigens and antibodies: Application to murine hepatitis virus. J. Immunol. Methods 73:177-188.

24. Fishman, P., Gass, J., Swoveland, P., Lavi, E., Highkin, M. and Weiss, S. (1985). Infection of the basal ganglia by a murine coronavirus. Science 229:877-879.

Positive Strand RNA Viruses, pages 423–436

RESPONSES OF PLANT CELLS TO VIRUS INFECTION WITH SPECIAL REFERENCE TO THE SITES OF RNA REPLICATION

R.I.B. Francki

Department of Plant Pathology, Waite Agricultural Research Institute, The University of Adelaide, Glen Osmond, South Australia, 5064.

ABSTRACT Responses of plants to virus infection at the whole plant, tissue and cellular levels are many and varied. One of the cellular responses which appears to be directly involved in virus replication is the development of numerous vesicles associated with cellular membrane systems. The vesicles can be associated with membranes of the endoplasmic reticulum, nuclei, chloroplasts, peroxisomes, mitochondria and tonoplasts. They usually contain fine fibrils which have been identified as ds-RNA. Evidence that the vesicles are the sites of viral RNA replication is discussed.

INTRODUCTION

The majority of viruses infecting plants (23 out of the 28 recognised taxonomic groups) have positive stranded RNA genomes. Most of these exert effects on the growth and development of their host plants resulting in a variety of disease symptoms (1-3). In recent years, very significant advances have been made in our understanding of the molecular events leading to virus synthesis. A voluminous literature has also accrued concerning the effects of virus infection on their hosts at the whole plant, tissue and cellular levels (2,4). However, most of these studies have been almost entirely descriptive and have contributed little to an understanding of how virus-infection leads to disease symptom development.

It is well established that the symptoms expressed by a plant following virus infection is the result of the interaction between the viral and host genomes modulated by the environment. This is a very complex interaction involving molecular events which affect the physiology as well as the morphology of the plants. Moreover, little is known as to which effects of infection are actually involved in virus synthesis and which are indirect consequences of infection. In this paper I wish to confine myself largely to discussing one cytopathic response to virus infection. The structures most consistently seen in virus infected cells, except for the virus particles themselves, are the numerous small vesicles which develop in association with different organelles depending on the identity of the infecting virus. I will examine the available evidence that the vesicles are the sites of RNA replication.

VIRUS-INDUCED VESICLES IN PLANT CELLS

Vesicle Development in the Cytoplasm.

Infection by many viruses results in the appearance of numerous small vesicles in the cytoplasm of plant cells (4, 5). In Comovirus and Nepovirus-infected cells which show very similar cytopathic effects, the vesicles usually aggregate together with some ribosomes and the endoplasmic reticulum into large vesicular inclusions (Fig. 1). They are usually located between, but separate from the nucleus, chloroplasts and mitochondria, and are not bounded by a membrane. The origin of the vesicles is not clear and it has been suggested that they are derived from membranes of the dictyosomes, endoplasmic reticulum or the outer nuclear lamella (5). Many of the vesicles contain fibrils which in the case of broad bean true mosaic virus, have been shown to consist of ds-RNA by enzyme cytochemical tests (6).

The most convincing evidence that the vesicular inclusions are involved in virus synthesis comes from the work with cowpea mosaic virus (CPMV) by van Kammen and his group (7). Results of organelle fractionation and autoradiographic experiments showed that the inclusions contained virus-specific ds-RNA and replicase bound to its endogenous template (8-10). It has also been demonstrated that in protoplasts infected only with CPMV B component (ie. without the M component which carries the genes for coat protein), B RNA is expressed and replicated, and

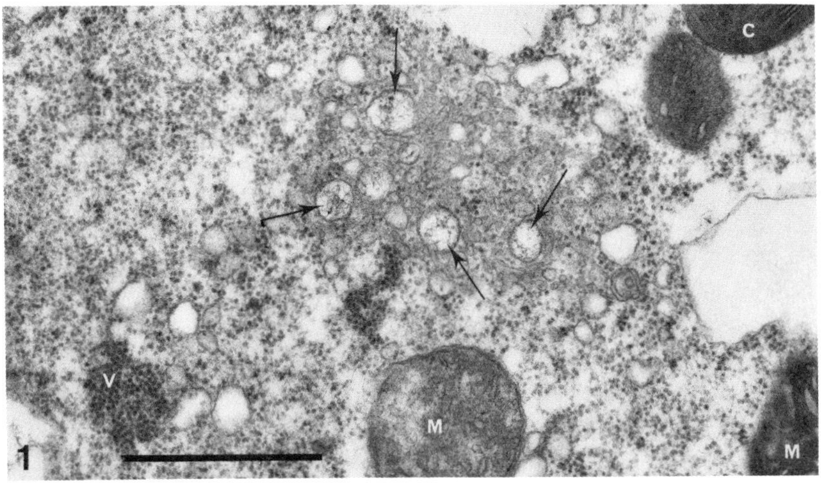

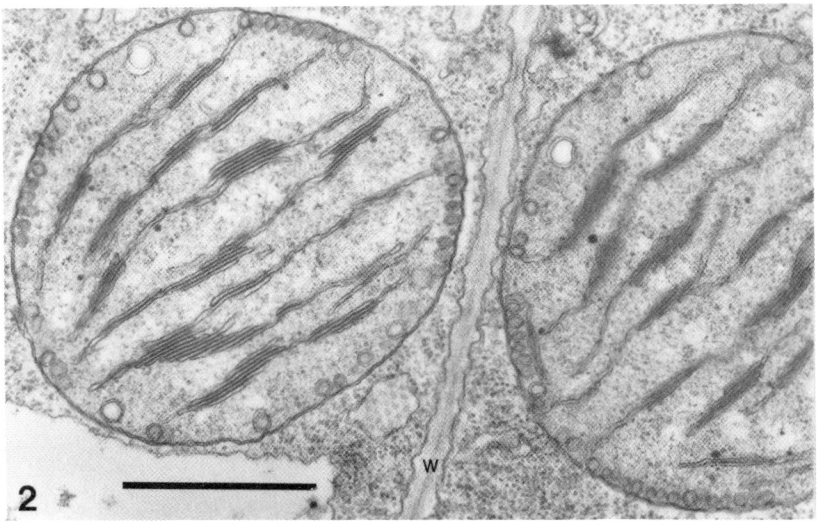

FIGURE 1. A cucumber leaf cell infected with tobacco ringspot virus (Nepovirus) showing a virus-induced cytoplasmic inclusion with vesicles containing electron-dense fibrils (arrows).

FIGURE 2. A Chinese cabbage leaf cell infected with turnip yellow mosaic virus (Tymovirus) showing vesiculation of the chloroplasts. (C = chloroplast; M = mitochondrion; V = virus particles; W = cell wall; Bar = 1μm).

normal vesicular inclusions develop (11, 12). This indicates that the vesicular inclusions are sites of viral RNA replication (7).

Somewhat similar cytoplasmic vesicles, although not usually organized into inclusions, have also been observed in cells infected by some Closteroviruses, Potyviruses and Dianthoviruses (4), and to a lesser extent some Ilarviruses and alfalfa mosaic virus (13). However, little effort has been made to study them in any detail although it has been suggested that they may be the sites of viral RNA synthesis.

Vesicles Associated with Nuclei.

Vesicles with fine fibrils have been seen both in the perinuclear spaces and cytoplasm of cells infected by some Luteoviruses, Bromoviruses and pea enation mosaic virus (PEMV) (4, 5, 13). In PEMV-infected cells, the vesicles are located predominantly in the perinuclear spaces having apparently developed from the inner nuclear membrane (14). The vesicles seen in the cytoplasm probably also originate from the nucleus, having been released into the perinuclear space and thence into the cytoplasm via the endoplasmic reticulum. Autoradiographic and biochemical experiments support the conclusion that the virus-specific ds-RNA and RNA-dependent RNA polymerase are associated with the nuclei of infected cells (15, 16). Furthermore, it was demonstrated that PEMV RNA could undergo the initial stages of replication in isolated pea nuclei (16).

Perinuclear and cytoplasmic vesicles have been observed in cells infected by several Luteoviruses, but detailed observations have only been reported on plants infected with beet western yellows virus (BWYV), barley yellow dwarf virus (BYDV) and potato leafroll virus (PLRV) (4, 5).

Luteoviruses are transmitted only by their aphid vectors and they are confined to phloem tissues which makes them difficult to work with. However, observations have been made on successively older leaves of BWYV-infected plants in an endeavour to examine cells at different stages of infection (17). Virus-induced vesicles were seen first in the cytoplasm and then in the perinuclear spaces. This was interpreted, rather boldly perhaps, to indicate that the vesicles develop from the endoplasmic reticulum and are then transported to the nuclei where virus synthesis takes place. It was concluded that a similar sequence of events took place in PLRV-infected cells (18).

Isolated protoplasts infected with two Bromoviruses, cowpea chlorotic mottle and brome mosaic viruses, were examined at various times after inoculation but there was no clear difference in the time of appearance of the vesicles in the cytoplasm and perinuclear spaces (19). However, the authors gained the impression that the structures were not derived from the lumen of the endoplasmic reticulum as suggested for BWYV (17), but rather that they arose by a budding process involving both membranes of the nuclear membrane.

Some perinuclear and cytoplasmic vesicles containing fibrils have been observed in cells infected with several definite and tentative members of the Sobemovirus group (20). However, much larger aggregates of fibrils were also seen in the cytoplasm free of any bounding membranes. In organelle fractionation experiments with velvet tobacco mottle virus-infected plants, it was shown that both the virus-specific ds-RNA and RNA-dependent RNA polymerase was detected almost exclusively in the membrane-free cytoplasmic fraction (20). This indicates that the viral RNA may be synthesized in the ground cytoplasm and raises the question of the significance of the cytoplasmic and perinuclear vesicles.

Chloroplast Vesicles.

The development of vesicles on the chloroplasts has been studied extensively in cells of plants infected with turnip yellow mosaic virus (TYMV; Fig 2). However, very similar structural changes appear to take place in cells infected with all other Tymoviruses which have been examined (4, 21). Detailed studies have established that the TYMV-induced chloroplast vesicles appear at an early stage of infection and the sequence of events in their development has been followed in detail (22, 23). The vesicles are formed as small invaginations of both chloroplast membranes with their necks remaining open to the cytoplasm (Fig 2). Observations on freeze-fractioned specimens indicate that there are no proteins within the vesicle membranes (24). Inside the vesicles, fibrillar osmiophilic material with the expected appearance of nucleic acid is often observed (22).

The vesicles are scattered singly or in groups over the surfaces of chloroplasts. At early stages of infection, endoplasmic reticulum can usually be seen near the vesicles but disappears with time to be replaced with what appear to be coat protein subunits. Later still, numerous virus particles can be observed between the chloroplasts which are by now swollen and clustered together (23, 25). This sequence of events suggests that the chloroplast vesicles are the sites of viral RNA replication and that the RNA is released into the cytoplasm to be encapsidated by viral protein synthesized in the cytoplasm.

Results of biochemical and autoradiographic experiments support the view that TYMV-induced chloroplast vesicles are the sites of RNA synthesis. RNA-dependent RNA synthesis was shown to be associated with a fraction containing the chloroplast bounding membranes with virus induced vesicles (26).

Some chloroplast vesicles have also been observed in plant cells infected with the Hordeivirus, barley stripe mosaic virus and galinsoga mosaic virus (GaMV). However, they have not been studied in detail (27-29). The significance of the vesicles in GaMV-infected cells is especially interesting because this virus also induces numerous mitochondrial vesicles which are discussed later.

Multivesicular bodies originating from peroxisomes.

Large multivesicular bodies appear to be characteristic of infection by most Tombusviruses (4, 5). When fully developed, they can be as large as chloroplasts, consisting of an outer membrane enclosing electron-dense, finely granular or crystalline matrix and many globose vesicles containing fibrilar material (Fig 3).

At first it was thought that multivesicular bodies are derived from membranes of the endoplasmic reticulum and dictyosomes (30) and later that they originated from, or were associated with the chloroplasts (31). However, there is now reliable evidence that multivesicular bodies in Tombusvirus-infected cells develop from peroxisomes (4,5). It has been demonstrated by cytochemical studies that in cells infected by several Tombusviruses, the multivesicular bodies contain catalase and glycolate oxidase, two enzymes usually associated with peroxisomes (32, 33). It was concluded that the multivesicular bodies are involved in virus replication because: (a) they do not occur in healthy plants or cells of infected plants which are not invaded by virus; (b) their vesicles contain ds-RNA which is absent in the peroxisomes of healthy plants; and (c) they appear in advance of virus particles (32).

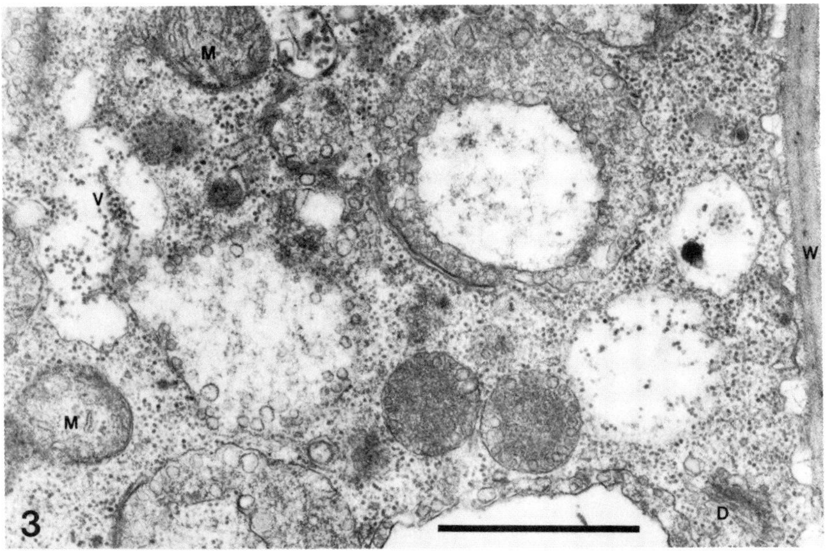

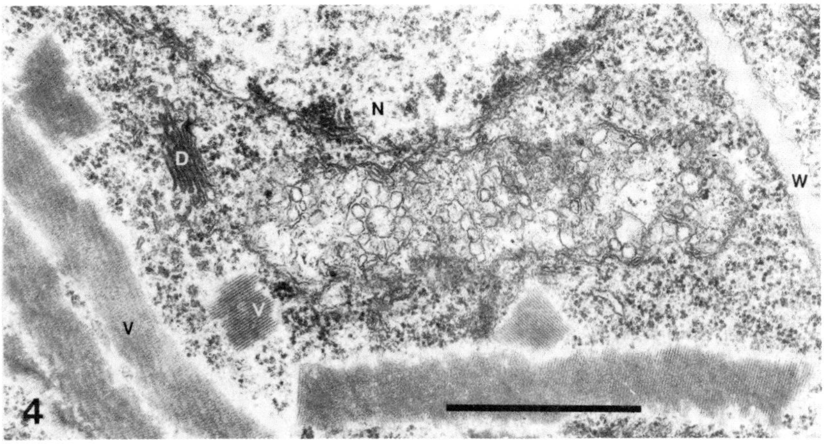

FIGURE 3. A leaf cell of *Datura stramonium* infected with tomato bushy stunt virus (Tombusvirus) showing peroxisomes at various stages of virus-induced vesiculation.

FIGURE 4. A cucumber leaf cell infected with cucumber green mottle mosaic virus (Tobamovirus) showing a mitochondrion with virus induced vesicles.
(W = cell wall; M = mitochondrion; V = virus particles; D = dictyosome; N = nucleus. Bar = 1μm)

Mitochondrial vesicles.

Plant cells infected by several viruses belonging to a number of different taxonomic groups develop multivesicular bodies resembling those induced by Tombusviruses which, however, develop from mitochondria (4, 5; Fig 4). The most thoroughly studied of these are cells infected with cucumber green mottle mosaic virus (CGMMV). The mitochondria are usually enlarged and mishapen, containing numerous vesicles about 50-70 nm in diameter located in the spaces between the outer and inner membranes of the mitochondria, including the intercristal spaces. Some of the vesicles can be seen to be open to the cytoplasm, the mouth being a continuation of the outer mitochondrial membrane. Many of the vesicles contain fibrils with the appearance of nucleic acid (34-36). The vesiculation of mitochondria precedes the appearance of virus particles indicating that the structures are involved in virus replication (35, 36).

Similar vesiculated mitochondria have been observed in cells infected with tobacco rattle virus defective in that it failed to produce coat protein (37). It was suggested that the mitochondria were the sites of virus RNA synthesis (37).

Mitochondrial vesiculation induced by two viruses with Tombusvirus-like particles, turnip crinkle virus and GaMV is also very similar to that in CGMMV-infected cells (29,38). In the case of GaMV, the massive vesiculation and distortion of the mitochondria was also accompanied by the development of a few chloroplast vesicles characteristic of Tymoviruses (29). On the other hand, the Tymovirus, clitoria yellow vein virus has been shown to induce some mitochondrial vesicles in addition to widespread chloroplast vesiculation (21).

Although their cytopathology has not been studied in detail, mitochondrial vesiculation has also been reported in cells infected with the Luteovirus, carrot red leaf virus (4) and dendrobium vein necrosis virus which is unclassified but has Closterovirus-like particles (39).

Vesiculation of tonoplasts.

Small tonoplast-associated vesicles have been observed in cells infected by all three Cucumoviruses (4, 13, 40). About 50-90nm in diameter, the vesicles protrude into the vacuole (Fig 5). Each vesicle is bounded by a membrane which in some micrographs can be seen to be continuous with the onoplast membrane and showing that its contents are in contact with the cytoplasm through a narrow neck. Some of the vesicles contained electron-dense fibrils which could be digested with ribonuclease in low but not in high salt buffers indicating that they were ds RNA (40).

Similar tonoplast vesiculation has also been seen in cells infected with tobacco necrosis virus, some Potex-viruses and the majority of the Tobamoviruses (4). Vesicles were observed in cells infected by six of the seven Tobamo-viruses examined although only a few were detected in tomato mosaic virus-infected cells (4). The only Tobamovirus which failed to induce any tonoplast vesicles was CGMMV (CV4 isolate) which, however, induced heavy mitochondrial vesiculation as already mentioned.

Tonoplast vesicles have also been observed in cells infected with carrot mottle, lettuce speckles and bean yellow vein viruses (41-43) whose virions have not been identified and the significance of the vesicles is not clear. It has been suggested that the vesicles in carrot mottle virus-infected cells may be the virions (41). However, it was concluded that in lettuce speckles virus-infected cells, they were not the virions but sites of infective ss-RNA synthesis which does not become encapsidated (42). Further work will be needed to determine their true identity and significance.

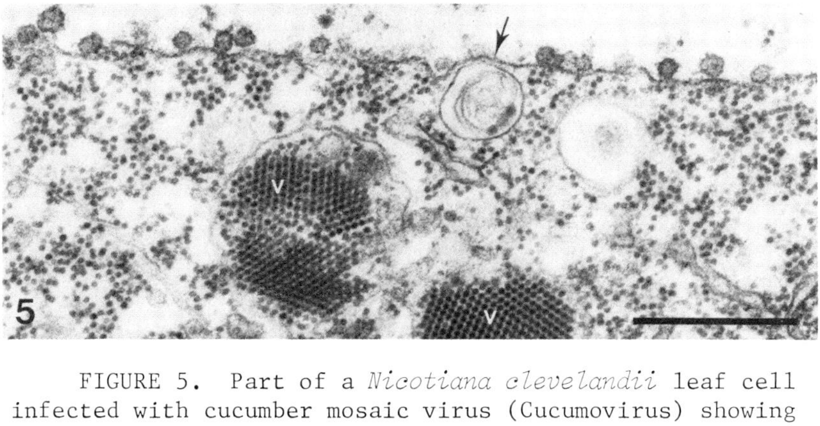

FIGURE 5. Part of a *Nicotiana clevelandii* leaf cell infected with cucumber mosaic virus (Cucumovirus) showing the tonoplast (arrow) with virus-induced vesicles. (V = virus particles; Bar = 0.5μm)

CONCLUSIONS

It is interesting that positive strand viruses, some with very similar genomic structure, induce vesicles at different locations in plant cells. The electron microscopic studies on these structures suggest that they are the sites of viral RNA replication which is also supported by the available biochemical data. At present we do not know how the virus recognises the membranes on which it induces the vesicles. The coat protein of the infecting virus cannot be responsible for recognition because isolated RNAs of all positive stranded viruses incite normal infections. It is also unlikely that any coat protein translated immediately following entry of the viral RNA into cells is involved because at least in the case of CPMV, normal vesiculated inclusions are produced in protoplasts inoculated with only B particles which do not carry the coat protein genes (12). This leaves as possible candidates, either the viral RNA itself or one or more of the non-structural virus-coded proteins. The latter seems more likely because there is evidence that in vertebrate cells infected with poliovirus, vesicles are induced by a virus-coded protein (44). In these cells, the vesicles and the cytoplasmic inclusions into which they are organized, are similar to those seen in plant cells infected with Nepoviruses, Comoviruses and Bromoviruses (4).

At present there is no evidence that some of the other virus-induced cytopathic effects are directly involved in virus synthesis. They include abnormalities of the chloroplasts and golgi bodies, development of myelin-like structures and tubules, accumulations of membranes in various configurations and cell wall outgrowths (4). It has been suggested that some of these may be the results of the plant's defence reactions to infection, but there is no conclusive evidence for this (45). Some structural changes to those associated with virus infection can also be observed in senescing cells suggesting that virus infection actually hastens plant senescence. Some of the physiological changes following infection such as increase in ethylene production (3), are also associated with senescence.

ACKNOWLEDGEMENT

I thank Dr. T. Hatta for the electron micrographs.

REFERENCES

1. Zaitlin M (1979). How viruses and viroids induce disease. In Horsfall JG, Cowling EB (eds): "Plant Disease, an Advanced Treatise" Vol IV, New York: Academic Press, p 257.
2. Matthews REF (1980). Host plant responses to virus infection. In Fraenkel-Conrat H, Wagner RR (eds): "Comprehensive Virology" Vol 16, New York: Plenum Press, p 297.
3. Matthews REF (1981). "Plant Virology" 2nd Ed. New York: Academic Press.
4. Francki RIB, Milne RG, Hatta T (1985). "Atlas of Plant Viruses" Vols I and II, Boca Raton: CRC Press.
5. Martelli GP, Russo M (1984). Use of thin sectioning for visualization and identification of plant viruses. Meth. Virol 8 : 143.
6. Hatta T, Francki RIB (1978). Enzyme cytochemical identification of single-stranded and double-stranded RNAs in virus-infected plant and insect cells. Virology 88 : 105.
7. Van Kammen A (1984). Expression of functions encoded on genomic RNAs of multiparticulate plant viruses. In Kurstak E (ed): "Control of Virus Diseases", New York: Marcel Dekker, p 301.
8. Assink AM, Swaans H, Van Kammen A (1973). The localization of virus-specific double-stranded RNA of cowpea mosaic virus in subcellular fractions of infected Vigna leaves. Virology 53 : 384.
9. De Zoeten GA, Assink AM, Van Kammen A (1974). Association of cowpea mosaic virus induced double-stranded RNA with a cytopathic structure in infected cells. Virology 59 : 341.
10. Zabel P, Weenen-Swaans H, Van Kammen A (1974). *In vitro* replication of cowpea mosaic virus RNA. I. Isolation and properties of the membrane-bound replicase. J Virol 14 : 1049.
11. Rezelman G, Goldbach R, Van Kammen A (1980). Expression of bottom component RNA of cowpea mosaic virus in cowpea protoplasts. J Virol 36 : 366.
12. Rezelman G, Franssen HJ, Goldbach RW, Ie TS, Van Kammen A (1982). Limits to the independence of bottom component RNA of cowpea mosaic virus. J Gen Virol 60 : 335.
13. Martelli GP, Russo M (1985). Virus-host relationships: Symptomatological and ultrastructural aspects. In Francki RIB (Ed): " The Plant Viruses : Polyhedral Virions with Tripartite Genomes", New York: Plenum Press, p163.

14. De Zoeten GA, Gaard G, Diez FB (1972). Nuclear vesiculation associated with pea enation mosaic virus-infected plant tissue. Virology 48 : 638.
15. De Zoeten GA, Powell CA, Gaard G, German TL (1976). *In situ* localisation of pea enation mosaic virus double-stranded ribonucleic acid. Virology 70 : 459.
16. Powell CA, De Zoeten GA (1977). Replication of pea enation mosaic virus RNA in isolated pea nuclei. Proc Natl Acad Sci USA 74 : 2919.
17. Esau K, Hoefert LL (1972). Development of infection with beet western yellows virus in the sugarbeet. Virology 48 : 724.
18. Shepardson S, Esau K, McCrum R (1980). Ultrastructure of potato leaf phloem infected with potato leafroll virus. Virology 105 : 379.
19. Burgess J, Motoyoshi F, Fleming EN (1974). Structural changes accompanying infection of tobacco protoplasts with two spherical viruses. Planta 117 : 133.
20. Francki RIB, Randles JW, Chu PWG, Rohozinski J, Hatta T (1985). Viroid-like RNAs incorporated in conventional virus capsids. In Maramorosch K, McKelvey JJ (eds): "Subviral Pathogens of Plants and Animals: Viroids and Prions, New York: Academic Press, p 265.
21. Lesemann DE (1977). Virus group-specific and virus-specific cytological alterations induced by members of the Tymovirus group. Phytopath Z 90 : 315.
22. Ushiyama R, Matthews REF (1970). The significance of chloroplast abnormalities associated with infection by turnip yellow mosaic virus. Virology 42 : 293.
23. Hatta, T, Matthews REF (1974). The sequence of early cytological changes in chinese cabbage leaf cells following systemic infection with turnip yellow mosaic virus. Virology 59 : 383.
24. Hatta T, Bullivant S, Matthews REF (1973). Fine structure of vesicles induced in chloroplasts of chinese cabbage leaves by infection with turnip yellow mosaic virus. J Gen Virol 20 : 37.
25. Hatta T, Matthews REF (1976). Sites of coat protein accumulation in turnip yellow mosaic virus-infected cells. Virology 73 : 1.
26. Lafléché D, Bové C, Dupant G, Mouches C, Astiev T, Garnier M, Bové JM (1972). Site of viral RNA replication in the cells of higher plants. TYMV (turnip yellow mosaic virus)-RNA synthesis on the chloroplast outer membrane system. Proc Fed Eur Biochem Soc 72:43.

27. Carrol TW (1970). Relation of barley stripe mosaic virus to plastids. Virology 42 : 1015.
28. McMullen CR, Gardner WS, Myers GA (1978). Aberrant plastids in barley leaf tissue infected with barley stripe mosaic virus. Phytopathology 68 : 317.
29. Hatta T, Francki RIB, Grivell CJ (1983). Particle morphology and cytopathology of galinsoga mosaic virus. J Gen Virol 64 : 687.
30. Russo M, Martelli GP (1972). Ultrastructural observations on tomato bushy stunt virus in plant cells. Virology 49 : 122.
31. Appiano A, Pennazio S, Redolfi P (1978). Cytological alterations in tissues of Gomphrena globosa plants systemically infected with tomato bushy stunt virus. J Gen Virol 40 : 277.
32. Russo M, Di Franco A, Martelli G (1983). The fine structure of cymbidium ringspot virus infections in host tissues. III. Role of peroxisomes in the genesis of multivesicular bodies. J Ultrastruct Res 82 : 52.
33. Martelli G, Di Franco A, Russo M (1984), The origin of multivesicular bodies in tomato bushy stunt virus-infected Gomphrena globosa plants. J. Ultrastruct Res 88 : 275.
34. Hatta T, Nakamoto T, Takagi Y, Ushiyama R (1971). Cytological abnormalities of mitochondria induced by infection with cucumber green mottle mosaic virus. Virology 45 : 292.
35. Hatta T, Ushiyama R (1973). Mitochondrial vesiculation associated with cucumber green mottle mosaic virus-infected plants. J Gen Virol 21 : 9.
36. Sugimura Y, Ushiyama R (1975). Cucumber green mottle mosaic virus infection and its bearing on cytological alterations in tobacco mesophyll protoplasts. J Gen Virol 29 : 93.
37. Harrison BD, Stefanac Z, Roberts IM (1970). Role of mitochondria in the formation of X-bodies in cells of *Nicotiana clevelandii* infected by tobacco rattle viruses. J Gen Virol 6 : 127.
38. Russo M, Martelli GP (1982). Ultrastructure of turnip crinkle and saguaro cactus virus-infected tissues. Virology 118 : 109.
39. Lesemann DE (1977). Long, filamentous virus-like particles associated with vein necrosis of Dendrobium phalaenopsis. Phytopath Z 89 : 330.

40. Hatta T, Francki RIB (1981). Cytopathic structures associated with tonoplasts of plant cells infected with cucumber mosaic and tomato aspermy viruses. J Gen Virol 53 : 343.
41. Murant AF, Roberts IM, Goold RA (1973). Cytopathological changes and extractable infectivity in *Nicotiana clevelandii* leaves infected with carrot mottle virus. J Gen Virol 21 : 269.
42. Falk BW, Morris TJ, Duffus JE (1979). Unstable infectivity and sedimentable ds-RNA associated with lettuce speckles mottle virus. Virology 96 : 239.
43. Cockbain AJ, Jones P (1981). Bean yellow vein-banding virus. Rothamsted Exp Stn Rep for 1980, Pt 1, p 189.
44. Bienz K, Egger D, Russer Y, Bossart W (1983). Intracellular distrubution of poliovirus proteins and the induction of virus-specific cytoplasmic structures. Virology 131 : 39.
45. Martelli GP (1980). Ultrastructural aspects of possible defence reactions in virus-infected plant cells. Microbiologia 3 : 369.

Positive Strand RNA Viruses, pages 437–452

STUDY ON VIRULENCE OF POLIOVIRUS TYPE 1 USING IN VITRO MODIFIED VIRUSES[1]

A. Nomoto[2], M. Kohara[2,3], S. Kuge[2], N. Kawamura[2], M. Arita[4], T. Komatsu[4], S. Abe[3], B. L. Semler[5], E. Wimmer[6], and H. Itoh[3*]

[2]Department of Microbiology, Faculty of Medicine, University of Tokyo, Bunkyo-ku, Tokyo 113, Japan
[3]Japan Poliomyelitis Research Institute, Higashi-murayama, Tokyo 189, Japan
[4]Department of Enteroviruses, National Institute of Health, Musashimurayama, Tokyo 190-12, Japan
[5]Department of Microbiology and Molecular Genetics, College of Medicine, University of California, Irvine, California, 92717, USA
[6]Department of Microbiology, School of Medicine, State University of New York, Stony Brook, New York 11794, USA

ABSTRACT Recombinant viruses between the virulent Mahoney and the attenuated Sabin strains of type 1 poliovirus were constructed in vitro. Biological tests including monkey neurovirulence test on these recombinants revealed genome loci that influenced biologically different phenotypes between the two parental strains.

[1]This work was supported in part by a grant from the Minstry of Education, Science, and Culture of Japan to A. Nomoto, by a grant from the Ministry of Health and Welfare of Japan to M. Arita, by a grant from the American Cancer Society (MV-183) to B. L. Semler, and by US Public Health Service grants AI 15122 and CA 28146 from the National Institute of Health to E. Wimmer.

*Deceased

INTRODUCTION

Poliovirus, the causative agent of poliomyelitis, is a human enterovirus that belongs to the Picornaviridae. The virus exists in three stable, antigenically distinct types, that is, type 1, type2, and type 3. To control the severe paralytic disease caused by poliovirus, attenuated strains of all three serotypes (Sabin 1, Sabin 2, and Sabin 3, respectively) have been isolated and used effectively as oral live vaccines (1-3). The single-stranded genomic RNA is enclosed in a naked, tightly packed icosahedral capsid consisting of 60 copies each of four proteins. Although humans are its only natural host, poliovirus can be transferred to monkeys, in which it also causes paralytic disease.

The genome of poliovirus is composed of approximately 7500 nucleotides, polyadenylylated at the 3' terminus (4) and covalently attached to a genome-linked protein (VPg) at the 5' terminus (5). This RNA is infectious in mammalian cells regardless of whether VPg is attached to the 5' end (6). In the host-cell cytoplasm, the viral RNA is translated into a single continuous polyprotein with molecular weight of 247000. The polyprotein is subsequently cleaved by proteases to form the viral structural and nonstructural proteins (7-9).

To date, cDNAs to the genomes of type 1 poliovirus(the virulent Mahoney and the attenuated Sabin 1 strains) (10-12) , type 2 poliovirus (the virulent Lansing and the attenuated Sabin 2 strains) (13, 14), and type 3 poliovirus (the virulent Leon and the attenuated Sabin 3 strains) (14-16) have been cloned using bacterial plasmids, and the total nucleotide sequences of these genomes have been determined.

The attenuated Sabin 1 strain was derived from the virulent Mahoney strain of type 1 poliovirus by multiple passages through host cells of nonhuman origin (17). In addition to their different potentials for causing disease, these two strains of virus differ in a number of biological characteristics. Some of these biological characteristics are used as in vitro marker tests to analyze the quality of batches of oral live vaccines (18). These include the sensitivity of viral multiplication to elevated temperatures (rct marker), the sensitivity of viral plaque-forming ability to low concentrations of sodium bicarbonate under agar overlay (d marker), and the size of plaques produced in infected monolayers of primate cells. Although much information has been gathered concerning the molecular biology of poliovirus, little is known about the molecular basis of attenuation. Similarly, the molecular events that influence the in vitro phenotypes of the Sabin strains have not been elucidated.

Here we summarize the results of monkey neurovirulence tests and in vitro marker tests of recombinant viruses of the Mahoney and Sabin 1 strains constructed in vitro. The data provide insight into the relationship between genome structure and attenuation of poliovirus type 1.

BIOLOGICAL AND STRUCTURAL DIFFERENCES BETWEEN THE MAHONEY AND THE SABIN 1 STRAINS

Major different biological characteristics between the Mahoney and the Sabin 1 strains are listed in TABLE 1. The Sabin 1 strain is considered to be a multi-step temperature -sensitive mutant and one of the steps appears to be in the assembly of viral capsid proteins into capsomeres (19). The temperature-sensitive phenotype of the Sabin strain is used as an in vitro marker to estimate the quality of live vaccines (TABLE 1). The rct marker test for this phenotype is one of the most reliable in vitro marker tests of poliovirus vaccines (18). The plaque-forming ability of the attenuated Sabin 1 strain is sensitive to reduced concentration of bicarbonate in the agar overlay, but that of the virulent Mahoney strain is not (TABLE 1). This characteristic of the Sabin strain is often used in the in vitro d marker test of the virus that may correlate with the attenuated phenotype. Plaque size is also one of the

TABLE 1. DIFFERENT PHENOTYPES BETWEEN THE MAHONEY AND THE SABIN 1 STRAINS OF POLIOVIRUS

Phenotype	Virulent Mahoney	Attenuated Sabin 1
Neurovirulence	strong	weak
Temperature-sensitivity (rct)	low	high
Bicarbonate conc. dependency (d)	low	high
Plaque size	large	small

marker tests used as an indication of the rate of viral multiplication, and it might correlate with viral virulence. Indeed, the plaque size of the Mahoney strain is larger than that of the Sabin 1 strain (TABLE 1). As with the attenuated phenotype, the molecular basis of these in vitro phenotype of the attenuated Sabin strain is unknown.

Different biological characteristics of the Mahoney and the Sabin 1 strains must be due to the differences in genome structures that resulted from the attenuation process used to create the Sabin 1 virus. Indeed, the elucidation of the total nucleotide sequence of the RNA genomes of both strains (7, 10, 12) has revealed 55 nucleotide substitutions (12, 14, 20) within the 7441 total heteropolymeric bases. These nucleotide changes were found to be scattered over the entire length of the genome (FIGURE 1), and result in 21 amino acid replacements within the viral polyprotein (12) (FIGURE 1).

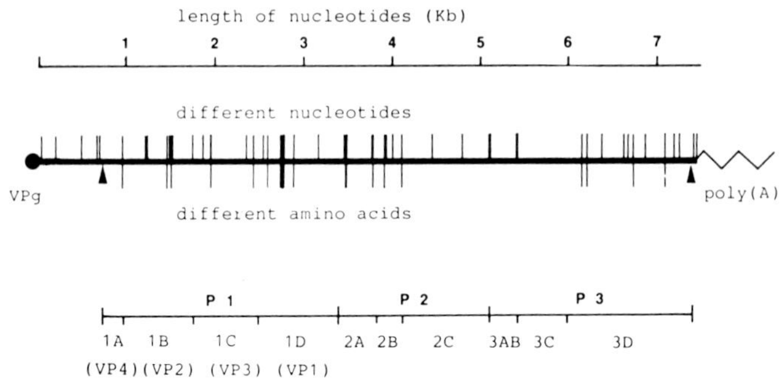

FIGURE 1. Location of nucleotide and amino acid differences between the Mahoney and Sabin 1 strains. Length of the entire genome of poliovirus type 1 is indicated at the top of the figure in kilobases (kb) from the 5' end. Gene organization is shown at the bottom of the figure. The positions of initiation and termination (▲) of viral polyprotein synthesis are indicated on the genome RNA. The locations of nucleotide and amino acid differences between the Mahoney and the Sabin 1 strains are indicated by lines over and under the genome RNA, respectively.

RECOMBINANT VIRUSES BETWEEN THE MAHONEY AND THE SABIN 1 STRAINS

Genetic recombinants between the neurovirulent and the attenuated poliovirus strains were isolated from infected tissue culture cells (21, 22). The genetic crossover in these recombinants was found to be located in the central region (P2 region) of the RNA genome (21, 22). Using such recombinants in tests of monkey neurovirulence, Agol et al. (21, 23) suggested that major determinants of neurovirulence reside in the 5'-terminal half of the poliovirus genome. The failure to isolate recombinants from infected cells with crossovers in regions other than in P2, probably because of the lack of suitable selectable markers, dose not allow a more precise identification of genomic sequences involved in attenuation.

Racaniello and Baltimore (24) have shown that a complete, cloned cDNA copy of the genome of the virulent Mahoney strain of type 1 poliovirus is infectious in mammalian cells. A similar clone of high specific infectivity was also constructed by Semler et al. (25). Omata et al. (26) isolated an infectious cDNA clone of the genome of the Sabin 1 vaccine strain and designated as pVS(1)IC-0(25). The availability of infectious clones of poliovirus RNA provides a molecular genetic approach for investigating the relationship between genome structure and function of viral genome using recombinant DNA technology. Thus, it is now possible to construct recombinant viruses between different strains and test the properties of recombinants. Indeed, Kohara et al. (27) and Omata et al. (20) constructed infectious recombinant cDNA clones from segments of the virulent Mahoney and attenuated Sabin 1 genomes.

Recently, Semler and Johnson (manuscript in preparation) have constructed a highly infectious cDNA clone of the Mahoney strain of type 1 poliovirus. In addition to a full-length cDNA copy of the type 1 Mahoney genome, this plasmid (pPVA55) contains the SV40 origin of replication as well as the promoter and coding region for the SV40 large T antigen. To obtain a highly infectious Sabin 1 clone, the Mahoney sequence in this plasmid was replaced by the Sabin 1 sequence. The cDNA clone of the Sabin 1 strain thus obtained was designated as pVS(1)IC-0(T) (28). Similarly, highly infectious Mahoney clone, designated as pVM(1)pDS306(T), was constructed by the replacement of the Mahoney sequence in pPVA55 with the Mahoney sequence in pVM(1)pDS306(25) (see refs. 20,27).

The total nucleotide sequences of the genomes of the

virulent Mahoney and the attenuated Sabin strains of type 1 poliovirus are known as described above. When the predicted amino acid sequences of the viral polyproteins between the Mahoney and the Sabin 1 strains were compared, a cluster of amino acid changes was observed in the NH_2-terminal half of capsid protein VP1 (12) (FIGURE 1). Since VP1 is not only the largest of the capsid proteins but also the most exposed surface protein of the virion (29), we initiated "allele replacement" experiments to determine how these amino acid changes might influence the biological properties of the virus (20, 27). To analyze the effect of amino acid changes, clustered predominantly in VP1, on the phenotype of the virus, we constructed recombinant cDNA clones in which Pst I fragments (nucleotide positions 1814 to 3421) of infectious cDNA clones pVS(1)IC-0(T) and pVM(1)pDS306(T) were replaced by the corresponding fragments of pVM(1)pDS306(T) and pVS(1)IC-0(T), respectively, and the infectious recombinant clones were designated as pVSM(1)IC-1a(T) and pVSM(1)IC-1b(T), respectively. The recovered viruses from AGMK cells transfected with pVS(1)IC-0(T), pVM(1)pDS306(T), pVSM(1)IC-1a(T), and pVSM(1)IC-1b(T) were designsted as PV1(Sab)IC-0, PV1(M)pDS306, PV1(SM)IC-1a, and PV1(SM)IC-1b, respectively(FIGURE 2).

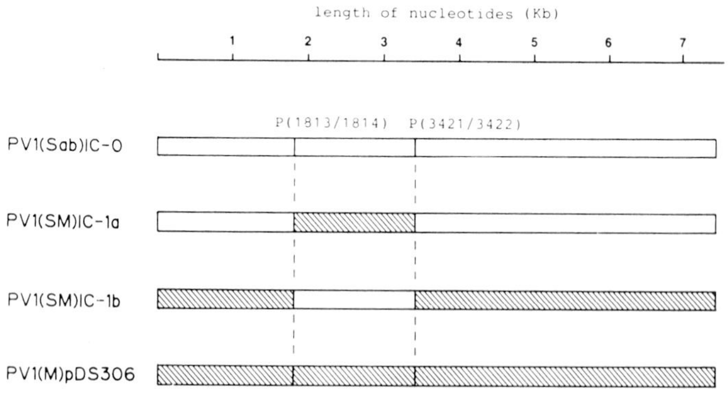

FIGURE 2. Genome structure of recombinant type 1 polio viruses. The expected genome structures of the recombinant viruses are shown by the combination of Sabin 1 (▭) and Mahoney (▨) sequences. P represents cleavage sites of the restriction enzyme Pst I. Numbers in parentheses following the restriction sites indicate nucleotide positions from the 5' end of the viral genome.

CHARACTERIZATION OF PV1(SM)IC-1a AND PV1(SM)IC-1b

Antigenicity

The segment exchanged in the recombinant viruses PV1(SM)IC-1a and PV1(SM)IC-1b codes for capsid protein VP1 and a portion of capsid protein VP3. All neutralizing monoclonal antibodies isolated so far for type 1 poliovirus have been found to recognize only antigenic sites that map in VP1 (30), although the immunogenic structures appear to be supported by amino acids in other loci of capsid proteins (31). It is therefore possible to use this property to detect intra-typic differences of the Mahoney and the Sabin 1 strains. Monoclonal antibodies specific to the Mahoney and the Sabin 1 strains were employed to test the antigenicity of PV1(SM)IC-1a and PV1(SM)IC-1b by an enzyme-linked immunosorbent assay procedure (see ref. 27). PV1(SM)IC-1a and PV1(SM)IC-1b were recognized only by neutralizing monoclonal antibodies specific to the Mahoney and the Sabin 1 strains, respectively (FIGURE 3). Furthermore, all recombinant viruses so far constructed that have the Mahoney sequence or the Sabin 1 sequence in the same genome locus showed Mahoney- or Sabin 1-specific antigenicity, respectively (20). These results clearly indicated that the strain specific antigenic determinant mapped in the genome locus represented by the segment of nucleotide positions 1814 to3421.

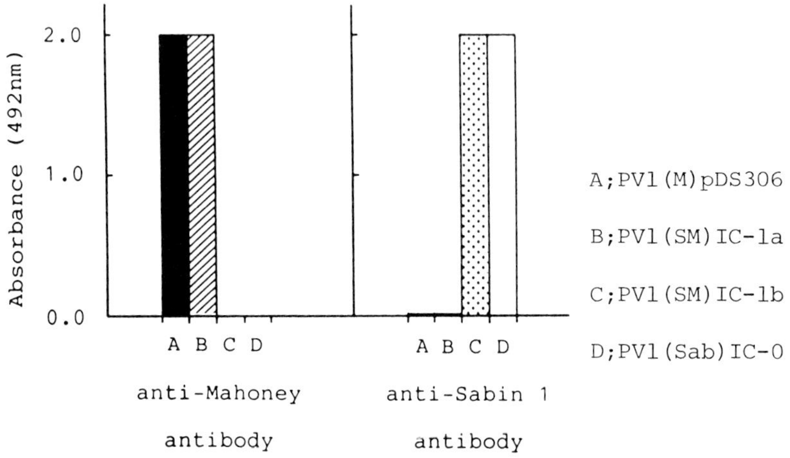

FIGURE 3. Emzyme-linked immunosorbent assay.

Neurovirulence

Recombinant viruses PV1(SM)IC-1a and PV1(SM)IC-1b were tested for their neurovirulence by being injected into cynomolgus monkeys intracerebrally as described previously (20). The resulting mean value of lesion scores of each virus is shown in TABLE 2. The recombinant virus PV1(SM) IC-1a showed low lesion score similar to that of the attenuated PV1(Sab)IC-0 parental virus. This result is compatible with our previous observation (20). PV1(SM)IC-1b, a reciprocal recombinant virus of PV1(SM)IC-1a, also showed very low lesion score. In contrast, PV1(M)pDS306 showed high lesion score.

TABLE 2. MONKEY NEUROVIRULENCE TESTS

Virus	Average lesion score	Incidence of paralysis[a]
PV1(Sab)IC-0	0.04	0 / 3
PV1(SM)IC-1a	0.05	0 / 3
PV1(SM)IC-1b	0.07	0 / 3
PV1(M)pDS306	2.48	3 / 4

[a]Number of monkeys paralyzed / number of monkeys injected.

Omata et al. (20) have indicated by allele replacement experiments that viral attenuation seems to be attributed to lowered efficiency in certain viral replication steps. According to this concept, the recombinant viruses PV1(SM) IC-1a and PV1(SM)IC-1b might have reduced efficiencies in certain viral replication steps. Replacement of one part of a viral coat protein by elements of the coat protein region of other strains that include high degree of amino acid variation might cause alterations in efficiency of capsid assembly or the correct folding required for the faithful processing of the precursor proteins. Indeed, a recombinant virus PV1(SM)IC-1a was observed to be an unstable virus (27). In any event, these reciprocal recombinant viruses were not able to be used for identification of genome regions influencing viral attenuation.

Reproductive capacity at different temperatures (rct) test

The rct marker test is based on the temperature sensitivity of multiplication of the poliovirus vaccine strains as described above, and is one of the most reliable tests among many in vitro biological marker tests to analyze the quality of a poliovirus live vaccine. Accordingly, the rct marker test was performed on PV1(SM)IC-1a and PV1(SM)IC-1b (TABLE 3). Both the recombinant viruses were quite temperature sensitive when compared with PV1(M)pDS306, although the sensitivity is a little less than that of PV1(Sab)IC-0 (TABLE 3). Thus, the temperature sensitivity of the recombinant viruses PV1(SM)IC-1a and PV1(SM)IC-1b appears to correlate well with their neurovirulence.

TABLE 3. REPRODUCTIVE CAPACITY AT DIFFERENT TEMPERATURES[a]

Virus	Log difference between temps (°C)		
	36 / 39	36 / 39.5	36 / 40
PV1(Sab)IC-0	5.13	>8.55	
PV1(SM)IC-1a	2.23	3.26	6.35
PV1(SM)IC-1b	3.02	5.93	>7.64
PV1(M)pDS306		0.07	0.48

[a]Values shown here are the logarithmic differences of virus titers obtained at indicated temperatures.

GENOME REGIONS INFLUENCING DIFFERENT PHENOTYPES BETWEEN THE MAHONEY AND THE SABIN 1 STRAINS

Omata et al. (20) have constructed seven different recombinant viruses including PV1(SM)IC-1a from the virulent Mahoney and the attenuated Sabin 1 parental strains of type 1 poliovirus by using infectious cDNA clones. Monkey neurovirulence tests using these recombinant viruses revealed that the loci influencing attenuation were spread over several areas of the viral genome. In vitro phenotypic marker tests (rct marker, d marker, and plaque size) were performed to identify the genome loci of these determinants and to investigate their correlation with attenuation. More recombinant viruses between both the strains have been constructed and tested for their biological characteristics (data not shown). These allele replacement experiments revealed the genome loci influencing the different phenotypes between the Mahoney and the Sabin 1 strains. The results are summarized in FIGURE 4.

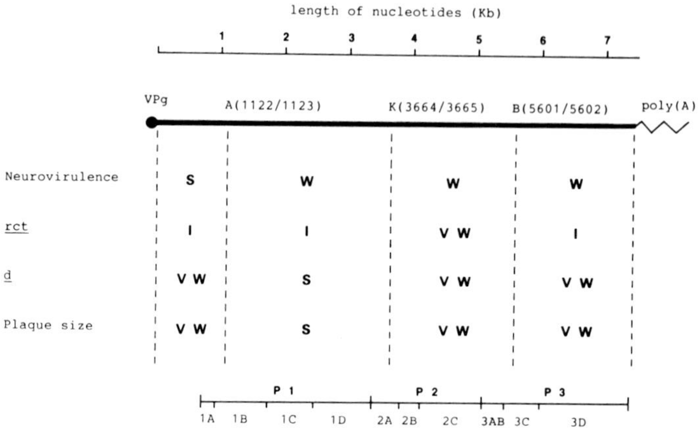

FIGURE 4. Summary of biological tests on recombinant viruses between the Mahoney and the Sabin 1 strains. Length of nucleotides from the 5' end of the genome is shown at the top of the figure in kilobases (kb). Gene organization is indicated at the bottom of the figure. A, K, and B represent cleavage sites of the restriction enzymes AatII, Kpn I, and Bgl II, respectively. Numbers in parentheses following the restriction sites indicate nucleotide positions from the 5' end of the viral genome. Extent of influence of genome region to each biological characteristic is roughly indicated by S(strong), I(intermediate), W(weak), or VW(very weak) in the corresponding genome region.

Strong determinant(s) influencing neurovirulence appear to reside in the nucleotide sequence upstream of the Aat II restriction site (nucleotide positions 1 to 1122). This genome region represents mainly the noncoding sequence, since the translation of the polyprotein commences at nucleotide position 743. Region 743 to 1122 harbors one nucleotide difference between the Mahoney and the Sabin 1, resulting in one amino acid change in VP4 (Ala ⟶ Ser)(12) (FIGURE 5). To know whether the nucleotide change or the amino acid change influences neurovirulence of type 1 poliovirus, we constructed a recombinant virus PV1(SM)IC-5a (FIGURE 5) in which only a genome sequence of Kpn I site (nucleotide position 71) to Ban II site (nucleotide position 909) of the Sabin 1 strain is replaced by the corresponding sequence of the Mahoney strain. This recombinant virus should have Mahoney-specific nucleotides at only four positions, that is, nucleotide positions 189, 480, 649, and 674 (FIGURE 5). Recombinant viruses PV1(SM)IC-4b and PV1(SM) IC-5a were tested for their neurovirulence by being injected into cynomolgus monkeys intraspinally. The results showed that both the strains have indistinguishable property of neurovirulence from each other (data not shown). These data clearly indicated that the determinant(s) influencing attenuation residedin the 5' noncoding sequence of the Sabin 1 genome. The mechanism for this phenomenon is unknown but could be a modulation in one or all of the following steps in viral proliferation: initiation of protein synthesis, RNA replication, or morphogenesis. Indeed, the genomes of the Mahoney and the Sabin 1 differ in their efficiencies in in vitro translation (32). A single base change in the 5' noncoding region of type 3 poliovirus has been observed to correlate with the attenuation phenotype of this strain (33).

It is of interest to know that a genome region of Aat II site(nucleotide position 1123) to Kpn I site(nucleotide position 3664) coding for mainly viral capsid proteins harbors only a weak determinant(s) in regard to phenotype of neurovirulence. It is therefore possible to construct more stable (and hence safer) live vaccines of type 2 and type 3 by the replacement of the capsid protein coding sequence of the Sabin 1 by the corresponding sequence of the type 2 and type 3 genome, since the Sabin 1 vaccine strain is the most stable virus among three serotype vaccine strains. Indeed, we recently succeeded in construction of a new candidate type 3 vaccine strain using cDNA clones of the Sabin 1 and Sabin 3 genomes.

The rct marker tests performed on many recombinant viruses between the Mahoney and Sabin 1 strains showed that the determinants of temperature sensitivity, like those of neurovirulence, were located across the entire poliovirus genome, and expressed fairly strong but not perfect correlation with attenuation (FIGURE 4).

Recombinant viruses with Sabin 1-derived capsid proteins showed a small-plaque phenotype, and their plaque-forming ability was strongly dependent on bicarbonate concentration, suggesting that these determinants map to

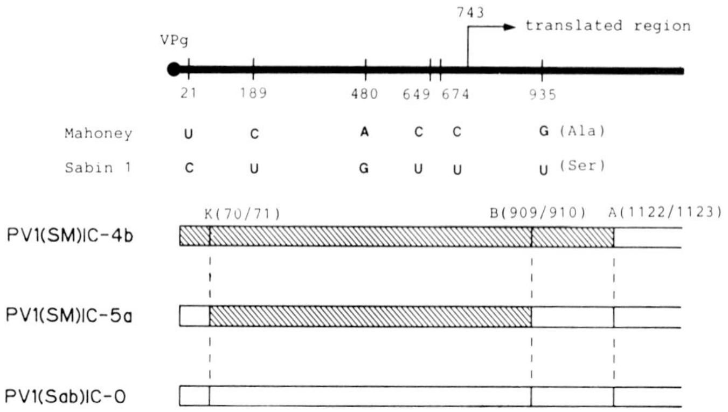

FIGURE 5. Genome structure of recombinant type 1 polioviruses. Genome RNA is shown at the top of the figure. Different nucleotides and amino acid between the Mahoney and Sabin 1 strains are indicated under the genome RNA. The expected genome structures of the recombinant viruses are shown by the combination of Sabin 1 (▭) and Mahoney (▨) sequences. K, B, and A represent cleavage sites of restriction enzymes Kpn I, Ban II, and Aat II, respectively. Numbers in parentheses following the restriction sites indicate nucleotide positions from the 5' end of the viral genome.

the genome region encoding the viral capsid proteins. Thus, the d marker and plaque size were found to be poor indicators of attenuation (FIGURE 4).

Our results suggest that the extent of viral multiplication in the central nervous system of monkeys might be one of the most important factors determining neurovirulence. Moreover, we conclude that the expression of the attenuated phenotype of the Sabin 1 strain of poliovirus is the result of several different biological characteristics, and that none of the in vitro phenotypic markers alone can serve as a good indicator of neurovirulence or attenuation.

ACKNOWLEDGMENTS

We thank Hiroshi Yoshikura and Hiroto Shimojo for many

enlightening suggestions and discussions. We also thank Nobuyuki Uchida for kind help in scoring the intensity of histological lesions.

REFERENCES

1. Sabin AB (1984). Strategies for elimination of poliomyelitis in different parts of the world with use of oral polio vaccine. Rev Infect Dis 6:S391.

2. Sabin AB (1985). Oral poliovirus vaccine: history of its development and use and current challenge to eliminate poliomyelitis from the world. J Infect Dis 151:420.

3. Melnick JL (1984). Live attenuated oral poliovirus vaccine. Rev Infect Dis 6: S323.

4. Yogo Y, Wimmer E (1972). Polyadenylic acid at the 3'-terminus of poliovirus RNA. Proc Natl Acad Sci USA 69: 1877.

5. Wimmer E (1982). Cell 28: 199.

6. Nomoto A, Kitamura N, Golini F, and Wimmer E (1977). Proc Natl Acad Sci USA 74: 5345.

7. Kitamura N, Semler BL, Rothberg PG, Larsen GR, Adler CJ, Dorner AJ, Emini EA, Hanecak R, Lee JJ, van der Werf S, Anderson CW, and Wimmer E (1981). Primary structure, gene organization and polypeptide expression of poliovirus RNA. Nature(London) 291: 547.

8. Hanecak R, Semler BL, Ariga H, Anderson CW, and Wimmer E (1984). Expression of a cloned gene segment of poliovirus in E. coli: evidence for autocatalytic production of the viral proteinase.

9. Toyoda H, Nicklin MJH, Murray MG, Anderson CW, Dunn JJ, Studier FW, and Wimmer E (1986). A second virus-encoded proteinase involved in ptoteolytic processing of poliovirus polyprotein. Cell in press.

10. Racaniello VR, and Baltimore D (1981). Molecular cloning of poliovirus cDNA and determination of the complete nucleotide sequence of the viral genome. Proc Natl Acad Sci USA 78: 4887.

11. Van der Werf S, Bregegere F, Kopecka H, Kitamura N, Rothberg PG, Kourilsky P, Wimmer E, and Girard M (1981). Molecular cloning of the genome of poliovirus type 1.

Proc Natl Acad Sci USA 78: 5983.

12. Nomoto A, Omata T, Toyoda H, Kuge S, Horie H, Kataoka Y, Genba Y, Nakano Y, and Imura N (1982). Complete nucleotide sequence of the attenuated poliovirus Sabin 1 strain genome. Proc Natl Acad Sci USA 79: 5793.

13. LaMonica N, Meriam C, and Racaniello V (1986). Mapping of sequences required for mouse neurovirulence of polio virus type 2 Lansing. J Virol 57: 515.

14. Toyoda H, Kohara M, Kataoka Y, Suganuma T, Omata T, Imura N, and Nomoto A (1984). Complete nucleotide sequences of all three poliovirus serotype genomes: implication for genetic relationship, gene function and antigenic determinants. J Mol Biol 174: 561.

15. Stanway G, Cann AJ, Hauptmann R, Hughes P, Clarke LD, Mountford RC, Minor PD, Schild GC, and Almond JW (1983). The nucleotide sequence of poliovirus type 3 Leon 12a_1b: comparison with poliovirus type 1. Nucleic Acids Res 11: 5629.

16. Stanway G, Hughes PJ, Mountford RC, Reeve P, Minor PD, Schild GC, and Almond JW (1984). Comparison of the complete nucleotide sequences of the genomes of the neurovirulent P3/Leon 37 and its attenuated Sabin vaccine derivative P2/Leon 12a_1b. Proc Natl Acad Sci USA 81: 1539.

17. Sabin AB, and Boulger LR (1973). History of Sabin attenuated poliovirus oral live vaccine strains. J Biol Stand 1: 115.

18. Nakano JH, Hatch MH, Thieme ML, and Nottay B (1978). Parameters for differentiating vaccine-derived and wild poliovirus strains. Prog Med Virol 24: 178.

19. Fiszman M, Reynier M, Bucchini D, and Girard M (1972). Thermosensitive block of the Sabin strain of poliovirus type 1. J Virol 10: 1143.

20. Omata T, Kohara M, Kuge S, Komatsu T, Abe S, Semler BL, Kameda A, Itoh H, Arita M, Wimmer E, and Nomoto A (1986). Genetic analysis of the attenuation phenotype of poliovirus type 1. J Virol in press.

21. Agol VI, Grachev VP, Drozdov SG, Kolesnikova MS, Kozlov VG, Ralph NM, Romanova LI, Tolskaya EA, Tyufanov AV, and Viktorova EG (1984). Construction and properties of intertypic poliovirus recombinants: first approximation

mapping of the major determinants of neurovirulence. Virology 136: 41.

22. Emini EA, Leibowitz J, Diamond DC, Bonin J, and Wimmer E (1984). Recombinants of Mahoney and Sabin strain poliovirus type 1: analysis of in vitro phenotype markers and evidence that resistance to guanidine maps in the non-structural proteins. Virology 137: 74.

23. Agol VI, Drozdov SG, Frolova MP, Grachev VP, Kolesnikova MS, Kozlov VG, Ralph NM, Romanova LI, Tolskaya EA, and Viktorova EQ (1985). Neurovirulence of the intertypic poliovirus recombinant v3/al-25: characterization of strains isolated from the spinal cord of diseased monkeys and evaluation of the contribution of the 3' half of the genome.

24. Racaniello VR, and Baltimore D (1981). Cloned poliovirus complementary DNA is infectious in mammalian cells. Science 214: 916.

25. Semler BL, Dorner AJ, and Wimmer E (1984). Production of infections poliovirus from cloned cDNA is dramatically increased by SV40 transcription and replication signals. Nucleic Acids Res 12: 5123.

26. Omata T, Kohara M, Sakai Y, Kameda A, Imura N, and Nomoto A (1984). Cloned infectious complementary DNA of the poliovirus Sabin 1 genome: biological and biochemical properties of the recovered virus. Gene 32: 1.

27. Kohara M, Omata T, Kameda A, Semler BL, Itoh H, Wimmer E, and Nomoto A (1985). In vitro phenotypic markers of a poliovirus recombinant canstructed from infectious cDNA clones of the neurovirulent Mahoney strain and the attenuated Sabin 1 strain. J Virol 53: 786.

28. Kohara M, Abe S, Kuge S, Semler BL, Komatsu T, Arita M, Itoh H, and Nomoto A (1986). An infectious cDNA clone of the poliovirus Sabin strain could be used as a stable repository and inoculum for the oral polio live vaccine. Virology in press.

29. Wetz K, and Habermehl KD (1979). Topographical studies on poliovirus capsid proteins by chemical modification and cross-linking with bifunctional reagents. J Gen Virol 44: 525.

30. Wimmer E, Jameson BA, and Emini EA (1984). Poliovirus antigenic sites and vaccines. Nature(London) 308: 19.

31. Diamond DC, Jameson BA, Bonin J, Kohara M, Abe S, Itoh H, Komatsu T, Arita M, Kuge S, Nomoto A, Osterhaus ADME, Crainic R, and Wimmer E (1985). Antigenic variation and resistance to neutralization in poliovirus type 1. Science 229: 1090.

32. Svitkin YV, Maslova SV, and Agol VI (1985). The genomes of attenuated and virulent poliovirus strains differ in their in vitro translation efficiencies. Virology 147: 243.

33. Evans DMA, Dunn G, Minor PD, Schild GC, Cann AJ, Stanway G, Almond JW, Currey K, and Maizel JV Jr (1985). Increased neurovirulence associated with a single nucleotide change in a noncoding region of the Sabin type 3 poliovaccine genome. Nature(London) 314: 548.

Positive Strand RNA Viruses, pages 453–466

CELLULAR RECEPTORS IN COXSACKIEVIRUS INFECTIONS[1]

Richard L. Crowell, Kuo-Hom Lee Hsu,
Maggie Schultz, and Burton J. Landau

Department of Microbiology and Immunology, Hahnemann
University School of Medicine, Philadelphia, PA 19102

ABSTRACT Cellular receptors are important determinants of virus tropism in the pathogenesis of human and animal diseases. This concept is derived from the observation that the presence or absence of specific receptors is the predominant determinant of the host range of a virus. In addition, the receptor specificity for the different species of picornaviruses places these viruses within sub-groups according to their original classification which was based on the type of disease and histopathology produced in humans or animals. This receptor specificity was determined by competition between viruses for a given receptor. More recently, monoclonal antibodies prepared against cellular receptors for picornaviruses have confirmed this receptor specificity. A receptor protein, Rp-a, of 49.5 kd which binds to group B coxsackieviruses (CB) has been isolated from HeLa cells. A rabbit antiserum prepared against the receptor preparation was found to block binding and protect cells against CB1, CB4, CB5, CB1-RD, and echovirus 6(E6), but not against poliovirus T1 or CB6. Surprisingly, this rabbit antiserum resembled the activity of a monoclonal antibody (2) which was active against receptors for CB1, CB3, CB5 and their respective RD variants, E6 and CA21, but not against CB2, CB4, CB6, polioviruses T1-3 or 12 other picornaviruses. These and other antibodies should

[1]This work was supported by a U.S. Public Health Service Research Grant AI-03771 from the National Institute of Allergy and Infectious Diseases

prove useful in determining the relationship betweeen the diverse receptors and their structures on different cell types. Because the CB-RD variant viruses may have acquired a second site for binding to an additional receptor and because CB3 was unique among the parental CB viruses for binding to receptors of the rat L_8 myogenic cell line, more than one type of receptor for binding virus may exist on different cell types which can influence the cellular host range and virus tropism. The distribution and function of these receptors on different cell types during stages of differentiation remains to be determined.

INTRODUCTION

The initial event for virus replication is attachment to specific receptors at the cell surface (1). Since the presence or absence of specific cellular receptors influences the host range of viruses, the identification of these structures is of importance for understanding viral tropism as an essential event in the pathogenesis of infection (2-6).

It is recognized that the receptor specificity for the different species of picornaviruses sorts these viruses according to their original classification based on disease and histopathology produced in humans and animals (Table 1).

TABLE 1
RECEPTOR FAMILIES FOR PICORNAVIRUSES BASED ON VIRUS COMPETITION FOR CELL RECEPTORS

Receptor Family	Reference
Poliovirus types 1-3	(7)
Coxsackievirus, group B, types 1-6	(7)
Coxsackievirus, group A, types 2 and 5; 13,15,18	(8-10)
Echovirus type 6	(7)
Human Rhinovirus types 1A, 2, 44, 49	(11-12)
HRV types 3,5,9,12,14,15,22,32,36,39,41 51,58,59,60,66,67,89, Cox A21	
FMDV types A_{12}119, O_{1B}, C_{3Res}; SAT_{1-3}	(13)
Cardioviruses	(13)

The polioviruses (PV) are well known agents of neurologic disease occurring only in man and sub-human primates. The group B coxsackieviruses (CB) cause pleurodynia, pancreatitis, meningitis, orchitis and myocarditis in man and a fatal paralysis of newborn mice in which necrosis of cells of the pancreas, fat, myocardium, brain and skeletal muscle occurs. The group A coxsackieviruses (CA) cause herpangina in man and a fatal skeletal muscle paralysis of newborn mice. The echoviruses (EV) are prominently associated with meningitis in man and do not produce disease in animals. The human rhinoviruses (HRV) cause the common cold syndrome and are without animal models of disease. The foot-and-mouth disease viruses (FMDV) cause severe disease in cattle and produce a fatal paralysis of newborn mice with myonecrosis. The cardioviruses of mice include the encephalomyocarditis virus (EMC), Maus Elberfeld virus (ME) and mengovirus (MV). These viruses are closely related serologically and cause fatal infections of adult mice. Each of the species of picornaviruses comprise multiple serotypes (14). Yet it is paradoxical that many of the serotypes (if not all in some cases) within each species share an affinity for binding to the same receptor on HeLa cells, but lack a common virion surface antigen (15). The recent elucidation of the topography of the picornavirus capsid (16,17) provides an apparent explanation for this paradox. X-ray crystallographic studies of HRV-14 (16) and of PV T1 (17) have revealed a "canyon" or cleft around the vertices of the virion which presumably serves as points of attachment for specific cellular receptors. The dimensions of these "canyons" shield the virion attachment sites from combining with antibodies, which might otherwise have recognized a common viral antigen for binding receptors within the receptor family (11). Our inability to obtain anti-idiotypic (anti-id) antibodies with specificity for the receptor for CB3 is further evidence that the virion attachment sites are shielded from antibody recognition. This information was obtained following the immunization of mice or rabbits with high titered CB3 neutralizing antiserum, prepared in the homologous species of animal. These negative results are similar to those found by others for CB4 (18) and FMDV (19). Thus, it is most unfortunate that the application of the "internal image" theory of antibody specificity to help isolate cellular receptors for picornaviruses, as has been done for reovirus type 3 (20), may not be possible.

Recently, some major advances have been made in the identification and isolation of cellular receptors for a number of non-enveloped viruses (Table 2). It is remarkable that each was isolated by use of a different strategy. For example, a receptor protein for the group B coxsackieviruses was isolated from detergent solubilized HeLa cells by use of labeled CB3 as an affinity surface (21). The virus-receptor complex (VRC) was then purified with methods normally used for virus purification. This procedure was made possible because the Triton X-100 (1%), sodium deoxycholate (0.5%) and sodium dodecylsulfate (1%) detergent mixture used to extract the receptor, did not dissociate the virus-receptor bond. The VRC sedimented more slowly (140 S) than native CB3 virions (155 S) when centrifuged on sucrose gradients. The purified VRC was iodinated, repurified, and analyzed by SDS-PAGE on slab gels. Only 1 protein, designated Receptor protein -a (Rp-a), of 49.5 Kd was found in addition to the virion capsid proteins. Rp-a was dissociated from the CB3 virions by heat (2 hr at 45°C) disruption of the virions and the iodinated Rp-a was shown to bind to unlabeled CB1 and CB3, but not to poliovirus T1. This procedure is not likely to be applicable to the isolation of other picornavirus receptors, since the detergents would dissociate the VRC.

The cellular receptor for the major group of HRV was isolated from HeLa cells by use of a receptor specific monoclonal antibody immunobilized on Affi-Gel (22, and elsewhere in this volume). Deoxycholate (0.3%) solubilized HeLa cell membranes were passed over the immunoaffinity column which was washed extensively, and the bound material eluted from the column with diethylamine, pH 11.5. Analysis by SDS-PAGE showed a major band of an apparent size of 90 Kd. The erythrocyte receptor for EMC virus has been shown to be glycophorin A, which binds virus in its multimeric form depending on the type and concentration of detergent used (23). Glycophorin A has been sequenced, treated with selected proteases, neuraminidase and reagents which block amino, carboxy and other reactive groups to locate the virus attachment site in the region of amino acid 35 to about 60 (25). One or more sialo-oligosaccharide side chains were found to be required for virus attachment.

As mentioned above, the receptor for reovirus type 3 has been isolated by using monoclonal and monospecific antireceptor antibodies (20). The strategy employed the use of a monoclonal anti-idiotype antibody prepared against a monoclonal antibody with specificity for the reovirus hemagglutinin (sigma-1 protein). The anti-id antibody represented an effective "internal image" for the reovirus HA domain that reacts with the cellular receptor for the

TABLE 2
STRATEGIES FOR ISOLATION AND IDENTIFICATION OF RECEPTORS FOR NONENVELOPED VIRUSES

Virus	Affinity Molecules	Receptor Mol. Wt ($x 10^3$)	Reference
Coxsackievirus B3	Virions (suspension)	49	21
Hu Rhinovirus 14	Monoclonal antibody(fixed)	90	22
Encephalomyocarditis	Virions (fixed)	31	23
Reovirus T3	Anti-Id antibody	67	20
Adenovirus T2	Virions (x-linked,fixed)	42	24

virus. Since a polyclonal rabbit anti-id antibody preparation was more efficient than the monoclonal antibody for immunoprecipitation of the detergent solubilized receptor, the polyclonal serum was used to purify the receptor. The receptor protein had an apparent size of 67 Kd, and was found to be identical to the beta-adrenergic hormone receptor (26). The cellular function(s) of receptors for picornaviruses remains unknown.

The receptor for adenovirus T2 is similar to the receptor for CB3 (11) in that virions and the adenovirus fiber antigen compete with CB3 for the same receptor. However, no direct biochemical comparisons of the respective purified receptor proteins have been made. Although a difference in the apparent molecular weights are known (22,24) this may only reflect differences in the carbohydrate composition between the two receptors. Further studies are needed to determine the relationship between these receptors, since adenovirus T2 neither replicates nor causes disease in mice, as does CB3.

RESULTS AND DISCUSSION

APPLICATION OF MONOCLONAL ANTIBODIES TO THE CHARACTERIZATION OF CELLULAR RECEPTORS

Campbell and Cords (27), were the first to obtain a monoclonal antibody with specificity for inhibiting attachment of picornaviruses to cells by blocking cellular receptors. Their IgM class antibody inhibited attachment of CB1, CB5 and CB6 to HeLa cells without blocking attachment of PV. These results stimulated the search for additional monoclonal antibodies which were reactive against cellular receptors for picornaviruses (28-31). As shown in Table 3 this search has been rewarding, although additional antibodies are needed to help determine the relationships among the different receptors for prototype viruses and those receptors that bind host range virus variants (32, 33).

TABLE 3
MONOCLONAL ANTIBODIES REACTIVE AGAINST CELLULAR RECEPTORS FOR PICORNAVIRUSES

Viruses Blocked	Cell Immunogen	Antibody Class	Reference
Coxsackievirus B1,B5,B6	HeLa	IgM	(27)
Coxsackievirus B1,B3,B5, Cox A21, Echo 6	HeLa	IgG2a	(31)
Polio T1, T2, T3	Hep 2c	IgG	(28)
Polio T1, T2, T3	HeLa	IgG1(Fab)	(29)
Human Rhino 78 types, Cox A11, A18, A21	HeLa	IgG1	(22,30)

In general the monoclonal antibodies to the different receptors show a high degree of specificity. These reagents help to confirm the existence of receptor families for the different species of picornaviruses, which were first recognized by virus competition (Table 1). However, the finding of a monoclonal antibody, designated Rmc CB3-CB3RD (31), which blocked receptors for some picornaviruses outside of the specific receptor family (i.e., CA21 and E6) was unexpected. In addition, the receptor for CB2, CB4 and CB6 was not blocked by this monoclonal antibody even though these viruses compete for a receptor on HeLa cells which binds all 6 group B coxsackievirus serotypes (34). Furthermore, the Rmc CB3-CB3RD antibody blocked attachment to the human rhabdomyosarcoma cell line (RD) of CB1-RD, CB3-RD, and CB5-RD variant viruses, which presumably use the same receptor on HeLa and RD cells as well as a second site on HeLa cells. This latter site is shared by the CB3 parental virus (32). Thus, this monoclonal antibody blocks not only the HeLa-specific receptor, but blocks the RD/HeLa receptor only on RD cells. In this regard, a sample of a monoclonal antibody obtained from Campbell and Cords (27) was found to block attachment to HeLa cells of CB3, but did not block attachment of CB3-RD to RD cells. This IgM-class monoclonal antibody, therefore, showed one of the antibody specificities for cellular receptors that was observed in the assay used for screening for monoclonal antibodies (Table 4). The dual receptor specificity of Rmc CB3-CB3RD was not the result of a mixture of antibodies, since fluids from each of 85 subclones of hybridoma cells derived from two consecutive subclonings retained the dual specificity. In addition, the hybridoma culture fluids were absorbed with HeLa cells or RD cells and titrations of the absorbed fluids revealed decreased titers for protection against CB3 on HeLa cells and for CB3-RD on RD cells, respectively.

TABLE 4

MONOCLONAL ANTIBODIES WITH THREE COXSACKIEVIRUS-CELL SPECIFICITIES OBSERVED IN SCREENING OF FUSION PRODUCTS[1]

Challenge Virus	Cell Line	Monoclonal Antibody
CB3	HeLa	Rmc CB3
CB3-RD	RD	Rmc CB3RD
CB3 and CB3-RD	HeLa and RD	Rmc CB3-CB3RD

[1] Numerous presumptive monoclonal antibodies with the indicated specificities were found in repeated screening assays.

The results observed for the Rmc CB3-CB3RD antibody may be explained by postulating that the antibody-combining site may exist on a subunit which is associated with various cellular receptor complexes for different enteroviruses (Fig. 1). Alternatively, the antibody site may be differentially expressed on the polypeptide chain which bears the receptor. Perhaps the capacity of the antibody to block the receptor site may occur by induced conformational effect or by steric hindrance. A schematic representation of some of the picornavirus receptor families is depicted in Fig. 1. Each receptor family is distinguished by the different shape of the virus attachment site (based on virus competition studies). The scheme also depicts two polypeptide chains forming the receptor (21). The combining site for the Rmc CB3-CB3RD antibody is shown on one of the polypeptides proximal to the virus attachment site. The development of additional monoclonal antibodies against the different cellular receptors for picornaviruses should be helpful to determine the relationships among the receptor proteins.

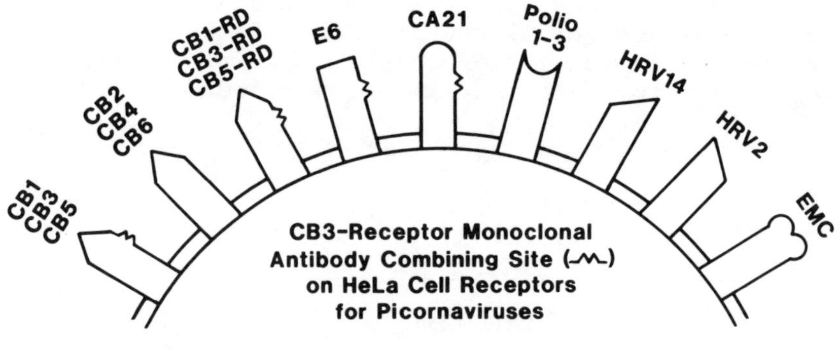

Figure 1. Schematic representation of some picornavirus receptor families and their relationship to postulated combining sites of Rmc CB3-CB3RD antibody. See text for details.

To study further the antigenic reactivities of the cellular receptor for the CB viruses, a rabbit was immunized with a purified virus-receptor complex (21). The polyclonal antiserum obtained showed specificity for blocking HeLa cell receptors against CB1, CB3, CB4, CB5, CB1-RD and E6, but not against PV T1, CB6 and CB3-RD (Table 5). The finding that the receptor for E6 was blocked by both the Rmc CB3-CB3RD antibody and the polyclonal antiserum prepared against the VRC was unexpected, since CB3 and E6 do not compete for a receptor on HeLa cells (7,31). Further studies are needed to determine the significance of this observation and to identify the cellular polypeptides which combine with the monoclonal and polyclonal antibody preparations. It will be important to learn whether the antiserums react with more than one polypeptide which may comprise the receptor complex as suggested previously (2,21). It is noteworthy to point out, however, that the VRC preparation used to immunize the rabbit (see above) was prepared using a modification of the method previously reported (21) to obtain larger amounts of the VRC. The VRC sample was passed through a Bio-Gel A-5 column following the metrizamide-sucrose step gradient centrifugation, and the CHAPS detergent was not used since iodination of the VRC was not needed. Plans are made to perform molecular cloning of the receptor gene in collaboration with R. Colonno, who is currently cloning the receptor gene for the major rhinovirus group.

Whereas, the monoclonal antibody preparations obtained for the cellular receptors for polioviruses are highly specific for blocking receptors for the 3 immunotypes of polioviruses (28,29), the monoclonal antibody which blocks

TABLE 5
COMPARATIVE RECEPTOR SPECIFICITY OF THE MONOCLONAL ANTIBODY Rmc CB3-CB3RD TO A POLYCLONAL RABBIT ANTISERUM PREPARED AGAINST PURIFIED CB3-RECEPTOR COMPLEX (VRC)

Challenge Virus	Receptors Blocked by Rmc CB3-CB3RD	Receptors Blocked by Anti-VRC
Cox B1, B3, B5	+	+
Cox B4	-	+
Cox B6	-	-
Cox B1-RD	+	+
Cox B3-RD	+	-
Echo 6	+	+
Polio T1	-	-

the receptor for the major group of HRV also blocks the receptor for CA11, CA18, CA21 (22,30). The reason for this latter cross reactivity is unknown and poses another interesting problem. In addition, it is unclear why the Rmc CB3-CB3RD antibody blocks the receptor on HeLa cells for CA21, but not that for HRV14 or CA18 (31). It is evident that the application of monoclonal antibody technology to the study of cellular receptors for picornaviruses will prove to be helpful in determining the relationships among the different receptor proteins. Furthermore, such antibodies with specificity for blocking host range viral mutants should prove useful in distinguishing those domains on the receptor which interact with complementary domains in the viral "canyons" which determine the host range variation (33).

RECEPTORS AS DETERMINANTS OF COXSACKIEVIRUS TROPISM IN PATHOGENESIS

The CB-RD virus variants have been found to be less virulent and to have altered tissue tropisms in mice as compared to their respective parental viruses (5,32). For example, a parental CB3 strain produced myocarditis in 8-week old male mice of C3H/HeJ, SJL/J and BALB/c inbred strains, whereas, no myocarditis was observed following infection with the CB3-RD variant. In contrast, CB3-RD infection of young SJL/J mice resulted in prominent skeletal muscle necrosis and inflammation followed by paralysis. This latter infection resembled the type of disease and histopathology associated with CA viruses. In tissue cultures of the differentiating rat L_8 myogenic cell line only CB3 strains, including CB3-RD, of the 6 CB parental serotypes replicated to levels of 10^6-10^7 PFU/ml of culture fluid. All of the other CB viruses were undetectable even after 3 or 4 sequential culture passages. Results of virus attachment studies revealed that CB3 virus attached to L_8 cells, whereas, CB1 did not (5). Reagan, et al (32) found that the RD cell line failed to support the growth of CB2 and CB4, yet these CB viruses share the same receptor specificity on HeLa cells (7,10,34). Perhaps differentiated or partially differentiated cell types express unique receptors that are shared by some, but not all of the CB viruses.

Based on the results of reciprocal virus competition studies and on the specificity of monoclonal antibodies for cellular receptors for picornaviruses, a logical case can be made for receptors as important determinants of cellular tropism in the pathogenesis of picornavirus infections. The grouping of these viruses into "receptor families" (7,10,11,34) according to the diseases produced in man and animals is unlikely to be due to chance (2) and yet we are a long way from identifying the distribution of specific receptors in tissues and organs of susceptible hosts. The detection of receptors in situ probably can best be performed by use of labeled monoclonal or monospecific antibodies, since picornaviruses as probes are too large and the specific activity of labeled virus is too low. In addition, no virus substructure or isolated polypeptide with specificity for binding to receptors is available for picornaviruses. We now have an explanation for this latter limitation if the virion attachment site is concealed at the bottom of the "canyon" (16,17) surrounding the vertices of the icosahedron. Future studies will need to specifically correlate the presence of receptors on those cells which are susceptible to infection. Susceptible cells can be identified by use of fluorescent antibodies to detect newly synthesized viral proteins, or better yet to use labeled cDNA probes for detection of viral RNA in infected cells (35). The identification of receptors may be more complicated if a virus can use more than one type of receptor for gaining entrance to the cell (32). Likewise, the expression of receptors on a given cell does not guarantee that the cell is in a state of susceptibility to the virus (8). Also, it is known that non-specific binding of virus must be adequately controlled (3). Thus, it will require more effort than would be predicted by cursory examination of this subject, to confirm the role of receptors as determinants of viral tropism. The use of host range virus variants in comparative studies with prototype viruses provide useful models to evaluate receptors in pathogenesis (32).

REFERENCES

1. Lonberg-Holm K, Crowell, RL (1986). Introduction and Overview. In Crowell RL, Lonberg-Holm K (eds.): "Virus Attachment and Entry Into Cells" Washington, DC: American Society for Microbiology, p 1.
2. Crowell RL, Landau BJ (1983). Receptors in the initiation of picornavirus infections. Comp Virology 18: 1.

3. Lonberg-Holm K (1981). Attachment of animal viruses to cells: an introduction. In Lonberg-Holm K, Philipson L (eds.): "Virus Receptors" Part 2. London: Chapman and Hall, Ltd. p 1.
4. Sharpe AH, Fields BN (1985). Pathogenesis of viral infections. Basic concepts derived from the reovirus model. N Eng J Med 312: 486.
5. Crowell RL, Reagan KJ, Schultz M, Mapoles JE, Grun JB, Landau BJ (1985). Cellular receptors as determinants of viral tropism. Banbury Report 22: "Genetically Altered Viruses and the Environment" Cold Spring Harbor, p 147.
6 Paulson JC (1985). Interactions of animal viruses with cell surface receptors. In Conn PM (ed.): "The Receptors" Vol 2, New York, Academic Press, p 131.
7. Crowell, RL (1966). Specific cell-surface alteration by enteroviruses as reflected by viral attachment interference. J Bacteriol 91: 198.
8. Schultz M, Crowell RL (1983). Eclipse of coxsackievirus infectivity: the restrictive event for a non-fusing myogenic cell line. J Gen Virol 64: 1725.
9. McLaren LC, Scaletti JV, James CG (1968). Isolation and properties of enterovirus receptors. In Manson LA (ed): "Biological Properties of the Mammalian Surface Membrane" Philadelphia, Wistar Institute Press, p 123.
10. Crowell RL (1976). Comparative generic characteristics of picornavirus-receptor interactions. In Beers RF Jr, Bassett EG (eds.): "Cell Membrane Receptors for Viruses, Antigens, and Antibodies, Polypeptide Hormones and Small Molecules" New York: Raven Press, p 179.
11. Lonberg-Holm K, Crowell RL, Philipson L (1976). Unrelated animal viruses share receptors. Nature (Lond.) 259: 679.
12. Abraham G, Colonno RJ (1984). Many rhinovirus serotypes share the same cellular receptor. J Virol 51: 340.
13. Sekiguchi K, Franke AJ, Baxt B (1982). Competition for cellular receptor sites among selected aphthoviruses. Arch Virol 74: 53.
14. Rueckert RR (1985). Picornaviruses and their replication. In Fields BN, et al (eds.) "Virology" New York: Raven Press, p. 705.
15. Katze MG, Crowell RL (1980). Immunologic studies of the group B coxsackieviruses by the sandwich enzyme-linked immunosorbent assay (ELISA) and immunoprecipitation. J Gen Virol 50: 357.

16. Rossmann MG, Arnold E, Erickson JW, Frankenberger EA, Griffith JP, Hecht H-J, Johnson JE, Kamer G, Luo M, Mosser AG, Rueckert RR, Sherry B, Vriend G (1985). Structure of a human common cold virus and functional relationship to other picornaviruses. Nature 317: 145.
17. Hogle JM, Chow M, Filman DJ (1985). Three-dimensional structure of poliovirus at 2.9 A resolution. Science 229: 1358.
18. McClintock PR, Prabhakar BS, Notkins AL (1986). Anti-idiotypic antibodies against anti-coxsackievirus type B4 monoclonal antibodies. In Crowell RL, Lonberg-Holm K (eds.): "Virus Attachment and Entry Into Cells" Washington, DC: American Society for Microbiology, p 36.
19. Baxt B, Morgan DO (1986). Nature of the interaction between foot-and-mouth disease virus and cultured cells. In Crowell RL, Lonberg-Holm K (eds): "Virus Attachment and Entry Into Cells" Washington DC: American Society for Microbiology, p 126.
20. Co MS, Gaulton GN, Fields BN, Greene MI (1985). Isolation and biochemical characterization of the mammalian reovirus type 3 cell-surface receptor. Proc Natl Acad Sci USA 82: 1494.
21. Mapoles JE, Krah DL, Crowell RL (1985). Purification of a HeLa cell receptor protein for group B coxsackieviruses. J Virol 55: 560.
22. Colonno RJ, Tomassini JE, Callahan PL, Long WJ (1986). Characterization of the cellular receptor specific for attachment of most human rhinovirus serotypes. In Crowell RL, Lonberg-Holm (eds.): "Virus Attachment and Entry into Cells" Washington DC: American Society for Microbiology, p 109.
23. Pardoe IU, Burness ATH (1981). The interaction of encephalomyocarditis virus with its erythrocyte receptor on affinity chromatography columns. J Gen Virol 57: 239.
24. Svensson U, Persson R, Everitt E (1981). Virus-receptor interactions in the adenovirus system. I. Identification of virion attachment proteins of the HeLa cell plasma membrane. J Virol 38: 70.
25. Allaway GP, Pardoe IU, Tavakkol A, Burness ATH (1986). Encephalomyocarditis virus attachment. In Crowell RL, Lonberg-Holm (eds.): "Virus Attachment and Entry Into Cells" Washington DC: American Society for Microbiology, p 116.

26. Co MS, Gaulton GN, Tominaga A, Homcy CJ, Fields BN, Greene MI (1985). Structural similarities between the mammalian beta-adrenergic and reovirus type 3 receptors. Proc Natl Acad Sci USA 82: 5315.
27. Campbell BA, Cords CE (1983). Monoclonal antibodies that inhibit attachment of group B coxsackieviruses. J Virol 48: 561.-564
28. Minor PD, Pipkin PA, Hockley D, Schild GC, Almond JW (1984). Monoclonal antibodies which block cellular receptors of poliovirus. Virus Res 1: 203. -212
29. Nobis P, Zibirre R, Meyer G, Kuhne J, Warnecke G, Koch G. (1985). Production of a monoclonal antibody against an epitope on HeLa cells that is the functional poliovirus binding site. J Gen Virol 66: 2563.-2569.
30. Colonno RJ, Callahan PL, Long WL (1986). Isolation of a monoclonal antibody that blocks attachment of the major group of human rhinoviruses. J Virol 57: 7.-12
31. Crowell, RL, Field AK, Schleif WA, Long WL, Colonno RJ, Mapoles JE, Emini EA (1986). Monoclonal antibody that inhibits infection of HeLa and rhabdomyosarcoma cells by selected enteroviruses through receptor blockade. J Virol 57: 438.-445.
32. Reagan KJ, Goldberg B, Crowell RL (1984). Altered receptor specificity of coxsackievirus B3 after growth in rhabdomyosarcoma cells. J Virol 49: 635.
33. LaMonica N, Meriam C, Racaniello VR (1986). Mapping of sequences required for mouse neurovirulence of poliovirus type 2 Lansing. J Virol 57: 515.
34. Crowell RL (1963). Specific viral interference in Hela cell cultures chronically infected with coxsackie B5 virus. J Bact 86: 517.
35. Kandolf R, Hofschneider PH (1985). Molecular cloning of the genome of a cardiotropic coxsackie B3 virus: Full length reverse-transcribed recombinant cDNA generates infectious virus in mammalian cells. Proc Natl Acad Sci USA 82: 4818.

Positive Strand RNA Viruses, pages 467–476

NUCLEIC ACID SEQUENCE ANALYSIS OF SINDBIS PATHOGENESIS AND PENETRATION MUTANTS[1]

Robert E. Johnston, Nancy L. Davis,
David F. Pence, Susan Gidwitz[2] and Frederick J. Fuller

Department of Microbiology
North Carolina State University
Raleigh, North Carolina 27695-7615

ABSTRACT Attenuated mutants of Sindbis virus (SB) have been isolated. These differed from wild-type with respect to virulence in neonatal mice, penetration of BHK cells in culture, and neutralization sensitivity to monoclonal antibodies R6 and R13. These three phenotypes are genetically linked as shown by characterization of virulent revertants selected in mice, mutants selected for fast penetration, and antibody selected variants. The mutation responsible for these phenotypes most probably resides at amino acid position 114 of glycoprotein E2. A serine to arginine substitution at this position was the only coding change detected in the glycoprotein genes of both the attenuated prototype, SB-RL, and the attenuated mutant selected for fast penetration, SB-FP. Reversion to serine at 114 was accompanied by reversion to the wild-type characteristics for each of the phenotypes. An antibody R6 selected second site suppressor mutation at E2 position 62 (asparagine to aspartic acid) also caused phenotypic reversion. Antibody R13 selected mutations clustered at E2 positions 96 and 159. However, only the mutations at 159 regained virulence in neonates. These results suggest a major role for glycoprotein E2 in both virulence in neonatal mice and penetration of cultured cells.

[1]This work was supported by NIH grant AI22186 and constitutes paper no. 10484 of the N. C. Agricultural Research Service, Raleigh, N. C. 27695-7601.
[2]Present address: Department of Medicine, University of North Carolina, Chapel Hill.

INTRODUCTION

Sindbis virus is the prototype of the alphavirus genus (1). The viral genome is a single strand of RNA containing 11,7Ø3 bases (strain HR) and encoding seven proteins (2). Three of these are present in Sindbis virions: a capsid protein, C (3ØK), which encloses the genome within an icosahedral nucleocapsid, and two envelope glycoproteins, E1 and E2 (5ØK and 45K, respectively), which are constituents of the lipoprotein envelope (3,4). In neonatal mice, Sindbis is extremely virulent. After subcutaneous (sc.) inoculation of virus dosages approaching a single PFU/mouse, mortality is 1ØØ%, and there are no survivors beyond 6 days post-infection (5,6). The virus initially establishes a viremia followed rapidly by invasion of the brain (7). Within 48 hr, brain virus titers exceeding $1Ø^9$ PFU/gm of tissue are found, and virus antigen is evident throughout the brain.

We have obtained mutants of Sindbis virus which display markedly reduced virulence in neonatal mice (8,9). These mutants were isolated after selection for rapid growth in baby hamster kidney (BHK) cells. The rationale for this procedure was based on the empirical finding that "adaptation" of viruses to cultured cells often reduces their virulence in animals. We reasoned that selection for rapid growth would increase the selective pressure for mutants which grew efficiently in BHK cells and therefore, increase the probability of isolating attenuated mutants. The prototype attenuated mutant, SB-RL (Sindbis-reduced latent period), was one of the attenuated mutants isolated in this way.

SB-RL does not invade the brain as rapidly as its virulent parent, strain AR339 (SB). At sc. doses less than $1Ø^2$ PFU/mouse, 8Ø-1ØØ% survival is observed, and up to 3Ø% survive at a sc. dose of $1Ø^6$ PFU/mouse. Of those animals which do die, only rarely does death occur prior to day 8 post-infection (J.E. Humphreys and R.E. Johnston, unpublished).

In addition to the *in vivo* phenotype of reduced virulence, SB-RL penetrates BHK cells at an accelerated rate compared to SB (defined as escape from extracellular neutralizing antibody) and is more sensitive to neutralization by two anti-E2 monoclonal antibodies, R6 and R13 (9). Analysis of other attenuated and virulent isolates indicated that these three phenotypes are linked genetically. This conclusion is supported by experiments in which a mutant selection was based on one of the three phenotypes, and the resulting strains were screened for the expression of the other two phenotypes (10). Virulent revertants of SB-RL, isolated from the brains of moribund SB-RL infected mice, also revert to the slower rate of penetration in BHK cells and reduced sensitivity to neutralization by R6 and R13 characteristic of the virulent parent, SB. SB-FP, a mutant of SB selected for the ability to rapidly penetrate BHK cells, displays reduced virulence in neonatal mice and increased sensitivity to neutralization by R6 and R13. Finally, an antibody escape mutant (RLmc6-2), selected from an SB-RL stock with R6, is virulent in neonatal mice and penetrates BHK cells at a rate similar to SB.

In this paper, we report the nucleotide sequence of the glycoprotein genes of these attenuated and virulent mutants. The results strongly suggest that a single amino acid substitution in glycoprotein E2 can have a profound pleiotropic effect on monoclonal antibody neutralization, penetration of cultured cells and virulence in neonatal mice.

METHODS

Virus Strains and Cell Culture.

SB (AR339) was supplied by H. R. Bose, University of Texas at Austin, was virulent in neonatal mice and was the parent of the prototype attenuated strain, SB-RL (for SB-reduced latent period), and SB-FP, a fast penetrating mutant (9). Virus stocks were maintained on baby hamster kidney (BHK) cells grown in Eagle's MEM supplemented with 10% donor calf serum and 10% tryptose phosphate broth. All stocks were tested for virulence phenotype in neonatal mice prior to extraction of RNA for sequencing.

Sequence Analysis of Genomic RNA.

Genomic RNA was extracted from purified virions (10) with 1% sodium dodecyl sulfate and phenol:chloroform (1:1) followed by chloroform alone. The RNA was concentrated by ethanol precipitation.

The oligonucleotide primers used for the E2 gene were synthesized either manually (New England Biolabs) or automatically using an Applied Biosystems DNA synthesizer. The El primers were the kind gift of Dennis T. Brown, University of Texas at Austin. The primers and their sequences are shown in Table 1. Viral gene sequence was obtained using the dideoxy chain termination method (11,12). Analysis of the sequence by computer utilized the SEQALIGN programs (13) and a hydropathicity program which was the kind gift of Carson Loomis, Duke University.

TABLE 1

Primers for RNA Sequencing

Primer	Complementary Nucleotides[a]	Primer Sequence (5' to 3')
E11[b]	11397-11410	CTGGTCGGATCATT
E12	11129-11142	TCGCGGTGCTAAAG
E13	10871-10885	CCGTATGAACAGTCC
E16	10635-10649	CATCGCTCCATATTC
E17	10389-10403	ATCTGCTGACAATTC
E21	9907-9920	CATGGTCTCGGTGA
E22	9722-9735	AGATGGTGTACACA
E24	9451-9464	ATGTATTACATTCG
E26	9166-9180	ATGATTCTTCCAGGT
E27	8955-8968	ACCGTTACGCTGTC
E28	8906-8919	TGTAGCTAAGCCTT

[a]Nucleotides numbered from the 5' end of the Sindbis HR genome (2).

[b]El primers were the gift of D. T. Brown, University of Texas.

RESULTS

Nucleotide Sequence Determination of Prototype Virulent and Attenuated Strains.

The nucleotide sequences of the glycoprotein genes of SB and SB-RL were determined using the dideoxy chain termination method (11,12). Reverse transcription of genomic RNA was initiated with the synthetic oligodeoxynucleotide primers listed in Table 1. In the regions encoding PE2 (the E2 precursor) and E1, SB-RL differed from SB at only two residues. At nucleotide 9779 (numbered from the 5' end of the viral genome by analogy with the sequence of the heat resistant strain, ref. 2) both U and C were detected in roughly equal proportions in the genome of SB. In SB-RL, as well as the other strains to be discussed below, U was found exclusively. This heterogeneity affected only the nucleotide sequence however, as either U or C at nucleotide 9779 would result in threonine at the amino acid level.

The only coding change found in the glycoprotein genes of SB-RL was at nucleotide 8972. The C in SB was changed to A in SB-RL resulting in a change from serine in SB to arginine at amino acid position 114 of the mature SB-RL glycoprotein E2 (Table 2). Because this was the only coding difference between SB and SB-RL, the mutation at nucleotide 8972 was a likely candidate for the mutation causing the pleiotropic phenotypic differences discussed above. This notion was supported by sequence results from other virulent and attenuated Sindbis strains. During the selection for SB-RL, several strains differing in virulence for neonatal mice were isolated by plaque-purification and characterized. The nucleotide sequence between nucleotides 8800 and 9145 was determined for five attenuated strains which displayed the fast penetration and efficient R6/R13 neutralization phenotypes. Three virulent, slowly penetrating and inefficiently neutralized strains also were sequenced in this region. The attenuated strains contained the SB-RL codon for arginine at E2 position 114, while the virulent strains had the SB codon for serine. The HR strain (gift of Dennis Brown) and an independently propagated AR339 strain (gift of Alan Schmaljohn) were both virulent in neonatal mice and were identical to SB at position 114.

TABLE 2

Mutations Affecting Sindbis Virulence

Virus	Phenotype	E2 Amino Acid Position 62	96	114	159
SB	Vir	AAC asn	UAC tyr	AGC ser	AAA lys
SB-RL	Att			AGA arg	
SB-FP	Att			AGA arg	
RLvrl-1	Vir			AGC ser	
RLmc6-2	Vir	GAC asp		AGA arg	
RLmc13-1 RLmc13-4	Vir			AGA arg	GAA glu
RLmc13-2 RLmc13-3	Att		CAC his	AGA arg	
RLmc13-5	Att		UUC phe	AGA arg	

Sequence Analysis of Revertant, Fast Penetrating and Antibody Selected Strains.

Four virulent revertants of SB-RL were isolated by plaque purification from the brains of three moribund, SB-RL infected mice (9). Sequencing in the 8800-9145 region revealed that all four revertants regained the SB serine codon at E2 amino acid position 114. The complete E1 and E2 sequence for one of the revertants (RLvrl-1) was determined, and no differences between it and SB were found

(Table 2), indicating that these were same site revertants.

From an SB parent stock, SB-FP was selected for rapid penetration of BHK cells. This independently selected mutant also exhibits the reduced virulence and increased R6/R13 neutralization sensitivity of SB-RL (9,10). Sequence analysis of the E1 and E2 genes of SB-FP revealed a single mutation at nucleotide 8972 identical to the coding change at E2 amino acid 114 found in SB-RL.

An antibody escape mutant (RLmc6-2), isolated from SB-RL with monoclonal antibody R6, reverted to wild-type with respect to both virulence in neonates and penetration of BHK cells (10). This mutant retained the SB-RL arginine at position 114, but had an additional mutation at base 8814 (A to G) resulting in the substitution of aspartic acid for asparagine at E2 position 62. Therefore, the coding change at amino acid 62 represents a second site or suppressor mutation which compensates for the primary mutation at position 114.

Monoclonal antibody R13 shares an antigenic site on E2 with R6. Escape mutants selected with this antibody also retained the arginine at 114, but had second site mutations which clustered at two other loci. Two mutants had a second site mutation at position 159 substituting glutamic acid for lysine. Both of these were virulent in neonatal mice. The other cluster of mutations was at position 96 where the tyrosine of SB-RL was replaced with histidine in two mutants and with phenylalanine in another. In contrast to the second site mutations at positions 62 and 159, neither of the replacements at position 96 caused reversion of SB-RL to virulence. As yet, none of the R13 selected mutants have been tested for penetration phenotype.

Selection of antibody escape mutants also allows provisional mapping of particular amino acid residues to specific epitopes. We have previously shown that R6 and R13 define an independently mutable, neutralizing E2 antigenic site accessible at the virion surface (10). Competition binding experiments indicated that this site, which we have designated E2c, was spatially distinct from two previously defined E2 sites (14). The positions of the antibody escape mutations selected with R6 and R13 suggest that E2 residues 62, 96 and 159 are constituents of the E2c antigenic site.

DISCUSSION

The successful selection of attenuated mutants by serial passage of virulent strains in cell culture implies the existence of genetic loci at which mutation simultaneously increases the efficiency of cell culture replication while decreasing the ability of the virus to cause overt disease in an animal. The evidence presented here strongly suggests that this was the case for the Sindbis virus - neonatal mouse model system. The arginine substitution in SB-RL for the wild-type serine at position 114 of glycoprotein E2 was the only coding change found in the genes for PE2 and E1. Phenotypic and sequence analysis of other virulent and attenuated isolates, of virulent revertants of SB-RL, and of an independently selected attenuated mutant, SB-FP, indicated that this single step mutation was responsible for reduced virulence in neonatal mice as well as rapid penetration of BHK cells. Although the complete genomes of SB and SB-RL were not sequenced, it is unlikely that a second mutation outside the glycoprotein genes was required for attenuation. Such an hypothesis would have required not only that this putative second mutation occur coincidentally with the mutation at E2 position 114 in at least two independent selection protocols but also that it undergo simultaneous reversion.

The sequence context of E2 position 114 and its conservation in two other alphaviruses suggests that it may be an important structural element of the virion spike. In SB, serine 114 is the central amino acid in a series of 18 hydrophobic and uncharged residues. The relative hydrophobicity of this region is conserved in Semliki Forest and Ross River viruses (15,16). Also, the tripeptide valine113-serine114-isoleucine115 is conserved in SB and Semliki Forest while Ross River has valine113-serine114-phenylalanine115. In addition to reducing the hydrophobicity of the region, an arginine substitution at position 114 would alter the local electrostatic and hydrogen bonding patterns within E2 and would introduce a large side chain in a presumably internal domain previously occupied by a much smaller residue. Therefore, introduction of a highly charged amino acid such as arginine into this hydrophobic region would be expected to have a significant effect on virion structure and/or stability.

An alteration in virion structure was indicated by the increased sensitivity of SB-RL and SB-FP to neutralization by monoclonal antibodies R6 and R13. R6 and R13 bind equally well to the surfaces of both SB and SB-RL, but they preferentially neutralize those virions containing the arginine for serine substitution at E2 position 114 (10). Another indication of a virion structural change derives from heat inactivation experiments. At 37C and 45C, comparable heat stability of SB and SB-RL was observed. At 51C however, those mutants containing arginine at E2 position 114 were considerably more heat sensitive than those having serine (Gidwitz and Johnston, unpublished).

In summary, we suggest that selection of attenuated mutants by passage in cell culture does rely on the existence of genetic loci at which mutation can simultaneously increase the efficiency of cell culture replication while reducing virulence in an animal host. Our data indicate that base 8972 of Sindbis virus, affecting E2 amino acid position 114, is one such locus. Substitution of an arginine at this site in SB-RL results in pleiotropic phenotypic changes including a reduction in virulence for neonatal mice, accelerated penetration of BHK cells and increased sensitivity to neutralization by monoclonal antibodies which recognize the E2c antigenic site.

ACKNOWLEDGMENTS

This work was supported by NIH grant AI22186 and the North Carolina Agricultural Research Service. The authors wish to thank William G. Dougherty for timely advice, especially regarding dideoxy chain termination sequencing, and Dennis T. Brown for the gift of the E1 primers.

REFERENCES

1. Andrewes CH, Pareira HG, Wildy P (1972). "Viruses of Vertebrates." London: Bailliere-Tindall, p. 76.
2. Strauss EG, Rice CM, Strauss JH (1984). Complete nucleotide sequence of the genomic RNA of Sindbis virus. Virology 133:92.
3. Strauss JH, Burge BW, Pfefferkorn ER, Darnell JE (1969). Identification of the membrane protein and 'core' protein of Sindbis virus. Proc Natl Acad Sci, USA 59:533.
4. Schlesinger MJ, Schlesinger S, Burge BW (1972). Identification of a second glycoprotein in Sindbis virus. Virology 47:539.

5. Reinarz ABG, Broome MG, Sagik BP (1971). Age-dependent resistance of mice to Sindbis virus infection: Viral replication as a function of host age. Infect Immun 3: 268.
6. Johnson RT, McFarland HF, Levy SE (1972). Age-dependent resistance to viral encephalitis: studies of infections due to Sindbis virus in mice. J Inf Dis 125:257.
7. Johnson RT (1965). Virus invasion of the central nervous system: A study of Sindbis virus infection in the mouse using fluorescent antibody. Am J Path 46:928.
8. Baric RS, Moore DB, Johnston RE (1980). In vitro selection of an attenuated variant of Sindbis virus. ICN-UCLA Symp Mol Cell Biol 18:685.
9. Olmsted RA, Baric RS, Sawyer BA, Johnston RE (1984). Sindbis virus mutants selected for rapid growth in cell culture display attenuated virulence in animals. Science 225:424.
10. Olmsted RA, Meyer WJ, Johnston RE (1986). Characterization of Sindbis virus epitopes important for penetration in cell culture and pathogenesis in animals. Virology 148:245.
11. Zimmern D, Kaesberg P (1978). 3' terminal nucleotide sequences of encephalomyocarditis virus RNA determined by reverse transcriptase and chain-terminating inhibitors. Proc Natl Acad Sci, USA 75:4257.
12. Ahlquist P, Dasgupta R, Kaesberg P (1981). Near identity of 3' RNA secondary structure in bromoviruses and cucumber mosaic virus. Cell 23:183.
13. Johnston RE, Mackenzie JM, Dougherty WG (1986). Assembly of overlapping DNA sequences by a program written in BASIC for 64K CP/M and MS-DOS IBM-compatible microcomputers. Nuc Acids Res 14:517.
14. Stec DS, Waddell A, Schmaljohn CS, Cole GA, Schmaljohn AL (1986). Antibody-selected variation and reversion in Sindbis virus neutralization epitopes. J Virol 57:715.
15. Garoff H, Frischauf AM, Simons K, Lehrach H, Delius H (1980). Nucleotide sequence of cDNA coding for Semliki Forest virus membrane glycoproteins. Nature (London) 288:235.
16. Dalgarno L, Rice CM, Strauss JH (1983). Ross River virus 26S RNA: Complete nucleotide sequence an deduced sequence of the encoded structural proteins. Virology 129:170.

Positive Strand RNA Viruses, pages 477–486

NATURAL DISTRIBUTION OF WILD TYPE 1 POLIOVIRUS GENOTYPES

Rebeca Rico-Hesse, Mark A. Pallansch
Baldev K. Nottay, and Olen Kew

Division of Viral Diseases, Center for Infectious Diseases
Centers for Disease Control, Atlanta, Georgia 30333

ABSTRACT Determination of the patterns of genomic variation of RNA viruses during natural replication has been a powerful method for establishing epidemiologic relationships among isolates. We have determined 150 bases of the genomic sequence encoding parts of the capsid protein VP1 and the noncapsid protein 2A for 45 wild polioviruses, obtained over a 30-year period from different geographic regions. Beyond revealing previously unrecognized links between outbreaks, these studies have provided a much broader view of the pathways of poliovirus transmission in nature.

Wild-type polioviruses are endemic to the developing countries of four continents and cause up to 400,000 cases of paralytic poliomyelitis annually (1,2). Introduction of wild strains from regions endemic for poliomyelitis has resulted in sporadic cases and limited outbreaks in developed countries, including the United States (3-5). Since a high proportion of poliovirus infections are asymptomatic, standard epidemiologic investigations may not reveal the origins of outbreak viruses. The sole conclusive evidence for relatedness of cases may derive from laboratory data (3,4). Such information is potentially critical to the implementation of effective control strategies.

Currently, poliovirus isolates are identified by their antigenic properties, using panels of neutralizing monoclonal antibodies (6-8), or by genotype, using RNase T_1 oligonucleotide fingerprinting (4,9,10). Each approach has important limitations. For example, the potential for polioviruses to undergo rapid antigenic variation during replication in humans can yield inconclusive results even when the best available monoclonal antibody panels are used (6). In contrast, oligonucleotide fingerprinting provides a rapid method for estimating genomic sequence relationships among RNA viruses, and can readily identify relationships among antigenic variants of the same serotype (3,8). However, quantitative estimates of relatedness are most reliable when the RNA genomes share $\geq$95% base sequence homology (11). Because the poliovirus genome evolves at a rate of about two base substitutions per week during epidemic transmission (3), fingerprinting is limited to recognizing relationships between isolates separated from the ancestral infection by no more than 3 to 5 years.

To overcome these limitations, we have used primer extension sequencing (12-14) to identify and compare poliomyelitis case isolates. Genomic sequencing was first used as an epidemiologic tool to monitor the transmission of influenza viruses (15,16). However, since the number of distinct influenza genotypes in simultaneous circulation is small, fingerprinting is generally sufficient for most epidemiologic studies. In contrast, many different genotypes of poliovirus currently exist in nature. Therefore, the capacity of genomic sequencing to reveal relationships among highly divergent strains has made this method especially suitable for tracking poliovirus isolates. Moreover, selected genomic regions can be specifically targeted for comparison, such that genomic regions having different rates of evolution may be independently examined.

We compared 45 wild-type 1 poliovirus isolates obtained from patients in five continents over a 30-year period. Type 1 strains were selected because they are associated with the greatest number of cases of paralytic disease worldwide. Nucleotide sequences were obtained for a genomic interval encoding the amino terminal region of the noncapsid protein 2A and the carboxy terminal region of VP1 (Figure 1). The VP1 sequences included those encoding a surface polypeptide loop thought to form a portion of a potential antibody binding site (17,18). This interval could be conveniently targeted because of the existence of a conserved site for primer binding adjacent to upstream sequences that are known to vary substantially across the three poliovirus serotypes (19). While many conserved sites exist along the poliovirus genome, they are most abundant within the noncapsid sequences, where active recombination has been detected among genetically

```
A. NUCLEOTIDE SEQUENCES:                                              VP1<--┐
                                                                            |
              3320                                                        3385
SABIN 1        GGCCCUGGAGUGGAUUACAAGGAUGGUACGCUUACACCCCUCUCCACCAAGGAUCUGACCACAUAU

USA/TX     70  --U-----G-----------------C--C-----U-----U--U-----A---U-A-----U---
USA/CT     72  --U-----G--------------------C-----U--U--U--------A---U-A-----U---
MEXICO     77  --U-----G-----------A--------C--C--U------------------U-------U---
USA/FL     77  --U-----G-----------A--------C--C--U------------------U-------C---
SALVADOR   81  --U-----G-----C--------------C--C--U-----U--------A---U---U---U---
HONDURAS   84  --U-----G-----C--U-----------C--C--------U------------U---U---U--C

BRAZIL     81  -----A--G--------U--A-----A--A-----U--A-----A--A-----CU-------G--C
ARGENTINA  82  -----A--G--U-----U--A-----G--A-----U--A-----A--------CU---U---G--C
DOM.REP.   82  -----A-----U--------A--C--A--A-----U--A-----A---------U---U---G--C
JAMAICA    82  -----A-----U--------A--C--A--A-----U--A-----A---------U---U---G--C
HAITI      85  -----A-----C--------A--C--A--A-----U--A-----A---------U-------U--C

NICARAGUA  77  -----------A-----------------U--CG-U-----U-----------C--A-----G---
NICARAGUA  78  -----------A-----------------U--CG-U-----U--U--------C--A-----G---

VENEZUELA  81  --U--G--G--A-----U--A-----C--U--CG-U---U-GA-------A--C--A--A-----C

TURKEY     77  -----A-----A--------A------U-U----G---UU-G-----------CU----A--C---
NETHER.    78  -----A-----A--------A--------U----G---UU-A-----------CU----A--C---
USA/MO     79  -----A-----A--------A--------U----G---UU-A-----------CU----A--C--C

INDIA      82  -----A--------------------G--A--C--C-----------U--A--C------------
```

```
B. DEDUCED AMINO ACID SEQUENCES:
                                                       VP1<--┐┌-->2A
                      273                                    ||                   20
SABIN 1                PPRAVAYYGPGVDYKDGTLTPLSTKDLTTYGFGHQNKAVYTAGYKICNYH

USA/TX           70    --------------------------------------------------
USA/CT           72    --------------------------------------------------
MEXICO           77    --------------------------------------------------
USA/FL           77    -------------------------------------------N------
EL SALVADOR      81    --------------------------------------------------
HONDURAS         84    R-------------------------------------------------

BRAZIL           81    --------------------------------------------------
ARGENTINA        82    --------------------------------------------------
DOMINICAN REP.   82    --------------------------------------------------
JAMAICA          82    --------------------------------------------------
HAITI            85    --------------------------------------------------

NICARAGUA        77    -------------------A------------------------------
NICARAGUA        78    -------------------A------------------------------

VENEZUELA        81    -------------------A--T---------------------------

TURKEY           77    -----------------S-A------------------------------
NETHERLANDS      78    -------------------A------------------------------
MISSOURI         79    -------------------A------------------------------

INDIA            82    AR-----------------------------Y-----R------------
```

Figure 1. A. Comparison of nucleotide sequences at the 3' end of the VP1 region of type 1 polioviruses of the North American, Caribbean, Nicaraguan, Venezuelan, and Indian genotypes. Sequencing was performed by extension of a synthetic DNA primer with reverse transcriptase in the presence of alpha-thio[^{35}S]dATP and dideoxynucleoside triphosphates (12,13,14). Nucleotide positions are numbered according to Nomoto, *et al.* (22). Nucleotide differences from Sabin 1 are shown, dashes indicate identities. The 66 nucleotides shown in each row constitute 44% of the total sequence information used for each strain to construct the phylogenetic tree in Figure 2.

B. Amino acid sequences deduced from the 150 bases of nucleotide sequence used for determination of genetic relatedness among type 1 isolates.

divergent polioviruses (20). Genetic exchanges appear to occur much less frequently within the capsid region; thus, we believe that the sequence differences among isolates observed in this study arose primarily by progressive fixation of mutations.

When the genomic sequences of the wild type 1 strains were compared with the corresponding sequences for the Sabin 1 vaccine strain, LSc 2ab (derived from the 1941 USA isolate, Mahoney), substantial divergence among isolates was found (Figure 1A). Over an interval of 150 nucleotides(60 from 2A and 90 from VP1), the wild isolates differed from Sabin 1 at 23 (15%) to 34 (23%) nucleotide positions, representing 44% to 60% of codons. The great majority (>98%) of mutations produced synonymous codons. The deduced polypeptide sequences encoded over this interval were identical to Sabin 1 for 19 of the 45 strains, and differed by up to 4 amino acid residues (8%) only for the 1982 isolate from India (Figure 1B). Thus, despite a high rate of fixation of mutations into the genome, polypeptide sequences were strongly conserved.

Rather than comparing sequences against those of an arbitrarily chosen reference strain, analyses can be refined and quantitated by performing pairwise comparisons between all sequences (21). Comparisons were made with a computer program written by one of the authors (MAP), and the results presented graphically in the form of a phylogenetic tree (Figure 2).

Sequence relatedness among isolates correlated better with geographic proximity than with time of isolation (Figure 2). These results suggest that genotypes characteristic of a particular endemic focus generally persist in the environment for many years, and are only infrequently displaced by genotypes from a different region. Thus, isolates from the United States (prevaccine period), Mexico, and El Salvador constitute a genotypic group, while isolates from the Caribbean region form a separate group. In all, at least 10 distinct genotypic groups (Caribbean, Nicaraguan, North American, Indian, Peruvian, Moroccan, Taiwanese, Middle Eastern, South African, Venezuelan) distinct genotypic groups were recognized. Although the close relationships among isolates within a group were generally unambiguous, the evolutionary relationships between different groups was difficult to asess, as the nucleotide base sequence divergence across groups (up to 22%, Figure 2) approached that observed across the three poliovirus serotypes (29%) in the same interval (19).

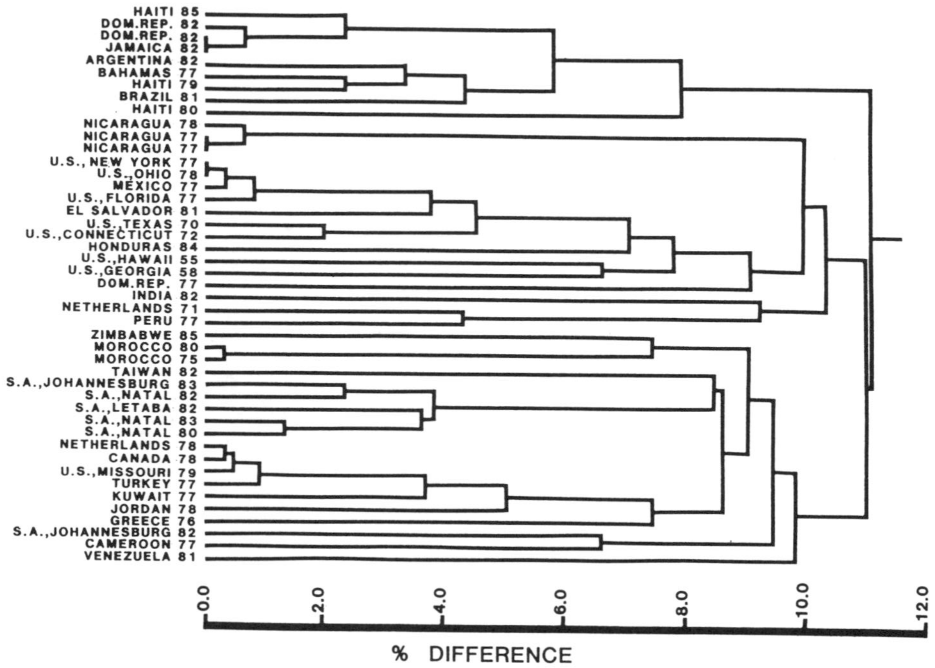

Figure 2. Phylogenetic tree (21) constructed according to sequence relatedness among wild type 1 polioviruses across the interval of nucleotides 3296 to 3445 (VP1-2A region). All pairwise comparisons were performed using a computer program that assigned each base substitution an equivalent statistical weight (M.A. Pallansch, unpublished). The per cent base sequence divergence between any two strains is twice the distance along the abscissa to the connecting node.

The phylogenetic tree bears some very interesting epidemiologic information. For example, since 1970 three outbreaks of poliomyelitis have occurred in the United States (Texas 1970, Connecticut 1972, Pennsylvania 1979), along with several sporadic cases, all attributable to wild type 1 strains (4). With the exception of the 1979 cases among the Amish of Pennsylvania, Missouri, and several other states (3), isolates from all other cases were closely related to polioviruses from Mexico, El Salvador, and Honduras (Figures 2 and 3). The 1979 isolate from Missouri is very closely related to isolates from Canada and the Netherlands, as had been shown previously by fingerprinting (3). The origin of the Netherlands epidemic virus was unknown, but a link to the Middle East was suspected (7). Interestingly, a close correspondence was found between the sequences of isolates from the Netherlands (1978) and Turkey (1977), strongly suggesting the existence of a direct link between poliomyelitis cases in the two countries. The isolate from Turkey is more closely related to the Netherlands virus than to other Middle Eastern isolates from Jordan (1978) and Kuwait (1977). These findings highlight the potential for wild polioviruses to be transmitted from any region of the world to susceptible persons in distant communities. Another example of this potential is illustrated by the wide distribution of a genotype from the Caribbean, Brazil, and Argentina. Most recent isolates from Haiti, the Dominican Republic, and Jamaica were closely related to each other (Figure 2). In fact, one 1982 isolate from the Dominican Republic was more similar to a 1982 Jamaican isolate than to another contemporary Dominican Republic isolate, a result fully consistent with earlier fingerprint data (4). On the other hand, the genetic similarities among isolates from the Caribbean, from Brazil, and from a 1982 epidemic in Argentina were quite unexpected. A common epidemiologic origin is suggested, presumably by recent importation from a single, as yet unidentified, endemic focus. It will be interesting to determine whether the range of this genotype has become more restricted following intensive immunization programs in Brazil and Argentina. While the range of the Caribbean genotype included the Bahamas, we have so far obtained no evidence for transmission of this genotype of poliovirus to susceptible persons in the United States (Figure 3).

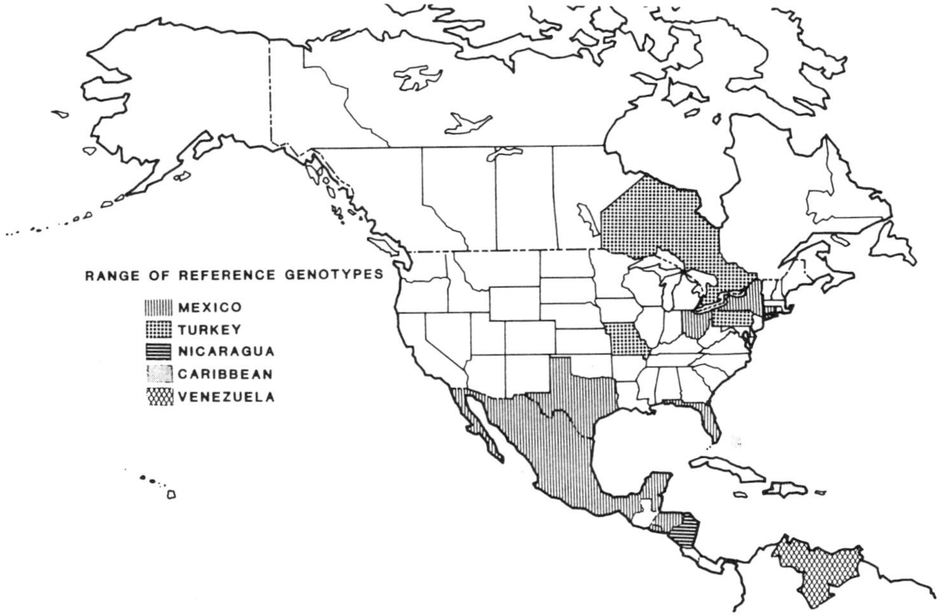

Figure 3. Range of different genotypes of type 1 poliovirus isolates from North and Central America, and the Caribbean (including Venezuela). Strains of the Middle Eastern genotype were isolated from patients in Turkey, Kuwait, Jordan, the Netherlands, four Canadian provinces, and six states in the USA (Figure 2; 3). All other USA wild type 1 isolates were related to strains from Mexico, El Salvador, and Honduras. The range of the Caribbean genotype extended from Haiti and the Dominican Republic to Jamaica, the Bahamas, Brazil, and Argentina. Type 1 isolates from Nicaragua and Venezuela appeared to represent two separate genotypes.

Although most of the epidemiologic transmission patterns revealed by these sequence studies are consistent with known patterns of human migration, some of the results were quite surprising. For example, it is not clear why the isolates from Nicaragua or Venezuela appear to be distinct from viruses obtained in neighboring countries. Neither do we understand why the Dominican Republic isolate from 1977 or the Haitian isolate from 1980 differ so markedly from each other and from other strains of the region. Such findings may imply that epidemiologically separate pockets of poliovirus transmission coexist in adjacent geographic areas. The dynamics of poliovirus circulation in these apparently exceptional situations may become better understood when a more comprehensive set of isolates is compared.

Our analyses can be extended both by sequencing a larger number of isolates and by generating more sequence information per strain. To test the effects of inclusion of differing amounts of sequence information from the VP1-2A genomic region, phylogenetic trees were constructed using 100, 150, and 245 nucleotides of sequence information for each member of a subset of strains. Inclusion of sequence information above 100 nucleotides did not significantly alter the structure of the phylogenetic tree. At least for this genomic region, differences in the evolution rate of functionally distinct domains tended to average out even across an interval of only 100 nucleotides, presumably because nearly all mutations were conservative. Thus, a reasonable picture of the natural distribution and transmission of wild type 1 polioviruses was obtained by sampling as little as 1.5% of the total genome. However, the results presented here did not provide a reliable quantitative measure of the extent of evolutionary divergence among major groups of type 1 strains. Consequently, we are comparing intervals within the generally conserved 5'-noncoding region to detect more distant relationships among polioviruses. It is hoped that by integration of sequence information obtained from different genomic regions of many strains, a detailed picture of the worldwide patterns of poliovirus transmission will become apparent.

REFERENCES

1. Assaad, F., and Ljungars-Esteves, K. (1984). World

overview of poliomyelitis: Regional patterns and trends. Rev. Infect. Dis. 6 (Supp. 2): S302-S307.

2. John, T.J. (1984). Poliomyelitis in India: Prospects and problems of control. Rev. Infect. Dis. 6 (Supp. 2): S439-S441.

3. Nottay, B.K., Kew, O.M., Hatch, M.H., Heyward, J.T., and Obijeski, J.F. (1981). Molecular variation of type 1 vaccine-related and wild polioviruses during replication in humans. Virology 108: 405-423.

4. Kew, O.M., and Nottay, B.K. (1984). Molecular epidemiology of polioviruses. Rev. Infect. Dis. 6 (Supp. 2): S499-S504.

5. Centers for Disease Control-Atlanta. (1985). Poliomyelitis-Finland. MMWR 34: 5-6.

6. Crainic, R., Couillin, P., Blondel, B., Cabau, N., Boue, A., and Horodniceanu, F. (1983). Natural variation of poliovirus neutralization epitopes. Infect. Immun. 41: 1217-25.

7. Osterhaus, A.D.M.E., van Wezel, A.L., Hazendonk, T.G., UytdeHaag, F.G.C.M., van Asten, J.A.A.M., and van Steenis, B. (1983). Monoclonal antibodies to polioviruses. Comparison of intratypic strain differentiation of poliovirus type 1 using monoclonal antibodies versus crossabsorbed antisera. Intervirology 20: 129-136.

8. Minor, P.D., Schild, G.C., Ferguson, M., Mackay, A., Magrath, D.I., John, A., Yates, P.J., and Spitz, M. (1982). Genetic and antigenic variation in type 3 polioviruses: Characterization of strains by monoclonal antibodies and T1 oligonucleotide mapping. J. Gen. Virol. 61: 167-176.

9. Clewley, J.P., and Bishop, D.H.L. (1982). Oligonucleotide fingerprinting of viral genomes. In: New Developments in Practical Virology (C.R. Howard, ed.) pp. 231-277. Alan R. Liss, New York.

10. Kew, O.M., Nottay, B.K., and Obijeski, J.F. (1984). Applications of oligonucleotide fingerprinting to the identification of viruses. In: Methods in Virology (K. Maramorosch and H. Koprowski, eds.) Vol. 8, pp. 41-84. Academic Press, New York.

11. Aaronson, R.P., Young, J.F., and Palese, P. (1982). Oligonucleotide mapping: Evaluation of its sensitivity by computer-simulation. Nucleic Acids Res. 10: 237-246.

12. Sanger, F., Nicklen, S., and Coulson, A.R. (1977).

DNA sequencing with chain-terminating inhibitors. Proc. Nat. Acad. Sci. USA 74: 5463-5467.

13. Zimmern, D., and Kaesberg, P. (1978). 3'-Terminal nucleotide sequence of encephalomyocarditis virus RNA determined by reverse transcriptase and chain-terminating inhibitors. Proc. Nat. Acad. Sci. USA 75: 4257-4261.
14. Biggin, M.D., Gibson, T.J., and Hong, G.F. (1983). Buffer gradient gels and ^{35}S label as an aid to rapid DNA sequence determination. Proc. Nat. Acad. Sci. USA 80: 3963-3965.
15. Ward, C.W. (1981). Structure of the influenza virus hemagglutinin. Curr. Topics Microbiol. Immunol. 94/95: 1-74.
16. Skehel, J.J., Daniels, R.S., Douglas, A.R., and Wiley, D.C. (1983). Antigenic and amino acid sequence variations in the hemagglutinin of type A influenza viruses recently isolated from human subjects. Bull. WHO 61: 671-676.
17. Hogle, J.M., Chow, M., and Filman, D.J. (1985). Three-dimensional structure of poliovirus at 2.9 A resolution. Science 229: 1353-1365.
18. Chow, M., Yabrov, R., Bittle, J., Hogle, J., and Baltimore, D. (1985). Synthetic peptides from four separate regions of the poliovirus type 1 capsid protein VP1 induce neutralizing antibodies. Proc. Nat. Acad. Sci. USA 82: 910-914.
19. Toyoda, H., Kohara, M., Kataoka, Y., Suganuma, T., Omata, T., Imura, N., and Nomoto, A. (1984). Complete nucleotide sequences of all three poliovirus serotype genomes: Implication for genetic relationship, gene function and antigenic determinants. J. Mol. Biol. 174: 561-585.
20. Kew, O.M., and Nottay, B.K. (1984). Evolution of the oral poliovaccines strains in humans occurs by both mutation and intramolecular recombination. In: Modern Approaches to Vaccines (R. Chanock and R. Lerner, eds.) pp. 357-362. Cold Spring Harbor Laboratory, Cold Spring Harbor, New York.
21. Fitch, W.M., and Margoliash, E. (1967). Construction of phylogenetic trees. Science 155: 279-284.
22. Nomoto, A., Omata, T., Toyoda, H., Kuge, S., Horie, H., Kataoka, Y., Genba, Y., Nakano, Y., and Imura, N. (1982). Complete nucleotide sequence of the attenuated poliovirus Sabin 1 strain genome. Proc. Nat. Acad. Sci. USA 79: 5793-5797.

Positive Strand RNA Viruses, pages 487–495

RHINOVIRUS DETECTION BY cDNA:RNA HYBRIDIZATION

W. Al-Nakib[1], G. Stanway[2*], M. Forsyth[1], P.J. Hughes[2], J.W. Almond[3] and D.A.J. Tyrrell[1]

MRC Common Cold Unit, Salisbury[1], Department of Biology, University of Essex[2] and Department of Microbiology, University of Reading[3], UK

ABSTRACT An M13 template comprising the first 800 nucleotides from the 5' end of human rhinovirus-14 was constructed. A ^{32}P-labelled probe was then generated by primer extension and used in cDNA:RNA hybridization experiments. Of the 35 human rhinoviruses so far investigated the probe hybridized with RNA from 32 (91%). The probe, however, did not react with RNA from other respiratory viruses such as influenza A and B viruses or coronavirus 229E, or extracts from uninfected tissue-culture fluids. The inclusion of inhibitors of RNase activity such as vanadyl-ribonucleoside-complexes (VRC) was found essential to avoid loss of sensitivity especially when virus was present at low concentration. Decreasing the concentration of formamide in the hybridization mixture was found to increase the hybridization signal for the heterologous rhinovirus serotype.

INTRODUCTION

The entire genome of human rhinovirus-14 (HRV-14) has recently been cloned and sequenced (1). The virus is a typical member of the picornaviridae family and possesses a single-stranded RNA genome of approximately 7200 nucleotides. The 5'-first 624 nucleotides are presumed to have a non-coding function. These are followed by the single open-reading frame of 6537 nucleotides, a short 3'

*This work was supported by an MRC grant no.G8324256CB

non-coding region (47 nucleotides) and a poly A tract. The 5' non-coding region exhibits quite a high degree of nucleotide sequence homology to those of related picornaviruses (1). We have therefore investigated the use of an M13 derived cDNA probe containing this region for the detection of homologous and heterologous rhinovirus RNA by hybridization.

MATERIALS AND METHODS

Construction of M13 Template

An HRV-14 clone, pAM 2, carries a cDNA insert corresponding to the first 2200 nucleotides of the genome. The cDNA is flanked by PstI sites and contains a unique Bgl II site at position 798. pAM 2 was digested with PstI and Bgl II and was ligated into M13 mp9 cut with PstI and Bam HI. After transfection and isolation of template DNA, the correct construct was identified by sequencing. This M13 molecule was then used as a template for the production of a labelled second strand complementary to the first 798 nucleotides of HRV-14 RNA (figure 1). This corresponds to the relatively conserved 5' non-coding region (position 1-624) and most of the sequence coding for VP4.

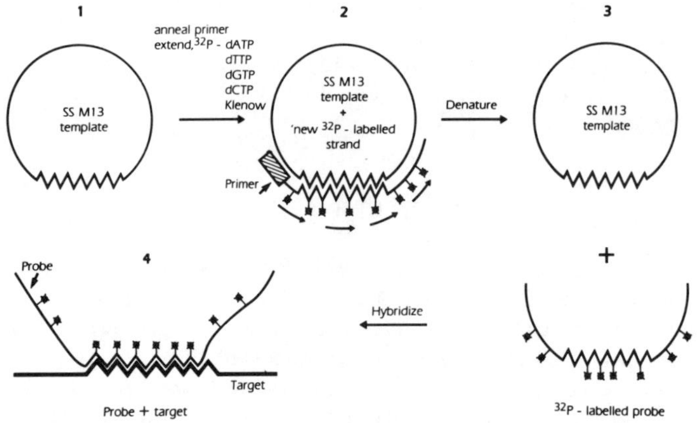

FIGURE 1. Preparation of cDNA probe.

cDNA Probe Preparation

To 2.5 µl of M13 template, 0.5 µl of distilled H_2O, 1 µl of primer (2 µg/ml, BRL) and 1 µl of TM buffer (100 mM Tris, 50 mM $MgCl_2$, pH 8.5) were added. The primer was annealed to the template (figure 1) by boiling in a water bath for three minutes and allowed to cool on the bench for 15 minutes. The mixture was then centrifuged for 15 seconds to bring down condensation. Two microlitres of ^{32}P-dATP (3000 µCi/m.mol), 8 µl of 1 mM dNTP mixture (dTTP, dGTP and dCTP in 50 mM Tris, 1 mM EDTA buffer, pH 8.0), 6 µl of 3 mM dithiotreitol and 8 U of the Klenow fragment of DNA polymerase 1 (BRL) were added and the mixture left on the bench for 30 minutes. The ^{32}P-labelled M13-probe was separated from unincorporated material by loading into a Sephadex G100 column.

Viral RNA Extraction and Hybridization

Details of these will be published elsewhere (Al-Nakib et al., manuscript in preparation). Briefly, samples containing stocks of rhinoviruses and other control viruses were prepared and treated with VRC (20 mM), sodium dodecyl sulphate and protease K. The RNA was extracted twice with a mixture of phenol/chloroform and precipitated with sodium acetate and ethyl alcohol. Viral RNA was then spotted onto nitrocellulose filters and these were pre-hybridized for 8 to 18 hours, hybridized overnight with ^{32}P-labelled cDNA probe (1×10^7 to 1×10^8 CPM/µg of DNA), washed and exposed to an x-ray film at -70° C.

RESULTS

Table 1 shows that our 5' end probe hybridized with 32 of the 35 (91%) human rhinoviruses so far investigated. Only HRV-13, 45 and 47 were negative. Furthermore, the probe also reacted with RNA from bovine rhinovirus EC11 and coxsackie A21 virus. In contrast, the probe did not cross-hybridize with RNA from any of our control virus panels; reovirus types 1 and 3, influenza viruses A, B, parainfluenza virus type 3 and coronavirus 229E or DNA from herpes simplex virus type 1 or control uninfected tissue-culture fluids. It also did not hybridize to the picornaviruses, calf rhinovirus SD1 and echovirus type 1.

TABLE 1
DETECTION OF RHINOVIRUSES BY cDNA:RNA HYBRIDIZATION USING HRV-14 ^{32}P-LABELLED PROBE

Hybridization reaction	Viruses
Positive	HRV-1A, 1B, 2, 3, 4, 5, 6, 7, 8, 9, 12, 14, 15, 16, 17, 18, 19, 23, 28, 29, 30, 31, 32, 48, 49, 64, 65, 72, 75, 77, 80, 83, EL, Bov RV EC11, Coxsackie A21
Negative	HRV-13, 45 and 47, calf RV SD1, Echo 1, Reoviruses 1 and 3, Influenza A and B, Parainfluenza 3, coronavirus 229E, Herpes simplex type 1

As can be seen from figure 2 the strength of the hybridization signal depends on the degree of homology in nucleic acid sequences in the 5' end of the genome.

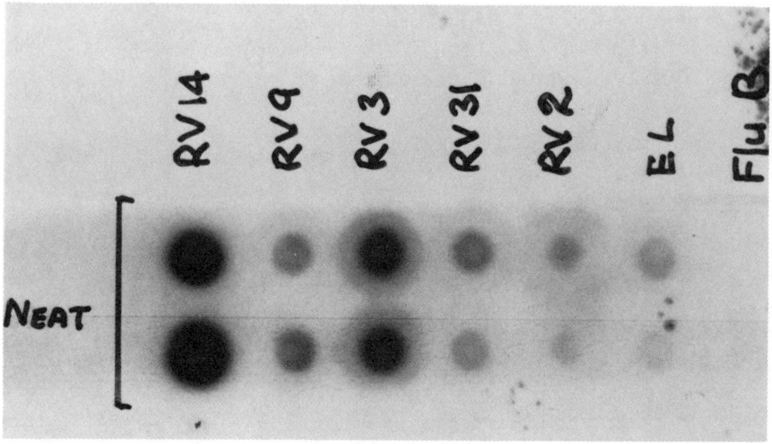

Thus HRV-3 appears to be more closely related to HRV-14 than are HRV-2, 9, 31 or EL, since it hybridizes to a greater extent while the titres, in terms of $TCID_{50}$/ml, were generally similar.

Table 2 demonstrates the importance of including vanadyl-ribonucleoside-complexes (VRC) in the extraction mixture. Generally, a stronger hybridization signal was obtained when VRC was included.

TABLE 2
EFFECT OF INCLUDING VRC AS INHIBITORS OF ENDOGENOUS RNase ACTIVITY ON THE DETECTION RATE OF RNA BY HYBRIDIZATION

	Virus dilution			
Formamide concentration	10^{-1}	10^{-2}	10^{-3}	HRV-14* in nasal washing
with VRC				
40%	++++	++	±	+
30%	++++	++	±	N.D.
20%	++++	+++	+	N.D.
10%	++++	+	-	N.D.
without VRC				
40%	++++	+	-	-
30%	++++	+	-	N.D.
20%	++++	++	±	N.D.
10%	++++	±	-	N.D.

*This is a clinical specimen from a volunteer who was infected with HRV-14 and who was subsequently shown by virus isolation to be excreting virus.

++++	very strong hybridization signal
+++	very good " "
++	good " "
+	positive " "
±	weak " "
-	negative " "
N.D.	not done

Furthermore, as was shown in table 2 this was particularly important when virus was detected at higher dilutions, i.e. 10^{-3}. This implied that in clinical specimens, such as nasal washings where virus concentration is expected to be relatively low, i.e. $< 10^2$ $TCID_{50}$/ml and where endogenous RNase activity is expected to be high, it would be imperative to include such inhibitors routinely. To test this hypothesis we have investigated the presence of viral RNA in a nasal washing from a volunteer who was infected with HRV-14 and who was subsequently shown to be excreting the virus. The RNA was extracted with and without VRC. Only in the presence of VRC were we able to detect a positive reaction (table 2).

Table 3 shows that it would be possible to increase the strength of the hybridization signals for the heterologous viruses, such as HRV-2, 3, 9, 31 and EL, by decreasing the concentration of formamide from 40% to 20% and hence decreasing the stringency of hybridization. This can be done without affecting the specificity of detection regarding control viruses such as influenza B or coronavirus 229E.

DISCUSSION

The data presented in this paper indicate the feasibility of constructing M13 templates incorporating specific regions of choice from the human rhinovirus genome, and then using these templates to generate labelled cDNA for use as a probe. The cDNA:RNA hybridization system used in this study was sensitive enough to detect rhinovirus specific RNA when virus was present at $10^{2.8}$ $TCID_{50}$/ml (data not presented) and proved to be extremely reproducible yielding consistent results.

Our results show that there is considerable cross-hybridization between the various human rhinoviruses using a probe constructed from the 5' non-coding region. Thus, 32 of 35 (91%) rhinoviruses examined were positive by hybridization. Only HRV-13, 45 and 47 were negative. We will, therefore, be re-examining these three rhinoviruses this time at higher concentrations. However, as was shown in figure 2, there was also quite a variation in the strength of the hybridization signals depending on the genomic homology in the 5' region of the genome. Rotbart et al. (2, 3) and Tracy (4) have also demonstrated considerable homology between the enteroviruses using cDNA

TABLE 3
EFFECT OF FORMAMIDE CONCENTRATION ON THE DETECTION RATE OF THE DIFFERENT HUMAN RHINOVIRUSES

Formamide concentration	Spot (5 µl) no.	Type of HRV and control virus							
		HRV-14	HRV-9	HRV-3	HRV-31	HRV-2	HRV EL	'Flu B	Coron. 229E
40%	1	++++	++	+++	+	–	–	–	–
	2	++++	++	+++	+	±	–	–	–
30%	1	++++	++	+++	++	+	+	–	–
	2	++++	++	+++	++	+	+	–	–
20%	1	++++	+++	++++	+++	++	±	–	–
	2	+++	+++	++++	+++	++	+	–	–

++++ very strong hybridization signal
+++ very good " "
++ good " "
+ positive hybridization signal
± weak " "
– negative " "

probes from polio and coxsackie viruses.

This study also provides evidence that extracted viral RNA is quite vulnerable to endogenous RNase activities leading to loss of RNA and a reduction in the sensitivity of detection. This effect is particularly important when virus is present in small concentrations, as the case may be in clinical specimens such as nasal washings. It is, therefore, important that inhibitors of RNase activity are included during all extraction procedures. We have used vanadyl-ribonucleoside-complexes (VRC) and found these to be quite satisfactory.

Although we routinely use 40% formamide in our hybridization reaction, we have investigated the effect of reduced concentration of formamide in the reaction mixture and whether this will amplify the hybridization signal at least for the heterologous rhinovirus serotype. As was shown in table 3, stronger hybridization signals were obtained for the heterologous viruses, i.e. HRV-2, 3, 9, 31 and EL, at 20% formamide when compared to 40% formamide. Hybridization at lower stringency conditions can therefore probably result in an improved detection rate of rhinoviruses and probably also other picornaviruses. This did not seem to affect the specificity of the reaction, at least for other respiratory viruses, such as influenza B and coronavirus 229E or uninfected tissue culture extracts.

We hope to extend our studies to include a larger series of human rhinoviruses and study in more detail the molecular relationship between the various human rhinoviruses. We also hope to evaluate further probes, that we have recently constructed, in the detection of human rhinoviruses with the ultimate objective of developing a clinical assay for the diagnosis of rhinovirus infections using cDNA:RNA hybridization.

REFERENCES

1. Stanway G, Hughes PJ, Mountford RC, Minor PD, Almond JW (1984). The complete nucleotide sequence of a common cold virus: human rhinovirus 14. Nucl Acid Res 12:7859.
2. Rotbart HA, Levin MJ, Villarreal LP (1984). Use of subgenomic poliovirus DNA hybridization probes to detect the major subgroups of enteroviruses. J Clin Microbiol 20:1105.

3. Rotbart HA, Levin MJ, Villarreal, Tracy SM, Semler BL, Wimmer E (1985). Factors affecting the detection of enteroviruses in cerebrospinal fluid with coxsackie virus B3 and poliovirus 1 cDNA probes. J Clin Microbiol 22:220.
4. Tracy S (1985). Comparison of genomic homologies in the coxsackie virus B group by use of cDNA:RNA dot-blot hybridization. J Clin Microbiol 21:371.

Positive Strand RNA Viruses, pages 497–507

SINGLE-CYCLE GROWTH KINETICS OF HEPATITIS A VIRUS IN BSC-1 CELLS

David A. Anderson, Stephen A. Locarnini, Bruce C. Ross, Anthony G. Coulepis, Bruce N. Anderson, and Ian D. Gust

Virus Laboratory, Fairfield Hospital, Yarra Bend Road, Fairfield 3078, Victoria, Australia

ABSTRACT Hepatitis A virus strain HM175 derived from persistently infected BSC-1 cell cultures was passaged at a low multiplicity of infection at 3 day intervals in BSC-1 cells. The single-cycle growth kinetics of HM175 were determined by radioimmunofocus assays for infectious virus and infectious RNA, indirect immunofluorescence for double-stranded RNA, in situ radioimmunoassay for viral antigen, and dot-blot hybridization of virus-related RNA with hepatitis A virus-specific cDNA probes. In contrast to the stable persistent infection from which the virus was isolated, BSC-1 cells infected with this HM175 pool developed a cytopathic effect in 2 to 3 days, which was prevented by preincubation of the virus with hepatitis A virus-specific monoclonal antibody K3-4C8. In cells infected at a multiplicity of 1 radioimmunofocus-forming unit per cell, an increase in intracellular infectious virus was first detected at 18 h, with a logarithmic phase lasting until 30 h, followed by a gradual increase to 51 h, at which time there was a yield of 20 radioimmunofocus-forming units per cell. Examination of the other parameters of infection studied suggests that RNA synthesis may be a rate-limiting step in the replication of HAV in BSC-1 cells.

This work was supported in part by a grant from the National Health and Medical Research Council. D.A.A. is the recipient of a Postgraduate Research Scholarship from the University of Melbourne.

INTRODUCTION

Hepatitis A virus (HAV) has been classified as an Enterovirus within the family Picornaviridae on the basis of biophysical and biochemical characteristics (1,2), but exhibits some growth properties atypical of this genus. Although many groups have been successful in propagating the virus in cell culture since the original report by Provost and Hilleman (3), HAV generally causes slow, persistent infections of susceptible cell cultures (3,4,5) with low yields of infectious HAV and HAV antigen (HAAg). The virus does not affect host-cell macromolecular synthesis (5,6) and is generally noncytopathic, making conventional infectivity assays untenable; accordingly, the study of the growth cycle of HAV in vitro has been difficult. However, with the development of a radioimmunofocus assay (RIFA) for infectious HAV in cell culture (7) it has become possible to study the growth cycle in detail. Recently, a one-step growth curve for HAV strain HAS-15, grown at a high multiplicity of infection (m.o.i.) in FRhK-4 cells has been reported, in which a lag phase of 20 h was observed prior to the production of large amounts of infectious HAV (8).

In an attempt to delineate the specific restrictive event(s) in the growth of HAV in vitro, we describe several parameters of virus replication under single-cycle growth conditions in BSC-1 cells (African green monkey kidney) using a cytopathic stock of HAV strain HM175 grown at a low m.o.i. to minimize the production of defective virus.

METHODS

Cells and Viruses

BSC-1 cells were cultivated in MEM containing 5% fetal calf serum (FCS). BSC-1 cells and BSC-1 cells persistently infected with HAV strain HM175 were obtained from S.M. Feinstone, National Institutes of Health, Bethesda, Md., and were subcultured each 7 to 28 days. HM175 was harvested from persistently infected cells in the 30th passage level in our laboratory by 3 cycles of freezing and thawing, followed by chloroform extraction, and was passaged a further 5 times at a m.o.i. of 0.01 radioimmunofocus-forming units (RFU) per cell at 3 day intervals at 34°. The infectious stock for growth curve analysis was prepared by removal of infected cells from roller bottles using 0.1% trypsin and 0.2% EDTA. The cells were washed with growth medium and pelleted at

5000g for 10 min. The cell pellet was resuspended in 5 volumes of 0.85% NaCl and subjected to 3 cycles of freezing and thawing followed by extraction with an equal volume of chloroform. After centrifugation at 15000g for 10 min, the aqueous phase was removed and Sarkosyl was added to a final concentration of 1%. Aliquots of 1 ml were layered on top of linear 10-50% sucrose gradients and centrifuged at 25000 rpm for 3 h at 18° in a Beckman SW41 rotor. Peak virus-containing fractions were pooled, diluted 1:4 in saline, and virus was pelleted at 40000 rpm for 5 h at 4° in an SW41 rotor. The virus-containing pellet was resuspended in MEM and extracted with an equal volume of chloroform, and stored at -70°.

Poliovirus type 1 (PV, Mahoney) was prepared similarily 24 h after infection, except that the cells and culture medium were pooled before extraction with chloroform.

Light-sensitive HM175 and PV were prepared after growth in the presence of 0.01% neutral red in the dark.

Infectivity Assays

The HAV-specific RIFA was modified from Lemon et al (7). BSC-1 cells were grown on 25 mm Lux "Thermanox" tissue culture coverslips (Miles Laboratories, Ill.) anchored in 30 mm petri dishes with a drop of 1% agar. Virus dilutions were prepared in MEM containing 2% FCS, while RNA prepared by the phenol-chloroform method (9) was heated at 100° for 1 min before dilution in MEM containing 0.05M Tris pH 7.4 and 0.5 mg of DEAE-dextran (Pharmacia) per ml. Cells were infected with 0.4 ml of virus dilutions at 34° for 1 h or RNA dilutions at 24° for 30 min, after which the inoculum was replaced with 2 ml of overlay medium (MEM containing 2% FCS and 0.5% agar). After incubation for 7 days, the coverslips were removed, washed in 0.85% NaCl, and fixed in acetone at 4° for 2 min. Coverslips were dried in air and inverted on 0.5 ml of 0.85% NaCl containing 2% FCS and 10^6 cpm of I-labelled monoclonal antibody K3-4C8 (10) for 1 h at 24° and were then washed in 0.85% NaCl, dried and exposed to Kodak X-Omat RP film overnight at -70° between 2 intensifying screens. PV and PV RNA were prepared and diluted as described for the RIFA, but plaques were visualized after 2 days incubation.

In Situ Radioimmunoassay

BSC-1 cell monolayers on 25 mm coverslips were fixed at appropriate times and stained and washed as described for the RIFA, and the bound radioactivity was measured in a gamma counter. The HAAg measured in this way is probably intact, intracellular virions.

Indirect Immunofluorescence

Cell monolayers were fixed in acetone and stained for double-stranded (ds) RNA or HAAg. Rabbit antiserum to dsRNA was a gift from R. Francki, University of Adelaide, South Australia, and was used at a concentration of 1 mg per ml. HAAg was stained using 1 µg of K3-4C8 per ml. Cells were then stained with fluorescein-labelled, anti-rabbit or anti-mouse IgGs, washed and examined using a Zeiss standard microscope and HBO 50 light source.

Dot-Blot Hybridization

RNA samples in distilled water were mixed with an equal volume of 20XSSC (0.33M NaCl, 0.3M sodium citrate, pH 7.0) and applied to nitrocellulose filters. Filters were then dried, baked at 80° for 2 h and probed as described (11) using 2×10^6 cpm of nick-translated HAV-specific cDNA clone pHM14 (Ross BC, Anderson BN, Coulepis AG, Chenoweth MP, Gust ID; submitted for publication)

Single-Cycle Growth Curves

BSC-1 cells in 30 mm dishes were infected with HAV strain HM175 at a m.o.i. of 1 RFU per cell or PV at a m.o.i. of 5 PFU per cell, for 1 h at 34°. Cells were then washed and fed with 2 ml of MEM containing no additions, or supplemented with either 2% FCS, 1 mM guanidine, or 0.5 mM zinc chloride. At various times, the medium was removed from HM175-infected cells, which were then dissolved in 0.2 ml of 1% NP40 for the extraction of virus or 2% SDS for the extraction of viral RNA as described. Poliovirus-infected cells were scraped into the medium, and adjusted with NP40 or SDS before all samples were stored at -70° until analysis. Parallel cultures were sampled for immunofluorescence and *in situ* RIA.

Penetration and Uncoating

Ten-fold dilutions of light-sensitive HM175 and PV were inoculated onto monolayers of BSC-1 cells for RIFA or plaque assay, respectively, at 4° for 1 h. Monolayers were then washed and fed with MEM containing 2% FCS (HAV penetration) or overlay medium (uncoating). At appropriate times, monolayers for the analysis of HAV penetration (resistance to neutralization) were refed MEM containing 10 µg of K3-4C8 per ml for 1 h at 4°, then washed 3 times and overlay added. Monolayers for the analysis of virus uncoating (resistance to inactivation by light) were exposed to four 150W lamps at a distance of 80 cm for 5 min.

RESULTS

Cytopathic Effects

When BSC-1 cells were infected with the HM175 stock used and maintained in MEM, a CPE was observed to develop in 2 to 3 days, in parallel with the accumulation of HAAg as detected by indirect immunofluorescence using monoclonal antibody K3-4C8 (fig. 1). Development of this CPE was abolished by preincubation of the virus with K3-4C8, and was delayed to 4 to 7 days in cultures maintained in MEM containing 2% FCS or 1 mM guanidine (results not shown)

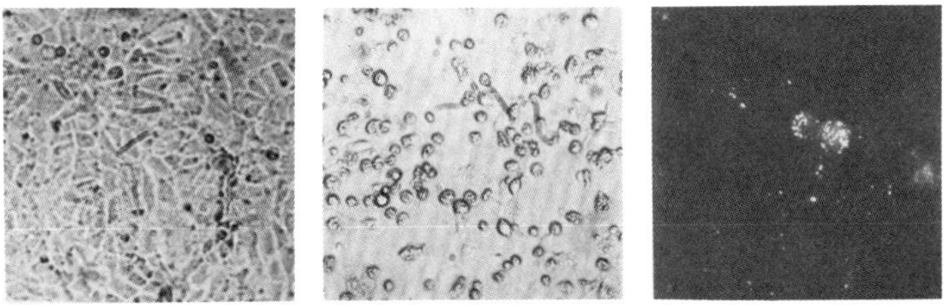

A B C

FIGURE 1. HM175-specific CPE and indirect immunofluorescence at 72 h. (A) uninfected (B) HM175-infected BSC-1 cells (X100). (C) HM175 indirect immunofluorescence (X320).

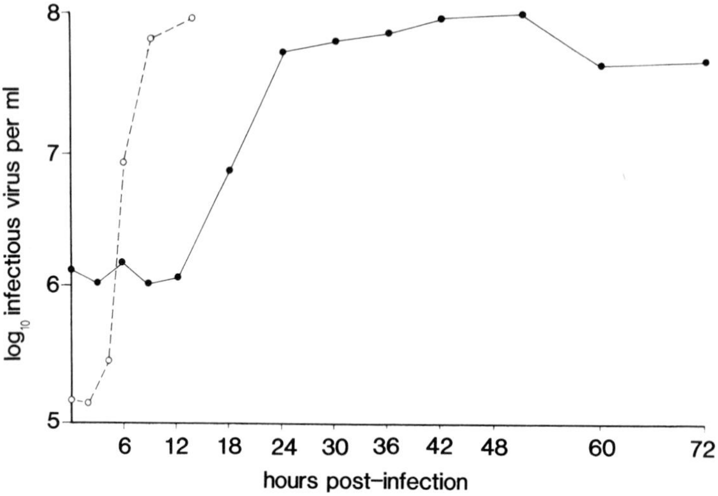

FIGURE 2. Single-cycle growth curves of HM175 (●——●) and PV (O---O).

Single-Cycle Growth Curves

In order to define the kinetics of HAV replication in BSC-1 cells, cultures were infected at a multiplicity of 1 RFU of HM175 or 5 PFU of PV and examined at intervals after infection.

The growth curves for infectious HM175 and PV are shown in fig. 2. In PV-infected cells, eclipse of the inoculum virus was evident at 0 to 2 h, reflecting uncoating of the virus, followed by logarithmic accumulation of mature virus to 9 h and a final yield of infectious virus at 15 h of 500 PFU per cell. In contrast, eclipse of the inoculum HM175 appeared to be incomplete, with at least 25% of the virus being recovered up to 9 h. From 12 to 24 h there was logarithmic virus growth, with a peak yield of infectious HM175 at 51 h of 20 RFU per cell.

The replication of HAV RNA was examined by the analysis of total intracellular HAV RNA and infectious RNA. Cultures were examined for total HM175-related RNA by dot-blot hybridization with HAV-specific cDNA probes (fig. 3). Dilutions of total RNA from inoculum virus ("i", representing 25% of the inoculum added to each culture), untreated

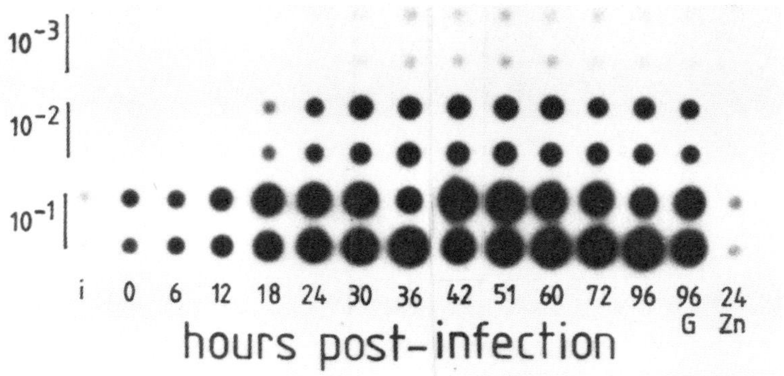

FIGURE 3. Dot-blot hybridization of RNA from HM175-infected BSC-1 cells under single-cycle growth conditions.

cultures, guanidine-treated cultures ("G"), and zinc chloride-treated cultures ("Zn") were probed as described. An increase in total HAV-related RNA in untreated cultures can be seen from 6 to 12 h, with a more rapid increase from 12 to 24 h. Zinc chloride abolished the replication of RNA, while guanidine appears to have had no effect.

The pattern of HM175 infectious RNA accumulation (fig. 4) was similar to that seen for infectious virions in replicate cultures (fig. 2), while in PV-infected cells a significant increase was seen from 0 to 4 h, during which time little mature virus was produced.

The accumulation of HAAg was determined using an in situ RIA for the detection of antigen in acetone-fixed monolayers (results not shown). An increase in the amount of labelled K3-4C8 bound was first detected at 18 h, and HAAg continued to accumulate at a constant rate until 51 h, when infected cell monolayers bound 43000 cpm, compared to 2000 cpm bound to uninfected cell monolayers.

Double-stranded RNA in infected cells was analyzed by indirect immunofluorescence using rabbit antiserum to dsRNA (fig. 5). In HM175-infected cells, dsRNA was first detectable at 24 h, by which time the majority of infectious virus (fig. 2) and infectious RNA (fig. 4) had been produced. In contrast, dsRNA was detected in PV-infected cells as early as 2 h, and the increase over the 9 h period studied was consistent with the accumulation of infectious virus and RNA in these cells.

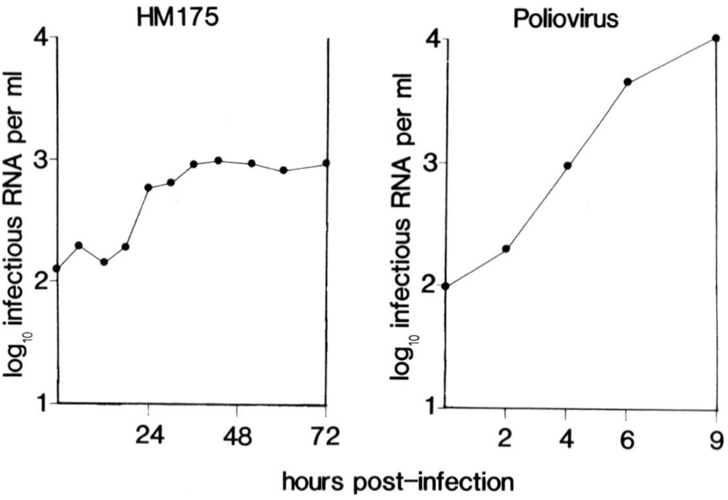

FIGURE 4. Accumulation of infectious RNA in BSC-1 cells infected with HM175 or PV under single-cycle growth conditions.

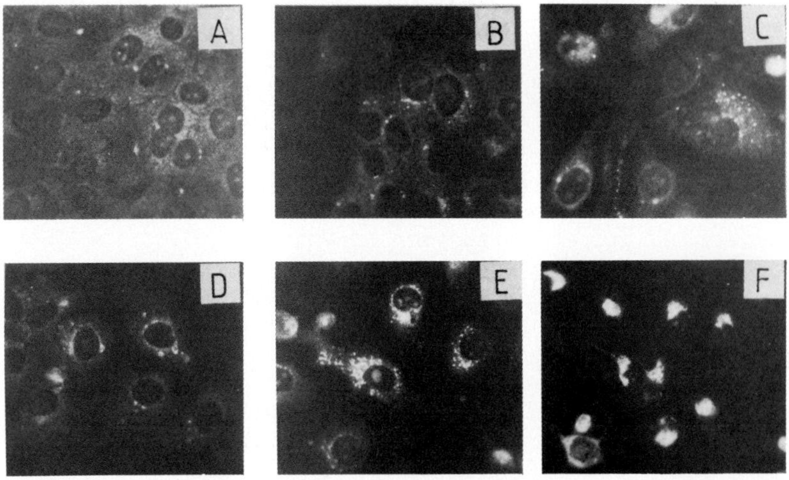

FIGURE 5. Indirect immunofluorescence of dsRNA in (A) mock-infected cells, (B and C) HM175-infected cells at 24 and 72 h, respectively, (D-F), PV-infected cells at 2 (D), 4 (E) and 9 h (F), respectively. X320.

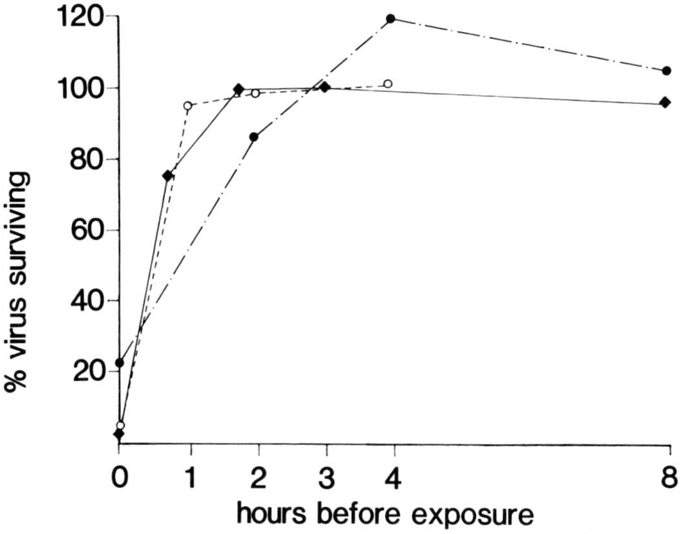

FIGURE 6. Penetration of HM175 (O---O) and uncoating of HM175 (●·-●) and PV (◆—◆) in BSC-1 cells at 34°.

Penetration and Uncoating of HM175 and PV.

The observation of a large amount of apparently uneclipsed HM175 at up to 9 h in single-cycle growth experiments (fig. 2) prompted us to study the rate of penetration and uncoating of HM175 in BSC-1 cells. However, in contrast to the observation of "uneclipsed" virus (fig. 2), when cells were infected with light-sensitive virus and exposed to either monoclonal antibody K3-4C8 (to measure penetration) or light (to measure uncoating), little difference was seen between the penetration and uncoating of HM175 and the uncoating of PV (fig. 6). The reason for these anomalous results is not clear.

DISCUSSION

This paper examines the single-cycle growth kinetics of HAV strain HM175 in an attempt to identify the restrictive events acting to limit the replication of this virus.

During this study, the development of a specific CPE was observed in HM175-infected cells. Another report has noted a

CPE in FRhK-4 cells infected with a fast-growing strain of HAV (12). This observation warrants further investigation because HAV is not cytocidal in vivo, and the HM175 studied here was isolated from a persistently infected cell line.

Under single-cycle growth conditions (fig. 2), HM175 demonstrated a lag phase of 12 h, followed by logarithmic growth to 24 h. Qualitatively similar results have been reported for HAV strain BGM in MRC-5 cells (13) and strain HAS-15 in FRhK-4 cells (8).

The modified RIFA described has, for the first time, allowed the direct demonstration of the infectivity of HAV RNA in vitro (fig. 4). In conjunction with dot-blot hybridization this has been used to examine the kinetics of RNA accumulation in infected cells.

Total HM175 RNA was seen to accumulate from 6 to 12 h, then increasing more rapidly (fig. 3), while infectious RNA (fig. 4) followed the same pattern as infectious virus. The lag phase for infectious RNA production was therefore sufficient to account for the lag phase in infectious virion production (fig. 2). This suggests a restriction at the level of RNA replication.

The low level of dsRNA detected during the productive phase of HM175 infection (fig. 5) indicates that there are relatively few replicative complexes in HM175-infected cells when compared to PV-infected cells, which also lends support to this hypothesis. It is interesting to note that the CPE develops after the yield of infectious virus has reached a plateau (fig. 2), but at a time when a considerable amount of dsRNA is present in the cell.

Wheeler and coworkers (8) suggested that HAV uncoating or penetration was a rate-limiting step, and our own observations of a 12 h lag phase combined with the recovery of apparently non-eclipsed virus from cells up to 9 h (fig. 2) also supported this hypothesis. However, an analysis of penetration and uncoating of light-sensitive HAV in comparison to PV (14; fig. 6) revealed no significant difference in the kinetics of uncoating for the two viruses.

In contrast, the kinetics of HM175 RNA accumulation suggest a level of restriction at this level. The mechanism of the apparent restriction of RNA replication is currently under investigation.

REFERENCES

1. Coulepis AG, Locarnini SA, Westaway EG, Tannock GA, Gust ID (1982). Biophysical and biochemical characterization of hepatitis A virus. Intervirology 18:107.

2. Gust ID, Coulepis AG, Feinstone SM, Locarnini SA, Moritsugu Y, Najera R, Siegl G (1983). Taxonomic classification of hepatitis A virus. Intervirology 20:1.
3. Provost PJ, Hilleman MR (1979). Propagation of human hepatitis A virus in cell culture in vitro. Proc Soc Exp Biol Med 160:213.
4. Vallbracht A, Hofmann L, Wurster KG, Flehmig B (1984). Persistent infection of human fibroblasts by hepatitis A virus. J gen Virol 65:609.
5. Gauss-Müller V, Deinhardt F (1984). Effect of hepatitis A virus infection on cell metabolism in vitro. Proc Soc Exp Biol Med 175:10.
6. Locarnini SA, Coulepis AG, Westaway EG, Gust ID (1981). Restricted replication of human hepatitis A virus in cell culture: intracellular biochemical studies. J Virol 37:216.
7. Lemon SM, Binn LN, Marchwicki RH (1983). Radioimmunofocus assay for quantitation of hepatitis A virus in cell cultures. J Clin Microbiol 17:834.
8. Wheeler CM, Fields HA, Schable CA, Meinke WJ, Maynard JE (1986). Adsorption, purification, and growth characteristics of hepatitis A virus strain HAS-15 propagated in fetal rhesus monkey kidney cells. J Clin Microbiol 23:434.
9. Coulepis AG, Tannock GA, Locarnini SA, Gust ID (1981). Evidence that the genome of hepatitis A virus consists of single-stranded RNA. J Virol 37:473.
10. Coulepis AG, Veal M, MacGregor A, Kornitschuk M, Gust ID (1985). Detection of hepatitis A virus and antibody by solid-phase radioimmunoassay and enzyme-linked immunosorbent assay with monoclonal antibodies. J Clin Microbiol 22:119.
11. Thomas PS (1980). Hybridization of denatured RNA and small DNA fragments transferred to nitrocellulose. Proc Natl Acad Sci USA 77:5201.
12. Venuti A, DiRusso C, del Grosso N, Patti A-M, Ruggeri F, De Stasio PR, Martiniello MG, Pagnotti P, Degener AM, Midulla M, Pana A, Perez-Bercoff R (1985). Isolation and molecular cloning of a fast-growing strain of human hepatitis A virus from its double-stranded replicative form. J Virol 56:579.
13. Siegl G, deChastonay J, Kronauer K (1984). Propagation and assay of hepatitis A virus in vitro. J Virol Methods 9:53.
14. Mandel B (1967). Relationship between penetration and uncoating of poliovirus in HeLa cells. Virology 31:702.

Positive Strand RNA Viruses, pages 509–518

HOST RANGE DETERMINANTS OF AVIAN RETROVIRUS ENVELOPE GENES[1]

Carol A. Bova and Ronald Swanstrom

Department of Biochemistry and
Lineberger Cancer Research Center
University of North Carolina
Chapel Hill, NC 27514

ABSTRACT The *env* gene product of avian retroviruses is allelic in the virus population and specifies which cell receptor is used to enter the host cell. We determined the nucleotide sequence of the region of the *env* gene that encodes the glycoprotein gp85 from subgroup A, B, and D viruses, and we compared the predicted amino acid sequences of these *env* genes to the previously reported sequences of subgroup B, C, and E *env* genes. Based on these comparisons, we draw the following conclusions: i) There are four variable regions, ranging in size from 9 to 52 amino acids, within the gp85 coding domain. ii) The percent homology within the variable regions ranges between 41% and 60% when comparing most of the different subgroup viruses; however, the average homology is increased to the 81% to 90% range when comparing viruses of the same subgroup. iii) The variable regions are flanked by conserved domains that are on average 95% homologous between the different subgroup viruses. Biological testing of molecular recombinants among A, B, C, D and E subgroup viruses indicates that the host range determinant defining subgroup specificity is encoded within the DNA that spans the variable regions.

[1]This work is supported by N.I.H. Grant R01 CA33147. Computer resources were provided by Bionet whose funding is from N.I.H. Grant # U41 RR-01685-02.

INTRODUCTION

The *env* gene of avian leukosis virus (ALV) and Rous sarcoma virus (RSV) encodes two proteins, gp85 and gp37, that are associated with the viral envelope (1,2). Individual isolates of ALV and RSV have been classified into five distinct subgroups, designated A through E. Subgroup assignments are determined by the ability to infect genetically defined chicken cells, by virus interference patterns, and by antibody neutralization of infectivity (3). Several lines of evidence indicate that the *env* gene encodes the host range determinants that define subgroup specificity.

The nucleotide sequences of the gp85 region of the *env* genes from different subgroup viruses have been compared (4,5,6,7). These comparisons revealed four distinct variable regions within the gp85 coding domain. A direct role for at least one of these variable regions has been shown by analysis of the host range mutant NTRE-4 (5,8).

In this report we present a comparison of the predicted amino acid sequence of five different *env* genes and an analysis of sequence homology. We also review our work on the host range of viruses generated from molecular recombinants.

RESULTS

Comparison of the predicted amino acid sequence of gp85 from the different *env* gene alleles. The amino termini of the mature gp85 and gp37 proteins have been determined (9) allowing the identification of the gp85 coding domain within the nucleotide sequence of the *env* gene. Figure 1 shows the predicted amino acid sequence of the mature gp85 for the following subgroup viruses: A, Schmidt-Ruppin strain of RSV, SR-A RSV (6); B, Rous associated virus 2, RAV-2 (6); C, Prague strain of RSV, Pr-C RSV (4); D, Schmidt-Ruppin strain of RSV, SR-D RSV (7); and E, Rous associated virus 0, RAV-0 (5,6). The sequences were aligned to allow the maximum homology between the virus isolates with the introduction of the fewest number of gaps.

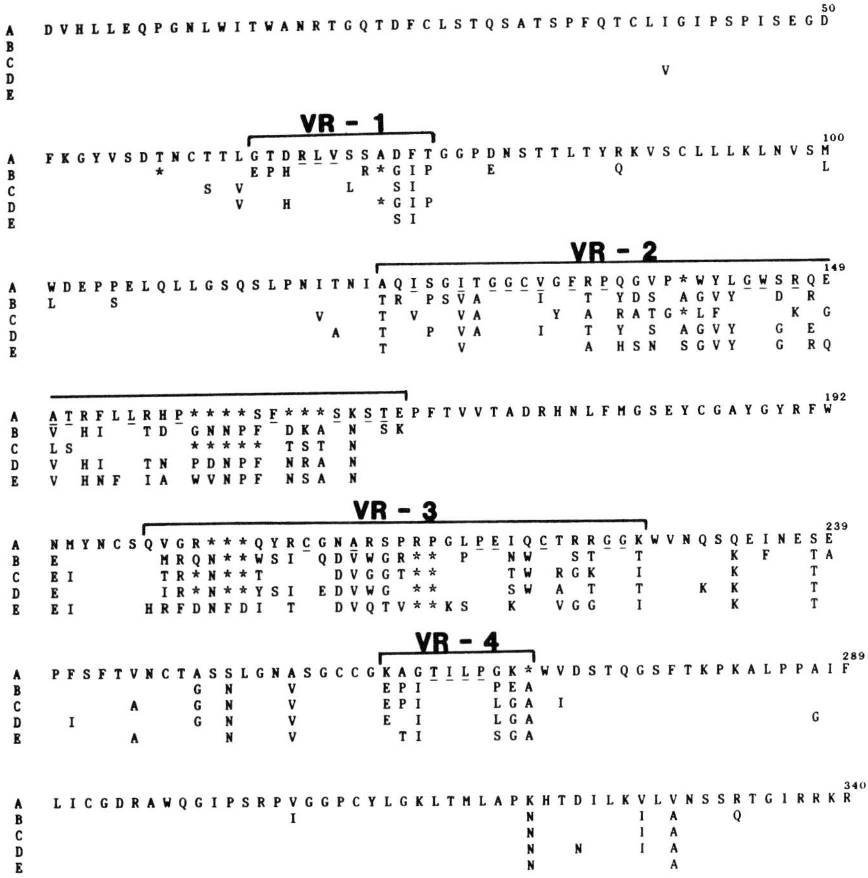

FIGURE 1. Amino acid sequence comparison of the predicted gp85 coding region of subgroup A, B, C, D, and E avian retrovirus *env* genes. The complete predicted amino acid sequence is shown for the subgroup A isolate, while only the differences for the remaining subgroup viruses as compared to the subgroup A virus are shown. The symbol (*) denotes the position of a gap inserted during the alignment. The four variable regions are shown in brackets. Framework amino acids, as described in the text, are underlined.

As noted previously (6) there are four variable regions, denoted VR-1 through VR-4, that are apparent when comparing the gp85 coding sequence of the different subgroup isolates. The positions of the variable regions within the SR-A RSV sequence are as follows: VR-1, 64 through 75; VR-2, 122 through 166; VR-3, 199 through 227; and VR-4, 261 through 269. In VR-1, the subgroup B virus is the most divergent. Within VR-2, subgroups B and D are the most homologous (73%) while subgroups B and C are the least homologous (37%). Subgroups B, D, and E contain eight amino acids not found at equivalent positions in the subgroup A sequence and six amino acids not found in the subgroup C sequence. The subgroup B and D viruses show the greatest homology in VR-3 (69%). VR-4 is most divergent for the subgroup A virus.

The individual domains flanking the variable regions are approximately 95% conserved between the different subgroup isolates. All cysteines are conserved between the subgroup viruses. There are only six differences in potential asparagine-linked glycosylation sites (10) when comparing all the different subgroup viruses. There is a loss of three potential glycosylation sites in the SR-A RSV sequence, the loss of a potential glycosylation site for SR-D RSV, and the gain of a potential glycosylation site at equivalent positions in VR-3 for both the subgroup B and D sequences.

Figure 2 is a compilation of the results of pairwise homology comparisons of individual conserved or variable domains between viruses of different subgroup classifications (Fig. 2A) or between viruses of the same subgroup classification (Fig. 2B). Figure 2A includes the SR-A RSV, RAV-2, Pr-C RSV, SR-D RSV, and RAV-0 sequences as shown in Figure 1. Figure 2B includes comparisons of RAV-2, MAV-2 (11), and Pr-B RSV (5) subgroup B sequences, and a comparison of SR-A RSV and RAV-1 (7) subgroup A sequences. Most of the conserved domains are greater than 90% homologous when comparing either viruses from different subgroups or viruses from the same subgroup. However, the comparison of variable regions from different subgroup viruses shows less homology (on average, 41% to 60%) than comparisons of viruses from the same subgroup where the homology is generally greater than 81%.

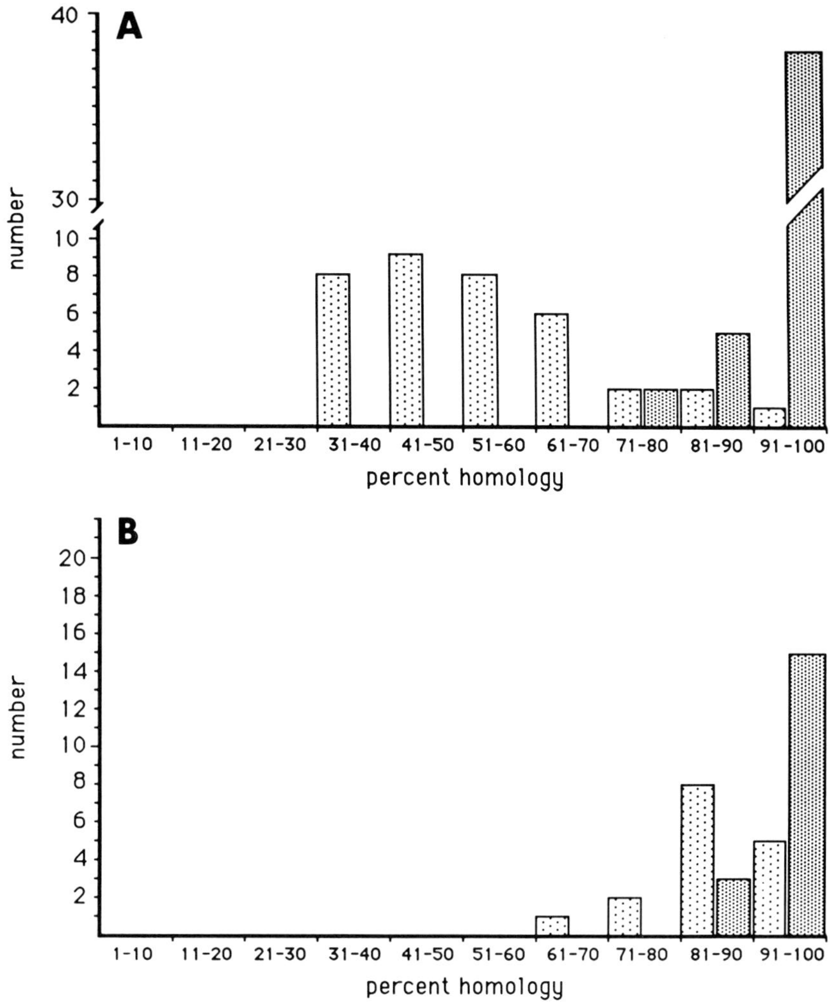

FIGURE 2. Each variable region (VR-1 through VR-4) and each flanking conserved region were compared between the different viruses (see text), and the percent homology for each comparison was calculated. The histogram shows the number of individual comparisons that fall within a range of percentage values.
Comparisons of individual variable regions: light stipple
Comparisons of individual conserved regions: heavy stipple
A. Comparisons of viruses from different subgroups.
B. Comparisons of viruses from the same subgroup.

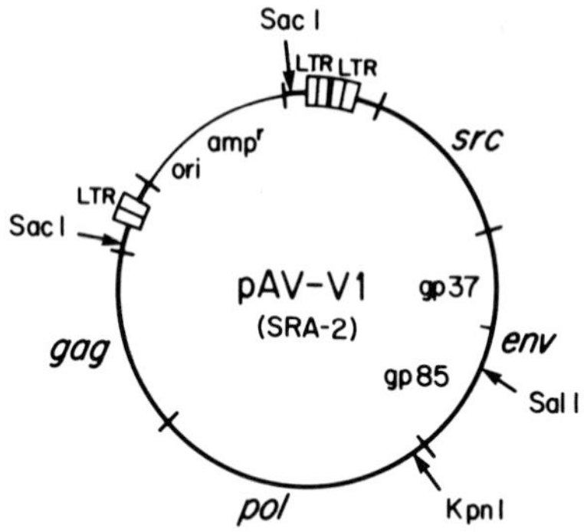

Figure 3. Map of pAV-V1.

Host range analysis of molecular recombinants. Molecular recombinants were constructed to analyze the role of the sequences within gp85 in defining subgroup specificity. Figure 3 shows the plasmid pAV-V1 derived from the SRA-2 DNA clone of SR-A RSV (12). Transfection of the plasmid into susceptible host cells yields SR-A RSV. To make recombinant viruses, the Kpn I to Sal I restriction fragment was removed from pAV-V1 and replaced by the equivalent fragment from other subgroup viruses as indicated: pAV-B01 from RAV-2; pAV-D01 from SR-D RSV; and pAV-E03 from RAV-0. pAV-C02 contains the Kpn I to Hind III fragment from Pr-C RSV which is slightly larger than the Kpn I to Sal I fragment used in the other constructions. The Kpn I to Sal I fragment encodes the last 64 amino acids of the *pol* gene, the signal sequence of the precursor glycoprotein, and sequences to codon position V_{271} of SR-A RSV gp85 (Fig. 1). pOJ2-1 (7), the plasmid giving rise to ANV-A virus, contains the SV-40 promoter and neomycin resistance gene in place of the v-*src* gene in pAV-V1. pANV-D, the plasmid giving rise to ANV-D virus, is equivalent to pOJ2-1 except that it contains the

Kpn I to Sal I fragment of SR-D RSV. Transfection of the plasmids containing the recombinant virus genomes into susceptible cells produced an active infection. Virus stocks were then used to infect the following cell types: C/O (chicken embryo fibroblast cells, CEF, susceptible to all subgroup viruses), C/E (CEF resistant to subgroup E viruses), C/ABE (CEF resistant to subgroups A, B, and E viruses), Q/BD (QT-6 cells, quail cells resistant to B and D subgroup viruses), and COS-7 cells (a transformed monkey cell line). Virus infection and spread was detected by assaying for the presence of reverse transcriptase activity in the medium of infected cells (6,13). ANV-A and ANV-D infectivity was measured by determining the number of G418-resistant cells after infection (7).

The results indicate that the host range of each recombinant virus changed from that of the parental subgroup A virus to that of the genome that donated the Kpn I to Sal I fragment (Table 1).

TABLE 1

	C/O	C/ABE	C/E	Q/BD	COS 7
SR-A	+	-	+	+	NT
AV-B01	+	-	+	-	NT
AV-C02	+	+	+	-	NT
AV-D01	+	-	NT	NT	NT
AV-E03	+	-	-	+	NT
ANV-A	+	NT	NT	+	-
ANV-D	+	NT	NT	-	+

NT-not tested

DISCUSSION

We have used analysis of the host range of molecular recombinant viruses to show that a 1.1 kb region of DNA bounded by the Kpn I and Sal I restriction endonuclease sites in the avian retrovirus genome defines subgroup specificity (i.e., ability of the virus to infect genetically defined chicken cells). Infection of chicken cells does not distinguish between the subgroup B and D viruses; however, the ANV-D recombinant virus was able to infect COS-7 cells, indicating that the determinant of mammal-tropism of the subgroup D viruses also lies within the Kpn I to Sal I fragment. Although the Kpn I to Sal I fragment contains *pol* sequences and the glycoprotein leader sequence in addition to gp85 coding sequences, we argue that the host range determinant lies within the coding sequences of gp85 since subgroup specificity is a cell surface phenomenon. These results are consistant with previous mapping studies (14,15) that link this region of the genome to subgroups specificity. A direct role for one region of gp85 (VR-3) has been shown by sequence analysis of a recombinant virus, NTRE-4 (5,8), that possesses an altered host range.

In an attempt to identify specific regions within the gp85 coding domain that may be involved in determining host range, we compared the predicted primary amino acid sequence of gp85 from molecularly cloned virus isolates representing each of five different subgroups (Fig. 1). Sequence alignment of the various subgroup viruses revealed a clustering of amino acid changes on which the division of the gp85 coding into variable and conserved regions is based. The results of pairwise comparisons of viruses from the same or different subgroup support the role of the variable regions in host range. In comparisons of viruses of different subgroup designation (Fig. 2A), the variable regions show much greater sequence divergence than in comparisons of viruses from the same subgroup. The average homology for the variable regions of viruses from the same subgroup is still lower (81 to 90%) than the average homology for the conserved domains from viruses of the same subgroup or from different subgroup classifications. The individual variable regions therefore can tolerate some sequence heterogeneity.

Subgroup B, D, and E viruses may utilize the same receptor in chickens (16). In the two large variable regions, the B, D and E subgroup viruses show greater similarity in both length and sequence among themselves than when compared to A and C. The B and D subgroup viruses are more similar to each other (73%) than to the subgroup E virus (58% and 63% respectively) in VR-2. In VR-3, the subgroup E virus is much less homologous (32% to 35%) to either the subgroup B or D virus than the subgroup B and D virus are to each other (69%). Thus, the similarity between subgroup B and D viruses detected using biological assays is mirrored in a higher level of homology in the amino acid sequence.

Secondary structure analysis of gp85 for each of the subgroup viruses results in similar preductions for protein secondary structure (6). This includes predictions for the variable regions. Review of the actual amino acids within the variable regions reveals positions along the primary sequence that are conserved between the different subgroup viruses or where only conservative changes (i.e., amino acids with equivalent side chains) have occured. These amino acids may serve as framework positions within the variable regions (underlined in Fig. 1). It is noteworthy that many of these conserved amino acids are cysteines, or prolines and glycines (disruptors of regular secondary structure). The secondary structure predictions are very similar between the subgroup viruses with the degree of hydrophobicity and length of looped regions dependent on the subgroup-specific sequence.

Sites on a protein surface recognized by the immune system as antigenic are often hydrophilic looped regions of the protein (17,18). The majority of the gp85 coding sequences is predicted to be composed of hydrophobic β-sheet topology with looped or turn regions spaced along the primary sequence. Since neutralizing antibodies are often type specific (19), hydrophilic looped regions within the variable regions may be antigenic and involved in virus interaction with the host cell receptor.

ACKNOWLEDGEMENTS

We thank Sylvia Gault for stenographic assistance.

REFERENCES

1. Duesberg, PH, Martin GS, Vogt PK (1970). *Virology* 41:631.
2. Bolognesi DP, Bauer H, Gelderblom H, Huper G (1972). *Virology* 47:551.
3. Vogt PK (1977). In Fraenkel-Conrat H, and Wagner R (eds): "Comprehensive Virology," New York: Plenum Press, Vol. 9, p. 341.
4. Schwartz DE, Tizard R, Gilbert W (1983). *Cell* 32:853.
5. Dorner AJ, Stoyle JP, Coffin JM (1985). *J. Virol.* 53:32.
6. Bova CA, Manfredi JP, Swanstrom R (1986). *Virology*, in press.
7. Bova CA, Olsen JC, Swanstrom R (1986). Manuscript in preparation.
8. Tsichlis PN, Conklin KF, Coffin JM (1980). *PNAS USA* 77:536.
9. Hunter E, Hill E, Hardwick M, Bhown A, Schwartz D, Tizard R (1983). *J. Virol.* 46:920.
10. Kornfield R, Kornfield S (1985). *Ann. Rev. Biochem.* 54:631.
11. Kan NC, Baluda MA, Papas T (1985). *Virology* 145:323.
12. DeLorbe WJ, Luciw PA, Goodman HM, Varmus HE, Bishop JM (1980). *J. Virol.* 36:50.
13. Tereba A, Murti KG (1977). *Virology* 80:166.
14. Joho RH, Billeter MA, Weissman C (1975). *PNAS USA* 72:4772.
15. Coffin JM, Champion M, Chabot F (1978). *J. Virol.* 28:972.
16. Weiss R (1982). In Weiss R, Teich N, Varmus H, Coffin J (eds) "RNA Tumor Viruses," New York: Cold Spring Harbor Laboratory, p.209.
17. Hopp TP, Woods KR (1980). *PNAS USA* 78:3824.
18. Westhof E, Altschuh D, Moras D, Bloomer AC, Mondragon A, Klug A, Van Regenmortel MHV (1984). *Nature* 311:123.
19. Vogt PK (1970). In Dutcher RM (ed): "Comparative Leukemia Research 1969," New York: Plenum Press, p.153.

Positive Strand RNA Viruses, pages 519–536

RUBELLA VIRUS ASSOCIATED WITH CYTOSKELETON (RUBELLA VACS) PARTICLES - REVEVANT TO SCRAPIE?

Diane Van Alstyne,[2] Marc DeCamillis,[3] Paul S. Sunga and Richard F. Marsh[3]

Quadra Logic Technologies Inc.
Vancouver, B.C., Canada V6H 3Z6

and

Department of Veterinary Science, University of Wisconsin-Madison, Madison, Wisconsin 53706

ABSTRACT Persistent rubella virus (RV) infection in the central nervous system (CNS) has been associated with[1-3] chronic, degenerative neurologic disorders where there is a long initial period of latency. However, the mechanism underlying RV persistence as it relates to CNS disease remains unclear. To further our ongoing[4-6] investigations of persistent rubella infection in the CNS we have produced hybridomas secreting monospecific antirubella antibodies. In the course of their characterization, it was observed that an immunosorbent column of monoclonal IgG linked to Sepharose beads could be employed to affinity-purify all RV proteins simultaneously from RV-infected L cell lysates. The viral proteins were eluted as rod-like structures averaging 10-20 x 50-100 nm and were shown to react positively with

[1]This work was supported by the Multiple Sclerosis Society of Canada and the B.C. Health Care Research Foundation (D.V.), and by grant NS 14822 from the National Institutes of Health, and the United States Department of Agriculture (R.F.M.)
[2]Present address: Quadra Logic Technologies Inc., 2660 Oak Street, Vancouver, B.C., Canada V6H 3Z6
[3]Present address: Department of Veterinary Science, 1655 Linden Drive, Madison, Wisconsin, 53706

anti-actin antibody. These structures were then shown to be associated with single-stranded (ss) RNA, rendering the RNA nuclease resistent. The rods were likely composed of RV associated with actin-containing structures, and have been designated rubella VACS (Virus Associated with Cytoskeleton) particles. These structures were furthur shown to be infectious and to demonstrate some unique biological properties unlike those of the virus stocks from which they were derived. Particles could also be affinity-purified from lysates of VACS-infected cells. These particles, designated VACS-2, were distinct from VACS-1 in that they were composed predominantly of a 27-30 Kd protein and reacted positively with anti-tubulin antibody. These *in vitro* data suggest that RV may possess an alternate mode of infection mediated by rubella VACS particles which share several of the unique characteristics of the rod-shaped structures associated with the subacute spongiform virus encephalopathies. One of these, the scrapie agent, has been propagated in hamster brain where the highest titers of infectivity co-purified with insoluble cytoskeletal proteins. These preparations, although relatively homogeneous in protein content, contained large amounts of heterogeneous low-molecular-weight RNA and treatment with RNase A resulted in a substantial reduction in scrapie infectivity. However, a similar response was observed to treatment with poly-L-lysine, indicating that the reduction in infectivity may be due to aggregation of the sample caused by the polycationic properties of RNase A rather than a hydrolytic effect on an essential RNA component of the scrapie agent.

INTRODUCTION

Rubella virus is classified in the family *Togaviridae*, notable for containing an extraordinarily high number of viruses which, like rubella, can invade the CNS[7]. Interest in RV stems largely from its induction of congenital malformations in humans and other animals, known collectively as the congenital rubella syndrome (CRS)[8]. Symptoms of CNS involvement can be expected to present in more than 80% of children affected with CRS[9,10]. In addition, other subtle neurological deficits have been associated with CRS of late onset in adolescents and young adults who exhibit few, if any, signs of the syndrome at birth[3,11]. This indicates that our present appreciation of the consequences of RV persistence in the CNS may still be incomplete. There is at present no clear understanding of a mechanism to account for the late onset of symptoms associated with RV persistence in the CNS. However, recent studies indicate that persistently-infected glioblasts in continuous culture restrict RV-directed synthesis of p24 and p30, two polypeptides associated with a productive infection in vitro[4]. Subsequent exposure of these infected blast cells to dibutyryl cyclic AMP (dB-cAMP) has been shown to promote glioblast maturation into astrocytes and oligodendrocytes[6] and a concomitant activation of rubella replication[5] long after the initial infection. This indicates that there is some essential cell-virus interaction which is prerequisite for the complete synthesis of progeny virions from glia in the CNS.

The scrapie agent produces a degenerative brain disease in sheep and is indistinguishable from similar neuropathic agents infecting man. While the physicochemical nature of the scrapie agent remains controversial, there is agreement that infectivity is highest in membrane fractions[12,13], especially in their detergent-soluble residues[14]. Previous attempts to detect differences in proteins[15] or lipids[16] in brain membranes from scrapie-infected or uninfected hamsters have been unsuccessful, but differences have been observed in nucleic acid content[17,18]. The finding of a unique low-molecular-weight RNA in scrapie-infected brain[18] has produced speculation that the scrapie agent may be a small nucleoprotein complex[19]. These studies further examine the association of scrapie infectivity with the cytoskeleton and test sensitivity to treatment with RNase A.

ISOLATION AND CHARACTERIZATION OF RUBELLA VIRUS ASSOCIATED WITH CYTOSKELETON (RUBELLA VACS) PARTICLES METHODS AND RESULTS

In order to investigate the precise nature of this cell-virus interaction operative in RV-infected glia, we have prepared monoclonal antibodies specific for RV polypeptides. One of these antibodies (C3-8C) has been found to be positive at a dilution of 10^{-5} in detecting RV antigens in infected cells using immunofluorescent labeling. It was found to be negative in the hemagglutination inhibition (HAI) and in the virus neutralization tests[20]. This antibody was linked to Sepharose 4B beads and used to affinity-purify RV antigens from infected L cell lysates. Polyacrylamide slab gel electrophoresis (PAGE) analysis of proteins in the column eluate illustrated in

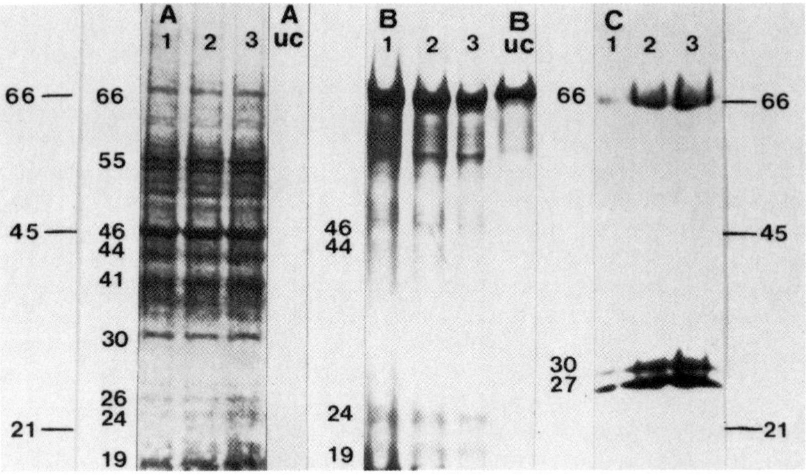

FIGURE 1. Polyacrylamide slab gel electrophoresis (PAGE) analysis of rubella proteins affinity-purified from infected cell lysates before (A) and after (C) treatment with 2% sodium dodecyl sulfate (SDS). Monoclonal antiRV antibody was prepared according to the procedure described by Koprowski et al.[35]. A total of 45.12 mg of monoclonal C3 antibody (IgG) were linked to 5.0 ml of cyanogen bromide-activated Sepharose 4B beads (Pharmacia). RV antigens were prepared from a polytroned lysate of L cells harvested from 50-175 cc^2 infected monolayers. Cells were suspended in 2.0 ml of phosphate buffered saline (PBS), containing protease inhibitors, lysed and centrifuged at 15K for 30 min in a Sorvall centrifuge. The supernatent was passed over the immunosorbent column and the bound material was eluted with 0.1 N HCl and neutralized with 3.0 M Tris buffer, pH 7.5, resulting in a final volume of 20 ml of antigen with an $A_{280m\mu}$= 0.249. PAGE analysis was carried out according to the procedure described by Laemmli[36], using a 10% discontinuous gel system. Fig. 1A illustrates all of the proteins isolated from infected cell lysates (lanes 1-3, 5-20 µl) and from uninfected control cell lysates (lane uc=uninfected control). The structural rubella proteins are contained in Fig. 1B. These were eluted from an affinity column containing monoclonal antibody E1-8F which is directed against an external antigen, permitting the affinity-purification of rubella virions from 500 ml of tissue culture medium removed from $5x10^8$ infected cells. Fig. 1B uc=uninfected control. Two ml of antigen shown in 1A were dissociated with 2% SDS (final concentration), placed in a boiling water bath

for 2 min, diluted ten fold with cold PBS and rechromatographed. A total of 13% of the total antigen was recovered (0.164 $A_{280m\mu}$ units) as column eluate which was subjected to PAGE. The proteins contained in affinity-purified antigen after SDS treatment using C3 antibody are illustrated in Fig. 1(C).

Figure 1A, revealed all RV polypeptides previously reported in infected L cells[4,5], most of which have been confirmed by other investigators[21-23]. Examination of the proteins in the Phillips 301 electron microscope revealed small, filamentous structures of varying lengths (50-100 nm), shown in Figure 2A and 2B. When these structures were dissociated with SDS and rechromatographed on the C3-8C column, two proteins could be identified following PAGE. These had molecular weights of 27 and 30 Kd. One additional 66 Kd protein was also observed and may have been a precursor which undergoes proteolytic cleavage to form some of the smaller rubella proteins[4,23], a replication strategy common to many of the Togaviruses[7]. An equivalent amount of uninfected L cell lysate was also subjected to affinity chromatography and no cell proteins sticking non-specifically to the Sepharose beads or cross-reacting with the bound IgG could be detected. These data suggest that the monoclonal antibody was specific for an epitope present on 2 proteins (27 and 30 Kd). Western blot analysis has confirmed and extended these data, showing that the monoclonal also cross-reacted with 2 proteins with molecular weights of 47 and 42 Kd. These observations are consistent with those of Oker-Blom[24] who has suggested that some smaller proteins may represent unglycosylated precursors of 2 envelope proteins which appeared virtually identical following tryptic peptide analysis. While the precise structural role of these proteins remains to be determined, the data indicate that the C3-8C antibody was directed against specific viral proteins and coincidentally adsorbed all of the other RV antigens and possibly some cell proteins which were associated with them on the structures illustrated in Figure 2A. To determine whether these structures were artifacts generated by lysing the cells in the presence of mature rubella virions, a mixing experiment was performed. Uninfected L cells were collected into 2 ml of PBS and mixed with 1 ml of a virus suspension containing 1×10^7 virions. The suspension was then lysed, centrifuged and the supernatent was passed over the C3 immunosorbent column as described. However, no detectable material with an absorbence at 280 mμ could be eluted from the column, suggesting that the structures were

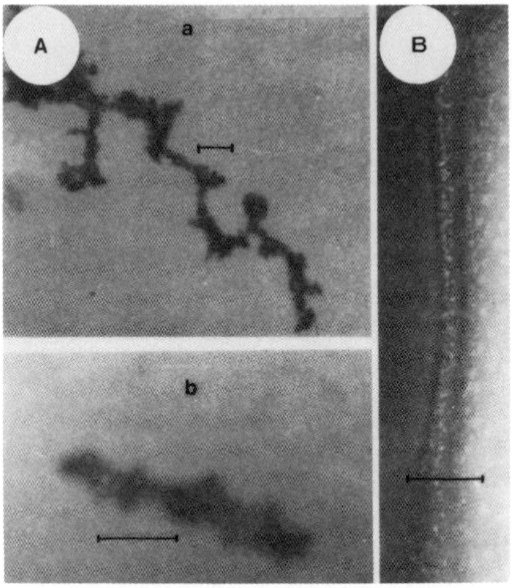

FIGURE 2. Electron micrographs illustrating VACS structures affinity-purified from (A) RV-infected L cells (VACS-1 particles) shown in A(a) as aggregated rods and in A(b) as a single rod, and from (B) VACS-1 infected L cells (VACS-2 particles). Structures in A were stained with uranyl acetate and in B with PTA on formvar-coated grids. Scale bars, 50 nm.

---

present in infected cells and were not generated as lysis artifacts.

The structures contained in Figure 2A appeared to have a tendency to aggregate as further evidenced by their continuous distribution throughout a linear 25-45% Renografin gradient. They were then characterized further. First, they were examined for the presence of single stranded viral RNA. L cells were infected and incubated in medium containing 2 μCi/ml ^{3}H-uridine (New England Nuclear). The structures were prepared as described and found to contain trichloroacetic acid (TCA) precipitable counts which were largely (88.4%) insensitive to the addition of RNase A (Sigma). However, if the structures were pre-treated with 2% SDS and placed in a boiling water bath for 2 minutes, then 91.2% of the TCA-precipitable counts were rendered nuclease sensitive (summarized in Table 1). This suggests that single-stranded RNA is associated with and stabilized by these structures. The origin of the RNA remains to be determined but the RV genome has been shown to be comprised of standard ssRNA with a molecular weight of $3x10^6$

TABLE 1

Trichloroacetic acid (TCA) precipitable ^{3}H-uridine counts associated with rubella VACS rods

SAMPLE	TCA PRECIPITABLE COUNTS	% OF SAMPLE 1
1. Untreated VACS	1053	100.0
2. VACS + RNase A	931	88.4
3. SDS-treated VACS + RNase A	93	8.8

Each 2 ml aliquot of rubella VACS particles ($A_{280\ m\mu}$=0.249) in PBS was precipitated and washed 3 times with cold 5% TCA followed by a final 95% ethanol wash, and the precipitates were retained on glass-fiber paper discs, which were counted after (1) no RNase treatment (2) addition of 10 μg RNase (Sigma) and 10 min incubation at 37^{0}C and (3) addition of SDS to a final concentration of 2%, heating in a boiling water bath for 2 min, 1/5 dilution with cold PBS, cooling and the final addition of 10 μg RNase and 10 min incubation. Each sample represents an average of triplicate samples.

daltons[25,26] as well as subgenomic, defective RNA species with variable molecular weights representing as little as 35% of the standard RNA[26].

Second, the structures were bound to microtiter dishes and the enzyme-linked immunosorbent assay (ELISA)[27] was performed employing anti-actin, anti-tubulin and C3-8C antibodies. The results are contained in Figure 3 which confirms that VACS-1 particles, affinity-purified from RV-infected L cells, bind antibodies directed to G-actin. In addition, some tubulin-containing fragments may also have been present as indicated by the binding of anti-tubulin antibody. Thus, the rod-shaped structures isolated using affinity column chromotography appeared to be composed of rubella virus proteins and possibly viral ssRNA associated with fragments of the host cytoskeleton, and have been designated rubella VACS (virus associated with cytoskeleton) particles.

The undissociated, affinity-purified rubella VACS-1 particles were then tested for infectivity in three ways. First, immunofluorescent antibody techniques were used to detect intracellular rubella antigens using C3-8C monoclonal antibody in

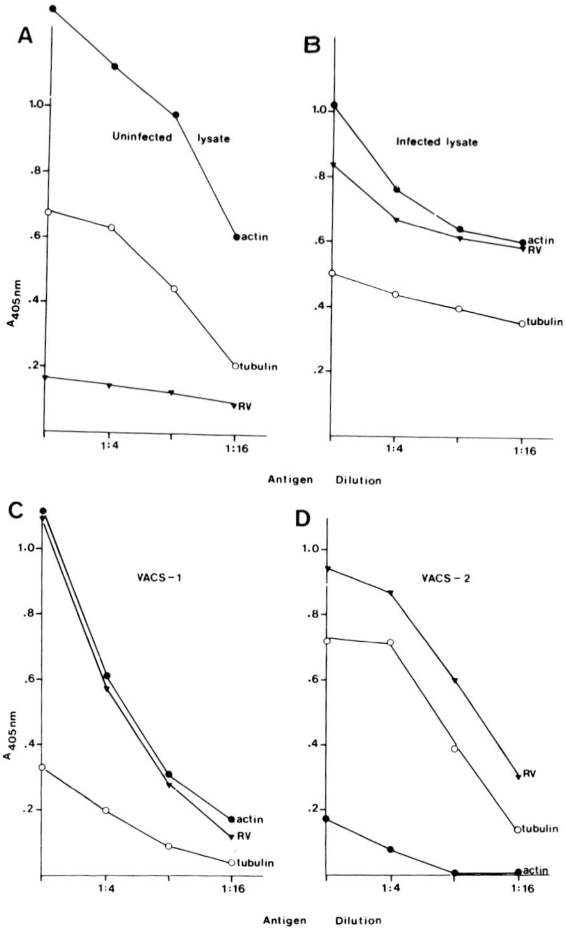

FIGURE 3. Relative concentrations of actin, tubulin and RV in uninfected (A) and RV-infected (B) L cell lysates and in affinity-purified VACS-1 (C) and VACS-2 (D) structures. The ELISA was performed according to the procedure described by Voller et al.[27] Polytroned lysates of 5 uninfected and 5 RV-infected 175 cc^2 L cell monolayers were diluted 1:10 in coating buffer ($A_{280\mu m}$=0.10) and duplicate 200 µl aliquots of each were adsorbed to microtiter plate wells (Dynatech Laboratories). A predetermined dilution of each of the following antibodies was added to each well after washing (rabbit anti-actin (Miles Laboratories), 1:50; rabbit anti-tubulin (a gift from V. Kalnins

Department of Anatomy, the University of Toronto), 1:50; and C3-8C, 1:4000. Antibody binding was measured using a previously determined 1:4000 dilution of goat anti-rabbit IgG and rabbit anti-mouse IgG (Quadra Logic Technologies Inc.) linked to alkaline phosphatase. The $A_{400 \mu m}$ was determined after 1 hour incubation at room temperature.

VACS-innoculated L cell monolayers after 5 sequential subcultures. The results are illustrated in Figure 4 which shows that both the rubella virus and the VACS-infected L cells expressed cytoplasmic fluorescence but the latter cells, in Fig. 4C, displayed altered morphologies, appearing larger and expressing longer cytoplasmic extensions. The pattern of fluorescent labeling varied somewhat in that the VACS-infected cultures contained an unusually high number (20%) of multinucleated cells. The morphological differences between RV- and VACS-infected cells were striking and could be observed within 2 days of infection. On continued subculturing of RV- and VACS-infected cells for 5 to 6 weeks, these differences became more pronounced. Persistently-infected L cells continued to produce rubella virions and to resemble those monolayers shown in Fig. 4B. The VACS-infected L cells exhibited progressively more bizarre shapes and increased cell lysis until these monolayers could no longer be subcultured. In view of the elapsed time between infection and the subsequent staining of the RV antigens, it is likely that the detection of rubella antigen in VACS-infected cells represented new synthesis rather than the uptake of the original inoculum. If VACS particles were treated with 2% SDS and placed in a boiling water bath for 2 min prior to inoculation of L cell monolayers, infectivity was lost and no cytoplasmic fluorescence or morphological changes were observed.

Second, RV and VACS-infected cultures were examined for the presence of rubella polypeptides. In each case, infected cells were collected, lysed, centrifuged and passed over an affinity column as previously described. Rod-like structures derived from each of the virus and VACS-infected cultures were eluted from the columns, dissociated and subjected to PAGE analysis on 10% gels. The structures obtained from RV-infected cells (called VACS-1 particles) were composed of all the major proteins associated with a productive rubella infection[4,23]. The structures obtained from the VACS-1-infected cells (called VACS-2 particles) were composed predominantly of at least one polypeptide with a molecular weight of 27-30 Kd. One additional band at 66 Kd may have represented a larger molecular

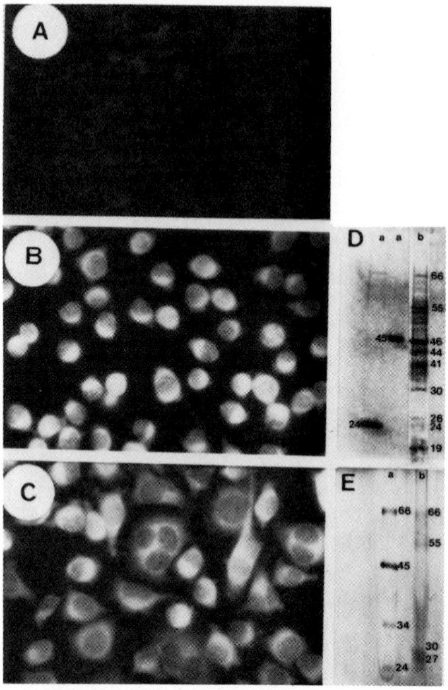

FIGURE 4. A comparison of intracellular RV-associated with cytoskeleton fragments in RV- and VACS-1-infected cells. A, B and C illustrate rubella-specific fluorescein-labeling in uninfected, RV- and VACS-1 infected L cells respectively (x200). The PAGE analysis of proteins affinity-purified from RV-infected and VACS-1-infected cells is shown in Fig. 4D and 4E respectively. Cells were infected with RV at an m.o.i. of 0.01 as described elsewhere[21], and prepared for labeling 4 days post-infection. A second monolayer composed of 10^5 cells on a glass coverslip was infected with 0.1 ml of VACS-1 ($A_{280m\mu}$=0.249), containing 40 µg protein. The suspension was added to the monolayer after removing the medium and was allowed to adsorb for 1 hour at 37°C. Five ml of medium were added to the dish. This infected monolayer was then subcultured 5 consecutive times splitting the cells 1:3 at 4 day intervals. Indirect immunofluorescence techniques were employed as described in detail elsewhere[5,6]. Ten 175 cc^2 flasks were employed for each PAGE analysis. Cells were harvested, lysed and subjected to affinity chromatography. Eluted antigens were dialyzed against distilled water, lyophylized, resuspended in sample buffer and analyzed on 10% PAGE slab gels. Molecular weight markers for gel calibration appear in (a) and VACS samples in lanes marked (b).

weight precursor protein which was present in Figs. 1A and 1C as well. A 55Kd protein was also observed and may have represented an unusual rubella precursor protein, or, in view of the binding of anti-tubulin antibody (Fig. 3D) and the rod-like nature of the VACS-2 particles, possibly 55 Kd tubulin subunits of microtubules[28]. This protein was also observed in Fig. 4D as a component of VACS-1 structures. Thus, VACS-1 particles have been shown to infect L cells and to direct the synthesis of at least one protein, with a molecular weight of 27-30 Kd, which affinity purifies with tubulin-containing fragments and possibly one additional precursor protein. If other RV proteins were present in VACS-1 infected cells they were not associated with the affinity-purified particles shown in Fig. 2B.

Third, attempts were made to detect infectious virions in culture fluid from VACS-1-infected cells. No virions could be detected by direct microscopic examination of culture media, nor were there any infectious hemadsorbing (Had) units greater than 10^2 (background) scored when culture fluid from VACS-infected monolayers was used to infect fresh L monolayers[21]. Further, when 10-175 cc^2 monolayers of VACS-infected cells were incubated in the presence of 2 µCi/ml ^{3}H-uridine for 4 days and standard centrifugation techniques were employed to detect ^{3}H-labeled virions with densities of 1.19 to 1.12 gm/ml[3], no virions could be detected[21,26]. It has previously been shown that RV generates defective interfering (DI) virions and that these DI particles may be present in all virus stocks[26]. However, it is unlikely that the VACS-mediated infectivity could be due to small numbers of defective virions present in the VACS suspensions since DI rubella virus promotes the reduced synthesis of all RV proteins and limited budding of virus from cells which are morphologically indistinguishable from uninfected controls. In contrast, the hallmark of a rubella-VACS-mediated infection is the pronounced alteration in cell morphology and the appreciable synthesis of only one (27-30 Kd) protein. This would indicate that the VACS particles exhibit properties unlike the virus stocks from which they were derived. The inability of VACS-infected cells to sustain the production of mature rubella virus would be consistent with the very limited synthesis of filament-associated viral proteins in these cells.

CONCLUSIONS

These data indicate that normal RV replication is intimately associated with intracellular cytoskeleton structures. Lysis of infected cells released actin and tubulin-containing fragments with bound viral proteins and ss RNA (possibly of viral origin), which exhibited virus-like, infectious characteristics. There have been various reports of other viral proteins associating with cytoskeletal components, but these observations have never been extended to include infectivity assays of such structures[29-33]. (It may be relevant to note, in this regard, that fragments of actin and tubulin have been shown to integrate into existing cytoskeletal structures following their microinjection into cells[34]). The data presented here indicate that rubella virus possesses an alternate mode of infection in vitro which is VACS-mediated and which results in a persistent infection characterized by (1) an altered host morphology (2) severely limited synthesis of viral proteins and (3) an undetectable release of progeny virions.

CHARACTERIZATION OF SCRAPIE INFECTIVITY METHODS AND RESULTS

Extraction of Cytoskeleton

Whole brains from endstage scrapie-infected or age-matched uninfected male hamsters were homogenized to 10% in a Ten Broeck glass grinder using buffer A (20mM Tris, 100 mM NaCl, 5mM EDTA, 1mM DTT, pH 7.8). After clarification at 500 x g/5°/10 min to remove nuclei and large cell fragments, the supernatant was centrifuged at 1000 x g/5°/30 min to obtain a plasma membrane enriched pellet[13]. These pellets were resuspended to 10% in buffer B (10mM Tris, 6mM 2-mercaptoethanol, 4mM $MgCl_2$, 1 mM EGTA, pH 7.8), then extracted with 1%NP40 at 4°C for 2 hr. The cytoskeletal fraction was pelleted at 12 K x g/5°/15 min and washed twice in buffer B to remove residual detergent. The washed pellets were resuspended to 10% in buffer B without $MgCl_2$ and turned overnight at 4°C.

After centrifugation at 12 K x g/5°/20 min, the supernatants containing soluble cytoskeletal proteins were removed and the pellets resuspended to 10% in buffer B without $MgCl_2$, then extracted with 4% N-lauroylsarcosine for 2 hr at 4°C followed by centrifugation at 20 K x g/5°/30 min. Five ml samples of these supernatants were fractionated on seven ml 20-40% Nycodenz^R gradients[14] (in buffer A containing 0.4M $(NH_4)_2SO_4$) centrifuged at 100 K x g/5°/16 hr. One and a half ml fractions were collected from the separated gradients and their densities measured before dialyzing at 4°C against 10 mM Tris, 2mM EDTA, pH 7.8 for 48-72 hr. A fraction containing a visible band of material sedimenting towards the bottom of the gradients at a density of 1.226 g/ml was found to contain a high level of scrapie infectivity and was selected for protein and nucleic acid characterization.

Proteins. Protein concentrations were determined by the dye-binding method of Bradford, then samples containing 20 μg of protein each were separated in a 9-18% denaturing polyacrylamide gel and stained with silver[15]. In Figure 5, the soluble cytoskeletal protein samples from both infected and uninfected hamster brains contain two major proteins which appear to be tubulin (55 Kd) and actin (43 Kd) based on their molecular weights and co-migration with these proteins. The 1.226 g/ml Nycodenz^R gradient fractions contain a major 50 Kd protein which we have identified as glial fibrillary acidic protein (GFAP) based on molecular weight, reaction with GFAP monoclonal antibody (Boehringer Mannheim, Cat. No. 814369) in immunoblot, and the formation of 10 nm filaments after incubation at 37°C in 150 mM KCl.

Nucleic Acids. Nucleic acids were quantitated by $A_{260m\mu}$ absorbance after phenol extraction, then 3' or 5'-end labeled and separated in SDS-polyacrylamide gels containing 8M urea[18]. Samples enriched for tubulin and actin cytoskeletal proteins contained 8-12 μg/ml of nucleic acid, those enriched for GFAP 30-40 μg/ml. Qualitatively, all samples were principally composed of heterogeneous low-molecular-weight RNA. A more complete description of the nucleic acids in these cytoskeletal extracts is *in preparation* (Judd Aiken *et al*.).

Infectivity and Treatment with RNase A

Table 2 reports the scrapie infectivity of each sample before and after treatment with RNase A (Sigma, Cat. No. R 4875) or poly-L-lysine (Sigma, Cat. No. P 9404). Both treatments reduced infectivity 2-3 $Log_{10}LD_{50}$.

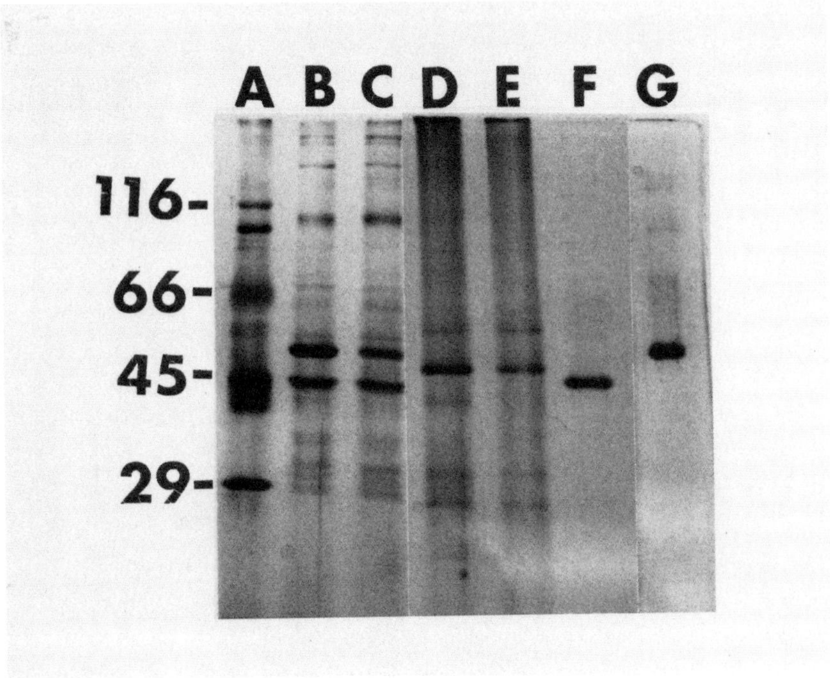

FIGURE 5. Lane A contains molecular weight markers indicated on left, lanes B and C soluble cytoskeletal proteins from infected (B) and uninfected (C) hamster brain, and lanes D (infected) and E (uninfected) insoluble proteins extracted with 4% N-lauroylsarcosine. Lane F contains 15 μg of rabbit muscle actin (Sigma, Cat. No. 2522) and lane G,7 μg of tubulin (a gift from Gary Borisy).

TABLE 2

SUSCEPTIBILITY OF SCRAPIE INFECTIVITY IN CYTOSKELETAL EXTRACTS TO TREATMENT[a] WITH RNASE A OR POLY-L-LYSINE

Sample	Untreated	RNase A (100 μg/ml)	Poly-L-lysine (100 μg/ml)
Tubulin-Actin enriched	7.8[b]	5.5	6.0
GFAP enriched	8.5	6.0	6.8

[a]37°C/2 hr with intermittent vortexing
[b]$Log_{10}LD_{50}$ per ml as measured by the method of incubation interval assay[14].

CONCLUSIONS

Scrapie infectivity is associated with the cytoskeleton, co-purifying to a greater degree with insoluble (GFAP) rather than soluble (tubulin and actin) cytoskeletal proteins. Infectivity in these preparations can be reduced by treatment with RNase A. However, treatment with poly-L-lysine produces a similar result indicating that the reduction in infectivity may be caused by aggregation of the sample due to the polycationic properties of RNase A rather than a hydrolytic effect on an essential RNA component.

ACKNOWLEDGMENTS

DV is grateful to George Spiegelman and Wayne Vogel for their advice and to Leo Smyrnis and Kathy Wong for expert technical assistance. RFM wishes to thank Gary Borisy for helpful discussions and the gift of purified tubulin.

REFERENCES

1. Holt S, Hudgins D, Krishman DR, Critchley EMR (1976). Diffuse myelitis associated with rubella vaccination. Brit Med J 30 Oct:1037.
2. Cremer NE, Oshiro LS, Weil ML, Lenette EJ, Itabashi HH, Carnay L (1979). Isolation of rubella virus from brain in chronic progressive panencephalitis. J gen Virol 29:143.
3. Callaghan N, Feely M, Walsh B (1977). Relapsing neurological disorder associated with rubella virus infection. J Neurol Neurosurg Psych 40:1117.
4. Pope DD, Van Alstyne D (1981). Evidence for restricted replication of rubella virus in rat glial cells in culture. Virol 113:776.
5. Van Alstyne D, Paty DW (1983). The effect of dibutyryl cyclic AMP on restricted replication of rubella virus in rat glial cells in culture. Virol 124:173.
6. Van Alstyne D, Smyrnis EM, Paty DW (1983). Differentiation of glioblasts from adult brain. Neurosci Letters 40:327.
7. Shope RE, Monath TP, Cooper LZ, et al. (1985). The Togaviruses (chap 39-43). In Fields BN (ed): "Virology", Raven Press, p 927.
8. Miller E, Cradock-Watson JE, Pollock TM (1982). Consequences of confirmed maternal rubella at successive stages of pregnancy. Lancet 2:781.
9. Desmond MM, Fisher ES, Vorderman AL, et al. (1978). The longitudinal course of congenital rubella encephalitis in non-retarded children. J Pediat 93:584.
10. DesmondMM, Wilson GS, Melnick Jl, et al. (1967). Congenital rubella encephalitis. J Pediat 71:311.
11. Levine JB, Berkowitz CD, St Geme JW Jr (1982). Rubella virus reinfection during pregnancy leading to late onset congenital rubella syndrome. J Pediat 100:589.
12. Millson GC, Hunter GD, Kimberlin RH (1971). An experimental examination of the scrapie agent in cell membrane mixtures. II. Association of scrapie activity with membrane fractions. J Comp Path 81:255.
13. Semancik JS, Marsh RF, Geelen JL, Hanson RP (1976). Properties of the scrapie agent - endomembrane complex from hamster brain. J Virol 18:693.
14. Marsh RF, Castle BE, Dees C, Wade WF (1984). Equilibrium density gradient centrifugation of the scrapie agent in Nycodenz. J gen Virol 65:1963.
15. Dees C, German TL, Wade WF, Marsh RF (1985). Characterization of proteins in membrane vesicles from scrapie-

infected hamster brain. J gen Virol 66:851.
16. Dees C, German TL, Wade WF, Marsh RF (1985). Characterization of lipids in membrane vesicles from scrapie-infected hamster brain. J gen Virol 66:861.
17. German TL, McMillan BC, Castle BF, Dees C, Wade WF, Marsh RF (1985). Comparison of RNA from health and scrapie-infected hamster brain. J gen Virol 66:839.
18. Dees C, McMillan BC, Wade WF, German TL, Marsh RF (1985). Characterization of nucleic acids in membrane vesicles from scrapie-infected hamster brain. J Virol 55:126.
19. Marsh RF (1986). The scrapie agent. In Semancik JS (ed): "Viroids and viroid-like pathogens", Baton Rouge: CRC Press Inc (in press).
20. Stewart GL, Parkman PD, Hopps HE, et al. (1967). Rubella virus hemagglutination inhibition test. New Eng J Med 276:554.
21. Van Alstyne D, Krystal G, Kettyls GD, Bohn EM (1981). Purification of rubella virus and determination of its polypeptide composition. Virol 108:491.
22. Waxham MN, Wolinsky JS (1983). Immunochemical identification of rubella virus hemagglutinin. Virol 126:194.
23. Chantler JK (1979). Rubella virus: intracellular polypeptide synthesis. Virol 98:275.
24. Oker-Blom C, Kalkkinen N, Kaariainen L, Pettersson RF (1983). Rubella virus contains one capsid protein and three envelope glycoproteins, E1, E2a and E2b. J Virol 46:964.
25. Hovi T, Vaheri A (1970). Infectivity and some physicochemical characteristics of rubella virus RNA. Virol 42:1.
26. Bohn EM, Van Alstyne D (1981). The generation of defective interfering rubella virus particles. Virol 111:549.
27. Voller A, Bidwell D, Bartlett A (1976). Enzyme linked immunosorbent assay, Chap 69. In Rose NR, Friedman H (eds): "Manual of Clinical Immunology", Am Soc Microbiol p 506.
28. Wisenberg R, Borisy G, Taylor EW (1968). The isolation and characterization of tubulin and the determination of subunit structure. Biochem 7:4466.
29. Lenk R, Penman S (1979). The cytoskeleton framework and poliovirus metabolism. Cell 16:289.
30. Krempien U, Schneider G, Hiller K, et al. (1981). Conditions for poxvirus-specific microvilli formation studied during synchronized virus assembly. Virol 113:556.
31. Rutter G, Mannweiler K (1977). Alterations of actin-containing structures in BHK-21 cells infected with NDV and VSV. J gen Virol 37:233.
32. Luftig RB, Weihing RR (1975). Adenovirus binds to rat

brain microtubules in vitro. J Virol 16:696.
33. Murti KG, Goorha RJ (1983). Interaction of frog virus-3 with the cytoskeleton. I. Altered organization of microtubules, intermediate filaments and microfilaments. J Cell Biol 96:1248.
34. Birchmeier W (1984). Cytoskeleton structure and function. TIBS 9:192.
35. Koprowski H, Gerhard W, Croce CM (1977). Production of antibodies against influenza virus by somatic cell hybrids between mouse myeloma and primed spleen cells. Proc Natl Acad Sc USA 74:2985.
36. Laemmli UK (1970). Cleavage of structural proteins during the assembly of the head of bacteriophage T4. Nature (London) 227:680.

VIII. STRATEGIES FOR CONTROL OF POSITIVE STRAND VIRUS DISEASES

Positive Strand RNA Viruses, pages 539–553

MOLECULAR BASIS OF ANTIGENICITY OF POLIOVIRUS

P D Minor, D M A Evans, M Ferguson, G C Schild, J W Almond, G Stanway

National Institute for Biological Standards and Control, Holly Hill, London NW3 6RB

INTRODUCTION

Polioviruses are members of the family picornaviridae. The virion consists of sixty copies of each of the four structural proteins VP1, VP2, VP3 and VP4 arranged with icosahedral symmetry about a messenger sense single stranded RNA genome approximately 7400 bases in length. The atomic structure of the virus has been resolved to 2.9Å (Hogle et al 1985) and the RNA sequences of the genomes of several strains have been determined (Kitamura et al 1981, Toyoda et al 1984, Cann et al 1984).

The virus occurs in three serotypes, designated types 1, 2 and 3 which show little antigenic cross reactivity in neutralisation tests although there is some cross protection between types 1 and 2 in animals. Vaccines, which may consist of either killed virus (Salk and Salk 1977) or live attenuated strains (Sabin and Boulger 1973) contain a single strain of each serotype, and have effectively controlled poliomyelitis in the developed world. Cases of vaccine associated disease (WHO 1981, Cann et al 1984) or possible vaccine failure (Leinikki et 1985, Magrath et al 1986) are very rare.

The high degree of efficacy of poliovirus vaccines based on a single strain of each serotype implies that wild poliovirus is not subject to antigenic drift of the kind described for influenza virus. The antigenic structure of the virus is thus of considerable interest from several points of view. A full understanding of the apparent antigenic stability of the virus would have a bearing on the development of vaccines against other virus diseases, and the definition of the antigenic sites involved in protective immunity is a preliminary step in the development of novel vaccines prepared by recombinant DNA technology or chemical synthesis.

IDENTIFICATION OF ANTIGENIC SITES INVOLVED IN THE NEUTRALISATION OF POLIOVIRUSES

While several strategies have been adopted for identifying the sites involved in the neutralisation of polioviruses, the work described here was based on the preparation of monoclonal antibodies with neutralising activity for the Sabin strains of polioviruses of types 1, 2 and 3 followed by the selection and characterisation of antigenic variants resistant to neutralisation (Ferguson et al 1984, Minor et al 1983, Evans et al 1983, Minor et al 1985, Minor et al 1986). Hybridoma lines were prepared from splenocytes obtained from Balb/c mice or Lewis rats, immunised in the initial experiments by intraperitoneal injection of purified, native preparations of poliovirus of types 1, 2 or 3. In later studies different routes of immunisation and different preparations of virus were employed to increase the spectrum of antibodies obtained.

Antigenic variants were selected from the Sabin strains of poliovirus by two cycles of plaque formation after treatment with neutralising monoclonal antibody before small working pools were grown for antigenic and genetic characterization. The experimental strategy was based on the assumption that in the main mutations affecting the reactions of antigens occur at the site of binding of the antibody, although mutations having an effect on the conformation of a distant amino acid sequence are possible. Early studies involved the neutralisation of type 3 poliovirus using a panel of monoclonal antibodies generated from mice immunised by intraperitoneal injection with native purified virus. The variants were intitially studied by examining their neutralisation by a panel of monoclonal antibodies. A total of sixteen different patterns of reaction were identified in a total of 129 plaques selected from the Sabin type 3 vaccine strain, P3 Leon $12a_1b$ (Table 1). The thirteen antibodies could be divided into two groups composed of 12 antibodies and one antibody respectively.

Table 1

Reactions of representatives of 129 antigenic mutant viruses selected from the Sabin type 3 strain Leon $12a_1b$ with monoclonal antibodies

Virus strain P3/Leon/37	25-1-14[a]	25-4-12[a]	27-4-4[a]	199	194	134	208	175	204	197	165	198	138
1	r	r	r	r	r		r	r	r	r	r	r	
2	r	r	r	r	r			r	r	r	r	r	
3	r	r	r	r	r	r	r	r	r	r	r	r	
4	r	r	r	r	r	r	r	r		r		r	
5	r	r	r	r	r		r		r	r	r		
6		r		r	r	r	r	r			r	r	
7		r										r	
8				r	r	r	r				r		
9				r	r	r	r	r	r		r	r	
10				r	r	r	r	r		r	r	r	
11				r			r			r	r		
12				r		r	r				r		
13				r	r		r	r		r	r		
14				r	r	r	r	r	r		r		
15						r	r				r	r	
16													r

Antibodies were raised by fusion of splenocytes from Lewis rats (a) or Balb/C mice immunised by intraperitoneal injection of native virus.

The fifteen mutants which were resistant to an antibody of the major group were invariably resistant to several other antibodies of the same group and one (mutant 3) was resistant to all antibodies of this group. In contrast the remaining mutant (mutant 16) was only resistant to neutralisation by the thirteenth antibody, 138, and 138 was able to neutralise all fifteen mutants of the other class. The antibodies of the panel therefore appeared to be recognising two distinct independently mutable antigenic sites, one of which was a target of more antibodies than the other.

The mutations were identified by primer extension sequencing of the genomic RNA of the mutants (Evans et al 1983) through regions which had been tentatively identified by other methods (Minor et al 1983, Minor et al 1985). Single point mutations leading to predicted amino acid changes were identified in all cases, and are summarised in Table 2 which also includes substitutions identified in comparable studies with the related type 3 strain P3/Leon/USA/1937. The major group of mutants had lesions in VP1 between residues 89 and 100 suggesting that this was the target recognised by the majority of the antibodies. By analogy with the three dimensional structure of the Mahoney strain of type 1 poliovirus (Hogle et al 1985) the sequence from residues 89 to 100 of type 3 poliovirus would be expected to be close to residues 166 and 253 which can also be changed in mutants affecting the reactions of the main group of antibodies. The subsidiary site recognised by antibody 138 was shown to include residues 286, 287 and 288 of VP1 (Table 2) and by analogy with Mahoney this sequence is distant from the major site in the three dimensional structure of the virus, consistent with its distinct and independently mutable properties.

In contrast to these results with type 3 poliovirus others working with type 1 poliovirus reported the existence of multiple mutational sites distinct from both those described here (Diamond et al 1985, Blondel et al 1986). As it was possible that these findings were due to differences in the methods used to obtain antibodies and mutants monoclonal antibodies specific for type 1 were derived using the same immunisation and screening protocols described above for type 3 and mutants of type 1 poliovirus selected and characterised. Two groups of variant were identified, one involving mutations in VP3

TABLE 2
Amino acid substitutions identified in antigenic variants resistant to monoclonal antibodies raised by intraperitoneal injection of native Sabin poliovirus strains into mice

Type 1 (Sabin strain)	Type 2 (Sabin strain)	Type 3 P3/Leon $12a_1b$ (Sabin strain)	P3/Leon/ USA/1937
a)	a)	a)	a)
$VP3_{58}$ S-C	$VP1_{94}$ A-T	$VP1_{89}$ E-K,G	$VP1_{91}$ D-N
$VP3_{59}$ A-Q	$VP1_{95}$ P-I,N	$VP1_{91}$ D-N	$VP1_{93}$ E-G
$VP3_{60}$ K-Q,R,T	$VP1_{97}$ K-E,Q	$VP1_{97}$ T-N,S	$VP1_{95}$ P-L
$VP3_{71}$ R-Q,W	$VP1_{98}$ R-L,S,C	$VP1_{98}$ R-W,G,Q	$VP1_{96}$ T-I,A,N
b)	$VP1_{99}$ A-P	$VP1_{99}$ A-T,V	$VP1_{97}$ T-N,I
$VP2_{169}$ S-P	$VP1_{174}$ K-E	$VP1_{100}$ Q-K,R	$VP1_{98}$ R-Q,W
$VP2_{170}$ P-S		$VP1_{166}$ K-M	$VP1_{99}$ A-T,V
$VP1_{220}$ S-L		$VP1_{253}$ T-A	$VP1_{100}$ Q-L,P,R
$VP1_{222}$ A-V		b)	
		$VP1_{286}$ R-K	
		$VP1_{287}$ N-D	
		$VP1_{288}$ N-D	

Amino acids are designated by the single letter code

at residues 58, 59, 60 and 71, and the other involving mutations in VP2 at residues 169 and 170, or in VP1 at residues 220 and 222 (Table 2). These findings were consistent with those reported for the type 1 strains Mahoney (Diamond et al 1985) and the Sabin type 1 strain (Blondel et al 1986). The mutations in VP2 and VP1 were close together in the three dimensional structure of the virus, and distinct from the mutations in VP3, which formed a separate cluster (Hogle et al 1985). Both were distant from the major sequence identified in type 3 poliovirus, which was identified as a major site in type 2 poliovirus by the same strategy of mutant selection (Table 2).

Comparable studies have been undertaken of the antigenic structure of rhinovirus 14 (Sherry and Rueckert 1985, Rossmann et al 1985). The results are strikingly similar, and have identified a major site in VP1 including residues 91 and 95 a complex subsidiary site including residues 72, 75, 78 and 203 of VP3 with residue 287 of VP1 and a complex site made up of residues 136,

158, 159, 161, 162 of VP2 and residue 210 of VP1. The location of these sequences in the three dimensional structure of the virion is comparable to their location in poliovirus type 1. However, in poliovirus type 3 the site including residues 89 to 100 of VP1 is highly immunodominant, such that of a total of twenty six antibodies with neutralising activity raised by intraperitoneal immunisation of mice with native virus only one was directed against another, subsidiary site. In contrast only one antibody specific for type 1 poliovirus has been described which is directed against this major site (Blondel et al 1983). The reason for the difference between the serotypes is unknown. However preliminary studies suggest that monoclonal antibodies against site 1 of poliovirus type 1 may be generated by immunising the mice by intrasplenic injection of native virus.

IDENTIFICATION OF SUBSIDIARY SITES IN POLIOVIRUS TYPE 3

In 1984 and 1985 a limited outbreak of poliomyelitis occurred in Finland, a country in which more than 95% of the population are recorded as having received inactivated poliovaccine (Leinikki et al 1985). Strains isolated from the outbreak were antigenically unusual in their reactions with both monoclonal antibodies and human and animal sera (Magrath et al 1986). Monoclonal antibodies were raised with a strain isolated during this outbreak, and one was found to have neutralising activity for a range of strains of poliovirus type 3, including the mutants described above and in Tables 1 and 2. This antibody was therefore possibly directed against a distinct site.

The immunodominant site of the Sabin strain of poliovirus type 3 contains an arginine which renders it sensitive to trypsin so that after digestion antibodies no longer bind to it (Fricks et al 1985, Icenogle et al submitted for publication). Monoclonal antibodies raised by immunisation of mice with trypsin treated virus were found to be able to neutralise all mutants having lesions within the immunodominant site of type 3. They were therefore possibly directed against a different site.

Mutants were selected with this new panel of monoclonal antibodies and characterised as before and shown to fall into two distinct groups. The mutations identified are summarised in Table 3.

TABLE 3
Location of amino acid substitutions in antigenic mutants of type 3 poliovirus selected with monoclonal antibodies against trypsin treated Sabin type 3 virus or P3/Finland/23127/85

Mutation
a) $VP2_{164}$ N - K,T
$VP2_{166}$ V - A
$VP2_{167}$ I - K
$VP2_{172}$ E - A,K
b) $VP1_{287}$ D - N
$VP3_{290}$ D - E
$VP3_{58}$ E - N
$VP1_{59}$ S - N,R
$VP3_{77}$ D - E
$VP3_{79}$ S - L

Amino acids are designated by the single letter code. Groups (a) and (b) were substitutions found in two distinct operationally defined antigenic sites

One group had lesions in VP1 between residues 287 and 290 (which includes the subsidiary site described above) or in VP3 at residues 58, 59, 77 or 79. These sites are homologous to those identified for type 1 except that the site including residues from VP2 has not yet been shown to include residues from VP1 for type 3, and the site including residues from VP3 has not been shown to include residues from VP1 in type 1. The location and nomenclature of the sites are summarised in Table 4. The immunodominant site of type 3 is designated site 1, the site including residues of VP1 and VP2 is designated site 2, its VP1 component being site 2a, and its VP2 component site 2b, and the site including residues of VP1 and VP3 is designated site 3, its VP1 component being site 3a, and its VP3 component site 3b. For type 3 approximately twice as many antibodies have been obtained against site 3 as against site 2, but for native virus site 1 is strongly immunodominant in both mice and rats. It is striking that trypsin treated type 3 poliovirus appears to be similar to native type 1 virus in the profile of antibodies it stimulates in mice.

TABLE 4
Location and occurrence of antigenic sites in poliovirus of serotypes 1,2 and 3

Site	Location(a)	Serotype
1	VP1; 89-100	2,3
2a	VP1; 220-222	1
2b	VP2; 165-171	1,3
3a	VP1; 286-290	3
3b	VP3; 58-60, 70, 71, 77, 79	1,3

(a) numbering is from the N terminus of the protein

SIGNIFICANCE OF ANTIGENIC SITES IN POLIOVIRUS INFECTIONS

Mutations are readily selected in in vitro systems in each of the three antigenic sites identified in poliovirus. Passage of mixtures of wild type and mutant virus and the general stability of the mutant virus on passage, suggest that the antigenic variants are not subjected to a selective disadvantage in vitro. The antigenic stability of the virus in the wild is therefore unexpected.

A series of isolates of type 3 poliovirus was obtained from vaccinees and their contacts to examine the antigenic stability of the Sabin strain of type 3 poliovirus in the course of replication in the gut. The results are summarised in Table 5, which shows the antigenic reactivity of strains isolated from four primary vaccinees (DM, KT1, KT2 and KT3) four strains isolated over a period of 14 months from a hypogamma-globulinaemic patient given monovalent type 3 polio-vaccine (H18095, HA,HD,HF) and twelve vaccine related strains from individuals of unknown vaccination history. Nine of the twenty strains showed evidence of antigenic variation in sites 2 and 3. Antigenic variation with respect to monoclonal antibodies has been previously reported for excreted strains of type 1 poliovirus (Crainic et al 1983, Kew and Nottay 1985). In contrast only two of the strains, H18095 and HA showed signs of antigenic variation in site 1. It was notable that H18095, isolated 3 months post infection, was resistant to many of the site 1 antibodies while HA, isolate six months later, was resistant to only one and HD and HF isolated at 13 and 14 months post infection were

Table 5 Reaction of antibodies with excreted strains of type 3 poliovirus by vaccinees

Vaccine	Site 1												Site 2				Site 3					
related strains	25-1-14	25-4-12	27-4-4	199	194	134	208	175	204	197	165	198	792	248	665	756	138	1023	868	557	1007	961
DM35														r	r	r						
DM66														r	r	r						r
KT1(30)														r	r	r		r	r	r	r	r
KT2(30)																						
KT3(30)																						
H18095(3)				r	r	r				r	r	r					r	r	r	r	r	
HA(9)											r			r			r	r	r	r	r	
HD(13)																	r	r	r	r	r	r
HF(14)																	r	r	r	r	r	r
9052																						
13533																						
13118																					r	r
21746																						
10196														r				r				
15521																	r	r			r	
5229																						
15282																r		r				
11340																						
16057																	r	r				
285																	r	r	r	r		r

r = resistant

MSAABO

neutralised by all of the site 1 specific antibodies. Oligonucleotide mapping indicated a steady accumulation of mutations resulting in about one large T1 oligonucleotide change per month (P Minor unpublished). The observation that variation in site 1 occurred but that the variant strain did not persist raises the possibility that variation in site 1 is selected against during replication in the gut.

Twenty four wild isolates of type 3 poliovirus predating the use of the Sabin vaccine strain were examined with the same panel of antibodies and the results are shown in Table 6. The site 1 antibody 165 failed to neutralise 22 of the strains and the site 1 antibody 197 failed to neutralise 9. The remaining 10 site 1 antibodies neutralised all strains except 11, 22 and 702. In general, therefore the site 1 antibodies were broadly reactive. In contrast site 3 antibodies were highly specific, such that only three strains could be shown to be neutralised by any of the antibodies. One site 2 antibody (792) was as broadly reactive as the site 1 antibodies, while the remainder failed to neutralise virtually all wild strains. In wild strains therefore site 1 was well conserved relative to the Sabin strain while sites 2 and 3 were not. This suggests that sites 2 and 3 may also be variable between wild strains of poliovirus type 3 and is consistent with the hypothesis that, in the human gut, sites 2 and 3 are subject to immune pressure while site 1 is not. One possible explanation for this is that site 1 is destroyed by the action of proteases in the gut in a similar way to its destruction by trypsin.

Faecal samples from vaccinees were treated with monoclonal antibodies to type 1 and type 2 poliovirus and with a site 1, type 3 specific monoclonal antibody. After adsorption of the virus to a cell sheet the inoculum and antibodies were washed off, and plaques allowed to develop. The data are summarised in Table 7 and show that virus with site 1 intact was well neutralised by this treatment, but trypsin treated virus or unpassaged virus from faecal samples were not. In contrast a site 3 type 3 specific monoclonal antibody neutralised all virus samples effectively. Virus eluted from plaques from virus preparations treated with the site 1 antibody was found to be sensitive to neutralisation by site 1 specific antibodies and was thus not antigenically variant.

Table 6 Reactions of antibodies with strains of type 3 poliovirus isolated before the use of Sabin vaccine

	Site 1												Site 2				Site 3					
Strain	25-1-14	25-4-12	27-4-4	199	194	134	208	175	204	197	165	198	792	248	665	756	138	1023	868	557	1007	961
1											r			r	r	r	r	r	r	r	r	r
2														r	r	r	r	r	r	r	r	r
5														r	r	r	r	r	r	r	r	r
7											r			r	r	r	r	r	r	r	r	r
8										r	r			r	r	r	r	r	r	r	r	r
9											r			r	r	r	r	r			r	r
10											r			r	r	r	r	r			r	r
11			r	r						r	r			r	r		r	r	r	r	r	r
19										r	r			r	r		r	r			r	r
21																r	r	r	r	r	r	r
22		r	r	r	r	r	r		r	r	r	r	r	r	r	r	r	r	r	r	r	r
41											r			r	r	r	r	r	r	r	r	r
694											r			r	r	r	r	r	r	r	r	r
699											r			r	r	r	r	r	r	r	r	r
702				r			r	r	r	r	r			r	r	r	r	r	r	r	r	r
710										r	r			r	r	r	r	r	r	r	r	r
715										r	r			r	r	r	r	r	r	r	r	r
730										r	r			r	r	r	r	r	r	r	r	r
77689										r	r		r	r	r	r	r	r			r	r
77731											r			r	r	r	r	r	r	r	r	r
77733											r			r	r	r	r	r	r	r	r	r
77750											r			r			r	r	r	r	r	r

r = resistant

TABLE 7
Sensitivity of unpassaged virus in faecal specimens to monoclonal antibodies

Virus preparation	Number of plaques formed		
	No antibody* to type 3	type 3 site 1 antibody	type 3 site 3 antibody
Native virus	18	0	0
Trypsin treated virus	52	63	0
DM 14	1	1	0
DM 28	57	51	0
DM 42	11	12	0
KT1 14	complete lysis	complete lysis	
KT1 28	11	14	0
KT2 14	complete lysis	complete lysis	0

* All samples were incubated with a type 1 and type 2 monoclonal antibody and the inoculum washed off thoroughly after adsorption

These findings suggest that the site on type 3 virus which is most potently immunogenic for mice and rats is destroyed during replication in the human gut, presumably by proteases. If this site is also immunogenic in man, then any virus in which the site was less susceptible to destruction would be vigorously selected against, as it would then be a target for additional neutralising antibodies. Studies with human sera suggest that site 1 is not so potent an immunogen in man as in the mouse (Icenogle et al 1986). On the other hand variation in site 1 appears to have been selected against in the isolates shown in Table 5. In addition isolates from the recent Finnish outbreak were shown to be antigenically somewhat variable within site 1, and to have lost the trypsin sensitive site (Magrath et al 1986). The human response to the different sites identified by the use of murine monoclonal antibodies is under further study.

CONCLUSION

The data reported here describe the identification of three antigenic sites on poliovirus which differ in their immunodominance between the three strains for reasons which are as yet unknown. The location of the sites is analagous to those on rhinovirus 14 described by others. The extent to which the human immune response to poliovirus parallels that of mice is not clear but there is evidence for an immune pressure upon the virus based upon antigenic variation, and it is possible that the site of replication of the virus has a strong influence on the antigenic variation observed.

REFERENCES

1. Blondel B, Crainic R, Fichot O, Dufraisse G, Cardrea A, Girard M, and Horaud F (1986). Mutations conferring resistance to neutralization with monoclonal antibodies in type 1 poliovirus can be located outside or inside the antibody binding site. Virology (in press).
2. Cann AJ, Stanway G, Hughes PJ, Minor PD, Evans DMA, Schild GC, Almond JW (1984). Reversion to virulence of the live attenuated Sabin type 3 oral poliovirus vaccine. Nucleic Acids Res. 12: 7787-7792.
3. Cranic R, Couillin P, Blondel B, Cabau N, Bouse A and Horodniceanu F (1983). Natural variation of poliovirus epitopes. Infection and Immunity 41: 1217-1225
4. Diamond DC, Jameson BA, Brown J, Kohara M, Abe S, Itoh H, Komatsu T, Arita M, Kuge S, Osterhaus ADME, Crainic R, Nomoto A, and Wimmer E. (1985) Antigenic variation and resistance to neutralization in poliovirus type 1. Science 1090-1093.
5. Evans DM, Minor PD, Schild GS, Almond JW (1983). Critical role of an eight amino acid sequence of VP1 in neutralization of poliovirus type 3. Nature 304: 459-462.
6. Ferguson M, Minor PD, Magrath DI, Qi, Yi-Hua, Spitz M and Schild GC (1984). Neutralization epitopes on poliovirus type 3 particles: an analysis using monoclonal antibodies. J.Gen.Virol 65: 197-201.
7. Fricks CE, Icenogle JP and Hogle JM (1985). Trypsin sensitivity of the Sabin strain of type 1 poliovirus: cleavage sites in virions and related particles. J.Virol 54: 856-
8. Hogle JM, Chow M, Filman DJ. (1985). The three dimensional structure of poliovirus at 2.9 A^o resolution. Science 229: 1358-1365.

9. Kew OM and Nottay BK (1985). Evolution of the oral polio vaccine strains in humans occurs by both mutation and intramolecular recombination in Modern Approaches to Vaccines: Molecular and Chemical Basis of Virus Virulence and Immunogenecity: pp 357-362. Edited by R.M. Chancok and R.A. Lerner. New York: Cold Spring Harbor Laboratory.
10. Kitamura N, Semler BL, Rothberg PG, Larsen GR, Adler CJ, Dorner AJ, Emini EA, Hanecak R, Lee JL, Van der Werf S, Anderson CW and Wimmer E (1981). Primary structure, gene organisation and polypeptide expression of poliovirus RNA. Nature 291: 547-553.
11. Leinikki PO, Pasternack A, Mustonen J, Tanuaripaa P, Hyoty H (1985) Paralytic poliomyelitis in Finland. Lancet II 507.
12. Magrath DI, Evans DMA, Ferguson M, Schild GC, Minor PD, Horaud F, Crainic R, Stenik M, and Hovi T. (1986). Antigenic and Molecular properties of type 3 poliovirus responsible for an outbreak of poliomyelitis in a vaccinated population. J.Gen.Virol. (in press).
13. Minor PD, Schild GC, Bootman J, Evans DMA, Ferguson M, Reeve P, Spitz M, Stanway G, Cann AJ, Hauptmann R, Clarke LD, Mountford RC and Almond JW (1983). Location and primary structure of a major antigenic site for poliovirus neutralization. Nature 301: 674-679.
14. Minor PD, Evans DMA, Ferguson M, Schild GC, Westrop G and Almond JW (1985). Principal and subsidiary antigenic sites of VP1 involved in the neutralization of poliovirus type 3. J gen Virol. 65: 1159-1165.
15. Minor PD, Ferguson M, Evans DMA, Almond JW, and Icenogle JP. (1986). Antigenic structure of polioviruses of serotypes 1, 2 and 3. J.Gen.Virol. (in press).
16. Rossmann MG, Arnold E, Erickson JW, Frankenberger EA, Griffith JP. Hecht HJ, Johnson JE, Kramer G, Ming Luo, Mosser AG, Rueckert RR, Sherry B, and Vriend G (1985) The structure of a human common cold virus (rhinovirus 14) and its functional relations to other picornaviruses. Nature 317: 145-153.
17. Sabin AB, and Boulger LR (1973). History of Sabin attenuated poliovirus oral live vaccine strains. J.Biol.Stand. 1: 115-118.

18. Salk J and Salk D (1977). Control of influenza and poliomyelitis with killed virus vaccines. Science 195: 8314-835.
19. Sherry B and Rueckert R (1985) Evidence for at least two dominant neutralization antigens on human rhinovirus 14. J Virol. 53: 137-143.
20. Toyoda H, Kohara M, Katadia Y, Suganuma T, Omata T, Imura N, and Nomoto A (1984). Complete nucleotide sequence of all three poliovirus serotype genomes. Implication for genetic relations map, gene function and antigenic determinants. J.Mol.Biol. 174: 561-585.
21. World Health Organization Report (1981). Markers of poliovirus strains isolated from cases temporararally associated with the use of live poliovirus vaccine: report on a WHO collaborative study. J.Biol.Stand. 9: 163-184.

Positive Strand RNA Viruses, pages 555–564

CONTROL OF FOOT-AND-MOUTH DISEASE: THE PRESENT POSITION AND FUTURE PROSPECTS.

Fred Brown

Wellcome Biotechnology Ltd
Beckenham
Kent
U.K.

ABSTRACT Foot-and-mouth disease is controlled either by slaughter of infected animals and those in close contact with them in those countries where the disease does not normally occur or by vaccination in those areas where the disease is endemic. Vaccination is complicated by the occurrence of the virus as seven serotypes and multiple sub-types within each serotype. Nevertheless, vaccines prepared by inactivation of virus grown in a variety of cell cultures are effective as demonstrated by the virtual eradication of the disease from Western Europe during the last 30 years. The difficulties encountered in controlling the disease in many countries are often associated with the poor quality and stability of the vaccines and the antigenic variation exhibited by the virus. The recent work on the chemistry of the virus particle and its antigenic sites holds out promise for a synthetic peptide vaccine which would be stable and provide protection against a wider spectrum of antigenic variants. Moreover it would be capable of being delivered by a delayed release mechanism which would allow booster "injections" to be given without the need to muster the animals.

INTRODUCTION

Foot-and-mouth disease is one of the most devastating diseases of farm animals, with productivity losses usually estimated to be as high as 25%. Indirect losses due to embargoes on trading can be even more devastating, particularly to those countries whose economy is heavily based on the export of farm animals and their products. The disease occurs in most parts of the world except Australia, Japan, North America and Western Europe. Control is by slaughter in those countries where the disease does not normally occur but in the endemic areas control is by vaccination. This is a major operation and an estimated 1.5 billion doses are given each year.

Vaccines are prepared from virus grown in a variety of cell cultures. The most commonly used are tongue epithelial cell fragments (1) and the baby hamster kidney cell line BHK21 (2). The culture fluid is then inactivated with an imine, adsorbed on to aluminium hydroxide gel and administered after mixing with low concentrations of saponin. In some instances the inactivated virus is given as an emulsion with incomplete Freund's adjuvant. About 10μg of inactivated virus particles as one injection will protect cattle against the severe challenge of 100,000 ID_{50} of homologous virus injected directly into the tongue, the test normally applied in the potency testing of foot-and-mouth disease vaccines.

Application of vaccines prepared in this way have essentially eradicated the disease from Western Europe. Since a comprehensive vaccination policy was introduced in most countries on that continent over the last 20 to 30 years, the number of cases has fallen from many thousands each year to a mere handful. However, the situation in other parts of the world is not nearly so encouraging. The reasons for this difference are not difficult to find. In the first place the vaccines are relatively unstable and there is a need to ensure that there is a cold chain which enables the potency of the vaccines to be retained until they are administered. It is often difficult to ensure that these conditions are achieved. Secondly, the veterinary services in many of the under developed countries have not yet reached the level of efficiency of those in Western

Europe so that the necessary administrative supervision is lacking. Probably the most important factor, however, is the problem posed by the antigenic variation of the virus. It occurs as seven serotypes, A, O, C, SAT 1, SAT 2, SAT 3, and Asia 1 which provide no cross-protection. This means that an animal which has recovered from infection with a virus belonging to one serotype is completely susceptible to infection with the remaining serotypes although being solidly immune to the homologous serotype. In addition there is a wide antigenic spectrum of viruses within an individual serotype, which often means that a vaccine which provides immunity against challenge infection with the homologous virus will not provide protection against other virus isolates within the same serotype. Consequently there is a need to monitor outbreak strains for their serological relatedness to the available vaccines. This is usually done by comparing the neutralizing activity of sera from animals which have received the available vaccine against the outbreak virus and homologous virus strain. The ratio between the two titres gives a measure of the protection to be expected. A high ratio indicates that good protection would be expected whereas a low value (e.g.0.1) would indicate that the available vaccine was not suitable for the outbreak.

Vaccines with a wide antigenic spectrum are clearly desirable so that the difficulties associated with intra-typic variation can be met. There are examples of viruses with these properties and a comparison of the amino acid sequences of antigenic sites of these viruses with those of viruses of the same sub-types with narrow antigenic spectra could provide the information necessary to overcome this problem.

Recent work has provided information on the antigenic sites of the virus and sequencing studies have allowed comparisons to be made between viruses belonging to different serotypes and sub-types. Together with the demonstration that synthetic peptides corresponding to one of the antigenic sites will elicit levels of neutralizing antibody which protect experimental animals against infection, the route to synthetic vaccines which are designed to evoke the required immune response seems to be open.

CHEMICAL STRUCTURE OF THE VIRUS

For the rational design of new vaccines it is necessary to identify and understand the molecular structure of the proteins which elicit the immune response. Foot-and-mouth disease virus harvests contain four virus specified particles: (a) the infectious 146S particle consisting of one molecule of ssRNA, mol.wt about $2.6x10^6$ and 60 copies of each of four proteins VP1-VP4; (b) a so called empty particle sedimenting at 75S, with the same size and shape as the infective particle but possessing no RNA and consisting only of the four proteins, but with VP2 and VP4 covalently bound; (c) a 12S sub-unit consisting of five copies of VP1,VP2 and VP3, and (d) the RNA polymerase. The major immunogenic component of the mixture is the virus particle (3) although the empty particles and 12S sub-unit possess some immunogenic activity. The empty particles are as efficient as the virus particles in absorbing neutralizing antibody and their immunogenic activity can be enhanced by fixing them with formaldehyde (4). This suggests that their low activity in eliciting neutralizing antibody is caused by their instability in the injected animal.

IDENTIFICATION OF IMMUNOGENIC PROTEIN

The immunogenic activity of the virus particle is dramatically reduced by lowering its pH below 7. This lowering of pH causes the particle to disrupt, releasing the virus RNA in an infectious form and leading to the formation of the 12S sub-unit referred to above and an aggregate of VP4. The immunogenic activity is also reduced by treating the virus particle with various proteolytic enzymes (5). These enzymes do not alter the morphology of the virus particle but protein VP1 is cleaved and the particles no longer react with IgM antibody. The reaction with IgG antibody is unimpaired.

These results suggested the VP1 contained the immunodominant site and this was supported by the evidence of Laporte et al(6) and subsequently of several other groups of workers that VP1 alone, separated from the virus particle, will elicit neutralizing antibody. Consequently attention has been focussed on this protein for all the subsequent work which has been done on the immunogenic activity of the virus.

IDENTIFICATION OF IMMUNOGENIC SITES ON PROTEIN VP1

By cleaving VP1 either in situ with different proteolytic enzymes or, after isolation, with cyanogen bromide, followed by fractionation and testing of the products, Strohmaier and his colleagues (7) predicted that immunogenic sites would be located at amino acid residues 146-154 and 200-213. Comparison of the amino acid sequences of VP1 from different serotypes and sub-types showed that these two sequences were variable in comparison with the remainder of the molecule. On the assumption that the serological variability can be correlated with sequence variability this information pinpointed the two regions as being immunogenic sites. These predictions have been confirmed by testing synthetic peptides of VP1 (8 and unpublished data). The hypervariable region 141-160 elicits high levels of neutralizing antibody and protects guinea pigs against experimental infection with the virus. The second region 200-213 is much less immunogenic although it does afford some protection. The remainder of the molecule appears to elicit little or no neutralizing antibody. These results have consequently focussed attention on the 141-160 region of VP1.

CHEMICAL BASIS FOR ANTIGENIC VARIATION.

Comparison of the specificity of the anti-141-160 peptide sera and anti-virus particle sera showed that the ratio of the neutralizing activities of the two sera against the heterologous and homologous viruses was similar. This result, obtained with sera from two sub-types A10 and A12 of virus of serotype A, provided further evidence that the 141-160 sequence is a dominant immunogenic site on the virus. More precise evidence has been obtained from a comparison of three antigenic variants which have been found in a virus isolate of sub-type 12. These were found to differ only at positions 148 and 153 on VP1 (9 and unpublished data). Using both anti-virus particle and anti-peptide sera to the 141-160 sequence, we have shown that a single amino acid change at either of these positions is sufficient to reduce the ratio of heterologous to homologous neutralization to below 0.1. In practical terms this would mean that a vaccine produced from one of the variants would not protect the animals from challenge with the other variants.

A similar if more complex situation exists with viruses of serotype O (10). We have found, contrary to expectation, that anti-peptide sera are more cross-reactive in neutralization tests with other viruses of the serotype than the corresponding anti-virus particle sera (Table 1). Sequencing studies have shown that even when the amino acids in positions 140-145 differed, anti-peptide serum to the 141-160 sequence of the Kaufbeuren and BFS viruses would neutralize the virus. However, in those examples where the anti-peptide serum did not neutralize a particular virus, the amino acid at position 148 also differed (Fig.1). With the virus of sub-type 6 the change is from Leu to Thr and in clone 10 of the Thailand isolate it is from Leu to Arg.

The Thailand virus provided another example of the presence of antigenic variants in the same isolate. Two virus clones were isolated which differed in the amino acid at position 148. This change was sufficient to influence dramatically the neutralization with antiserum against the anti-peptide serum.

FUTURE PROSPECTS

The demonstration that synthetic peptides will elicit neutralizing antibodies which protect animals against experimental infection has provided the opportunity to tailor-make vaccines which provide immunity against a highly variable virus. The cross-reactivity of the antiserum against the 141-160 peptide with heterologous strains of the virus may have important practical implications for vaccine design. The 145-151 amino acid sequence is highly conserved in the viruses we have tested and one could postulate a critical role for the 146-160 sequence in the production of cross-reactivity antibody. The Arg Gly Asp sequence at positions 145,146 and 147 may also have some significance in the furnishing of a universal vaccine. This sequence occurs in all the viruses so far examined, with the exception of the sub-type A10 virus. With this virus the sequence is Ser Gly Asp at the same positions. By inserting an Arg residue in the A10 peptide between Ser and Gly residues the antibody which it then elicited cross-neutralized some of the A12 viruses, whereas the "natural" A10 anti-peptide antiserum does not neutralize these viruses.

Most killed vaccines are given as multiple injections, suitably spaced so that an effective booster response is obtained. The prospect of a vaccine which is composed of a product which is stable at body temperature opens up another method for giving multiple doses at intervals. Guinea pigs which have received a primary dose of conventional inactivated vaccine or the peptide produce high levels of neutralizing antibody when they receive a secondary inoculation of peptide (11). The stability of the peptide in the dry state allows it to be incorporated in a delayed release mechanism from which it can be released at a pre-selected time. Such a possibility would allow booster "injections" to be given without the need to muster the animals.

Exploration of these avenues will provide information which should enable us to take peptide vaccines out of the laboratory and into the field. The prospect is exciting and the possibility of a vaccine which is defined in precise chemical terms is a goal which is capable of attainment.

FIGURE 1

Amino acid sequences of the 141-160 region of VP1 of several isolates of foot and mouth disease virus, serotype O.

Virus	141									150
Kaufbeuren B64	Val	Pro	Asn	Leu	Arg	Gly	Asp	Leu	Gln	Val
Kaufbeuren B7	.	.	.	.	.	.	.	.	.	.
BFS 1848	.	.	.	.	.	.	.	.	.	.
BFS 1860	.	.	.	.	.	.	.	.	.	.
OV1 (Subtype 6)	.	.	.	Val	.	.	.	Thr	.	.
Hong Kong	Met	Ser	.	Val	.	.	.	.	.	.
Indonesia 7/83	Thr	Thr	.	Val	.	.	.	.	.	.
Thailand 1/80										
Clone 10	Leu	Thr	.	Val	.	.	.	Arg	.	.
Clone 2	Leu	Thr	.	Val	.	.	.	.	.	.

	151									160
Kaufbeuren B64	Leu	Ala	Gln	Lys	Val	Ala	Arg	Thr	Leu	Pro
Kaufbeuren B7	.	.	.	.	.	.	.	.	.	.
BFS 1848	.	.	.	.	.	.	.	.	.	.
BFS 1860	.	.	.	.	.	.	.	.	.	.
OV1 (Subtype 6)	.	Asp	.	.	.	Ser	.	Ala	.	.
Hong Kong	.	Thr	.	.	Ala	Ser	.	Ala	.	.
Indonesia 7/83	.	.	.	.	Ala	Ala	.	.	.	.
Thailand 1/80										
Clone 10	.	.	.	.	Ala	.	.	Pro	.	.
Clone 2	.	.	.	.	Ala	.	.	Pro	.	.

.= No change from Kaufbeuren B64 sequence.

TABLE 1
CROSS NEUTRALIZATION BETWEEN ISOLATES BELONGING TO FOOT-AND-MOUTH DISEASE VIRUS, SEROTYPE O

Virus	BFS 1860 Serum Anti-virus particle	BFS 1860 Serum Anti-peptide 141-160
Kaufbeuren B64	0.2	0.9
Kaufbeuren B7	0.2	0.6
BFS 1848	0.2	1.0
BFS 1860	1.0	1.0
OV1 Sub-type 6	<0.1	<0.1
Hong Kong	0.5	1.0
Indonesia 7/83	0.2	0.9
Thailand 1/80 Clone 2	ND	0.8
Clone 10	0.1	<0.1

Values expressed as

$$\frac{\text{neutralization of heterologous virus}}{\text{neutralization of homologous virus}}$$

REFERENCES

1 Frenkel HS,(1947). La culture du virus de la fievre aphteuse sur l'epithelium de la langue des bovides. Bull Off Int Epizoot 28:155.

2. Mowat GN, Chapman WG (1962). Growth of foot-and-mouth disease virus in a fibroblast cell line derived from hamster kidneys. Nature 194:253.

3. Brown F,Crick J (1959). Application of agar gel diffusion analysis to a study of the antigenic structure of inactivated vaccines prepared from the virus of foot-and-mouth disease. J Immunol 82:444.

4. Rowlands DJ,Sangar DV,Brown F (1975). A comparative chemical and serological study of the full and empty particles of foot-and-mouth disease virus. J Gen Virol 26:227.
5. Wild TF, Burroughs JN,Brown F (1969) Surface structure of foot-and-mouth disease virus. J Gen Virol 4:313.
6. Laporte J, Grosclaude J, Wantyghem J, Bernard S, Rouze P (1973). Neutralization en culture cellulaire du pouvoir infectieux du virus de la fievre aphteuse par des serums provenant de porcs immunise a l'aide d'une proteine virale purifee. Compt rend Acad Sci 276:3399.
7. Stohmaier K, Franze R, Adam K-H (1982). Localization and characterisation of the antigenic protein of the foot-and-mouth disease virus protein. J Gen Virol 59:295.
8. Bittle JL,Houghten RA, Alexander H, Shinnick TM, Sutcliffe JG, Lerner RA,Rowlands DJ, Brown F (1982). Protection against foot-and-mouth disease by immunization with a chemically synthesised peptide predicted from the viral nucleotide sequence. Nature 298:30.
9. Rowlands DJ,Clarke BE,Carroll AR,Brown F,Nicholson BH,Bittle JL,Houghten RA, Lerner RA,(1983). Chemical basis of antigenic variation in foot-and-mouth disease virus. Nature 30:694.
10. Ouldridge EJ, Parry NR,Barnett PV,Bolwell C,Rowlands DJ,Brown F (1986) I. Comparison of the structure of the major anitgenic site of foot-and-mouth disease viruses of two different serotypes. Vaccines 86 ed F Brown,RM Chanock,RA Lerner, Cold Spring Harbor Laboratory (in press).
11. Francis MJ,Fry CM,Rowlands DJ,Brown F,Bittle JL,Houghten RA,Lerner RA (1985). Immunological priming with synthetic peptides of foot-and-mouth disease virus. J Gen Virol 66:2347.

Positive Strand RNA Viruses, pages 565–574

STATUS OF HEPATITIS A VACCINES

by Robert H. Purcell

Hepatitis Viruses Section, Laboratory of Infectious Diseases
National Institute of Allergy and Infectious Diseases
National Institutes of Health, Bethesda, MD 20892

ABSTRACT Hepatitis A virus is an important human pathogen and there is a continuing need for a hepatitis vaccine. The virus is a unique picornavirus and the only human hepatitis virus to be isolated in cell culture. Its genome has been cloned and sequenced. Experimental inactivated and live hepatitis vaccines have been developed and are in early clinical trials. Recombinant DNA technology may lead to their improvement, or alternatively, to their replacement with expressed or synthetic viral antigens.

INTRODUCTION and DISCUSSION

Hepatitis A virus (HAV) is one of at least five recognized human hepatitis viruses (the others are hepatitis B virus, hepatitis delta virus, at least one blood-borne non-A, non-B hepatitis virus and at least one enterically transmitted non-A, non-B hepatitis virus). Although hepatitis A virus generally causes less severe disease than certain of the other hepatitis viruses and never causes chronic disease, it is the cause of significant human illness and lost productivity (1). Approximately 25% of clinical hepatitis cases in developed countries and up to 80% of such cases in certain developing countries are diagnosed as hepatitis A. In the United States, 38% of the cases reported to the Centers for Disease Control are diagnosed as hepatitis A and, until recently, this was the most frequently diagnosed type of hepatitis. However, the falling incidence of hepatitis A in the United States and the rising incidence of hepatitis B combined to drop hepatitis A to second place for the first time in 1983 (2).

This falling incidence of hepatitis A is part of a long-term trend in all developed countries and is thought to be the result of improved public and personal hygiene.

However, HAV continues to be highly endemic in underdeveloped countries and, like other enterically transmitted viruses, infects virtually 100% of the population by age 5 years. Such infections, occurring in infancy and early childhood, are usually unrecognized. As sanitary conditions improve and more infants escape infection, susceptible older children and, eventually, young adults are at increasing risk of more severe clinical disease. Thus, such countries experience a paradoxical increase in clinical type A hepatitis. In the United States at present, high-risk populations include the military and staff of day-care centers.

Thus, there will be a continuing need for prophylaxis against hepatitis A in developing countries throughout the world and in high-risk populations in developed countries. Passive immunoprophylaxis (immune serum globulin or gamma globulin) is effective but of only temporary benefit and active immunization (vaccination) would be much more practical for those experiencing repeated or continuous exposure.

The Virus

Hepatitis A virus was first visualized in 1973 (3) and first isolated in cell culture in 1979 (4). Its host range is limited to man and several species of primates. The most useful of these from the standpoint of research have been the chimpanzee, certain species of marmoset monkeys and the owl monkey. Cell cultures susceptible to infection with HAV include primary, diploid and continuous cells of primate origin. Isolation of HAV in vitro is very difficult and growth of the virus is relatively poor, especially in early passages, resulting in passage intervals measured in weeks and poor yield of virus (5). HAV does not produce a cytopathic effect in culture (with one or two possible exceptions) and must be identified indirectly by serologic means.

HAV has been classified as a picornavirus (6). It has been further subclassified as enterovirus 72 (7), although there is considerable evidence that HAV is unique among the picornaviruses and not closely related to any of the recognized genera (8). For example, HAV shares little homology (of nucleotides or amino acids) when compared with any of the other picornaviruses that have been cloned and sequenced and it does not hybridize even under nonstringent conditions, when compared with a variety of other picornaviruses. HAV more closely resembles the cardioviruses than it does the enteroviruses but is no more closely related to the former than cardioviruses and enteroviruses are related to each other. Although the organization of the genome of HAV is similar to that of enteroviruses, the relative sizes of the capsid proteins, especially VP4, are somewhat different and the post-translational cleavage sites are less like those of enteroviruses than some other picornaviruses.

Hepatitis A Vaccines

Because hepatitis A virus can be isolated and serially propagated in cell culture, and because it does not cause chronic or malignant disease or serious sequellae, the "classical" approaches to vaccine development (live attenuated and inactivated whole-virus vaccines) are feasible, as well as newer approaches based upon recombinant DNA technology.

The feasibility of killed whole-virus vaccines was demonstrated when HAV was extracted from the liver of experimentally infected marmosets, inactivated with formalin and used to vaccinate other marmosets. This crude "vaccine" was fully inactivated and immunogenic: marmosets vaccinated with it did not develop hepatitis but were protected when subsequently challenged with live HAV (9).

Serial passage of HAV in cell culture has led to the development of several strains of more rapidly growing virus that attain relatively high titers _in vitro_. Such strains may yield sufficient viral antigen to make a killed hepatitis A vaccine economically feasible and several experimental vaccines have been developed and are being evaluated (10-13). Most are produced in diploid cell lines that are approved for vaccine development. Preliminary results indicate that such killed vaccines are highly immunogenic and capable of stimulating protective antibody in marmosets and chimpanzees. Phase I clinical trials in man have begun with some of these vaccines.

Live attenuated HAV vaccines are also in the developmental stage. The virulence of HAV appears to be rapidly attenuated by serial passage in tissue culture or in lower primates (13-16). As few as 10 serial passages of one strain of HAV (HM-175) in primary African green monkey kidney cell culture resulted in partial attenuation of virulence for chimpanzees but not marmosets. Similar results have been obtained with another strain of HAV (CR-326) in studies conducted at the Merck Sharp and Dohme research laboratories by Provost and, more recently, by Emini, and their colleagues. The more rapid attenuation of the virulence of HAV for chimpanzees than for marmosets raises important questions about suitability of various primate species as models of human hepatitis. Chimpanzees are closer evolutionary relatives of man than marmosets or owl monkeys and this suggests that the former would be more appropriate for evaluating candidate vaccines, but chimpanzees are also used for such experiments at a younger biological age than monkeys (chimpanzees used for such studies are generally pre-pubescent whereas monkeys are usually adult animals) and it is well recognized that infection of man with HAV produces a less severe infection in infants and young children than in older children or adults. It is therefore important to establish objective criteria for attenuation of HAV in primates and to identify objective markers of attenuation so that candidate vaccine strains can be identified and differentiated from wild type virus. Quantitative data for diminished replication of these strains of HAV has been obtained in chimpanzees and marmosets (17,18, R. Karron, et al., in preparation). Since shedding of HAV is often greater in children than adults (19), diminished shedding of a candidate vaccine strain in young chimpanzees suggests that the virus is, indeed, attenuated and not simply producing a milder clinical response in these young animals.

A second problem in evaluating HAV replication in primates stems from uncertainty as to the sites of replication of this virus in man and other primates. Previous attempts to demonstrate an enteric site of HAV replication in marmosets infected by the oral route were unsuccessful and the coincident detection of HAV antigen in the liver, bile and intestinal contents of experimentally infected marmosets suggested that the bulk of excreted virus originated in the liver (20,21). Evidence for at least a primary site of enteric replication comes from a) the knowledge that HAV is an enterically-transmitted virus, b) the detection of HAV, albeit at low titer, in the oropharynyx of patients with hepatitis A and, c) the detection of IgA class anti-HAV, presumably secretory antibody, in the feces of such patients. Direct evidence of enteric replication of HAV was recently reported in marmosets that were infected with HAV (22). HAV antigen was detected by immunofluorescence in the mucosa of the duodenum of one of two immunosuppressed marmosets and in a marmoset that was not immunosuppressed. These animals had been infected intravenously. It is not known whether reported enteric replication is unique to marmosets or characteristic also of chimpanzees and man; comparable experiments are difficult to perform in higher primates. However, marmosets appear to be more susceptible to infection with HAV (or at least clinical expression of disease) than chimpanzees (and possibly man) and expression of viral antigen in the liver and excretion in the stool is more prolonged also. Sabin has described a spectrum of susceptibility to poliovirus infection in primates: man is the most susceptible to enteric infection and least susceptible to infection of the central nervous system by poliovirus whereas monkeys (principally the macaque species) are least susceptible to enteric replication and most susceptible to central nervous system involvement; chimpanzees are intermediate in both cases (23). It is not clear whether a similar spectrum exists for HAV and different primate species. Furthermore, nothing is known about the receptor sites for binding of HAV to cells or if the putative receptor site for one species of primate is identical to that for another. These are important questions because chimpanzees (and possibly marmosets) are relatively difficult to infect orally (15), whereas epidemiologic evidence would suggest that man is easily infected by the enteric route but quantitative data are limited. It would be preferable to administer live attenuated HAV orally, perhaps as a tetravalent vaccine with the polioviruses. However, intramuscular administration of one such vaccine is being considered (17).

Future Developments

Future development of hepatitis A vaccines, beyond the inactivated and attenuated vaccines currently being prepared, will likely depend upon recombinant DNA technology to improve existing vaccines or to supplant them with more defined preparations. Recombinant DNA technology is likely to have little effect on improvement of killed whole-virus vaccines unless the regions of the genome responsible for improved growth of the virus in cell culture can be identified and chemically modified or similar regions transferred from attenuated polio virus strains to the HAV genome. The latter carries the risk of inadvertently transferring the capacity for broadened host range and pathogenicity for other organ systems.

Improvement of live attenuated vaccine strains by recombinant DNA technology is a distinct possibility, however. Such an approach is being actively pursued for the Sabin vaccine strains of polio virus by sequencing the wild-type and attenuated strains of each serotype and determining those nucleotide changes that are responsible for altered virulence (24-26). This approach offers a possibility of precisely determining the molecular basis of virulence and introducing modifications to the genome, such as small deletions, that will minimize the chances of reversion to virulence by single base mutations.

This approach is being explored with the HM-175 strain of HAV. The wild-type parent virus has been completely cloned and sequenced and the sequence compared with other partial or completely sequenced genomes of other HAV isolates (J. Cohen, et al., in preparation). There was a high degree of homology among the different isolates; the number of single base changes was no greater than between wild-type and attenuated strains of the same serotype of polio virus. In most of these, the mutations were conservative third base changes. The attenuated candidate vaccine strain of HAV derived from wild type HM-175 is also being cloned and sequenced and this will be compared with the sequence of the wild-type parent as well as with sequences of other HAV isolates to determine a) the changes that appear to cause attenuation and b) the changes that are associated with growth in cell culture. If attempts, now in progress, to generate an infectious cDNA or RNA clone of HAV are successful (J. Cohen, et al., unpublished), we will be able to make _in vitro_ changes in the genome and prepare hybrid genomes for further characterization of the molecular basis of virulence and attenuation for this virus. One outcome of such studies might be the development of host-range mutants that replicate only in the enteric tract and not in the liver.

Direct application of recombinant DNA technology to the development of nonvirion HAV vaccines is less likely in the immediate future. Like other picornaviruses, the single large open reading frame of the HAV genome is probably translated into a single large polyprotein that is cleaved posttranslationally by viral proteases into a large number of structural and nonstructural proteins (27-29). From the standpoint of immunogenicity, the structural proteins VP1-VP4 are the most important. Although epitopes that stimulate neutralizing antibody have been identified on VP1, VP2 and VP3 of other picornaviruses (30,31), especially enteroviruses, all currently available neutralizing monoclonal antibodies to HAV appear to be directed to a single region of VP1 (32). Furthermore, as is generally characteristic of other picornaviruses, the epitopes that stimulate neutralizing antibody are conformational and not linear epitopes: they exist in the intact virion but not in the individual denatured proteins. Therefore, a recombinant HAV vaccine that is expressed in yeast or mammalian cells would probably have to be capable of autocleavage and self-assembly to be an efficient immunogen. The success achieved with expressed FMDV vaccine has not been matched with other picornaviruses to date (33).

Similarly, hepatitis A vaccines prepared from synthetic peptides will be difficult to develop, and for the same reasons (34). However, generation of synthetic peptides representing different regions of the capsid proteins have been useful adjuncts to the use of monoclonal antibodies in mapping the important epitopes of HAV, as has been done for other picornaviruses. If methods can be found to stabilize synthetic peptides (or expressed capsid proteins) in the three-dimensional configuration that they assume in the intact capsid (30,31), useful synthetic vaccines may be possible.

REFERENCES

1. Lemon S (1985). Type A viral hepatitis. New developments in an old disease. N Engl J Med 313:1059.
2. Centers for Disease Control (1985). ACIP: Recommendations for protection against viral hepatitis. MMWR 34:313.
3. Feinstone SM, Kapikian AZ, Purcell RH (1973). Hepatitis A: detection by immune electron microscopy of a virus-like antigen associated with acute illness. Science 182:1026.
4. Provost PJ, Hilleman MR (1979). Propagation of human hepatitis A virus in cell culture in vitro. Proc Soc Exp Biol Med 160:213.
5. Daemer RJ, Feinstone SM, Gust ID, Purcell RH (1981). Propagation of human hepatitis A virus in African green

monkey kidney cell culture: primary isolation and serial passage. Infect Immun 32:388.
6. Gust ID, Coulepis Ag, Feinstone SM, Locarnini SA, Moritsugu Y, Najera R, Siegl G (1983). Taxonomic classification of hepatitis A virus. Intervirology 20:1.
7. Melnick JL (1982). Classification of hepatitis A virus as enterovirus type 72 and of hepatitis B virus as hepadnavirus type 1. Intervirology 18:105.
8. Ticehurst JR (1986). Hepatitis A virus: clones, cultures, and vaccines. Semin Liver Dis (In Press).
9. Provost PJ, Hilleman MR (1978). An inactivated hepatitis A virus vaccine prepared from infected marmoset liver. Proc Soc Exp Biol Med 159:201.
10. Binn LN, Bancroft WH, Lemon SM, Marchwicki RH, LeDuc JW, Trahan CJ, Staley EC, Keenan CM (1986). Preparation of a prototype inactivated hepatitis A virus vaccine from infected cell cultures. J Infect Dis 153:749.
11. Flehmig B, Haage A, Pfisterer M (1986). Immunogenicity of a cell-produced killed hepatitis A virus vaccine. Abstracts of the 86th Annual Meeting of the American Society of Microbiology, p 17.
12. Shimojo H (1986). Development of inactivated hepatitis A vaccine in Japan. Seventh Joint Conference on Hepatitis: The United States - Japan Cooperative Medical Science Program; Session III.
13. Emini EA, Provost PJ, Hughes JV (1986). Towards the development of a hepatitis A vaccine. Seventh Joint Conference on Hepatitis: The United States - Japan Cooperative Medical Science Program; Session III.
14. Provost PJ, Conti PA, Giesa PA, Banker FS, Buynak EB, McAleer WJ, Hillman MR (1983). Studies in chimpanzees of live, attenuated hepatitis A vaccine candidates. Proc Soc Exp Biol Med 172:357.
15. Purcell RH, Feinstone SM, Ticehurst JR, Daemer RJ, Baroudy BM (1984). Hepatitis A virus. In Vyas GN, Dienstag JL, Hoofnagle JH (eds): "Viral Hepatitis and Liver Disease," Orlando, Florida: Grune & Stratton, p 9.
16. Bradley DW, Schable CA, McCaustland KA, Cook EH, Mercy BL, Fields HA, Ebert JW, Wheeler C, Maynard JE (1984). Hepatitis A virus: growth characteristics of in vivo and in vitro propagated wild and attenuated virus strains. J Med Virol 14:373.

17. Provost PJ, Buynak EB, McLean AA, Hilleman MR, Scolnick EM (1984). Progress toward a live attenuated human hepatitis A virus vaccine. In Vyas GN, Dienstag JL, Hoofnagle JH (eds): "Viral Hepatitis and Liver Disease," Orlando, Florida: Grune & Stratton, p 467.
18. Feinstone SM, Daemer RD, Gust ID, Purcell RH (1983). Live attenuated vaccine for hepatitis A. In Papavangelou G, Hennessen W (eds): "Viral Hepatitis: Standardization in Immunoprophylaxis of Infections by Hepatitis Vaccines," Basel, S Karger, p 429.
19. Tassoupoulos NC, Papaevangelou GJ, Ticehurst JR, Purcell RH (1986). Fecal excretion of hepatitis A virus in sporadic cases of hepatitis A in Greece and in chimpanzees experimentally infected with Greek strains. J Infect Dis, In Press.
20. Mathiesen LR, Moller AM, Purcell RH, London WT, Feinstone SM (1980). Hepatitis A virus in the liver and intestine of marmosets after oral inoculation. Infect Immun 28:45.
21. Krawczynski KK, Bradley DW, Murphy BL, Ebert JW, Anderson TE, Doto IL,Nowoslawski A, Duermeyer W, Maynard JE (1981). Pathogenetic aspects of hepatitis A virus infection in enterally inoculated marmosets. Am J Clin Pathol 76:698.
22. Karayiannis P, Jowett T, Enticott M, Moore D, Pignatelli M, Brenes F, Scheuer PJ, Thomas HC (1986). Hepatitis A virus replication in tamarins and host immune response in relation to pathogenesis of liver cell damage. J Med Virol 18:261.
23. Sabin AB (1956). Pathogenesis of poliomyelitis. Science 123:1151.
24. Schild GC, Minor PD, Evans DMA, Ferguson M (1984). Molecular basis for the antigenicity and virulence of poliovirus type 3. In Chanock RM, Lerner RA (eds): "Modern Approaches to Vaccines," New York, Cold Spring Harbor Laboratory, p. 27.
25. Almond JW, Westrop GD, Cann AJ, Stanway G, Evans DMA, Minor PD, Schild GC (1985). Attenuation and reversion to neurovirulence of the Sabin poliovirus type-3 vaccine. In Lerner RA, Chanock RM, Brown F (eds): "Vaccines 85: Molecular and Chemical Basis of Resistance to Parasitic, Bacterial, and Viral Diseases," New York, Cold Spring Harbor Laboratory, p. 271.
26. Omata T, Kohara M, Abe S, Itoh H, Toshihiko K, Arita M, Semler BL, Winner E, Shusuke K, Atsuko K, Nomoto A

(1985). Construction of recombinant viruses between Mahoney and Sabin strains of type-1 poliovirus and their biological characteristics. In Lerner RA, Chanock RM, Brown F (eds): "Vaccines 85: Molecular and Chemical Basis of Resistance to Parasitic, Bacterial, and Viral Diseases," New York, Cold Spring Harbor Laboratory, p 279.

27. Ticehurst JR, Rancaniello VR, Baroudy BM, Baltimore D, Purcell RH, Feinstone SM (1983). Molecular cloning and characterization of hepatitis A virus cDNA. Proc Natl Acad Sci USA 80:5885.
28. Baroudy BM, Ticehurst JR, Miele TA, Maizel JV, Purcell RH, Feinstone SW (1985). Sequence analysis of hepatitis A virus cDNA coding for capsid proteins and RNA polymerase. Proc Natl Acad Sci USA 82:2143.
29. Najarian R, Caput D, Gee W, Potter SJ, Renard A, Merryweather J, Van Nest G, Dina D (1985). Primary structure and gene organization of human hepatitis A virus. Proc Natl Acad Sci USA 82:2627.
30. Hogle JM, Chow M, Filman DJ (1985). Three-dimensional structure of poliovirus at 2.9 A resolution. Science 229:1358.
31. Rossman MG, Arnold E, Erickson JW, Frankenberger Ea, Griffith JP, Hechgt H.-J., Johnson JE, Kamer G, Luo M, Mosser Ag, Rueckert RR, Sherry B Vriend G (1985). Structure of a human common cold virus and functional relationship to other picornaviruses. Nature 317:145.
32. Lemon SM, Stapleton JT, Jansen RW (1986). A single dominant neutralization immunogenic site on hepatitis A virus is defined by virus variants resistant to monoclonal antibody-mediated neutralization. Seventh Joint Conference on Hepatitis: The United States-Japan Cooperative Medical Science Program; Session II.
33. Kleid DG, Yansura D, Small B, Dowbenko D, Moore DM, Grubman MJ, McKercher PD, Morgan DO, Robertson BH, Bachrach HL (1981). Cloned viral protein vaccine for foot-and-mouth disease: responses in cattle and swine. Science 214:1125.
34. Bittle JL, Houghten RA, Alexander H, Shinnick TM, Sulcliffe JG, Lerner RA, Rowlands DJ, Brown F (1982). Protection against foot-and-mouth disease by immunization with a chimically synthesized peptide predicted from the viral nucleotide sequence. Nature 298:30.

Positive Strand RNA Viruses, pages 575–585

TWO CONTRASTING TYPES OF HOST RESISTANCE OF POTENTIAL USE FOR CONTROLLING PLANT VIRUSES[1]

B.D. Harrison*, H. Barker*, D.C. Baulcombe†, M.W. Bevan†, and M.A. Mayo*

*Scottish Crop Research Institute, Invergowrie, Dundee DD2 5DA, UK

†Plant Breeding Institute, Trumpington, Cambridge CB2 2LQ, UK

ABSTRACT The particle antigen of potato leafroll virus (PLRV) was detected by fluorescent antibody staining only in phloem companion cells of infected potato plants. In resistant potato genotypes, as compared with susceptible ones, fewer companion cells in leaves were stained but the intensity of staining was unaffected, PLRV content measured by ELISA was decreased by 90-95%, the potency of infected leaves as sources of virus for aphids was decreased by about 95% and the susceptibility of plants to infection by aphid-borne virus was decreased by about 80%. These changes may all have the same underlying mechanism: impaired PLRV transport within the phloem system and/or decreased intrinsic susceptibility of companion cells to infection.

In a contrasting approach, tobacco plants were transformed with DNA copies of the satellite RNA of cucumber mosaic virus (CMV), using the Agrobacterium Ti plasmid system, in an attempt to modify their reaction to infection and/or their potency as virus sources when infected. In uninoculated transformed plants, small amounts of RNA were transcribed from the inserted satellite sequences but, when the plants were infected with a satellite-free culture of CMV, large amounts of biologically active unit-size molecules of satellite RNA were produced, in some instances at the expense of CMV genomic RNA. This satellite RNA was packaged in virus-like particles having a CMV protein

[1]Supported in part by the Rockefeller Foundation

coat and was perpetuated as a component of the CMV culture. Such modified CMV cultures caused less severe symptoms in Nicotiana clevelandii than did satellite-free CMV.

INTRODUCTION

One of the most desirable ways of controlling plant virus diseases is to breed virus-resistant plants. This paper describes our recent work on two contrasting types of host resistance to plant viruses in two different taxonomic groups, potato leafroll luteovirus (PLRV) and cucumber mosaic cucumovirus (CMV). PLRV has a single-stranded RNA genome in one piece of about 6000 nt packaged in 25 nm-diameter isometric shells composed of a protein of molecular weight about 25000 (1,2). This RNA is of positive sense, is covalently linked to a protein of molecular weight about 7000 and lacks any substantial polyadenylate tract (3). PLRV is not sap transmissible. It is maintained from year to year in potato tubers and is spread from plant to plant by aphids, especially Myzus persicae, in which it can be retained for more than 10 days, apparently without multiplying (4,5). In plants, PLRV seems to be confined to phloem tissue (6). CMV, in contrast, infects most plant tissues, and is transmitted by several species of aphid in the non-persistent manner as well as by inoculation with sap (7). Its single-stranded, positive sense RNA genome is in three capped pieces, RNA-1 (3389 nt), RNA-2 (3035 nt) and RNA-3 (2193 nt), which are packaged in separate 30 nm diameter isometric particles, RNA-3 along with a subgenomic fragment (RNA-4; 1027 nt) that is derived from it (8,9). In addition, some cultures contain single-stranded satellite RNA molecules of 300-400 nt which do not share any substantial nucleotide sequence with CMV genomic RNA (10). Satellite RNA is produced only in CMV-infected cells, and in some instances interferes with CMV replication (11). Some kinds of satellite RNA ameliorate the symptoms of CMV infection and others increase disease severity (12).

RESISTANCE TO PLRV

Most previous work on this topic has centred on breeding potato cultivars with resistance to infection. Genotypes have been screened by exposing them, usually in field plots, to aphids that have acquired the virus from infectors of a standard cultivar. This procedure has enabled genotypes with enhanced resistance to infection to be selected and it was concluded that this type of resistance is under the control of several genes (13). More recently the leaves of several potato genotypes, selected for resistance to PLRV, were found to produce only low concentrations of PLRV antigen when infected (14). The same effect was observed in plants with primary or secondary infection, whether inoculated by grafting or by aphids and whether grown in the glasshouse or field. Most of these genotypes developed only mild symptoms, although one was very severely affected.

Restricted Distribution of PLRV in Resistant Potatoes

When PLRV antigen content, measured by ELISA (14), in different parts of resistant and susceptible genotypes was compared in several experiments, large differences (6- to 50-fold) were found in leaf lamina and mid-vein tissue (Table 1). The effect was expressed to similar extents in young leaves and in the more mature leaves most favoured by *Myzus persicae*. Differences between antigen concentrations in the above-ground stems were less and those between below-ground tissues were least (2- to 4-fold). Similar results were obtained with plants grown in the glasshouse at 15-20°C (Table 1) and with plants grown in the field (15). In resistant genotypes, PLRV accumulated least in the leaves and most in the basal parts of the plants.

When tissue sections were treated with fluorescein-labelled antibody to PLRV particles, as described elsewhere (15), staining was confined to phloem companion cells, which fluoresced brightly. In leaf blades and petioles, the intensity of staining of companion cells was similar in resistant and susceptible cultivars but many fewer cells were stained in tissue from resistant than in tissue from susceptible genotypes. Moreover, in resistant plants the stained companion cells were confined to the internal phloem whereas in susceptible cultivars they occurred in both internal and external phloem. The difference between resistant and susceptible cultivars in numbers of stained cells was proportional to the difference in PLRV content, measured by ELISA. These results indicate that, in resistant cultivars, PLRV is more restricted in distribution within the phloem system than it is in susceptible cultivars. Indeed even in susceptible cultivars, most phloem cells appear to be virus-free. PLRV seems to multiply mainly, or perhaps solely, in companion

TABLE 1
CONCENTRATIONS OF POTATO LEAFROLL VIRUS IN TISSUES OF A RESISTANT AND A SUSCEPTIBLE POTATO CULTIVAR

Tissue	Virus concentration (ng/g fresh wt) in Maris Piper (susceptible)	Pentland Crown (resistant)	Ratio: susceptible/resistant
Young leaf[a]			
Lamina	600	95	6.3
Mid-vein	1150	75	15.3
Petiole	1208	193	6.3
Mature leaf[b]			
Lamina	1975	130	15.2
Mid-vein	2150	300	7.2
Petiole	1650	350	4.7
Stem (above ground)	1175	450	2.6

[a]Fourth leaf from shoot tip
[b]Twelfth leaf from shoot tip

cells, suggesting that sieve tubes may serve principally as routes of transport of PLRV inoculum between infected and uninfected companion cells. In individual companion cells, the extent of virus accumulation does not depend on the resistance status of the cultivar.

Resistant Cultivars as Sources and Receptors of Aphid-borne PLRV

As might be expected from their low virus content, resistant potato cultivars are much poorer sources of PLRV for aphids than susceptible cultivars. For example, in one series of tests the virus content (estimated by ELISA) of

Myzus persicae caged for 4 days on infected leaves of resistant cultivars was about a twentieth of that (2.8 ng/insect) of similar aphids caged on susceptible cultivars. Moreover, in a trial using young field-grown plants, only 3% of the aphids from a resistant cultivar transmitted the virus to Physalis floridana test plants, whereas 58% of those from a susceptible cultivar did so (15).

In further tests, two resistant and two susceptible cultivars were compared for infectibility by aphid-borne inoculum. The results (Table 2) show that the susceptible cultivars were substantially more likely to become infected

TABLE 2

INFECTION OF POTATO GENOTYPES WITH POTATO LEAFROLL VIRUS INOCULATED BY APHIDS (MYZUS PERSICAE)

Potato genotype	% plants infected using	
	1 aphid	3 aphids
Kerr's Pink (susceptible)	82	100
Bintje (moderately susceptible)	58	66
Pentland Crown (resistant)	8	42
G7445(1) (resistant)	0	18

than the resistant ones. Thus, if transmission from resistant cultivars is 20 times less probable than from susceptible cultivars and transmission to resistant cultivars is 5 times less probable than to susceptible cultivars, transmission from plant to plant of resistant cultivars should be about 100 times less common than with susceptible cultivars.

It is clear that breeders' trials, in which potato genotypes were screened for resistance to infection with PLRV, have unwittingly selected genotypes in which the virus reaches only low concentrations. However, both properties could have the same basis, which may be a decreased distribution of virus within the phloem system. In resistant genotypes, this could result in decreased probabilities, both of PLRV particles reaching a companion cell after their inoculation by aphids into a sieve tube, and of their transport from an initially infected companion cell to uninfected ones, the next stage in establishment of a systemic infection. Alternatively, companion cells of resistant genotypes may be intrinsically less readily infectible than those of susceptible cultivars. Further work is needed to test the validity of these rival hypotheses.

Because we have studied a type of PLRV resistance that, without being selected for, has been incorporated into some potato cultivars, its effectiveness under commercial conditions is already known, and no resistance-breaking strain of the virus has been reported. However, the convenience of this resistance for use in plant breeding programmes is limited by the lack of information on the nature of its genetic control.

EFFECTS OF TRANSFORMING PLANTS WITH DNA COPIES OF CUCUMBER MOSAIC VIRUS SATELLITE RNA

Discovery that the occurrence of some kinds of satellite RNA in CMV cultures lessens disease severity (12) has led to the deliberate introduction of such virus cultures into Chinese pepper crops, apparently with beneficial effects on crop yields (16). It therefore seemed possible that transforming plants with DNA copies of a benign satellite RNA might produce genotypes in which CMV replication or disease symptoms are lessened, and satellite-free isolates are converted to less virulent, satellite-containing ones that would be spread to other plants by aphids. Some progress towards testing these ideas (17) can now be reported.

The procedures used to produce tobacco plants transformed with copies of satellite RNA (17) can be summarized as follows. Satellite RNA from a French isolate of CMV (18) was converted to complementary DNA (cDNA), cloned and sequenced. Two head-to-tail concatemeric cDNA molecules were then constructed and inserted into vector plasmid pT7-2. These inserts consisted of one (pT104) or two (pT105) units of the complete satellite nucleic acid sequence (335 nt) with incomplete units at the 5' end. The cDNA molecules were then transferred into an expression vector (Rok 1) based on the Ti plasmid system of Agrobacterium tumifaciens. Rok 1 is derived from the binary vector Bin19 (17,19). In Rok 1/104 and Rok 1/105, the above cDNA sequences were placed between the promoter of cauliflower mosaic virus 35*S* RNA and the terminator fragment of the nopaline synthase gene of *Agrobacterium* T-DNA in such a way that RNA transcripts would contain 10 nucleotides from the promoter and 200 nucleotides from the terminator as well as the inserted sequence. Transformed tobacco plants (cv. Samsun NN) were then regenerated from *Agrobacterium*-infected leaf discs (17) and were propagated as stem cuttings.

The transformed plants were shown by Southern blot analysis to contain the satellite DNA sequences. RNA transcripts of the expected sizes were found in unfractionated, and in polyadenylated, leaf RNA by Northern blot analysis but no unit length molecules of satellite RNA were detected. However, when the plants were inoculated with a satellite-free isolate of CMV, large amounts of unit length and some double length molecules of satellite RNA were produced in inoculated and in systemically infected

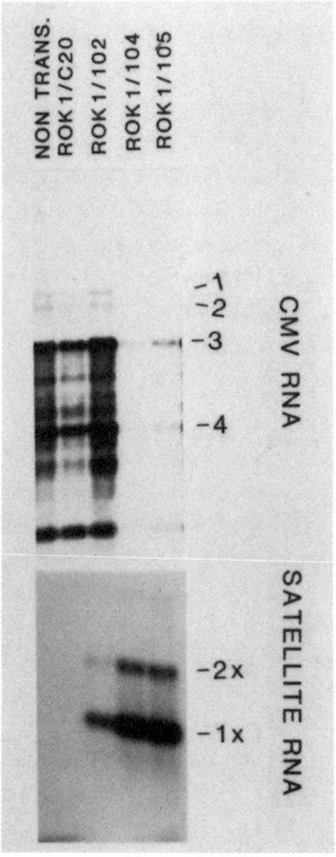

FIGURE 1. Northern blot analysis of RNA extracted from systemically infected leaves of different tobacco plants inoculated with satellite-free CMV. Each sample was probed for CMV RNA 3 and RNA 4 (upper panel) and for satellite RNA (lower panel). The positions of CMV RNA 1-4 and of unit-length (1x) and double-length (2x) satellite RNA are marked. The leaves were from non-transformed Samsun NN plants (NON TRANS), or plants transformed with the CMV coat protein gene (ROK1/C20), unit length satellite DNA (ROK1/102), or Rok 1/104 or Rok 1/105.

leaves. In systemically infected leaves, there was a tendency for the production of satellite RNA to be accompanied by decreased accumulation of CMV genomic RNA (Fig. 1). In contrast, no satellite RNA was detected in nontransformed Samsun NN tobacco plants or in plants transformed with other viral nucleotide sequences, whether or not these plants were inoculated with satellite-free CMV.

When sap from inoculated or systemically infected leaves of the satellite-transformed plants was inoculated to Nicotiana clevelandii plants, satellite RNA was again produced and CMV particles purified from these plants contained satellite RNA. Furthermore, whereas N. clevelandii plants inoculated with satellite-free CMV developed persistent mosaic symptoms and remained stunted, plants inoculated with sap from the CMV-infected transformed plants at first developed a systemic mosaic but later produced symptomless leaves and did not remain stunted (Fig. 2). This series of experiments therefore established several important points:

1. Tobacco plants transformed with CMV satellite DNA produced modest amounts of satellite RNA transcripts of the expected sizes but not unit-length satellite RNA.

2. When the transformed plants were inoculated with satellite-free CMV, large amounts of unit-length satellite RNA were produced and the accumulation of CMV RNA in systemically infected leaves seemed to be decreased.

3. This unit-length satellite RNA was transmitted to N. clevelandii plants as a component of the virus culture, and symptoms in N. clevelandii were ameliorated as a result.

4. Acquisition by CMV cultures of satellite RNA expressed from the nuclear genome of the host plant illustrates a possible way in which viruses may have acquired satellite RNA species, and carries implications for the evolution of plant viruses in general.

This work on plants genetically engineered so as to contain CMV satellite nucleic acid sequences in their nuclear genome is still at an early stage and much remains to be done. For example, it seems very possible that the results obtained could also be relevant to satellite RNA species associated with other plant viruses, particularly the smaller satellites that decrease virus genome replication and affect symptom expression (20). Further work is also needed to assess the risks

FIGURE 2. Comparison of effects on Nicotiana clevelandii of CMV cultures from different sources. Inocula were: middle, satellite-free CMV cultured in non-transformed tobacco; left and right, CMV that had acquired satellite RNA during propagation in tobacco plants transformed with Rok 1/105 and Rok 1/104, respectively.

involved in growing satellite-transformed cultivars commercially. These include the possibility that the sequences inserted in the host genome might mutate to a form that generates a virulent satellite, and the risk that a satellite, although benign in the transformed cultivar, might be virulent in other cultivars or crop species. Nevertheless, enough has already been done to suggest that, with the aid of genetic engineering methods, plant virus satellite nucleic acids can be used to man's advantage.

REFERENCES

1. Takanami Y, Kubo S (1979). Enzyme-assisted purification of two phloem-limited plant viruses: tobacco necrotic dwarf and potato leafroll. J Gen Virol 44:153.
2. Rowhani A, Stace-Smith R (1979). Purification and characterization of potato leafroll virus. Virology 98:45.
3. Mayo MA, Barker H, Robinson DJ, Tamada T, Harrison BD (1982). Evidence that potato leafroll virus RNA is positive-stranded, is linked to a small protein and does not contain polyadenylate. J Gen Virol 59:163.

4. Harrison BD (1984). Potato leafroll virus. CMI/AAB Descriptions of Plant Viruses No. 291.
5. Tamada T, Harrison BD (1981). Quantitative studies on the uptake and retention of potato leafroll virus by aphids in laboratory and field conditions. Ann Appl Biol 98:261.
6. Arai K, Doi Y, Yora K, Asuyama H (1969). Electron microscopy of the potato leafroll virus in leaves of three kinds of host plants and the partial purification of the virus. Ann Phytopath Soc Japan 35:10.
7. Francki RIB, Mossop DW, Hatta T (1979). Cucumber mosaic virus. CMI/AAB Descriptions of Plant Viruses No. 213.
8. Rezaian MA, Williams RHV, Gordon KJH, Gould AR, Symons RH (1984). Nucleotide sequence of cucumber-mosaic-virus RNA 2 reveals a translation product significantly homologous to corresponding proteins of other viruses. Eur J Biochem 143:277.
9. Rezaian MA, Williams RHV, Symons RH (1985). Nucleotide sequence of cucumber mosaic virus RNA 1. Eur J Biochem 150:331.
10. Diaz-Ruiz JR, Kaper JM (1977). Cucumber mosaic virus-associated RNA 5. III. Little nucleotide sequence homology between CARNA 5 and helper RNA. Virology 80:204.
11. Kaper JM, Tousignant ME (1977). Cucumber mosaic virus-associated RNA 5. I. Role of host plant and helper strain in determining amount of associated RNA 5 with virions. Virology 80:186.
12. Waterworth HE, Kaper JM, Tousignant ME (1979). CARNA 5, the small cucumber mosaic virus-dependent replicating RNA, regulates disease expression. Science 204:845.
13. Cockerham G (1945). Some genetical aspects of resistance to potato viruses. Ann Appl Biol 32:280.
14. Barker H, Harrison BD (1985). Restricted multiplication of potato leafroll virus in resistant potato genotypes. Ann Appl Biol 107:205.
15. Barker H, Harrison BD (1986). Restricted distribution of potato leafroll virus antigen in resistant potato genotypes and its effect on transmission of the virus by aphids. Ann Appl Biol (in press).
16. Tien P, Chang XH (1984). Vaccination of pepper with cucumber mosaic virus isolates attenuated with satellite RNA. Proc 6th Int Congr Virol Sendai:379.

17. Baulcombe DC, Saunders GR, Bevan MW, Mayo MA, Harrison BD (1986). Expression of biologically active viral satellite RNA from the nuclear genome of transformed plants. Nature (in press).
18. Jacquemond M, Lot H (1981). L'ARN satellite du virus de la mosaique du concombre. I. Comparaison de l' aptitude a induire la necrose de la tomate d'ARN satellites isoles de plusieurs souches du virus. Agronomie 1:927.
19. Bevan MW (1984). Binary Agrobacterium vectors for plant transformation. Nucl Acids Res 12:8711.
20. Murant AF, Mayo MA (1982). Satellites of plant viruses. A Rev Phytopath 20:49.

Index

Actin and rubella virus, 525, 526, 530
  and scrapie, 532–533
Acylation. *See* Aminoacylation
Adenovirus T2, 457
Agrobacterium tumefaciens Ti plasmid system, 580–581
Alphavirus
  defective interfacing RNAs, conservation, 242, 245
  promoters, cf. brome mosaic virus subgenomic RNA4 internal promoter, 332–334
  see also specific alphaviruses
Alphavirus plus and minus strand RNA synthesis, 251–258
  CEF cells, 256
  cistrons, 252, 253, 255
  cf. coronavirus, 253–254
  minus strand polymerase, 252, 254, 255
  nonstructural proteins, 252, 255–257
  plus strand polymerase, 252, 254
  cf. poliovirus, 254
  replication complex, 257–258
  Semliki Forest virus, 251
    ts mutants, 252–258
  Sindbis virus, 251
    heat-resistant strain, 252, 254
  transcription model, 257–258
Alphavirus replication, proteolytic processing of polyproteins, 209–217
  capsid protein as serine protease, 210–212
  glycoprotein cleavage, 213–215
  Ross River virus, 216, 216, 217
  Semliki Forest virus, 212, 213, 216
  Sindbis virus, 209–213, 216
    nonstructural protein, cleavage, 214–217
    O'Nyong-nyong virus, 216–217
  tryptophan residue, 211
  Venezuelan equine encephalitis virus, 212, 213
Aminoacyl acceptor arm
  BMV, tRNA-like region mutational analysis, 321–322
  TMV RNA, 301–303, 306, 312
Aminoacylation
  Sindbis virus gene expression in *S. cerevisiae*, 189
  turnip yellow mosaic virus, 149, 153–154
  see also Brome mosaic virus, tRNA-like region, 3′ region mutational analysis; Tobacco mosaic virus, aminoacylable RNA, histidine-accepting, 3′ noncoding region spatial folding
β-Annulus, turnip crinkle virus assembly/disassembly, 391–394
Antibodies. *See* Monoclonal antibodies
Anticodon stem C, BMV, 320, 321
Antigenic sites, neutralizing, poliovirus, 86–89
  monoclonal release mutants, 86–88
  synthetic peptides, 88–89
  see also Poliovirus, molecular basis of antigenicity
Antigenic variation
  flavivirus, 120–121
  FMDV, 557, 559–560
Anti-sense leader RNA inhibition, MHV leader RNA-primed transcription, 288–290, 295
Antiviral drug binding, rhinovirus 14, 70–74
Aphthoviruses, genome structure, 27, 29–32
Arildone, anti-rhinovirus 14, 70
Arteritis. *See* Equine arteritis virus, multiple subgenomic mRNAs in gene expression
Arthropod, 137
Assembly
  poliovirus, 84–85
  rhinovirus 14, 67–68
  see also Moloney murine leukemia virus assembly, $Pr65^{gag}$ myristylation

site; Tobacco mosaic virus co-translational disassembly of filamentous nucleocapsids; Tobacco mosaic virus, encapsidation initiation sites on non-virion RNA species; Turnip crinkle virus, assembly/disassembly mechanisms
Attachment blocking. *See* Rhinovirus, human, attachment-blocking monoclonal antibody
AUA sequence, TMV, 305, 308, 310
AUG codon
  cowpea mosaic virus, 196
  flavivirus, 265
  MHV, 130
AU-rich region, brome mosaic virus subgenomic RNA4 internal promoter, 330, 332–334
Avian leukemia virus. *See* Retrovirus, avian, *env* envelope genes, host range determinants

Baby hamster kidney (BHK) cells, 140, 468
Bacteriophage. *See* Phage *entries*
Barley stripe mosaic virus cf. TMV, aminoacylation, 307–310, 312
β-Barrel, 4
  poliovirus, 80–83, 88
  rhinovirus 14, 59, 62, 63, 66–68, 72
  turnip crinkle virus assembly/disassembly, 380
B component RNA, cowpea mosaic virus, 195–206, 424
Beet western yellow virus, 426, 427
BHK cells, 140, 468
Broad bean mottle virus, pseudoknots, 308, 310
Bromelain-solubilized influenza X:31 hemagglutinin (BHA), 104, 106–111
Brome mosaic virus, 6, 55
  cf. TMV, aminoacylation, 300, 302, 307–309, 313
Brome mosaic virus subgenomic RNA4, internal promoter, mutational analysis, 327–335
  cf. alphavirus promoters, 332–334
  AU-rich region, 330, 332–334
  cistrons, 328
  coat protein, 327
  generation from RNA3, 328
  cf. other viruses, 332
  phage SP6 DNA-dependent RNA polymerase, 328, 329, 331, 334
  poly(A) stretch, 330, 332
  replicase, 328, 330, 331, 334
  restriction map, 329
  tripartite genome, 327
Brome mosaic virus, tRNA-like region, 3′ region mutational analysis, 317–324
  aminoacyl acceptor arm, 321–322
  aminoacylation, tyrosine-specific, 318–324
    in vitro, 320–322
  anticodon stem C, 320, 321
  pseudoknot, 322
  RNA-dependent RNA-polymerase (replicase), 318
    cf. Qβ replicase, 318
    replicase promoter, 318, 324
    template activity, 322–324
  substrate for nucleotidyl transferase, 318
BSC-1 cells. *See* Hepatitis A virus (HM175), single-cycle growth kinetics in BSC-1 cells

Calcium binding sites, turnip crinkle virus assembly/disassembly, 383, 384, 391
CA21, MHV leader RNA-primed transcription, 287
Canyon(s)
  Coxsackievirus infection, cellular receptors, 455, 462, 463
  rhinovirus 14, 61, 67, 72
    receptor binding site, 69–70
Capsid protein
  alphavirus replication, proteolytic processing of polyproteins, 210–212
  cowpea mosaic virus, 197
  Semliki Forest virus, recombinant DNA, 352, 359
    -p62 cleavage, 352–355
  Sindbis virus
    gene expression in *Saccharomyces cerevisiae*, 184, 186, 189
    glycoproteins, 366
  subunits, poliovirus, 80–89
  see also Nucleocapsid (N) protein, MHV; Tobacco mosaic virus, encapsida-

tion initiation sites on non-virion RNA species
Carbohydrates, flavivirus, 119–120
Cardiovirus
genome structure, 27, 29, 30, 32
hepatitis A virus as, 567
Cassettes, cDNA, 375
CAT gene, insertion, Sindbis virus, 247, 248
CEF cells, 244, 256, 515
Chain elongation rate, TMV, 161–162
Chicken embryo fibroblasts, 244, 256, 515
Chimpanzee
hepatitis A virus, 567–569
rhinovirus attachment-blocking monoclonal antibody, 98–100
Chloroplast
TMV-infected cells, 400, 404
virus-induced vesicles, 427–428
Cistrons
alphavirus plus and minus strand RNA synthesis, 252, 253, 255
brome mosaic virus subgenomic RNA4 internal promoter, 328
Cleavage, proteolytic. *See* Proteolytic cleavage
CNS virulence, rubella virus associated with cytoskeleton particles, 520–521; *see also* MHV-JHM (MHV-4), molecular determinants of CNS virulence; Neurovirulence
Coat protein
brome mosaic virus subgenomic RNA4 internal promoter, 327
TMV, 305, 313
turnip crinkle virus assembly/disassembly, 387, 389, 392
turnip yellow mosaic virus, gene, 150, 151, 156
Co-infecting viruses, exchange, MHV leader RNA-primed transcription, 287–288, 295
Coliphage. *See* Phage *entries*
Competition
binding studies, rhinovirus attachment-blocking monoclonal antibody, 97
between short-chained RNA species, 14–16
Complementary strands
interaction, short-chained RNA species, 12, 14
MDV-1 RNA cf. microvariant RNA replication, reassociation, 41–42
Computers
RNA5 program, 263, 265
SEQALIGN program, 470
Conformational change, acid-induced, influenza X:31 hemagglutinin, 103–104, 107, 109–111
Congenital rubella syndrome, 520
Consensus sequence, Shine-Dalgarno, 166
Conservation, defective interfacing, alphaviruses, 242, 245
Coronavirus (Coronaviridae), 410
cf. alphavirus plus and minus strand RNA synthesis, 253–254
replication, cf. equine arteritis virus, 142–144
see also specific coronaviruses
COS-7 cells, 515, 516
Cotranslation. *See* Tobacco mosaic virus cotranslational disassembly of filamentous nucleocapsids
Cowpea chlorotic mottle virus, pseudoknots, 310
Cowpea mosaic virus, 424, 432
Cowpea mosaic virus genome expression, polyprotein processing, 195–206
AUG codon, 196
bipartite genome strategy, 196–199
cf. picornaviruses, 198–199, 201
replication, 197, 204
capsid proteins, 197
genetic map, 196
24K protein, 197–200, 202–206
proteolytic cleavage sites, 199–200, 205, 206
RNA, M and B components, 195–206, 424
viral proteins in infected protoplasts, 202–204
VPg, 196–199, 204–206
linkage to RNA 5′ end, 201–202
Coxsackievirus, 94, 95
genome structure, 25–33
Coxsackievirus infection, cellular receptors, 453–463

canyons, 455, 462, 463
determine tropism in pathogenesis, 462–463
HeLa cells, 455, 456, 458–462
mAbs, 456, 458–462
cf. other picornaviruses, 454–455, 458, 460, 462–463
RD rhabdomyosarcoma cell line, 459, 461, 462
virus-receptor complex, 456, 461
Cross-reaction, HRV-14 and other picornaviruses, 489, 494
Crystallography, X-ray, 79, 104, 149, 455
turnip crinkle virus, 383
Cucumber green mottle mosaic virus, 429–431
Cucumber mosaic virus, 431
pseudoknots, 310
satellite RNA, host resistance, viral control, 576
transformation of tobacco plants with DNA copies, 580–583
Cucurbita maxima, 18S ribosomal RNA, 401–403
Cytoplasm, virus-induced vesicles, 424–426
Cytoskeleton. *See* Rubella virus associated with cytoskeleton (VACS) particles

Datura stramonium, 429
Defective interfering genome technique. *See* Sindbis virus, replication and packing sequences in defective interfering RNAs
Deletion analysis, Sindbis virus, 246
Demyelination, MHV-JHM, 410, 411, 415–419
Developing countries
hepatitis A vaccine, 566
poliovirus wild type 1 distribution, 477
DHFR gene, Semliki Forest virus, recombinant DNA, 354–356, 360
Disassembly. *See* Tobacco mosaic virus cotranslational disassembly of filamentous nucleocapsids; Turnip crinkle virus, assembly/disassembly mechanisms
Disulfide
bonds, Sindbis virus gene expression in *Saccharomyces cerevisiae*, 187–190
-linked glycoprotein, influenza X:31 hemagglutinin, 104
HA1, 104, 110, 111
HA2, 104, 109–111
d marker, poliovirus type 1, Sabin cf. Mahoney, 438, 439, 445, 447–448
DNA, complementary (cDNA)
cassettes, 375
probe, HRV-14, 489, 490
see also Rhinovirus, human 14 (HRV-14) detection, cDNA:RNA hybridization; Tobacco mosaic virus, genomic-length cDNA clone construction
DNA polymerase I, 19
Docking
Semliki Forest virus, 358, 360
turnip crinkle virus, 394
Dot-blot hybridization, hepatitis A virus, 500, 502, 503
Drosophila melanogaster, 152
Drug binding, rhinovirus 14, 70–74; *see also* Rhinovirus, human 14 (HRV-14), structure and function

Echovirus cf. Coxsackievirus, cellular receptors, 455, 460, 461
EIS. *See* Tobacco mosaic virus, encapsidation initiation sites on non-virion RNA species
Electron density map, rhinovirus 14, 62, 64
WIN 51711, 71, 72
Electron microscopy, TMV, 161, 164, 171
Electrophoresis, SDS-PAGE
rubella virus associated with cytoskeleton particles, 522–523, 528
Sindbis virus gene expression in *Saccharomyces cerevisiae*, 184, 186, 188
ELISA
poliovirus type 1, Sabin cf. Mahoney, 443
potato leaf roll virus, 577, 578
rubella virus associated with cytoskeleton particles, 525, 526
Elongation, poliovirus-specific RNA polymerase, 275–276
Encephalitis virus
St. Louis, 262–265
Venezuelan equine, 212, 213, 369, 371

see also Flavivirus glycoprotein, epitope mapping
Encephalomyocarditis virus
cf. Coxsackievirus, cellular receptors, 455–457, 460
genome structure, 25–33
Endoglycosidase F, Sindbis virus gene expression in *Saccharomyces cerevisiae*, 184, 186
Endoplasmic reticulum, 210, 221
rough (RER), MHV-A59, 340, 344
Semliki Forest virus, recombinant DNA, 352, 358–359
p62 luminal domain translocation across, 354–356
Endosomes, influenza X:31 hemagglutinin, 103–104, 107, 109–111
Enterovirus, 498
genome structure, 26, 32
Envelope genes. *See* Retrovirus, avian, *env* envelope genes, host range determinants
Epitope mapping. *See* Flavivirus glycoprotein, epitope mapping
Equine arteritis virus, Venezuelan, multiple subgenomic mRNAs in gene expression, 137–144
BHK cells, 140
nonarthropod-borne togavirus, 137
positive-strandedness, 143
proteins, 137–139, 141
replication strategy, cf. coronavirus, 142–144
MHV, 143–144
Equine encephalitis virus, Venezuelan, 212, 213, 369, 371
Escherichia coli
lysates, TMV, 164–165
RNA polymerase, 52
Evolution
and genome structure, 3–6
MDV-1 RNA cf. microvariant RNA replication, 42

Fibroblasts, chicken embryo, 244, 256, 515
Fidelity, poliovirus-specific RNA polymerase, 275–276
Filamentous nucleocapsids. *See* Tobacco mosaic virus cotranslational disassembly of filamentous nucleocapsids
Fingerprinting. *See under* Poliovirus wild type 1 genotypes, worldwide distribution
Flavivirus glycoprotein, epitope mapping, 113–121
antigenic variation, 120–121
glycosylation, 114
carbohydrate role, 119–120
hemagglutinin, 114
mAbs, domains A, B, C, 115–116, 118, 120
interactions, 116–117
structural characterization, 117–119
SDS, 117, 118
structural proteins, 114
C, 114
E, 114, 115, 118, 120
M, 114
see also specific flaviviruses
Flavivirus replication, 261–271
conservation of sequences, 264
minus strand RNA, 261, 266
nonstructural proteins, 261, 266
plus strand RNA synthesis, 261, 266
polymerases, viral proteins as, 261, 266, 267, 269
proteolytic processing of polyproteins, 209–210, 215
genome translation, 217
nonstructural protein cleavages, 219–221
structural protein cleavages, 217–219
yellow fever virus, 209, 217
replication complexes, 268, 270, 271
St. Louis encephalitis virus, 262–265
signal sequences, 262–265
AUG codon, 265
loops, 263, 265
plus strand, 3′ terminal, 262–263, 265, 270
plus strand, 5′ terminal, 264–265
West Nile virus strain E101, 262–270
host involvement, 268–270
yellow fever virus, 262–264
FMDV. *See* Foot-and-mouth disease virus
Foot-and-mouth disease virus, 177, 558
antigenic variation, 557, 559–560
chemical structure, 558–560

control, 555–563
economic effects, 556
synthetic peptides, 559–561
Thailand, 560
vaccination, 556–557, 560–561
genome structure, 25–33
immunogenic protein, identification, 558–559
pH, 558
cf. rhinovirus 14, 62, 70, 73, 74
serotype O, 560, 563
vaccine, 571
VP1, 558, 559, 562
-VP4, 448, 558
Footprinting, turnip yellow mosaic virus, 154
Formamide concentration, HRV-14, 493–494

Gene expression. *See* Cowpea mosaic virus genome expression, polyprotein processing; Equine arteritis virus, multiple subgenomic mRNAs in gene expression; Sindbis virus structural gene expression in *Saccharomyces cerevisiae*; Turnip yellow mosaic virus, gene expression
Genome
bipartite
cowpea mosaic virus, 196–199, 201, 204
tobraviruses, 310
maps, picornaviruses, 30
structure, and evolution, 3–6; *see also* Picornaviruses, genome structure
translation, flavivirus replication, proteolytic processing of polyproteins, 217
tripartite, brome mosaic virus subgenomic RNA4 internal promoter, 327
GFAP, rubella virus associated with cytoskeleton particles, 531, 533
Glioblasts, rubella virus-infected, 521
α-Globin gene, 354–356, 360
Glycine, Moloney murine leukemia virus assembly, 228–229, 231, 232, 234
Glycophorin A, 456
Glycoproteins
avian retrovirus *env* genes, host range determinants
gp 37, 510
gp 85, 510–513, 516, 517
cleavage, alphavirus replication, proteolytic processing of polyproteins, 213–215
E1 membrane, MHV-A59, 340, 343–344
mAb, 344
E2 spike, MHV-A59, 340, 344–346, 348
HA spike, influenza X:31 hemagglutinin, 103–104
MHV-JHM, molecular determinants of CNS virulence
E1, 411
E2 spike, 411, 413, 419
Sindbis virus pathogenesis and penetration mutants, 469–475
see also Flavivirus glycoprotein, epitope mapping; Sindbis virus strain 339 glycoproteins, structure–function relationships
Glycosylation
flavivirus, 114
carbohydrate role, 119–120
Sindbis virus gene expression in *Saccharomyces cerevisiae*, 184–185, 189, 190
Golgi, 210, 214, 221–222
MHV-A59, 340, 344, 345
Sindbis virus
gene expression in *Saccharomyces cerevisiae*, 185, 188
glycoproteins, 367, 370
G protein, VSV, 185, 187, 189, 190, 247, 248
Growth kinetics. *See* Hepatitis A virus (HM175), single-cycle growth kinetics in BSC-1 cells

Hairpins
MDV-1 RNA cf. microvariant RNA replication, 37, 42
MHV leader RNA-primed transcription, 294
TMV, aminoacylable RNA, 301–304, 306–309
see also Loops
HAV antigen, hepatitis A virus, growth kinetics in BSC-1 cells, 498, 501, 503

Heat-resistant variants, Sindbis virus glycoproteins, 252, 254, 372
HeLa cells
  Coxsackievirus infection, cellular receptors, 455, 456, 458–462
  host factor, poliovirus RNA, 274, 276–282
  rhinovirus attachment-blocking monoclonal antibody, 94, 96, 98, 99
Hemagglutinin, flavivirus, 114; *see also* Influenza X:31 hemagglutinin, hydrophobic probe photolabeling
Hepatitis A vaccine, 565–571
  cf. cardiovirus, 567
  high-risk populations, 566
  HM-175 strain, 570
  killed vaccines, 567, 569–570
  live attenuated, 567, 568, 570
  lost productivity, 565
  nonhuman primates, 567–569
  cf. other forms of hepatitis, 565
  picornavirus, 566–567, 571
  recombinant DNA technology, 569–570
  underdeveloped countries, 566
  VP1-VP4, 571
Hepatitis A virus, genome structure, 25–33
Hepatitis A virus (HM175), single-cycle growth kinetics in BSC-1cells, 497–506
  cytopathic effects, 501, 505, 506
  dot-blot hybridization, 500, 502, 503
  Enterovirus, family Picornaviridae, 498
  HAV antigen, 498, 501, 503
  indirect immunofluorescence, 500, 503, 504
  mAb, 500, 501, 505
  penetration and uncoating, 501, 505, 506
  cf. poliovirus, 499–506
  radioimmunofocus assay, 498–501, 506
  RIA in situ, 500, 503
  dsRNA, 500, 503, 504, 506
Hepatitis virus, murine. *See* MHV *entries*
Histidine. *See* Tobacco mosaic virus, aminoacylable RNA, histidine-accepting, 3′ noncoding region spatial folding
HM175. *See* Hepatitis A virus (HM175), single-cycle growth kinetics in BSC-1 cells
Host specificity, MHV-A59, 339, 341, 346–348; *see also* Coxsackievirus infection, cellular receptors; Plant virus control, use of host resistance; Retrovirus, avian, *env* envelope genes, host range determinants
Hot spots, 17
HRV. *See* Rhinovirus, human *entries*
Hybridization, dot-blot, HAV, 500, 502, 503; *see also* Rhinovirus, human 14 (HRV-14) detection, cDNA:RNA hybridization
Hydrophobicity, Semliki Forest virus, recombinant DNA, 357, 358, 360, 361; *see also* Influenza X:31 hemagglutinin, hydrophobic probe photolabeling

eIF2 β subunit, 277
IgG, anti-measles, and MS, 410
Immunoaffinity isolation, rhinovirus attachment-blocking monoclonal antibody, 97–98
Immunogenic sites, neutralization, rhinovirus 14, 68–69
Indirect immunofluorescence, hepatitis A virus in BSC-1 cells, 500, 503, 504
Influenza X:31 hemagglutinin, hydrophobic probe photolabeling, 103–111
  acid-induced conformational change, endosomes, 103–104, 107, 109–111
  bromelain-solubilized (BHA), 104, 106–111
  disulfide-linked glycoproteins, 104
    HA1, 104, 110, 111
    HA2, 104, 109–111
  HA spike glycoprotein, 103–104
  liposomes, 107–111
Initiation, poliovirus RNA, 276; *see also* Tobacco mosaic virus, encapsidation initiation sites on non-virion RNA species
Insertion of foreign sequences, Sindbis virus, 246–249
Interferon, 277
Intermediate translation complexes, TMV, 161, 166, 168, 172, 173, 177, 178

JHM strain. *See* MHV-JHM (MHV-4), molecular determinants of CNS virulence

Latent period, reduced, Sindbis virus mutants, 468, 469, 471–475
LDV, 138, 139
Leader-primed RNA transcription. *See* MHV leader RNA-primed transcription
Leukemia virus, avian, *env* genes, host range determinants, 510; *see also* Moloney murine leukemia virus assembly, $Pr65^{gag}$ myristylation site
Liposomes, influenza X:31 hemagglutinin, 107–111
Loops
  flavivirus, 263, 265
  poliovirus, 80–83, 85, 86, 88
  turnip crinkle virus, 385, 393
  see also Hairpins
Lysates, TMV
  E. coli, 164, 165
  rabbit reticulocyte, 161–163, 165, 167, 176

Mahoney poliovirus. *See* Poliovirus *entries*
Marmosets, hepatitis A virus, 567–569
Maus Elberfeld virus, 455
M component RNA, cowpea mosaic virus, 195–205, 424
MDV-1 RNA, 14, 16
MDV-1 RNA cf. microvariant RNA replication, secondary structure formation, 35–42
  and evolution, 42
  hairpin structures, 37, 42
  nucleotide sequence and structure, 37
  Qβ replicase, 35–36, 38–40
  reassociation of complementary strands in the replication complex, 41–42
  RNA synthesis kinetics, 38–39
Measles, 410
Membrane, cell
  glycoprotein, E1, MHV-A59, 340, 343–344
    mAb, 344
  Moloney murine leukemia virus assembly, 228, 234, 235
  protein (M)
    MHV, 127, 129, 131
  see also Plasma membrane
Mengo virus, 59, 60, 63, 73, 74
MHV-A59 protein processing and virion assembly, 339–348
  alkalinity, 345
  E1 membrane glycosylation, 340, 343–344
    mAb, 344
  E2 spike glycoprotein, 340, 344–346, 348
  Golgi, 340, 344, 345
  interaction with target cell plasma membrane receptors, 339, 340, 345–348
    virus overlay protein blot assay, 346, 347
  cf. MHV-JHM, 341
  N protein, 340–343
    RNA overlay protein blot assay, 342
  RER, 340, 344
  specificity, host and tissue, 339, 341, 346–348
  ts mutants, 342
MHV, genomic organization, 127–132
  cDNA clone nucleotide sequences, 129
  M protein, 127, 129, 131
  N protein, 127, 129, 131
  mRNA, 127–131
    AUG initiator codon, 130
    discontinuous transcription, 128
    ORFs, 129–130, 132
  S protein, 127, 129, 131
  viral RNA polymerase, 128, 130, 132
MHV-JHM (MHV-4), molecular determinants of CNS virulence, 409–419
  demyelination, 410–411, 415–419
  E1 gp, 411
  E1 spike gp, 411, 413, 419
  histopathology, 414–418
  mAbs, 411, 412
  MBP, 414, 419
  cf. MHV-A59 ts mutant, 411, 414
  N nucleocapsid protein, 411
  resistance gene, 410
MHV leader RNA-primed transcription, 285–296
  A59, 291, 292, 295
  B1 and CA21, 287
  3′ end, 295
  5′-end sequences, molecular cloning, 293–295

hairpin loop, 294
RNA polymerase gene, 293–295
free exchange between co-infecting viruses, 287–288, 295
high-frequency and multiple RNA recombination, 290–293, 295
genetic map, 291
ts mutants, 291, 295
inhibition by anti-sense leader RNA, 288–290, 295
JHM, 291, 292, 294, 295
cf. picornavirus, 286
RNA polymerase, 285
mRNAs, 285–287, 295
cf. transcriptional pausing in Qβ and T7, 296
MHV replication strategy, cf. equine arteritis virus, 142–144
Minus strand RNA, flavivirus, 261, 266; *see also* Alphavirus plus and minus strand RNA synthesis
Mitochondria, virus-induced vesicles, 429, 430
MNV-11 RNA, 13, 16, 18
Moloney murine leukemia virus assembly, $Pr65^{gag}$ myristylation site, 227–235
cf. avian virus, 235
and cell membrane, 228, 234, 235
glycine, 228, 229, 231, 232, 234
lack of reverse transcriptase in mutants, 232
synthesis by mutants, 229–231
Monkey, cynomolgus, poliovirus type 1, 444, 447, 448; *see also* Hepatitis A virus (HM175), single-cycle growth kinetics in BSC-1 cells
Monoclonal antibodies
attachment-blocking, HRV, 93–101
Coxsackievirus infection, cellular receptors, 456, 458–462
flavivirus, 115–118, 120
interactions, 116–117
hepatitis A virus, growth kinetics in BSC-1 cells, 500, 501, 505
MHV-JHM, molecular determinants of CNS virulence, 411, 412
poliovirus, molecular basis of antigenicity, 540–544, 547–548
sensitivity of unpassaged virus in fecal specimens, 548, 550
poliovirus type 1, 443, 477–478
rubella virus associated with cytoskeleton particles, 522, 523, 525
Sindbis virus, 367, 368, 371–375, 468, 469, 472–474
Monoclonal release mutants, poliovirus, 86–88
Monster particles, turnip crinkle virus assembly/disassembly, 389, 390, 393
Mouse hepatitis virus. *See* MHV *entries*
Multiple sclerosis, anti-measles IgG, 410
Murine leukemia virus. *See* Moloney murine leukemia virus assembly, $Pr65^{gag}$ myristylation site
Mutagenesis, site-directed, Semliki Forest virus, recombinant DNA, 357; *see also* Brome mosaic virus subgenomic RNA4, internal promoter, mutational analysis; Brome mosaic virus, tRNA-like region, 3′ region mutational analysis
Myelin and MHV-JHM
basic protein (MBP), 414, 419
demyelination, 410, 411, 415–419
Myristylation. *See* Moloney murine leukemia virus assembly, $Pr65^{gag}$ myristylation site
Myzus persicae (aphid) vector, 576–580

*nde*I site, TMV cDNA clone, 49, 51, 52
Neonatal mice, Sindbis virus pathogenesis and penetration mutants, 468, 469, 475
Neurovirulence
poliovirus type 1, Sabin cf. Mahoney, 444, 447, 448
rubella virus associated with cytoskeleton particles, 520–521
Sindbis virus mutants, 374–375
see also MHV-JHM (MHV-4), molecular determinants of CNS virulence
Nicotiana
clevelandii, 431, 576, 582–583
tabacum, 52, 163, 169, 170, 172, 173, 176, 177
Noncapsid protein 2A, poliovirus wild type 1 genotypes, worldwide distribution, 478, 481

Nonstructural proteins
alphavirus, 214–217, 252, 255–257
flavivirus, 217–221, 261, 266
N protein. *See* Nucleocapsid (N) protein
NTRE-4 mutant, avian retrovirus, 510, 516
Nuclei-associated virus-induced vesicles, 426–427
Nucleocapsid (N) protein, MHV, 127, 129, 131
-A59, phosphorylated, 340–343
-JHM, molecular determinants of CNS virulence, 411
see also Tobacco mosaic virus cotranslational disassembly of filamentous nucleocapsids
Nucleotidyl transferase, BMV, 318

O'Nyong-nyong virus replication, 216, 217
Open reading frames, MHV, 129–130, 132

Packing sequences. *See* Sindbis virus, replication and packing sequences in defective interfering RNAs
Papovavirus, 410
Paramyxovirus, 214, 215
Penetration, hepatitis A virus, growth kinetics in BSC-1 cells, 501, 505, 506; *see also* Sindbis virus pathogenesis and penetration mutants, nucleic acid sequence analysis
Peptides
6kD, Semliki Forest virus, recombinant DNA, 352, 359–362
synthetic
FMDV, 559–561
poliovirus, 88–89
Peroxisomes, 428–429
pH
FMDV, 558
MHV-A59, 345
TMV, 162, 163, 166–168
turnip crinkle virus assembly/disassembly, 383–386
Phage Qβ, 35–36, 38–40, 42, 155, 268
cf. BMV, tRNA-like region mutational analysis, 318
cf. MHV leader RNA-primed transcription, 296
pause during chain elongation, replicase, 39–40
see also under Short-chained RNA species, replication and selection kinetics
Phage SP6 DNA-dependent RNA polymerase, 328, 329, 331, 334
Phage T7 cf. MHV leader RNA-primed transcription, 296
Phaseolus vulgaris, 52–53
Photolabeling. *See* Influenza X:31 hemagglutinin, hydrophobic probe photolabeling
Phylogenetic tree construction, poliovirus wild type 1, 480–482
Phylogeny, picornaviruses, 32–33
Physalia floridana, 579
Picornaviruses (Picornaviridae), 498, 566–567, 571
cellular receptors, 454–455, 458, 460, 462–463
cf. cowpea mosaic virus, bipartite genome, 198–199, 201
cross-reaction with HRV-14, 489, 494
cf. MHV leader RNA-primed transcription, 286
TMV, aminoacylable RNA, 312
see also specific viruses
Picornaviruses, genome structure, 25–33
genomic maps, 30
processing map, 26
ssRNA genome, 25
sequence comparisons and reorganized phylogeny, 32–33
subgroups, 26–32
viral proteins, 29–32
viral RNA, 27–29
Pisum sativum, TMV, 173–175
Plant cell response to infection, RNA replication sites, 423–432
beet western yellow viruses, 426, 427
cowpea mosaic virus, 424, 432
cucumber green mottle mosaic virus, 429–431
Datura stramonium, 429
multivesicular bodies, peroxisomes, 428–429
Nicotiana clevelandii, cucumber mosaic virus, 431

potato leaf roll virus, 426
RNA-dependent RNA polymerase, 427
dsRNA, 424, 427, 428, 431
turnip yellow mosaic virus, 425, 427
vesicles
chloroplast, 427–428
mitochondrial, 429, 430
tonoplasts, 430, 431
virus-induced in cytoplasm, 424–426
virus-induced, nucleus-associated, 426–427
Plants, RNA recombination, 6
Plant virus(es)
cf. poliovirus, 85–86
pseudoknots, 307–312
see also specific plants and viruses
Plant virus control, use of host resistance, 575–583
Agrobacterium tumefaciens, Ti plasmid system, 580–581
cucumber mosaic virus satellite RNA, 576
transformation of tobacco plants with DNA copies, 580–583
ELISA, 577, 578
Myzus persicae, 576–580
Nicotiana clevelandii, 576, 582–583
Physalia floridana, 579
potato leaf roll virus, 576–580
Plaque size, poliovirus type 1, Sabin cf. Mahoney, 445, 447, 448
Plasma membrane
Semliki Forest virus, recombinant DNA, 352
Sindbis virus glycoproteins, 367, 370
Plasmid system, *Agrobacterium tumefaciens*, host resistance, viral control, 580–581
Plus strand RNA synthesis, flavivirus, 261, 266; *see also* Alphavirus plus and minus strand RNA synthesis
Poliovirus, 55, 198, 268
cf. alphavirus plus and minus strand RNA synthesis, 254
cf. Coxsackievirus infection, cellular receptors, 455, 458, 460
genome structure, 25–33
cf. hepatitis A virus, growth kinetics in BSC-1 cells, 499–506
cf. rhinovirus 14, 59, 62, 68, 73, 74, 85, 543, 551
Poliovirus, Mahoney type 1, structure, 79–89
assembly, 84–85
capsid protein subunits, 80–83
alpha-carbon models, 81, 87
β-barrel, 80–83, 88
loops, 80–86, 88
VP1, 80–89
VP2, 80–88
VP3, 80–84, 87, 88
VP4, 80–84
neutralizing antigenic sites, 86–89
monoclonal release mutants, 86–88
synthetic peptides, 88–89
cf. plant viruses, 85–86
virion structure, 83–84
Poliovirus, molecular basis of antigenicity, 539–551
antigenic sites
identification and neutralizaion, 540–544
and infectivity, 546–550
cf. HRV-14, 543, 551
mAbs, 540–544, 547–548
sensitivity of unpassaged virus in fecal specimens, 548, 550
primer extension sequencing, 542
Sabin strains, 540, 543, 545, 549
subsidiary sites, 544–547, 549, 550
types 1-3, 539, 540, 542, 544, 546, 548–551
VP1, VP2, VP3, 539, 542–545
VP4, 539
Poliovirus RNA replication, 273–282
HeLa cell host factor, 274, 276–282
protein kinase, 277
terminal uridylyl transferase (TUT), 277
initiation, 276
membrane bound replication complex, 274, 276
poly(A) sequence, 3′ terminal, 274, 275, 280
poliovirus-specific RNA polymerase, 274, 280, 281
elongation, 275–276
fidelity, 275–276

isolation with poly(A):oligo(U) template:primer, 274, 277, 278, 280
purification, 277
product mRNA, size, 278–281
VPg, 274, 276, 278–281
Poliovirus type 1, Sabin cf. Mahoney, in vitro modification, 437–448
d marker, 438, 439, 445, 447–448
genome regions influencing phenotypic differences, 445–448
plaque size, 445, 447, 448
neurovirulence in cynomolgus monkeys, 444, 447, 448
nucleotide and amino acid differences, 440
poly(A) 3′ terminus, 438
PV1(SM)IC-1a and -1b, characterization, 443–445
antigenicity, ELISA, mAbs, 443
rct marker, 438, 439, 445, 447
recombinant viruses, 440–442, 446–448
genome structure, 442,448
cf. types 2 and 3, 438, 447
VP1, 442, 443
VP4, 446
VPg, 438
Poliovirus wild type 1 genotypes, worldwide distribution, 477–484
antigenic variation, 477–478
case relatedness, 477
developing countries, 477
mAbs, 477–478
pairwise comparisons, phylogenetic tree construction, 480–482
RNase T_1 oligonucleotide fingerprinting, 477, 478, 482, 484
noncapsid protein 2A, 478, 481
primer extension sequencing, 478, 479
VP1, 3′-end, 478–481
cf. Sabin 1 vaccine strain, 480
U.S., 482, 483
Polyadenylation
:oligo(U) isolation, poliovirus-specific RNA polymerase, 274, 277, 278, 280
sequence, 3′ terminal, poliovirus RNA, 274, 275, 280
stretch, Brome mosaic virus subgenomic RNA4 internal promoter, 330, 332
3′ terminus, poliovirus type 1, Sabin cf. Mahoney, 438
TMV, aminoacylable RNA, 308
Polymerases
alphavirus plus and minus strand RNA synthesis
minus, 252, 254, 255
plus, 252, 254
flavivirus proteins as, 261, 266, 267, 269
see also specific polymerases
Polyomavirus VP1, 212
Polyprotein processing. *See* Alphavirus replication, proteolytic processing of polyproteins; Cowpea mosaic virus genome expression, polyprotein processing
Positive strand viruses, 5, 143
Potato leaf roll virus, 426
host resistance, viral control, 576–580
Potato virus X cf. TMV, 172–174
$Pr65^{gag}$. *See* Moloney murine leukemia virus assembly, $Pr65^{gag}$ myristylation site
Primates, nonhuman, hepatitis A virus, 567–569
Primer
extension sequencing, poliovirus, 478, 479, 542
Sindbis virus pathogenesis and penetration mutants, 470, 471
see also MHV leader RNA-primed transcription
Progressive multifocal leukoencephalopathy, papovavirus, 410
Promoter. *See* Brome mosaic virus subgenomic RNA4, internal promoter, mutational analysis
Protein kinase, HeLa cell, poliovirus and, 277
Proteolytic cleavage
cowpea mosaic virus, 199–200, 205–206
flavivirus, 217–221
turnip crinkle virus assembly/disassembly, 385, 386
see also Alphavirus replication, proteolytic processing of polyproteins
Protoplasts, infected, cowpea mosaic virus, 200–204

Pseudoknots
BMV, 322
TMV, 300–306, 312, 313
cf. other plant viruses, 307–312
pseudo-pseudoknot, 311
Pst Isite, TMV cDNA clone, 49–53

Qβ. *See* Phage Qβ
Quasispecies, short-chained RNA, 17–18
mutation frequency, 17

Rabbit reticulocyte lysate system, TMV, 161–163, 165, 167, 176
Radioimmunoassay in situ, hepatitis A virus, 500, 503
Radioimmunofocus assay, hepatitis A virus, 498–501, 506
rct marker, poliovirus type 1, 438, 439, 445, 447
RD rhabdomyosarcoma cell line, 459, 461, 462
Receptors. *See* Coxsackievirus infection, cellular receptors
Recombinant DNA technology, hepatitis A vaccine, 569–570; *see also* Semliki Forest virus, structural protein synthesis, recombinant DNA
Reovirus HA, internal image, 456, 457
Replicase
BMV, tRNA-like region mutational analysis, 318
promoter, 318, 324
template activity, 322–324
brome mosaic virus subgenomic RNA4 internal promoter, 328, 330, 331, 334
phage Qβ, 10–11, 18–20
TMV, aminoacylable RNA, 299, 313
turnip yellow mosaic virus, 153, 155
see also RNA polymerase
Replication
complex
alphavirus plus and minus strand RNA synthesis, 257–258
poliovirus RNA, 274, 276
cowpea mosaic virus, 197, 204
equine arteritis virus cf. coronavirus, 142–144
MHV, 143–144
hepatitis A virus, 568–569
sites. *See* Plant cell response to infection, RNA replication sites
see also Alphavirus replication, proteolytic processing of polyproteins; Flavivirus replication; MDV-1 RNA cf. microvariant RNA replication, secondary structure formation; Poliovirus RNA replication; Short-chained RNA species, replication and selection kinetics; Sindbis virus, replication and packing sequences in defective interfering RNAs
Reporter genes, Semliki Forest virus, recombinant DNA, 354–356, 360
Resistance. *See* Plant virus control, use of host resistance
Restriction map
brome mosaic virus subgenomic RNA4 internal promoter, 329
tobacco mosaic virus cDNA clone construction, 51
Retrovirus, 215
type C mammalian, 228, 234
cf. avian, 235
see also Moloney murine leukemia virus assembly, $Pr65^{gag}$ myristylation site
Retrovirus, avian, *env* envelope genes, host range determinants, 509–517
A–C subgroups compared, 516–517
ALV, 510
chicken embryo fibroblasts, 515
COS-7 cells, 515, 516
glycoproteins
gp 37, 510
gp 85, 510–513, 516–517
host range analysis of molecular recombinants, 514–516
host range mutant NTRE-4, 510, 516
cf. RSV, 510, 512, 514
Reverse transcriptase-lacking mutants, Moloney murine leukemia virus, 232
Rhabdomyosarcoma cell line RD, 459, 461, 462
Rhinovirus, human
cellular receptors, cf. Coxsackievirus, 455–458, 460, 462

genome structure, 26–27, 29, 32, 25–33
cf. poliovirus, 85
Rhinovirus, human, attachment-blocking monoclonal antibody, 93–101
characterization, 97
chimpanzee model, 98–100
competition binding, 97
HeLa cells and membranes, 94, 96, 98, 99
immunoaffinity isolation, 97–98
listed of HRVs protected/not protected from, 96
normal cellular function, 98
Rhinovirus, human 14 (HRV-14)
attachment-blocking monoclonal antibody, 101
molecular basis of antigenicity, cf. poliovirus, 543, 551
Rhinovirus, human 14 (HRV-14) detection, cDNA:RNA hybridization, 487–494
cross-reaction to other picornaviruses, 489, 494
formamide concentration, 493–494
M13 template construction, 488, 492
5′ noncoding region, 487, 488, 490, 492
cDNA probe, 489, 490
vanadyl-ribonucleoside complexes, 491–492, 494
inhibition of endogenous RNase, 491, 494
Rhinovirus, human 14 (HRV-14), structure and function, 59–74
antiviral drug binding, 70–74
assembly, 67–68
canyon, 61, 67, 72
as receptor binding site, 69–70
cf. FMDV, 62, 70, 73–74
cf. Mengo virus, 59, 60, 63, 73, 74
neutralizing immunogenic sites, 68–69
cf. poliovirus, 59, 62, 68, 73, 74
sequence alignment, 68
cf. Southern bean mosaic virus, 60, 62, 64–68
structure, 60–67
β-barrel, 59, 62, 63, 66–68, 72
electron density map, 62, 74
FMDV loop, 67, 72
VP1, 59–73
VP2, 59–62, 64–68
VP3, 59–73
VP4, 61, 62, 67, 68, 70
cf. TBSV, 64–68
Ribosomes, TMV
18S, 400–403
70S, 165
80S, 161, 178
RNA
double-stranded (dsRNA), 42, 427
hepatitis A virus, growth kinetics in BSC-1 cells, 500, 503, 505, 506
flavivirus, minus and plus strands, 261, 266
messenger (mRNA)
MHV leader RNA-primed transcription, 285–287, 295
poliovirus RNA, 278–281
Semliki Forest virus, recombinant DNA, 352, 360
see also Equine arteritis virus, multiple subgenomic mRNAs in gene expression
overlay protein blot assay, MHV-A59, 342
packaging, turnip crinkle virus assembly/disassembly, 382
picornaviruses, 27–29
-protein complex, turnip crinkle virus assembly/disassembly, 387–392
recombination, evolution and genome structure, 5–6
plants, 6
satellite, cucumber mosaic virus, 576
transformation of tobacco plants with DNA copies, 580–583
single-stranded (ssRNA)
phage Qβ, 36, 37, 42
picornaviruses, 25
synthesis. *See* Alphavirus plus and minus strand RNA synthesis
transfer (tRNA), -like structures
brome mosaic virus, 317–324
TMV, aminoacylable RNA, 300–308, 312, 313
turnip yellow mosaic virus, 151–155
translation stop signal, turnip yellow mosaic virus, 150–153

see also MDV-1 RNA cf. microvariant RNA replication, secondary structure formation; Rhinovirus, human 14 (HRV-14) detection, cDNA:RNA hybridization; Short-chained RNA species, replication and selection kinetics; Transcription; Translation
RNA5 computer program, 263, 265
RNA polymerase, 19
  E. coli, and tobacco mosaic virus, 52
  MHV, 128, 130, 132
    gene, 293–295
    leader RNA-primed transcription, 285
  poliovirus-specific, 274–281
  SP6 DNA-dependent, 328, 329, 331, 334
  TMV, aminoacylable RNA, 299, 313
  see also Replicase
RNase
  inhibition of, HRV-14, 491, 494
  rubella virus associated with cytoskeleton particles, 531, 533
  T_1 oligonucleotide fingerprinting, poliovirus wild type 1 genotypes, worldwide distribution, 477, 478, 482, 484
Rod formation, TMV encapsidation initiation sites, 398
Ross River virus
  replication, proteolytic processing of polyproteins, 213, 216, 217
  cf. Sindbis virus
    glycoproteins, 371, 374
    pathogenesis and penetration mutants, E2 glycoprotein, 474
Rous sarcoma virus, 510, 512, 514
Rubella, congenital, 520
Rubella virus associated with cytoskeleton (VACS) particles, 519–533
  CNS infection, 520–521
  glioblasts, persistently infected, 521
  isolation and characterization, 521–530
    actin, 525, 526, 530
    ELISA, 525, 526
    mAb, 522, 523, 525
    SDS-PAGE, 522–523, 528
    TCA precipitation, 524, 525
    tubulin, 525, 526, 529, 530
  and normal RV replication, 529, 530
  and scrapie, 521, 530–533
    actin and tubulin, 532, 533
    cytoskeleton extraction, 530–531
    GFAP, 531, 533
    infectivity, 531, 533
    RNase A, 531, 533

Saccharomyces cerevisiae. *See* Sindbis virus structural gene expression in *Saccharomyces cerevisiae*
St. Louis encephalitis virus, replication, 262–265
Sarcoma virus, Rous, 510, 512, 514
Scrapie. *See under* Rubella virus associated with cytoskeleton (VACS) particles
SDS, flavivirus, 117–119; *see also* Electrophoresis, SDS-PAGE
Secondary structures. *See* MDV-1 RNA cf. microvariant RNA replication, secondary structure formation
Selection kinetics. *See* Short-chained RNA species, replication and selection kinetics
Semliki Forest virus
  plus and minus strand RNA synthesis, 251
    ts mutants, 252–258
  replication, proteolytic processing of polyproteins, 212, 213, 216
  cf. Sindbis virus
    glycoproteins, 367, 371
    pathogenesis and penetration mutants, E2 glycoprotein, 474
Semliki Forest virus, structural protein synthesis, recombinant DNA, 351–362
  C protein, 359
    C–p62 cleavage, 352–355
  ER, 352, 358–359
  E1 synthesis, reinitiation of polypeptide chain translocation, 358–361
    docking, 358, 360
    membrane binding, 361–362
    SRP-dependent, 360–361
  hydrophobicity, 357, 358, 360, 361
  6kD peptide, 352, 359–362
  p62, 352, 359, 361, 362
    luminal domain translocation across ER, 354–356

stop-translocation signal, 356–358
signal peptide, 355–356
plasma membrane, 352
reporter genes, 354–356, 360
26S mRNA, 352, 360
site-directed mutagenesis, 357
SEQALIGN program, 470
Sequence alignment, rhinovirus 14, 68; *see also* Sindbis virus pathogenesis and penetration mutants, nucleic acid sequence analysis
SERF, TMV encapsidation initiation sites, 399–400
Serine protease, capsid protein as, 210–212
Shine-Dalgarno consensus sequence, 166
Short-chained RNA species, replication and selection kinetics, 9–20
competition between species, 14–16
interaction of complementary strands, 12, 14
MDV-1 RNA, 14, 16
MNV-11 RNA, 13, 16, 18
phage Qβ replicase
cf. DNA polymerase I and RNA polymerase, 19
GTP, 11, 19
template specificity, 10–11
phage Qβ without template instruction, 18–20
quasispecies, 17–18
mutation frequency, 17
replication kinetics, 11–13
SV-11 RNA, 11
Signal peptide, Semliki Forest virus, recombinant DNA, 355–356
SRP, 360–361
Sindbis virus
heat-resistant strain, 252, 254, 372
plus and minus strand RNA synthesis, 251
replication, proteolytic processing of polyproteins, 209–213, 216
Sindbis virus pathogenesis and penetration mutants, nucleic acid sequence analysis, 467–475
BHK cells, 468
fast-penetrating (SB-FP) mutant, 469, 472–474
glycoproteins
E1, 472–473
E2, amino acid substitution, 469, 471–475
E2 spike, 474
cf. SFV and Ross River virus, 474
mAbs, 468, 469, 472–474
mutations affecting virulence, 472
neonatal mice, virulence in, 468, 469, 475
primers, 470, 471
reduced latent period (SB-RL) mutant, 468, 469, 471–475
SEQALIGN computer program, 470
Sindbis virus, replication and packing sequences in defective interfering RNAs, 241–249
conservation among alphaviruses, 242, 245
insertion of foreign sequences, 246–249
CAT gene, 247, 248
VSV G protein, 247
3′-terminus, 244–245, 249
5′-terminus, 245–249
deletion analysis, 246
transcription and transfection, 243–246
chicken embryo fibroblasts, 244
Sindbis virus strain 339 glycoproteins, structure–function relationships, 365–375
and capsid protein C, 366
E1, 366–371, 373
E2, 366–374
E3, 366, 374
glycosylation, 367, 370
Golgi, 367, 370
mAbs to antigenic variants, 367, 368, 371–375
neurovirulence mutants, 374–375
plasmalemma, 367, 370
cf. Ross River virus, 371, 374
cf. SFV, 367, 371
ts mutants, 367, 368, 374, 375
gp transport, 370–371
cf. VEEV, 369, 371
Sindbis virus structural gene expression in *Saccharomyces cerevisiae*, 183–190
capsid protein, 184, 186, 189
disulfide bonds, 187–190

E1, 184, 185, 187–189
E2, 184, 186, 188
endoglycosidase F, 184, 186
glycosylation, 184, 185, 189, 190
Golgi, 185, 188
no acylation, 189
p62, 184–185, 187–189
SDS/PAGE, 184, 186, 188
Southern bean mosaic virus, 60, 62, 64–68, 381
Spatial folding. *See* Tobacco mosaic virus, aminoacylable RNA, histidine-accepting, 3′ noncoding region spatial folding
Spike glycoproteins
influenza X:31 hemagglutinin, 103–104
MHV
-A59, 340, 344–346, 348
-JHM, 411, 413, 419
Sindbis virus, 474
v-*src* gene, 514
SRP-dependent E1 synthesis, Semliki Forest virus, 360–361
Stop codons, TMV, 305, 308, 311, 312
Striposomes, TMV, 161, 166, 168, 172, 173, 177, 178
Structure–function relationships. *See* Sindbis virus strain 339 glycoproteins, structure–function relationships
Subacute sclerosing panencephalitis, measles virus, 410
Subgenomic RNA. *See* Brome mosaic virus subgenomic RNA4, internal promoter, mutational analysis
Subsidiary sites, poliovirus, molecular basis of antigenicity, 544–550
Surface (S) protein, MHV, 127, 129, 131
SV-11, RNA, 11

TCA precipitation, rubella virus associated with cytoskeleton particles, 524–525
TCV. *See* Turnip crinkle virus, assembly/disassembly mechanisms
Temperature-sensitive mutants
MHV
-A59, 342
leader RNA-primed transcription, 291, 295
Semliki Forest virus, 252–258
Sindbis virus glycoproteins, 367, 368, 374, 375
gp transport, 370–371
Template
activity, BMV, tRNA-like region mutational analysis, 322–324
construction, M13, HRV-14, 488, 492
phage Qβ, 10–11, 18–20
:primer poly(A):oligo(U) isolation, poliovirus-specific RNA polymerase, 274, 277, 278, 280
Thailand, foot-and-mouth disease, 560
Tick-borne encephalitis virus. *See* Flavivirus glycoprotein, epitope mapping
TID. *See* Influenza X:31 hemagglutinin, hydrophobic probe photolabeling
Ti plasmid system, *Agrobacterium tumefaciens*, 580–581
Tissue specificity, MHV-A59, 339, 341, 346–348
TMV. *See* Tobacco mosaic virus *entries*
Tobacco mosaic virus, aminoacylable RNA, histidine-accepting, 3′ noncoding region spatial folding, 299–313
aminoacyl acceptor arm, 301–303, 306, 312
AUA sequence, 305, 308, 310
cf. barley stripe mosaic virus, 307–310, 312
cf. BMV, 300, 302, 307–309, 313
coat protein, 305, 313
hairpins, 301–304, 306–309
cf. picornaviruses, 312
poly(A), 308
pseudoknots, 300–306, 312–313
cf. other plant viruses, 307–312
pseudo-pseudoknots, 311
tRNA-like structures, 300–308, 312, 313
stop codons, 305, 308, 311, 312
cf. TYMV, 300, 302
cf. valine and tyrosine aminoacylation, 300, 308
virus-specific RNA-dependent RNA polymerase (replicase), 299, 313
Tobacco mosaic virus cotranslational disassembly of filamentous nucleocapsids, 159–178
chain elongation rate, 161–162

chloroplasts, infected cells, 165
early events of infection, 160–161
E. coli lysates, 164, 165
EM, 161, 164, 171
intermediate translation complexes (striposomes), 161, 166, 168, 172, 173, 177, 178
in vivo, 166, 169–172, 177, 178
longevity, 163, 164
Nicotiana tabacum, 163, 169, 170, 172, 173, 176, 177
cf. other plant viruses, 166, 172–177
pH, 162, 163, 166–168
rabbit reticulocyte lysate system, 161–163, 165, 167, 176
ribosomes
70S, 165
80S, 161, 172, 178
translocation inhibition, 168
Tobacco mosaic virus, encapsidation initiation sites on non-virion RNA species, 397–404
chloroplast DNA transcripts, EIS on, 400, 404
non-functional, 399–400
SERF, 399–400
18S ribosomal RNA, 400–403
rod formation, 398
Cucurbita maxima, 401–403
termini, 398, 399
Tobacco mosaic virus, genomic-length cDNA clone construction, 47–55
cDNA synthesis, 49–52
*nde*I site, 49, 51, 52
*Pst*I site, 49–53
E. coli RNA polymerase, 52
infectivity, 52–54
progeny virus characteristics, 54
restriction map, 51
Tobacco rattle virus, 310–311, 334
cf. TMV, 172–174
Tobraviruses, bipartite genome, 310
Togaviruses (Togaviridae), 114, 138, 242, 246, 520
alphavirus replication, proteolytic processing of polyproteins, 209
see also specific togaviruses
Tomato bushy stunt virus, cf.
rhinovirus 14, 64–68
turnip crinkle virus assembly/disassembly, 381–386
Tonoplasts, vesiculation, 430, 431
Transcription
alphavirus plus and minus strand RNA synthesis, 257–258
discontinuous, MHV, 128
Sindbis virus, 243–246
see also MHV leader RNA-primed transcription
Transfection, Sindbis virus, 243–246
Translation
complexes, intermediate (striposomes), TMV, 161, 166, 168, 172, 173, 177, 178
flavivirus replication, proteolytic processing of polyproteins, 217
inhibition, TMV, 168
see also Tobacco mosaic virus cotranslational disassembly of filamentous nucleocapsids
Translocation, polypeptide chain, E1 synthesis reinitiation, Semliki Forest virus, 358–361
Tropism in pathogenesis, Coxsackievirus infection, cellular receptors, 462–463
Tryptophan, 211
Tubulin, rubella virus associated with cytoskeleton particles, 525, 526, 529, 530
and scrapie, 532–533
Turnip crinkle virus, assembly/disassembly mechanisms, 379–394
coat protein, 387, 389, 392
expansion at neutral to alkaline pH, 383–386
Ca^{2+} binding sites, 383, 384, 391
crystallography, 383
proteolytic cleavage, 385, 386
cf. TBSV, 383–386
looping, 385, 393
monster particles, 389, 390, 393
reassembly, 387–394
β-annulus, 391–394
docking, 394
RNA-protein complex, 387–392
cf. Southern bean mosaic virus, 381
structure, 380–383

β-barrel, 380
RNA packaging, 382
cf. TBSV, 381, 382
Turnip yellow mosaic virus, 425, 427
Turnip yellow mosaic virus, gene expression, 149–150
aminoacylation, 149, 153–154, 300, 302
tRNA-like structure, 151–155
footprinting, 154
wheat germ EF-1α, 155
RNA translation, 150–153
coat protein gene, 150, 151, 156
RNA replicase, 153, 155
stop signal, 151–152
structure, 150
Tymovirus group, 150, 153
Tyrosine aminoacylation, 300, 308
BMV, tRNA-like region mutational analysis, 318–324

Uncoating, hepatitis A virus, 501, 505, 506
United States, poliovirus wild type 1 distribution, 482, 483
Uridylyl transferase, terminal, poliovirus RNA, HeLa host, 277

Vaccine, foot-and-mouth disease, 556–557, 560–561, 571
instability, 556
see also Hepatitis A vaccine
VACS. *See* Rubella virus associated with cytoskeleton (VACS) particles
Valine aminoacylation, cf. TMV aminoacylable RNA, 300, 308
Vanadyl-ribonucleoside complexes, HRV-14, 491–492, 494
Venezuelan equine encephalitis virus
replication, proteolytic processing of
polyproteins, 212, 213
cf. Sindbis virus glycoproteins, 369, 371
Vesicles. *See* Plant cell response to infection, RNA replication sites
Vesicular stomatitis virus (VSV) G protein, 185, 187, 189, 190, 247, 248
Vesiculation of tonoplasts, 430, 431
Virion structure, poliovirus, 83–84
Virus overlay protein blot assay, MHV-A59, 346, 347

West Nile virus strain E101, replication, 262–270
host involvement, 268–270
Wheat germ EF-1α, turnip yellow mosaic virus, 155
WIN 51711, rhinovirus 14, 70–74
WIN 52084, rhinovirus 14, 72, 73

Xenopus, 154, 403
X-ray crystallography, 79, 104, 149, 455

Yeast. *See* Sindbis virus structural gene expression in *Saccharomyces cerevisiae*
Yellow fever virus replication, 209, 217, 262–264
Yellow mosaic virus. *See* Turnip yellow mosaic virus *entries*